SCHAUM'S
outlines

Probability, Random Variables, and Random Processes

Probability, Random Variables, and Random Processes

Second Edition

Hwei P. Hsu, Ph.D.

Schaum's Outline Series

New York   Chicago   San Francisco   Lisbon   London   Madrid
Mexico City   Milan   New Delhi   San Juan   Seoul
Singapore   Sydney   Toronto

The McGraw·Hill Companies

HWEI P. HSU received his B.S. from National Taiwan University and his M.S. and Ph.D. from Case Institute of Technology. He has published several books which include *Schaum's Outline of Analog and Digital Communications* and *Schaum's Outline of Signals and Systems.*

Schaum's Outline of
PROBABILITY, RANDOM VARIABLES, AND RANDOM PROCESSES

3 4 5 6 7 8 9 CUS/CUS 1 9 8 7 6 5 4 3 2 1

ISBN: 978-0-07-163289-8
MHID: 0-07-163289-1

This publication is designed to provide accurate and authoritative information in regard to the subject matter covered. It is sold with the understanding that neither the author nor the publisher is engaged in rendering legal, accounting, securities trading, or other professional services. If legal advice or other expert assistance is required, the services of a competent professional person should be sought.

> —*From a Declaration of Principles Jointly Adopted by a Committee of the American Bar Association and a Committee of Publishers and Associations*

Library of Congress Cataloging-in-Publication Data

Hsu, Hwei P. (Hwei Piao), 1930-
Schaum's outline's of probability, random variables & random processes / Hwei Hsu.– 2nd ed.
    p. cm. – (Schaum's outlines)
  Includes index.
  Rev. ed. of: Schaum's outline of theory and problems of probability, random variables, and random processes / Hwei Hsu. c1997.
  ISBN 0-07-163289-1 (alk. paper)
  1. Probabilities—Outlines, syllabi, etc. 2. Random variables—Outlines, syllabi, etc. 3. Stochastic processes—Outlines, syllabi, etc. I. Hsu, Hwei P. (Hwei Piao), 1930- Schaum's outline of theory and problems of probability, random variables, and random processes. II. Title.

  QA273.25.H78 2010
  519.2—dc22   2010016677

Preface to The Second Edition

The purpose of this book, like its previous edition, is to provide an introduction to the principles of probability, random variables, and random processes and their applications.

The book is designed for students in various disciplines of engineering, science, mathematics, and management. The background required to study the book is one year of calculus, elementary differential equations, matrix analysis, and some signal and system theory, including Fourier transforms. The book can be used as a self-contained textbook or for self-study. Each topic is introduced in a chapter with numerous solved problems. The solved problems constitute an integral part of the text.

This new edition includes and expands the contents of the first edition. In addition to refinement through the text, two new sections on probability-generating functions and martingales have been added and a new chapter on information theory has been added.

I wish to thank my granddaughter Elysia Ann Krebs for helping me in the preparation of this revision. I also wish to express my appreciation to the editorial staff of the McGraw-Hill Schaum's Series for their care, cooperation, and attention devoted to the preparation of the book.

HWEI P. HSU
Shannondell at Valley Forge, Audubon, Pennsylvania

Preface to The First Edition

The purpose of this book is to provide an introduction to the principles of probability, random variables, and random processes and their applications.

The book is designed for students in various disciplines of engineering, science, mathematics, and management. It may be used as a textbook and/or a supplement to all current comparable texts. It should also be useful to those interested in the field of self-study. The book combines the advantages of both the textbook and the so-called review book. It provides the textual explanations of the textbook, and in the direct way characteristic of the review book, it gives hundreds of completely solved problems that use essential theory and techniques. Moreover, the solved problems are an integral part of the text. The background required to study the book is one year of calculus, elementary differential equations, matrix analysis, and some signal and system theory, including Fourier transforms.

I wish to thank Dr. Gordon Silverman for his invaluable suggestions and critical review of the manuscript. I also wish to express my appreciation to the editorial staff of the McGraw-Hill Schaum Series for their care, cooperation, and attention devoted to the preparation of the book. Finally, I thank my wife, Daisy, for her patience and encouragement.

Hwei P. Hsu
Montville, New Jersey

Contents

CHAPTER 1

Probability

1.1 Introduction

The study of probability stems from the analysis of certain games of chance, and it has found applications in most branches of science and engineering. In this chapter the basic concepts of probability theory are presented.

1.2 Sample Space and Events

A. Random Experiments:

In the study of probability, any process of observation is referred to as an *experiment*. The results of an observation are called the *outcomes* of the experiment. An experiment is called a *random experiment* if its outcome cannot be predicted. Typical examples of a random experiment are the roll of a die, the toss of a coin, drawing a card from a deck, or selecting a message signal for transmission from several messages.

B. Sample Space:

The set of all possible outcomes of a random experiment is called the *sample space* (or *universal set*), and it is denoted by S. An element in S is called a *sample point*. Each outcome of a random experiment corresponds to a sample point.

EXAMPLE 1.1   Find the sample space for the experiment of tossing a coin (*a*) once and (*b*) twice.

(*a*) There are two possible outcomes, heads or tails. Thus:

$$S = \{H, T\}$$

where H and T represent head and tail, respectively.

(*b*) There are four possible outcomes. They are pairs of heads and tails. Thus:

$$S = \{HH, HT, TH, TT\}$$

EXAMPLE 1.2   Find the sample space for the experiment of tossing a coin repeatedly and of counting the number of tosses required until the first head appears.

Clearly all possible outcomes for this experiment are the terms of the sequence $1, 2, 3, \ldots$ Thus:

$$S = \{1, 2, 3, \ldots\}$$

Note that there are an infinite number of outcomes.

EXAMPLE 1.3 Find the sample space for the experiment of measuring (in hours) the lifetime of a transistor.

Clearly all possible outcomes are all nonnegative real numbers. That is,

$$S = \{\tau : 0 \leq \tau \leq \infty\}$$

where τ represents the life of a transistor in hours.

Note that any particular experiment can often have many different sample spaces depending on the observation of interest (Probs. 1.1 and 1.2). A sample space S is said to be *discrete* if it consists of a finite number of sample points (as in Example 1.1) or countably infinite sample points (as in Example 1.2). A set is called *countable* if its elements can be placed in a one-to-one correspondence with the positive integers. A sample space S is said to be *continuous* if the sample points constitute a continuum (as in Example 1.3).

C. Events:

Since we have identified a sample space S as the set of all possible outcomes of a random experiment, we will review some set notations in the following.

If ζ is an element of S (or belongs to S), then we write

$$\zeta \in S$$

If ζ is not an element of S (or does not belong to S), then we write

$$\zeta \notin S$$

A set A is called a *subset* of B, denoted by

$$A \subset B$$

if every element of A is also an element of B. Any subset of the sample space S is called an *event*. A sample point of S is often referred to as an *elementary event*. Note that the sample space S is the subset of itself: that is, $S \subset S$. Since S is the set of all possible outcomes, it is often called the *certain event*.

EXAMPLE 1.4 Consider the experiment of Example 1.2. Let A be the event that the number of tosses required until the first head appears is even. Let B be the event that the number of tosses required until the first head appears is odd. Let C be the event that the number of tosses required until the first head appears is less than 5. Express events A, B, and C.

$$A = \{2, 4, 6, \ldots\}$$
$$B = \{1, 3, 5, \ldots\}$$
$$C = \{1, 2, 3, 4\}$$

1.3 Algebra of Sets

A. Set Operations:

1. Equality:
Two sets A and B are equal, denoted $A = B$, if and only if $A \subset B$ and $B \subset A$.

2. Complementation:
Suppose $A \subset S$. The *complement* of set A, denoted $\bar{A}$, is the set containing all elements in S but not in A.

$$\bar{A} = \{\zeta : \zeta \in S \text{ and } \zeta \notin A\}$$

3. Union:

The *union* of sets A and B, denoted $A \cup B$, is the set containing all elements in either A or B or both.

$$A \cup B = \{\zeta : \zeta \in A \text{ or } \zeta \in B\}$$

4. Intersection:

The *intersection* of sets A and B, denoted $A \cap B$, is the set containing all elements in both A and B.

$$A \cap B = \{\zeta : \zeta \in A \text{ and } \zeta \in B\}$$

5. Difference:

The *difference* of sets A and B, denoted $A \setminus B$, is the set containing all elements in A but not in B.

$$A \setminus B = \{\zeta : \zeta \in A \text{ and } \zeta \notin B\}$$

Note that $A \setminus B = A \cap \bar{B}$.

6. Symmetrical Difference:

The *symmetrical difference* of sets A and B, denoted $A \, \Delta \, B$, is the set of all elements that are in A or B but not in both.

$$A \, \Delta \, B = \{\zeta : \zeta \in A \text{ or } \zeta \in B \text{ and } \zeta \notin A \cap B\}$$

Note that $A \, \Delta \, B = (A \cap \bar{B}) \cup (\bar{A} \cap B) = (A \setminus B) \cup (B \setminus A)$.

7. Null Set:

The set containing no element is called the *null set*, denoted $\varnothing$. Note that

$$\varnothing = \bar{S}$$

8. Disjoint Sets:

Two sets A and B are called *disjoint* or *mutually exclusive* if they contain no common element, that is, if $A \cap B = \varnothing$.

The definitions of the union and intersection of two sets can be extended to any finite number of sets as follows:

$$\bigcup_{i=1}^{n} A_i = A_1 \cup A_2 \cup \cdots \cup A_n$$

$$= \{\zeta : \zeta \in A_1 \text{ or } \zeta \in A_2 \text{ or } \cdots \zeta \in A_n\}$$

$$\bigcap_{i=1}^{n} A_i = A_1 \cap A_2 \cap \cdots \cap A_n$$

$$= \{\zeta : \zeta \in A_1 \text{ and } \zeta \in A_2 \text{ and } \cdots \zeta \in A_n\}$$

Note that these definitions can be extended to an infinite number of sets:

$$\bigcup_{i=1}^{\infty} A_i = A_1 \cup A_2 \cup A_3 \cup \cdots$$

$$\bigcap_{i=1}^{\infty} A_i = A_1 \cap A_2 \cap A_3 \cap \cdots$$

In our definition of event, we state that every subset of S is an event, including S and the null set $\varnothing$. Then

$$S = \text{the certain event}$$
$$\varnothing = \text{the impossible event}$$

If A and B are events in S, then

$$\bar{A} = \text{the event that } A \text{ did not occur}$$
$$A \cup B = \text{the event that either } A \text{ or } B \text{ or both occurred}$$
$$A \cap B = \text{the event that both } A \text{ and } B \text{ occurred}$$

Similarly, if $A_1, A_2, \ldots, A_n$ are a sequence of events in S, then

$$\bigcup_{i=1}^{n} A_i = \text{the event that at least one of the } A_i \text{ occurred}$$

$$\bigcap_{i=1}^{n} A_i = \text{the event that all of the } A_i \text{ occurred}$$

9. Partition of S:

If $A_i \cap A_j = \varnothing$ for $i \neq j$ and $\bigcup_{i=1}^{k} A_i = S$, then the collection $\{A_i; 1 \leq i \leq k\}$ is said to form a *partition* of S.

10. Size of Set:

When sets are countable, the *size* (or *cardinality*) of set A, denoted $|A|$, is the number of elements contained in A. When sets have a finite number of elements, it is easy to see that size has the following properties:

    (*i*)   If $A \cap B = \varnothing$, then $|A \cup B| = |A| + |B|$.

    (*ii*) $|\varnothing| = 0$.

    (*iii*) If $A \subset B$, then $|A| \leq |B|$.

    (*iv*) $|A \cup B| + |A \cap B| = |A| + |B|$.

Note that the property (*iv*) can be easily seen if A and B are subsets of a line with length $|A|$ and $|B|$, respectively.

11. Product of Sets:

The *product* (or *Cartesian product*) of sets A and B, denoted by $A \times B$, is the set of ordered pairs of elements from A and B.

$$C = A \times B = \{(a, b): a \in A, b \in B\}$$

Note that $A \times B \neq B \times A$, and $|C| = |A \times B| = |A| \times |B|$.

EXAMPLE 1.5   Let $A = \{a_1, a_2, a_3\}$ and $B = \{b_1, b_2\}$. Then

$$C = A \times B = \{(a_1, b_1), (a_1, b_2), (a_2, b_1), (a_2, b_2), (a_3, b_1), (a_3, b_2)\}$$
$$D = B \times A = \{(b_1, a_1), (b_1, a_2), (b_1, a_3), (b_2, a_1), (b_2, a_2), (b_2, a_3)\}$$

B. Venn Diagram:

A graphical representation that is very useful for illustrating set operation is the Venn diagram. For instance, in the three Venn diagrams shown in Fig. 1-1, the shaded areas represent, respectively, the events $A \cup B, A \cap B$, and $\bar{A}$. The Venn diagram in Fig. 1-2(*a*) indicates that $B \subset A$, and the event $A \cap \bar{B} = A \backslash B$ is shown as the shaded area. In Fig. 1-2(*b*), the shaded area represents the event $A \triangle B$.

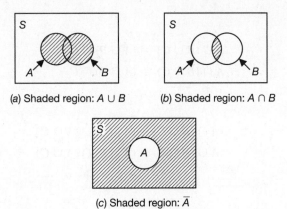

(a) Shaded region: $A \cup B$ (b) Shaded region: $A \cap B$

(c) Shaded region: $\bar{A}$

Fig. 1-1

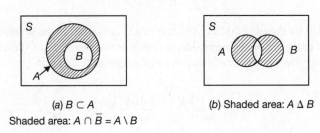

(a) $B \subset A$ (b) Shaded area: $A \, \Delta \, B$

Shaded area: $A \cap \bar{B} = A \setminus B$

Fig. 1-2

C. Identities:

By the above set definitions or reference to Fig. 1-1, we obtain the following identities:

$$\bar{S} = \varnothing \tag{1.1}$$

$$\bar{\varnothing} = S \tag{1.2}$$

$$\bar{\bar{A}} = A \tag{1.3}$$

$$S \cup A = S \tag{1.4}$$

$$S \cap A = A \tag{1.5}$$

$$A \cup \bar{A} = S \tag{1.6}$$

$$A \cap \bar{A} = \varnothing \tag{1.7}$$

$$A \cup \varnothing = A \tag{1.8}$$

$$A \cap \varnothing = \varnothing \tag{1.9}$$

$$A \setminus B = A \cap \bar{B} \tag{1.10}$$

$$S \setminus A = \bar{A} \tag{1.11}$$

$$A \setminus \varnothing = A \tag{1.12}$$

$$A \, \Delta \, B = (A \cap \bar{B}) \cup (\bar{A} \cap B) \tag{1.13}$$

The union and intersection operations also satisfy the following laws:

Commutative Laws:

$$A \cup B = B \cup A \tag{1.14}$$

$$A \cap B = B \cap A \tag{1.15}$$

Associative Laws:

$$A \cup (B \cup C) = (A \cup B) \cup C \tag{1.16}$$

$$A \cap (B \cap C) = (A \cap B) \cap C \tag{1.17}$$

Distributive Laws:

$$A \cap (B \cup C) = (A \cap B) \cup (A \cap C) \tag{1.18}$$

$$A \cup (B \cap C) = (A \cup B) \cap (A \cup C) \tag{1.19}$$

De Morgan's Laws:

$$\overline{A \cup B} = \bar{A} \cap \bar{B} \tag{1.20}$$

$$\overline{A \cap B} = \bar{A} \cup \bar{B} \tag{1.21}$$

These relations are verified by showing that any element that is contained in the set on the left side of the equality sign is also contained in the set on the right side, and vice versa. One way of showing this is by means of a Venn diagram (Prob. 1.14). The distributive laws can be extended as follows:

$$A \cap \left(\bigcup_{i=1}^{n} B_i \right) = \bigcup_{i=1}^{n} (A \cap B_i) \tag{1.22}$$

$$A \cup \left(\bigcap_{i=1}^{n} B_i \right) = \bigcap_{i=1}^{n} (A \cup B_i) \tag{1.23}$$

Similarly, De Morgan's laws also can be extended as follows (Prob. 1.21):

$$\overline{\left(\bigcup_{i=1}^{n} A_i \right)} = \bigcap_{i=1}^{n} \bar{A}_i \tag{1.24}$$

$$\overline{\left(\bigcap_{i=1}^{n} A_i \right)} = \bigcup_{i=1}^{n} \bar{A}_i \tag{1.25}$$

1.4 Probability Space

A. Event Space:

We have defined that events are subsets of the sample space S. In order to be precise, we say that a subset A of S can be an event if it belongs to a collection F of subsets of S, satisfying the following conditions:

(*i*) $S \in F$ $\hspace{6cm}$ (1.26)

(*ii*)   if $A \in F$, then $\bar{A} \in F$ $\hspace{4.5cm}$ (1.27)

(*iii*)   if $A_i \in F$ for $i \geq 1$, then $\displaystyle\bigcup_{i=1}^{\infty} A_i \in F$ $\hspace{3cm}$ (1.28)

The collection F is called an *event space*. In mathematical literature, event space is known as *sigma field* (σ-field) or σ-algebra.

Using the above conditions, we can show that if A and B are in F, then so are

$$A \cap B, A \backslash B, A \, \Delta \, B \text{ (Prob. 1.22).}$$

EXAMPLE 1.6 Consider the experiment of tossing a coin once in Example 1.1. We have $S = \{H, T\}$. The set $\{S, \varnothing\}, \{S, \varnothing, H, T\}$ are event spaces, but $\{S, \varnothing, H\}$ is not an event space, since $\overline{H} = T$ is not in the set.

B. Probability Space:

An assignment of real numbers to the events defined in an event space F is known as the *probability measure P*. Consider a random experiment with a sample space S, and let A be a particular event defined in F. The probability of the event A is denoted by $P(A)$. Thus, the probability measure is a function defined over F. The triplet (S, F, P) is known as the *probability space*.

C. Probability Measure

a. Classical Definition:

Consider an experiment with equally likely finite outcomes. Then the classical definition of probability of event A, denoted $P(A)$, is defined by

$$P(A) = \frac{|A|}{|S|} \tag{1.29}$$

If A and B are disjoint, i.e., $A \cap B = \varnothing$, then $|A \cup B| = |A| + |B|$. Hence, in this case

$$P(A \cup B) = \frac{|A \cup B|}{|S|} = \frac{|A| + |B|}{|S|} = \frac{|A|}{|S|} + \frac{|B|}{|S|} = P(A) + P(B) \tag{1.30}$$

We also have

$$P(S) = \frac{|S|}{|S|} = 1 \tag{1.31}$$

$$P(\overline{A}) = \frac{|\overline{A}|}{|S|} = \frac{|S| - |A|}{|S|} = 1 - \frac{|A|}{|S|} = 1 - P(A) \tag{1.32}$$

EXAMPLE 1.7 Consider an experiment of rolling a die. The outcome is

$$S = \{\zeta_1, \zeta_2, \zeta_3, \zeta_4, \zeta_5, \zeta_6\} = \{1, 2, 3, 4, 5, 6\}$$

Define:
  A: the event that outcome is even, i.e., $A = \{2, 4, 6\}$
  B: the event that outcome is odd, i.e., $B = \{1, 3, 5\}$
  C: the event that outcome is prime, i.e., $C = \{1, 2, 3, 5\}$

Then $P(A) = \dfrac{|A|}{|S|} = \dfrac{3}{6} = \dfrac{1}{2}, \quad P(B) = \dfrac{|B|}{|S|} = \dfrac{3}{6} = \dfrac{1}{2}, \quad P(C) = \dfrac{|C|}{|S|} = \dfrac{4}{6} = \dfrac{2}{3}$

Note that in the classical definition, $P(A)$ is determined *a priori* without actual experimentation and the definition can be applied only to a limited class of problems such as only if the outcomes are finite and equally likely or equally probable.

b.   Relative Frequency Definition:

Suppose that the random experiment is repeated n times. If event A occurs $n(A)$ times, then the probability of event A, denoted $P(A)$, is defined as

$$P(A) = \lim_{n \to \infty} \frac{n(A)}{n} \tag{1.33}$$

where $n(A)/n$ is called the relative frequency of event A. Note that this limit may not exist, and in addition, there are many situations in which the concepts of repeatability may not be valid. It is clear that for any event A, the relative frequency of A will have the following properties:

1. $0 \le n(A)/n \le 1$, where $n(A)/n = 0$ if A occurs in none of the n repeated trials and $n(A)/n = 1$ if A occurs in all of the n repeated trials.

2.   If A and B are mutually exclusive events, then

$$n(A \cup B) = n(A) + n(B)$$

and

$$\frac{n(A \cup B)}{n} = \frac{n(A)}{n} + \frac{N(B)}{n}$$

$$P(A \cup B) = \lim_{n \to \infty} \frac{n(A \cup B)}{n} = \lim_{n \to \infty} \frac{n(A)}{n} + \lim_{n \to \infty} \frac{N(B)}{n} = P(A) + P(B) \tag{1.34}$$

c.   Axiomatic Definition:

Consider a probability space (S, F, P). Let A be an event in F. Then in the *axiomatic* definition, the probability $P(A)$ of the event A is a real number assigned to A which satisfies the following three *axioms*:

Axiom 1: $P(A) \ge 0$ $\tag{1.35}$

Axiom 2: $P(S) = 1$ $\tag{1.36}$

Axiom 3: $P(A \cup B) = P(A) + P(B)$ if $A \cap B = \varnothing$ $\tag{1.37}$

If the sample space S is not finite, then axiom 3 must be modified as follows:

Axiom 3′:   If $A_1, A_2, \ldots$ is an infinite sequence of mutually exclusive events in S ($A_i \cap A_j = \varnothing$ for $i \neq j$), then

$$P\left(\bigcup_{i=1}^{\infty} A_i\right) = \sum_{i=1}^{\infty} P(A_i) \tag{1.38}$$

These axioms satisfy our intuitive notion of probability measure obtained from the notion of relative frequency.

d.   Elementary Properties of Probability:

By using the above axioms, the following useful properties of probability can be obtained:

1. $P(\bar{A}) = 1 - P(A)$ $\tag{1.39}$

2. $P(\varnothing) = 0$ $\tag{1.40}$

3. $P(A) \le P(B)$ if $A \subset B$ $\tag{1.41}$

4. $P(A) \le 1$ $\tag{1.42}$

5. $P(A \cup B) = P(A) + P(B) - P(A \cap B)$ $\tag{1.43}$

6. $P(A \backslash B) = P(A) - P(A \cap B)$ $\tag{1.44}$

7.   If $A_1, A_2, \ldots, A_n$ are n arbitrary events in S, then

$$P\left(\bigcup_{i=1}^{n} A_i\right) = \sum_{i=1}^{n} P(A_i) - \sum_{i \neq j} P(A_i \cap A_j) + \sum_{i \neq j \neq k} P(A_i \cap A_j \cap A_k)$$

$$- \cdots (-1)^{n-1} P(A_1 \cap A_2 \cap \cdots \cap A_n) \tag{1.45}$$

where the sum of the second term is over all distinct pairs of events, that of the third term is over all distinct triples of events, and so forth.

8. If $A_1, A_2, \ldots, A_n$ is a finite sequence of mutually exclusive events in S ($A_i \cap A_j = \emptyset$ for $i \neq j$), then

$$P\left(\bigcup_{i=1}^{n} A_i\right) = \sum_{i=1}^{n} P(A_i) \tag{1.46}$$

and a similar equality holds for any subcollection of the events.

Note that property 4 can be easily derived from axiom 2 and property 3. Since $A \subset S$, we have

$$P(A) \leq P(S) = 1$$

Thus, combining with axiom 1, we obtain

$$0 \leq P(A) \leq 1 \tag{1.47}$$

Property 5 implies that

$$P(A \cup B) \leq P(A) + P(B) \tag{1.48}$$

since $P(A \cap B) \geq 0$ by axiom 1.

Property 6 implies that

$$P(A \backslash B) = P(A) - P(B) \qquad \text{if } B \subset A \tag{1.49}$$

since $A \cap B = B$ if $B \subset A$.

1.5 Equally Likely Events

A. Finite Sample Space:

Consider a finite sample space S with n finite elements

$$S = \{\zeta_1, \zeta_2, \ldots, \zeta_n\}$$

where ζ_i's are elementary events. Let $P(\zeta_i) = p_i$. Then

1. $0 \leq p_i \leq 1 \qquad i = 1, 2, \ldots, n$ \hfill (1.50)

2. $\displaystyle\sum_{i=1}^{n} p_i = p_1 + p_2 + \cdots + p_n = 1$ \hfill (1.51)

3. If $A = \displaystyle\bigcup_{i \in I} \zeta_i$, where I is a collection of subscripts, then

$$P(A) = \sum_{\zeta_i \in A} P(\zeta_i) = \sum_{i \in I} p_i \tag{1.52}$$

B. Equally Likely Events:

When all elementary events $\zeta_i (i = 1, 2, \ldots, n)$ are equally likely, that is,

$$p_1 = p_2 = \cdots = p_n$$

then from Eq. (1.51), we have

$$p_i = \frac{1}{n} \qquad i = 1, 2, \ldots, n \tag{1.53}$$

and

$$P(A) = \frac{n(A)}{n} \tag{1.54}$$

where $n(A)$ is the number of outcomes belonging to event A and n is the number of sample points in S. [See classical definition (1.29).]

1.6 Conditional Probability

A. Definition:

The *conditional probability* of an event A given event B, denoted by $P(A|B)$, is defined as

$$P(A|B) = \frac{P(A \cap B)}{P(B)} \qquad P(B) > 0 \tag{1.55}$$

where $P(A \cap B)$ is the joint probability of A and B. Similarly,

$$P(B|A) = \frac{P(A \cap B)}{P(A)} \qquad P(A) > 0 \tag{1.56}$$

is the conditional probability of an event B given event A. From Eqs. (1.55) and (1.56), we have

$$P(A \cap B) = P(A|B) \, P(B) = P(B|A)P(A) \tag{1.57}$$

Equation (1.57) is often quite useful in computing the joint probability of events.

B. Bayes' Rule:

From Eq. (1.57) we can obtain the following *Bayes' rule*:

$$P(A|B) = \frac{P(B|A)P(A)}{P(B)} \tag{1.58}$$

1.7 Total Probability

The events $A_1, A_2, \ldots, A_n$ are called *mutually exclusive and exhaustive* if

$$\bigcup_{i=1}^{n} A_i = A_1 \cup A_2 \cup \cdots \cup A_n = S \quad \text{and} \quad A_i \cap A_j = \emptyset \qquad i \neq j \tag{1.59}$$

Let B be any event in S. Then

$$P(B) = \sum_{i=1}^{n} P(B \cap A_i) = \sum_{i=1}^{n} P(B|A_i)P(A_i) \tag{1.60}$$

which is known as the *total probability* of event B (Prob. 1.57). Let $A = A_i$ in Eq. (1.58); then, using Eq. (1.60), we obtain

$$P(A_i | B) = \frac{P(B|A_i)P(A_i)}{\displaystyle\sum_{i=1}^{n} P(B|A_i)P(A_i)} \tag{1.61}$$

Note that the terms on the right-hand side are all conditioned on events A_i, while the term on the left is conditioned on B. Equation (1.61) is sometimes referred to as *Bayes' theorem*.

1.8 Independent Events

Two events A and B are said to be (*statistically*) *independent* if and only if

$$P(A \cap B) = P(A)P(B) \tag{1.62}$$

It follows immediately that if A and B are independent, then by Eqs. (1.55) and (1.56),

$$P(A \mid B) = P(A) \quad \text{and} \quad P(B \mid A) = P(B) \tag{1.63}$$

If two events A and B are independent, then it can be shown that A and $\bar{B}$ are also independent; that is (Prob. 1.63),

$$P(A \cap \bar{B}) = P(A)P(\bar{B}) \tag{1.64}$$

Then

$$P(A | \bar{B}) = \frac{P(A \cap \bar{B})}{P(\bar{B})} = P(A) \tag{1.65}$$

Thus, if A is independent of B, then the probability of A's occurrence is unchanged by information as to whether or not B has occurred. Three events A, B, C are said to be independent if and only if

$$\begin{aligned} P(A \cap B \cap C) &= P(A)P(B)P(C) \\ P(A \cap B) &= P(A)P(B) \\ P(A \cap C) &= P(A)P(C) \\ P(B \cap C) &= P(B)P(C) \end{aligned} \tag{1.66}$$

We may also extend the definition of independence to more than three events. The events $A_1, A_2, ..., A_n$ are independent if and only if for every subset $\{A_{i1}, A_{i2}, ... A_{ik}\}$ $(2 \leq k \leq n)$ of these events,

$$P(A_{i1} \cap A_{i2} \cap \cdots \cap A_{ik}) = P(A_{i1}) P(A_{i2}) \cdots P(A_{ik}) \tag{1.67}$$

Finally, we define an infinite set of events to be independent if and only if every finite subset of these events is independent.

To distinguish between the mutual exclusiveness (or disjointness) and independence of a collection of events, we summarize as follows:

1. $\{A_i, i = 1, 2, ..., n\}$ is a sequence of mutually exclusive events, then

$$P\left(\bigcup_{i=1}^{n} A_i\right) = \sum_{i=1}^{n} P(A_i) \tag{1.68}$$

2.   If $\{A_i, i = 1, 2, \ldots, n\}$ is a sequence of independent events, then

$$P\left(\bigcap_{i=1}^{n} A_i\right) = \prod_{i=1}^{n} P(A_i)$$

(1.69)

and a similar equality holds for any subcollection of the events.

SOLVED PROBLEMS

Sample Space and Events

1.1.   Consider a random experiment of tossing a coin three times.

   (*a*)   Find the sample space S_1 if we wish to observe the exact sequences of heads and tails obtained.

   (*b*)   Find the sample space S_2 if we wish to observe the number of heads in the three tosses.

   (*a*)   The sampling space S_1 is given by

$$S_1 = \{HHH, HHT, HTH, THH, HTT, THT, TTH, TTT\}$$

   where, for example, HTH indicates a head on the first and third throws and a tail on the second throw. There are eight sample points in S_1.

   (*b*)   The sampling space S_2 is given by

$$S_2 = \{0, 1, 2, 3\}$$

   where, for example, the outcome 2 indicates that two heads were obtained in the three tosses. The sample space S_2 contains four sample points.

1.2.   Consider an experiment of drawing two cards at random from a bag containing four cards marked with the integers 1 through 4.

   (*a*)   Find the sample space S_1 of the experiment if the first card is replaced before the second is drawn.

   (*b*)   Find the sample space S_2 of the experiment if the first card is not replaced.

   (*a*)   The sample space S_1 contains 16 ordered pairs (i, j), $1 \leq i \leq 4$, $1 \leq j \leq 4$, where the first number indicates the first number drawn. Thus,

$$S_1 = \begin{cases} (1,1) & (1,2) & (1,3) & (1,4) \\ (2,1) & (2,2) & (2,3) & (2,4) \\ (3,1) & (3,2) & (3,3) & (3,4) \\ (4,1) & (4,2) & (4,3) & (4,4) \end{cases}$$

   (*b*)   The sample space S_2 contains 12 ordered pairs (i, j), $i \neq j$, $1 \leq i \leq 4$, $1 \leq j \leq 4$, where the first number indicates the first number drawn. Thus,

$$S_2 = \begin{cases} (1,2) & (1,3) & (1,4) \\ (2,1) & (2,3) & (2,4) \\ (3,1) & (3,2) & (3,4) \\ (4,1) & (4,2) & (4,3) \end{cases}$$

1.3. An experiment consists of rolling a die until a 6 is obtained.

    (*a*) Find the sample space S_1 if we are interested in all possibilities.

    (*b*) Find the sample space S_2 if we are interested in the number of throws needed to get a 6.

    (*a*) The sample space S_1 would be

$$S_1 = \{6,$$
$$16, 26, 36, 46, 56,$$
$$116, 126, 136, 146, 156, \ldots\}$$

where the first line indicates that a 6 is obtained in one throw, the second line indicates that a 6 is obtained in two throws, and so forth.

    (*b*) In this case, the sample space S_2 is

$$S_2 = \{i: i \geq 1\} = \{1, 2, 3, \ldots\}$$

where i is an integer representing the number of throws needed to get a 6.

1.4. Find the sample space for the experiment consisting of measurement of the voltage output v from a transducer, the maximum and minimum of which are $+5$ and -5 volts, respectively.

A suitable sample space for this experiment would be

$$S = \{v: -5 \leq v \leq 5\}$$

1.5. An experiment consists of tossing two dice.

    (*a*) Find the sample space S.

    (*b*) Find the event A that the sum of the dots on the dice equals 7.

    (*c*) Find the event B that the sum of the dots on the dice is greater than 10.

    (*d*) Find the event C that the sum of the dots on the dice is greater than 12.

    (*a*) For this experiment, the sample space S consists of 36 points (Fig. 1-3):

$$S = \{(i, j): i, j = 1, 2, 3, 4, 5, 6\}$$

where i represents the number of dots appearing on one die and j represents the number of dots appearing on the other die.

    (*b*) The event A consists of 6 points (see Fig. 1-3):

$$A = \{(1, 6), (2, 5), (3, 4), (4, 3), (5, 2), (6, 1)\}$$

    (*c*) The event B consists of 3 points (see Fig. 1-3):

$$B = \{(5, 6), (6, 5), (6, 6)\}$$

    (*d*) The event C is an impossible event, that is, $C = \varnothing$.

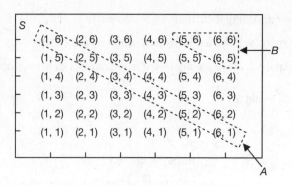

Fig. 1-3

1.6. An automobile dealer offers vehicles with the following options:

    (*a*)   With or without automatic transmission

    (*b*)   With or without air-conditioning

    (*c*)   With one of two choices of a stereo system

    (*d*)   With one of three exterior colors

If the sample space consists of the set of all possible vehicle types, what is the number of outcomes in the sample space?

The tree diagram for the different types of vehicles is shown in Fig. 1-4. From Fig. 1-4 we see that the number of sample points in S is $2 \times 2 \times 2 \times 3 = 24$.

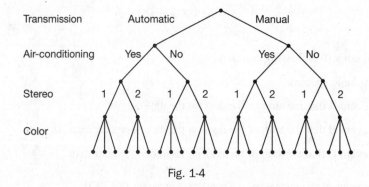

Fig. 1-4

1.7. State every possible event in the sample space $S = \{a, b, c, d\}$.

There are $2^4 = 16$ possible events in S. They are $\varnothing$; $\{a\}$, $\{b\}$, $\{c\}$, $\{d\}$; $\{a, b\}$, $\{a, c\}$, $\{a, d\}$, $\{b, c\}$, $\{b, d\}$, $\{c, d\}$; $\{a, b, c\}$, $\{a, b, d\}$, $\{a, c, d\}$, $\{b, c, d\}$; $S = \{a, b, c, d\}$.

1.8. How many events are there in a sample space S with n elementary events?

Let $S = \{s_1, s_2, \ldots, s_n\}$. Let Ω be the family of all subsets of S. (Ω is sometimes referred to as the *power set* of S.) Let S_i be the set consisting of two statements, that is,

$$S_i = \{\text{Yes, the } s_i \text{ is in; No, the } s_i \text{ is not in}\}$$

Then Ω can be represented as the Cartesian product

$$\Omega = S_1 \times S_2 \times \cdots \times S_n$$
$$= \{(s_1, s_2, \ldots, s_n) : s_i \in S_i \text{ for } i = 1, 2, \ldots, n\}$$

Since each subset of S can be uniquely characterized by an element in the above Cartesian product, we obtain the number of elements in Ω by

$$n(\Omega) = n(S_1)n(S_2) \cdots n(S_n) = 2^n$$

where $n(S_i)$ = number of elements in $S_i = 2$.

An alternative way of finding $n(\Omega)$ is by the following summation:

$$n(\Omega) = \sum_{i=0}^{n} \binom{n}{i} = \sum_{i=0}^{n} \frac{n!}{i!\,(n-i)!}$$

The proof that the last sum is equal to 2^n is not easy.

Algebra of Sets

1.9. Consider the experiment of Example 1.2. We define the events

$$A = \{k : k \text{ is odd}\}$$
$$B = \{k : 4 \le k \le 7\}$$
$$C = \{k : 1 \le k \le 10\}$$

where k is the number of tosses required until the first H (head) appears. Determine the events $\bar{A}, \bar{B}, \bar{C}$, $A \cup B, B \cup C, A \cap B, A \cap C, B \cap C$, and $\bar{A} \cap B$.

$$\bar{A} = \{k : k \text{ is even}\} = \{2,\ 4,\ 6, \ldots\}$$

$$\bar{B} = \{k : k = 1,\ 2,\ 3 \text{ or } k \ge 8\}$$

$$\bar{C} = \{k : k \ge 11\}$$

$$A \cup B = \{k : k \text{ is odd or } k = 4, 6\}$$

$$B \cup C = C$$

$$A \cap B = \{5,\ 7\}$$

$$A \cap C = \{1, 3, 5, 7, 9\}$$

$$B \cap C = B$$

$$\bar{A} \cap B = \{4,\ 6\}$$

1.10. Consider the experiment of Example 1.7 of rolling a die. Express

$$A \cup B, A \cap C, \bar{B}, \bar{C}, B\backslash C, C\backslash B, B \, \Delta \, C.$$

From Example 1.7, we have $S = \{1, 2, 3, 4, 5, 6\}, A = \{2, 4, 6\}, B = \{1, 3, 5\}$, and $C = \{1, 2, 3, 5\}$. Then

$$A \cup B = \{1, 2, 3, 4, 5, 6\} = S, \quad A \cap C = \{2\}, \quad \bar{B} = \{2, 4, 6\} = A, \quad \bar{C} = \{4, 6\}$$

$$B\backslash C = B \cap \bar{C} = \{\varnothing\}, \quad C\backslash B = C \cap \bar{B} = \{2\},$$

$$B \, \Delta \, C = (B\backslash C) \cup (C\backslash B) = \{\varnothing\} \cup \{2\} = \{2\}$$

1.11. The sample space of an experiment is the real line express as

$$S = \{v: -\infty < v < \infty\}$$

(*a*) Consider the events

$$A_1 = \left\{v: 0 \le v < \frac{1}{2}\right\}$$

$$A_2 = \left\{v: \frac{1}{2} \le v < \frac{3}{4}\right\}$$

$$A_i = \left\{v: 1 - \frac{1}{2^{i-1}} \le v < 1 - \frac{1}{2^i}\right\}$$

Determine the events

$$\bigcup_{i=1}^{\infty} A_i \qquad \text{and} \qquad \bigcap_{i=1}^{\infty} A_i$$

(*b*) Consider the events

$$B_1 = \left\{v: v \le \frac{1}{2}\right\}$$

$$B_2 = \left\{v: v \le \frac{1}{4}\right\}$$

$$\vdots$$

$$B_i = \left\{v: v \le \frac{1}{2^i}\right\}$$

Determine the events

$$\bigcup_{i=1}^{\infty} B_i \qquad \text{and} \qquad \bigcap_{i=1}^{\infty} B_i$$

(*a*) It is clear that

$$\bigcup_{i=1}^{\infty} A_i = \{v: 0 \le v < 1\}$$

Noting that the A_i's are mutually exclusive, we have

$$\bigcap_{i=1}^{\infty} A_i = \varnothing$$

(*b*) Noting that $B_1 \supset B_2 \supset \cdots \supset B_i \supset \cdots$, we have

$$\bigcup_{i=1}^{\infty} B_i = B_1 = \left\{v: v \le \frac{1}{2}\right\} \qquad \text{and} \qquad \bigcap_{i=1}^{\infty} B_i = \{v: v \le 0\}$$

1.12. Consider the switching networks shown in Fig. 1-5. Let A_1, A_2, and A_3 denote the events that the switches s_1, s_2, and s_3 are closed, respectively. Let A_{ab} denote the event that there is a closed path between terminals a and b. Express A_{ab} in terms of A_1, A_2, and A_3 for each of the networks shown.

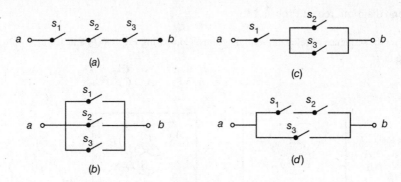

Fig. 1-5

(a) From Fig. 1-5(a), we see that there is a closed path between a and b only if all switches s_1, s_2, and s_3 are closed. Thus,

$$A_{ab} = A_1 \cap A_2 \cap A_3$$

(b) From Fig. 1-5(b), we see that there is a closed path between a and b if at least one switch is closed. Thus,

$$A_{ab} = A_1 \cup A_2 \cup A_3$$

(c) From Fig. 1-5(c), we see that there is a closed path between a and b if s_1 and either s_2 or s_3 are closed. Thus,

$$A_{ab} = A_1 \cap (A_2 \cup A_3)$$

Using the distributive law (1.18), we have

$$A_{ab} = (A_1 \cap A_2) \cup (A_1 \cap A_3)$$

which indicates that there is a closed path between a and b if s_1 and s_2 or s_1 and s_3 are closed.

(d) From Fig. 1-5(d), we see that there is a closed path between a and b if either s_1 and s_2 are closed or s_3 is closed. Thus

$$A_{ab} = (A_1 \cap A_2) \cup A_3$$

1.13. Verify the distributive law (1.18).

Let $s \in [A \cap (B \cup C)]$. Then $s \in A$ and $s \in (B \cup C)$. This means either that $s \in A$ and $s \in B$ or that $s \in A$ and $s \in C$; that is, $s \in (A \cap B)$ or $s \in (A \cap C)$. Therefore,

$$A \cap (B \cup C) \subset [(A \cap B) \cup (A \cap C)]$$

Next, let $s \in [(A \cap B) \cup (A \cap C)]$. Then $s \in A$ and $s \in B$ or $s \in A$ and $s \in C$. Thus, $s \in A$ and $(s \in B$ or $s \in C)$. Thus,

$$[(A \cap B) \cup (A \cap C)] \subset A \cap (B \cup C)$$

Thus, by the definition of equality, we have

$$A \cap (B \cup C) = (A \cap B) \cup (A \cap C)$$

1.14. Using a Venn diagram, repeat Prob. 1.13.

Fig. 1-6 shows the sequence of relevant Venn diagrams. Comparing Fig. 1-6(*b*) and 1.6(*e*), we conclude that

$$A \cap (B \cup C) = (A \cap B) \cup (A \cap C)$$

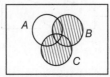 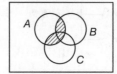

(*a*) Shaded region: $B \cup C$ (*b*) Shaded region: $A \cap (B \cup C)$

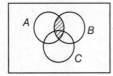 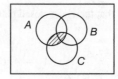

(*c*) Shaded region: $A \cap B$ (*d*) Shaded region: $A \cap C$

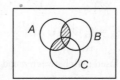

(*e*) Shaded region: $(A \cap B) \cup (A \cap C)$

Fig. 1-6

1.15 Verify De Morgan's law (1.24)

$$\overline{A \cap B} = \overline{A} \cup \overline{B}$$

Suppose that $s \in \overline{A \cap B}$, then $s \notin A \cap B$. So $s \notin$ {both A and B}. This means that either $s \notin A$ or $s \notin B$ or $s \notin A$ and $s \notin B$. This implies that $s \in \overline{A} \cup \overline{B}$. Conversely, suppose that $s \in \overline{A} \cup \overline{B}$, that is either $s \in \overline{A}$ or $s \in \overline{B}$ or $s \in$ {both $\overline{A}$ and $\overline{B}$}. Then it follows that $s \notin A$ or $s \notin B$ or $s \notin$ {both A and B}; that is, $s \notin A \cap B$ or $s \in \overline{A \cap B}$. Thus, we conclude that $\overline{A \cap B} = \overline{A} \cup \overline{B}$.

Note that De Morgan's law can also be shown by using Venn diagram.

1.16. Let A and B be arbitrary events. Show that $A \subset B$ if and only if $A \cap B = A$.

"If" part: We show that if $A \cap B = A$, then $A \subset B$. Let $s \in A$. Then $s \in (A \cap B)$, since $A = A \cap B$. Then by the definition of intersection, $s \in B$. Therefore, $A \subset B$.

"Only if" part: We show that if $A \subset B$, then $A \cap B = A$. Note that from the definition of the intersection, $(A \cap B) \subset A$. Suppose $s \in A$. If $A \subset B$, then $s \in B$. So $s \in A$ and $s \in B$; that is, $s \in (A \cap B)$. Therefore, it follows that $A \subset (A \cap B)$. Hence, $A = A \cap B$. This completes the proof.

1.17. Let A be an arbitrary event in S and let $\varnothing$ be the null event. Verify Eqs. (1.8) and (1.9), i.e.

(*a*) $A \cup \varnothing = A$ (1.8)

(*b*) $A \cap \varnothing = \varnothing$ (1.9)

(*a*) $A \cup \varnothing = \{s: s \in A \text{ or } s \in \varnothing\}$

But, by definition, there are no $s \in \varnothing$. Thus,

$$A \cup \varnothing = \{s: s \in A\} = A$$

(*b*) $A \cap \varnothing = \{s: s \in A \text{ and } s \in \varnothing\}$

But, since there are no $s \in \varnothing$, there cannot be an s such that $s \in A$ and $s \in \varnothing$. Thus,

$$A \cap \varnothing = \varnothing$$

Note that Eq. (1.9) shows that $\varnothing$ is mutually exclusive with every other event and including with itself.

1.18. Show that the null (or empty) set $\varnothing$ is a subset of every set A.

From the definition of intersection, it follows that

$$(A \cap B) \subset A \quad \text{and} \quad (A \cap B) \subset B \tag{1.70}$$

for any pair of events, whether they are mutually exclusive or not. If A and B are mutually exclusive events, that is, $A \cap B = \varnothing$, then by Eq. (1.70) we obtain

$$\varnothing \subset A \quad \text{and} \quad \varnothing \subset B \tag{1.71}$$

Therefore, for any event A,

$$\varnothing \subset A \tag{1.72}$$

that is, $\varnothing$ is a subset of every set A.

1.19. Show that A and B are disjoint if and only if $A \backslash B = A$.

First, if $A \backslash B = A \cap \bar{B} = A$, then

$$A \cap B = (A \cap \bar{B}) \cap B = A \cap (\bar{B} \cap B) = A \cap \varnothing = \varnothing$$

and A and B are disjoint.

Next, if A and B are disjoint, then $A \cap B = \varnothing$, and $A \backslash B = A \cap \bar{B} = A$.

Thus, A and B are disjoint if and only if $A \backslash B = A$.

1.20. Show that there is a distribution law also for difference; that is,

$$(A \backslash B) \cap C = (A \cap C) \backslash (B \cap C)$$

By Eq. (1.8) and applying commutative and associated laws, we have

$$(A \backslash B) \cap C = (A \cap \bar{B}) \cap C = A \cap (\bar{B} \cap C) = (A \cap C) \cap \bar{B}$$

Next,

$$(A \cap C) \backslash (B \cap C) = (A \cap C) \cap (\overline{B \cap C}) \qquad \text{by Eq. (1.10)}$$
$$= (A \cap C) \cap (\bar{B} \cup \bar{C}) \qquad \text{by Eq. (1.21)}$$

$$= [(A \cap C) \cap \bar{B}] \cup [(A \cap C) \cap \bar{C}] \qquad \text{by Eq. (1.19)}$$

$$= [(A \cap C) \cap \bar{B}] \cup [A \cap (C \cap \bar{C})] \qquad \text{by Eq. (1.17)}$$

$$= [(A \cap C) \cap \bar{B}] \cup [A \cap \varnothing] \qquad \text{by Eq. (1.7)}$$

$$= [(A \cap C) \cap \bar{B}] \cup \varnothing \qquad \text{by Eq. (1.9)}$$

$$= (A \cap C) \cap \bar{B} \qquad \text{by Eq. (1.8)}$$

Thus, we have

$$(A \backslash B) \cap C = (A \cap C) \backslash (B \cap C)$$

1.21. Verify Eqs. (1.24) and (1.25).

(*a*) Suppose first that $s \in \overline{\left(\bigcup_{i=1}^{n} A_i \right)}$; then $s \notin \left(\bigcup_{i=1}^{n} A_i \right)$.

That is, if s is not contained in any of the events A_i, $i = 1, 2, \ldots, n$, then s is contained in $\bar{A}_i$ for all $i = 1, 2, \ldots, n$. Thus

$$s \in \bigcap_{i=1}^{n} \bar{A}_i$$

Next, we assume that

$$s \in \bigcap_{i=1}^{n} \bar{A}_i$$

Then s is contained in $\bar{A}_i$ for all $i = 1, 2, \ldots, n$, which means that s is not contained in A_i for any $i = 1, 2, \ldots, n$, implying that

$$s \notin \bigcup_{i=1}^{n} A_i$$

Thus, $$s \in \overline{\left(\bigcup_{i=1}^{n} A_i \right)}$$

This proves Eq. (1.24).

(*b*) Using Eqs. (1.24) and (1.3), we have

$$\overline{\left(\bigcup_{i=1}^{n} \bar{A}_i \right)} = \bigcap_{i=1}^{n} A_i$$

Taking complements of both sides of the above yields

$$\bigcup_{i=1}^{n} \bar{A}_i = \overline{\left(\bigcap_{i=1}^{n} A_i \right)}$$

which is Eq. (1.25).

Probability Space

1.22. Consider a probability space (S, F, P). Show that if A and B are in an event space (σ-field) F, so are $A \cap B, A\backslash B$, and $A \triangle B$.

By condition (*ii*) of F, Eq. (1.27), if $A, B \in F$, then $\bar{A}, \bar{B} \in F$. Now by De Morgan's law (1.21), we have

$$\overline{A \cap B} = \bar{A} \cup \bar{B} \in F \qquad \text{by Eq. (1.28)}$$

$$A \cap B = \overline{\overline{A \cap B}} \in F \qquad \text{by Eq. (1.27)}$$

Similarly, we see that

$$A \cap \bar{B} \in F \qquad \text{and} \qquad \bar{A} \cap B \in F$$

Now by Eq. (1.10), we have

$$A\backslash B = A \cap \bar{B} \in F$$

Finally, by Eq. (1.13), and Eq. (1.28), we see that

$$A \triangle B = (A \cap \bar{B}) \cup (\bar{A} \cap B) \in F$$

1.23. Consider the experiment of Example 1.7 of rolling a die. Show that $\{S, \varnothing, A, B\}$ are event spaces but $\{S, \varnothing, A\}$ and $\{S, \varnothing, A, B, C\}$ are not event spaces.

Let $F = \{S, \varnothing, A, B\}$. Then we see that

$$S \in F, \bar{S} = \varnothing \in F, \bar{\varnothing} = S \in F, \bar{A} = B \in F, \bar{B} = A \in F$$

and

$$S \cup \varnothing = S \cup A = S \cup B = S \in F, \varnothing \cup A = A \cup \varnothing = A \in F,$$

$$\varnothing \cup B = B \cup \varnothing = B \in F, A \cup B = B \cup A = S \in F$$

Thus, we conclude that $\{S, \varnothing, A, B\}$ is an event space (σ-field).

Next, let $F = \{S, \varnothing, A\}$. Now $\bar{A} = B \notin F$. Thus $\{S, \varnothing, A\}$ is not an event space. Finally, let $F = \{S, \varnothing, A, B, C\}$, but $\bar{C} = \{2, 6\} \notin F$. Hence, $\{S, \varnothing, A, B, C\}$ is not an event space.

1.24. Using the axioms of probability, prove Eq. (1.39).

We have

$$S = A \cup \bar{A} \qquad \text{and} \qquad A \cap \bar{A} = \varnothing$$

Thus, by axioms 2 and 3, it follows that

$$P(S) = 1 = P(A) + P(\bar{A})$$

from which we obtain

$$P(\bar{A}) = 1 - P(A)$$

1.25. Verify Eq. (1.40).

From Eq. (1.39), we have

$$P(A) = 1 - P(\bar{A})$$

Let $A = \varnothing$. Then, by Eq. (1.2), $\bar{A} = \varnothing = S$, and by axiom 2 we obtain

$$P(\varnothing) = 1 - P(S) = 1 - 1 = 0$$

1.26. Verify Eq. (1.41).

Let $A \subset B$. Then from the Venn diagram shown in Fig. 1-7, we see that

$$B = A \cup (\bar{A} \cap B) \qquad \text{and} \qquad A \cap (\bar{A} \cap B) = \varnothing \tag{1.74}$$

Hence, from axiom 3,

$$P(B) = P(A) + P(\bar{A} \cap B)$$

However, by axiom 1, $P(\bar{A} \cap B) \geq 0$. Thus, we conclude that

$$P(A) \leq P(B) \qquad \text{if } A \subset B$$

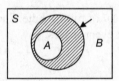

Shaded region: $\bar{A} \cap B$

Fig. 1-7

1.27. Verify Eq. (1.43).

From the Venn diagram of Fig. 1-8, each of the sets $A \cup B$ and B can be represented, respectively, as a union of mutually exclusive sets as follows:

$$A \cup B = A \cup (\bar{A} \cap B) \qquad \text{and} \qquad B = (A \cap B) \cup (\bar{A} \cap B)$$

Thus, by axiom 3,

$$P(A \cup B) = P(A) + P(\bar{A} \cap B) \tag{1.75}$$

and

$$P(B) = P(A \cap B) + P(\bar{A} \cap B) \tag{1.76}$$

From Eq. (1.76), we have

$$P(\bar{A} \cap B) = P(B) - P(A \cap B) \tag{1.77}$$

Substituting Eq. (1.77) into Eq. (1.75), we obtain

$$P(A \cup B) = P(A) + P(B) - P(A \cap B)$$

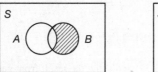

Shaded region: $\bar{A} \cap B$ Shaded region: $A \cap B$

Fig. 1-8

1.28. Let $P(A) = 0.9$ and $P(B) = 0.8$. Show that $P(A \cap B) \geq 0.7$.

From Eq. (1.43), we have

$$P(A \cap B) = P(A) + P(B) - P(A \cup B)$$

By Eq. (1.47), $0 \leq P(A \cup B) \leq 1$. Hence,

$$P(A \cap B) \geq P(A) + P(B) - 1 \tag{1.78}$$

Substituting the given values of $P(A)$ and $P(B)$ in Eq. (1.78), we get

$$P(A \cap B) \geq 0.9 + 0.8 - 1 = 0.7$$

Equation (1.78) is known as *Bonferroni's inequality*.

1.29. Show that

$$P(A) = P(A \cap B) + P(A \cap \bar{B}) \tag{1.79}$$

From the Venn diagram of Fig. 1-9, we see that

$$A = (A \cap B) \cup (A \cap \bar{B}) \quad \text{and} \quad (A \cap B) \cap (A \cap \bar{B}) = \varnothing \tag{1.80}$$

Thus, by axiom 3, we have

$$P(A) = P(A \cap B) + P(A \cap \bar{B})$$

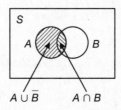

$A \cup \bar{B}$ $A \cap B$

Fig. 1-9

1.30. Given that $P(A) = 0.9$, $P(B) = 0.8$, and $P(A \cap B) = 0.75$, find (*a*) $P(A \cup B)$; (*b*) $P(A \cap \bar{B})$; and (*c*) $P(\bar{A} \cap \bar{B})$.

(*a*) By Eq. (1.43), we have

$$P(A \cup B) = P(A) + P(B) - P(A \cap B) = 0.9 + 0.8 - 0.75 = 0.95$$

(*b*)   By Eq. (1.79) (Prob. 1.29), we have

$$P(A \cap \bar{B}) = P(A) - P(A \cap B) = 0.9 - 0.75 = 0.15$$

(*c*)   By De Morgan's law, Eq. (1.20), and Eq. (1.39) and using the result from part (*a*), we get

$$P(\bar{A} \cap \bar{B}) = P(\overline{A \cup B}) = 1 - P(A \cup B) = 1 - 0.95 = 0.05$$

1.31. For any three events A_1, A_2, and A_3, show that

$$P(A_1 \cup A_2 \cup A_3) = P(A_1) + P(A_2) + P(A_3) - P(A_1 \cap A_2)$$
$$- P(A_1 \cap A_3) - P(A_2 \cap A_3) + P(A_1 \cap A_2 \cap A_3) \qquad (1.81)$$

Let $B = A_2 \cup A_3$. By Eq. (1.43), we have

$$P(A_1 \cup B) = P(A_1) + P(B) - P(A_1 \cap B) \qquad (1.82)$$

Using distributive law (1.18), we have

$$A_1 \cap B = A_1 \cap (A_2 \cup A_3) = (A_1 \cap A_2) \cup (A_1 \cap A_3)$$

Applying Eq. (1.43) to the above event, we obtain

$$P(A_1 \cap B) = P(A_1 \cap A_2) + P(A_1 \cap A_3) - P[(A_1 \cap A_2) \cap (A_1 \cap A_3)]$$
$$= P(A_1 \cap A_2) + P(A_1 \cap A_3) - P(A_1 \cap A_2 \cap A_3) \qquad (1.83)$$

Applying Eq. (1.43) to the set $B = A_2 \cup A_3$, we have

$$P(B) = P(A_2 \cup A_3) = P(A_2) + P(A_3) - P(A_2 \cap A_3) \qquad (1.84)$$

Substituting Eqs. (1.84) and (1.83) into Eq. (1.82), we get

$$P(A_1 \cup A_2 \cup A_3) = P(A_1) + P(A_2) + P(A_3) - P(A_1 \cap A_2) - P(A_1 \cap A_3)$$
$$- P(A_2 \cap A_3) + P(A_1 \cap A_2 \cap A_3)$$

1.32. Prove that

$$P\left(\bigcup_{i=1}^{n} A_i\right) \leq \sum_{i=1}^{n} P(A_i) \qquad (1.85)$$

which is known as *Boole's inequality*.

We will prove Eq. (1.85) by induction. Suppose Eq. (1.85) is true for $n = k$.

$$P\left(\bigcup_{i=1}^{k} A_i\right) \leq \sum_{i=1}^{k} P(A_i)$$

Then
$$P\left(\bigcup_{i=1}^{k+1} A_i\right) = P\left[\left(\bigcup_{i=1}^{k} A_i\right) \cup A_{k+1}\right]$$

$$\leq P\left(\bigcup_{i=1}^{k} A_i\right) + P(A_{k+1}) \qquad \text{[by Eq.(1.48)]}$$

$$\leq \sum_{i=1}^{k} P(A_i) + P(A_{k+1}) = \sum_{i=1}^{k+1} P(A_i)$$

Thus, Eq. (1.85) is also true for $n = k + 1$. By Eq. (1.48), Eq. (1.85) is true for $n = 2$. Thus, Eq. (1.85) is true for $n \geq 2$.

1.33. Verify Eq. (1.46).

Again we prove it by induction. Suppose Eq. (1.46) is true for $n = k$.

$$P\left(\bigcup_{i=1}^{k} A_i\right) = \sum_{i=1}^{k} P(A_i)$$

Then

$$P\left(\bigcup_{i=1}^{k+1} A_i\right) = P\left[\left(\bigcup_{i=1}^{k} A_i\right) \cup A_{k+1}\right]$$

Using the distributive law (1.22), we have

$$\left(\bigcup_{i=1}^{k} A_i\right) \cap A_{k+1} = \bigcup_{i=1}^{k} (A_i \cap A_{k+1}) = \bigcup_{i=1}^{k} \varnothing = \varnothing$$

since $A_i \cap A_j = \varnothing$ for $i \neq j$. Thus, by axiom 3, we have

$$P\left(\bigcup_{i=1}^{k+1} A_i\right) = P\left(\bigcup_{i=1}^{k} A_i\right) + P(A_{k+1}) = \sum_{i=1}^{k+1} P(A_i)$$

which indicates that Eq. (1.46) is also true for $n = k + 1$. By axiom 3, Eq. (1.46) is true for $n = 2$. Thus, it is true for $n \geq 2$.

1.34. A sequence of events $\{A_n, n \geq 1\}$ is said to be an *increasing sequence* if [Fig. 1-10(a)]

$$A_1 \subset A_2 \subset \cdots \subset A_k \subset A_{k+1} \subset \cdots \tag{1.86a}$$

whereas it is said to be a *decreasing sequence* if [Fig. 1-10(b)]

$$A_1 \supset A_2 \supset \cdots \supset A_k \supset A_{k+1} \supset \cdots \tag{1.86b}$$

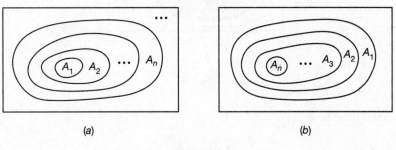

(a)   (b)

Fig. 1-10

If $\{A_n, n \geq 1\}$ is an increasing sequence of events, we define a new event A_∞ by

$$A_\infty = \lim_{n \to \infty} A_n = \bigcup_{i=1}^{\infty} A_i \tag{1.87}$$

Similarly, if $\{A_n, n \geq 1\}$ is a decreasing sequence of events, we define a new event A_∞ by

$$A_\infty = \lim_{n \to \infty} A_n = \bigcap_{i=1}^{\infty} A_i \tag{1.88}$$

Show that if $\{A_n, n \geq 1\}$ is either an increasing or a decreasing sequence of events, then

$$\lim_{n \to \infty} P(A_n) = P(A_\infty) \tag{1.89}$$

which is known as the *continuity theorem of probability*.

If $\{A_n, n \geq 1\}$ is an increasing sequence of events, then by definition

$$\bigcup_{i=1}^{n-1} A_i = A_{n-1}$$

Now, we define the events $B_n, n \geq 1$, by

$$B_1 = A_1$$
$$B_2 = A_2 \cap \bar{A}_1$$
$$\vdots$$
$$B_n = A_n \cap \bar{A}_{n-1}$$

Thus, B_n consists of those elements in A_n that are not in any of the earlier A_k, $k < n$. From the Venn diagram shown in Fig. 1-11, it is seen that B_n are mutually exclusive events such that

$$\bigcup_{i=1}^{n} B_i = \bigcup_{i=1}^{n} A_i \text{ for all } n \geq 1, \text{ and } \bigcup_{i=1}^{\infty} B_i = \bigcup_{i=1}^{\infty} A_i = A_\infty$$

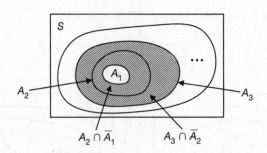

Fig. 1-11

Thus, using axiom 3', we have

$$P(A_\infty) = P\left(\bigcup_{i=1}^{\infty} A_i\right) = P\left(\bigcup_{i=1}^{\infty} B_i\right) = \sum_{i=1}^{\infty} P(B_i)$$

$$= \lim_{n \to \infty} \sum_{i=1}^{n} P(B_i) = \lim_{n \to \infty} P\left(\bigcup_{i=1}^{n} B_i\right)$$

$$= \lim_{n \to \infty} P\left(\bigcup_{i=1}^{n} A_i\right) = \lim_{n \to \infty} P(A_n) \tag{1.90}$$

Next, if $\{A_n, n \geq 1\}$ is a decreasing sequence, then $\{\bar{A}_n, n > 1\}$ is an increasing sequence. Hence, by Eq. (1.89), we have

$$P\left(\bigcup_{i=1}^{\infty} \bar{A}_i\right) = \lim_{n \to \infty} P(\bar{A}_n)$$

From Eq. (1.25),

$$\bigcup_{i=1}^{\infty} \bar{A}_i = \overline{\left(\bigcap_{i=1}^{\infty} A_i\right)}$$

Thus,

$$P\left[\overline{\left(\bigcap_{i=1}^{\infty} A_i\right)}\right] = \lim_{n \to \infty} P(\bar{A}_n) \tag{1.91}$$

Using Eq. (1.39), Eq. (1.91) reduces to

$$1 - P\left(\bigcap_{i=1}^{\infty} A_i\right) = \lim_{n \to \infty} [1 - P(A_n)] = 1 - \lim_{n \to \infty} P(A_n) \tag{1.92}$$

Thus,

$$P\left(\bigcap_{i=1}^{\infty} A_i\right) = P(A_\infty) = \lim_{n \to \infty} P(A_n)$$

Combining Eqs. (190) and (1.92), we obtain Eq. (1.89).

Equally Likely Events

1.35. Consider a telegraph source generating two symbols, dots and dashes. We observed that the dots were twice as likely to occur as the dashes. Find the probabilities of the dots occurring and the dashes occurring.

From the observation, we have

$$P(\text{dot}) = 2P(\text{dash})$$

Then, by Eq. (1.51),

$$P(\text{dot}) + P(\text{dash}) = 3P(\text{dash}) = 1$$

Thus,

$$P(\text{dash}) = \frac{1}{3} \quad \text{and} \quad P(\text{dot}) = \frac{2}{3}$$

1.36. The sample space S of a random experiment is given by

$$S = \{a, b, c, d\}$$

with probabilities $P(a) = 0.2, P(b) = 0.3, P(c) = 0.4$, and $P(d) = 0.1$. Let A denote the event $\{a, b\}$, and B the event $\{b, c, d\}$. Determine the following probabilities: (*a*) $P(A)$; (*b*) $P(B)$; (*c*) $P(\bar{A})$; (*d*) $P(A \cup B)$; and (*e*) $P(A \cap B)$.

Using Eq. (1.52), we obtain

(a) $P(A) = P(a) + P(b) = 0.2 + 0.3 = 0.5$

(b) $P(B) = P(b) + P(c) + P(d) = 0.3 + 0.4 + 0.1 = 0.8$

(c) $\bar{A} = \{c, d\}; P(A) = P(c) + P(d) = 0.4 + 0.1 = 0.5$

(d) $A \cup B = \{a, b, c, d\} = S; P(A \cup B) = P(S) = 1$

(e) $A \cap B = \{b\}; P(A \cap B) = P(b) = 0.3$

1.37. An experiment consists of observing the sum of the dice when two fair dice are thrown (Prob. 1.5). Find (a) the probability that the sum is 7 and (b) the probability that the sum is greater than 10.

(a) Let ζ_{ij} denote the elementary event (sampling point) consisting of the following outcome: $\zeta_{ij} = (i, j)$, where i represents the number appearing on one die and j represents the number appearing on the other die. Since the dice are fair, all the outcomes are equally likely. So $P(\zeta_{ij}) = \frac{1}{36}$. Let A denote the event that the sum is 7. Since the events ζ_{ij} are mutually exclusive and from Fig. 1-3 (Prob. 1.5), we have

$$P(A) = P(\zeta_{16} \cup \zeta_{25} \cup \zeta_{34} \cup \zeta_{43} \cup \zeta_{52} \cup \zeta_{61})$$
$$= P(\zeta_{16}) + P(\zeta_{25}) + P(\zeta_{34}) + P(\zeta_{43}) + P(\zeta_{52}) + P(\zeta_{61})$$
$$= 6\left(\frac{1}{36}\right) = \frac{1}{6}$$

(b) Let B denote the event that the sum is greater than 10. Then from Fig. 1-3, we obtain

$$P(B) = P(\zeta_{56} \cup \zeta_{65} \cup \zeta_{66}) = P(\zeta_{56}) + P(\zeta_{65}) + P(\zeta_{66})$$
$$= 3\left(\frac{1}{36}\right) = \frac{1}{12}$$

1.38. There are n persons in a room.

(a) What is the probability that at least two persons have the same birthday?

(b) Calculate this probability for $n = 50$.

(c) How large need n be for this probability to be greater than 0.5?

(a) As each person can have his or her birthday on any one of 365 days (ignoring the possibility of February 29), there are a total of $(365)^n$ possible outcomes. Let A be the event that no two persons have the same birthday. Then the number of outcomes belonging to A is

$$n(A) = (365)(364) \cdots (365 - n + 1)$$

Assuming that each outcome is equally likely, then by Eq. (1.54),

$$P(A) = \frac{n(A)}{n(S)} = \frac{(365)(364) \cdots (365 - n + 1)}{(365)^n} \tag{1.93}$$

Let B be the event that at least two persons have the same birthday. Then $B = \bar{A}$ and by Eq.(1.39), $P(B) = 1 - P(A)$.

(b) Substituting $n = 50$ in Eq. (1.93), we have

$$P(A) \approx 0.03 \quad \text{and} \quad P(B) \approx 1 - 0.03 = 0.97$$

(c) From Eq. (1.93), when $n = 23$, we have

$$P(A) \approx 0.493 \quad \text{and} \quad P(B) = 1 - P(A) \approx 0.507$$

That is, if there are 23 persons in a room, the probability that at least two of them have the same birthday exceeds 0.5.

1.39. A committee of 5 persons is to be selected randomly from a group of 5 men and 10 women.

   (*a*) Find the probability that the committee consists of 2 men and 3 women.

   (*b*) Find the probability that the committee consists of all women.

   (*a*) The number of total outcomes is given by

$$n(S) = \binom{15}{5}$$

It is assumed that "random selection" means that each of the outcomes is equally likely. Let *A* be the event that the committee consists of 2 men and 3 women. Then the number of outcomes belonging to *A* is given by

$$n(A) = \binom{5}{2}\binom{10}{3}$$

Thus, by Eq. (1.54),

$$P(A) = \frac{n(A)}{n(S)} = \frac{\binom{5}{2}\binom{10}{3}}{\binom{15}{5}} = \frac{400}{1001} \approx 0.4$$

   (*b*) Let *B* be the event that the committee consists of all women. Then the number of outcomes belonging to *B* is

$$n(B) = \binom{5}{0}\binom{10}{5}$$

Thus, by Eq. (1.54),

$$P(B) = \frac{n(B)}{n(S)} = \frac{\binom{5}{0}\binom{10}{5}}{\binom{15}{5}} = \frac{36}{429} \approx 0.084$$

1.40. Consider the switching network shown in Fig. 1-12. It is equally likely that a switch will or will not work. Find the probability that a closed path will exist between terminals *a* and *b*.

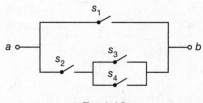

Fig. 1-12

Consider a sample space *S* of which a typical outcome is (1, 0, 0, 1), indicating that switches 1 and 4 are closed and switches 2 and 3 are open. The sample space contains $2^4 = 16$ points, and by assumption, they are equally likely (Fig. 1-13).

Let $A_i, i = 1, 2, 3, 4$ be the event that the switch s_i is closed. Let A be the event that there exists a closed path between a and b. Then

$$A = A_1 \cup (A_2 \cap A_3) \cup (A_2 \cap A_4)$$

Applying Eq. (1.45), we have

$$
\begin{aligned}
P(A) &= P[A_1 \cup (A_2 \cap A_3) \cup (A_2 \cap A_4)] \\
&= P(A_1) + P(A_2 \cap A_3) + P(A_2 \cap A_4) \\
&\quad - P[A_1 \cap (A_2 \cap A_3)] - P[A_1 \cap (A_2 \cap A_4)] - P[(A_2 \cap A_3) \cap (A_2 \cap A_4)] \\
&\quad + P[A_1 \cap (A_2 \cap A_3) \cap (A_2 \cap A_4)] \\
&= P(A_1) + P(A_2 \cap A_3) + P(A_2 \cap A_4) \\
&\quad - P(A_1 \cap A_2 \cap A_3) - P(A_1 \cap A_2 \cap A_4) - P(A_2 \cap A_3 \cap A_4) \\
&\quad + P(A_1 \cap A_2 \cap A_3 \cap A_4)
\end{aligned}
$$

Now, for example, the event $A_2 \cap A_3$ contains all elementary events with a 1 in the second and third places. Thus, from Fig. 1-13, we see that

$$
\begin{aligned}
n(A_1) &= 8 \qquad n(A_2 \cap A_3) = 4 \qquad n(A_2 \cap A_4) = 4 \\
n(A_1 \cap A_2 \cap A_3) &= 2 \qquad n(A_1 \cap A_2 \cap A_4) = 2 \\
n(A_2 \cap A_3 \cap A_4) &= 2 \qquad n(A_1 \cap A_2 \cap A_3 \cap A_4) = 1
\end{aligned}
$$

Thus,

$$P(A) = \frac{8}{16} + \frac{4}{16} + \frac{4}{16} - \frac{2}{16} - \frac{2}{16} - \frac{2}{16} + \frac{1}{16} = \frac{11}{16} \approx 0.688$$

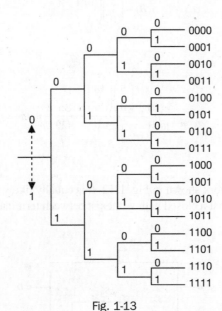

Fig. 1-13

1.41. Consider the experiment of tossing a fair coin repeatedly and counting the number of tosses required until the first head appears.

(*a*) Find the sample space of the experiment.

(*b*) Find the probability that the first head appears on the *k*th toss.

(*c*) Verify that $P(S) = 1$.

(a) The sample space of this experiment is

$$S = \{e_1, e_2, e_3, \ldots\} = \{e_k : k = 1, 2, 3, \ldots\}$$

where e_k is the elementary event that the first head appears on the kth toss.

(b) Since a fair coin is tossed, we assume that a head and a tail are equally likely to appear. Then $P(H) = P(T) = \frac{1}{2}$. Let

$$P(e_k) = p_k \qquad k = 1, 2, 3, \ldots$$

Since there are 2^k equally likely ways of tossing a fair coin k times, only one of which consists of $(k-1)$ tails following a head we observe that

$$P(e_k) = p_k = \frac{1}{2^k} \qquad k = 1, 2, 3, \ldots \tag{1.94}$$

(c) Using the power series summation formula, we have

$$P(S) = \sum_{k=1}^{\infty} P(e_k) = \sum_{k=1}^{\infty} \frac{1}{2^k} = \sum_{k=1}^{\infty} \left(\frac{1}{2}\right)^k = \frac{\frac{1}{2}}{1 - \frac{1}{2}} = 1 \tag{1.95}$$

1.42. Consider the experiment of Prob. 1.41.

(a) Find the probability that the first head appears on an even-numbered toss.

(b) Find the probability that the first head appears on an odd-numbered toss.

(a) Let A be the event "the first head appears on an even-numbered toss." Then, by Eq. (1.52) and using Eq. (1.94) of Prob. 1.41, we have

$$P(A) = p_2 + p_4 + p_6 + \cdots = \sum_{m=1}^{\infty} p_{2m} = \sum_{m=1}^{\infty} \frac{1}{2^{2m}} = \sum_{m=1}^{\infty} \left(\frac{1}{4}\right)^m = \frac{\frac{1}{4}}{1 - \frac{1}{4}} = \frac{1}{3}$$

(b) Let B be the event "the first head appears on an odd-numbered toss." Then it is obvious that $B = \bar{A}$. Then, by Eq. (1.39), we get

$$P(B) = P(\bar{A}) = 1 - P(A) = 1 - \frac{1}{3} = \frac{2}{3}$$

As a check, notice that

$$P(B) = p_1 + p_3 + p_5 + \cdots = \sum_{m=0}^{\infty} p_{2m+1} = \sum_{m=0}^{\infty} \frac{1}{2^{2m+1}} = \frac{1}{2} \sum_{m=0}^{\infty} \left(\frac{1}{4}\right)^m = \frac{1}{2}\left(\frac{1}{1 - \frac{1}{4}}\right) = \frac{2}{3}$$

Conditional Probability

1.43. Show that $P(A \mid B)$ defined by Eq. (1.55) satisfies the three axioms of a probability, that is,

(a) $P(A \mid B) \geq 0$ (1.96)

(b) $P(S \mid B) = 1$ (1.97)

(c) $P(A_1 \cup A_2 \mid B) = P(A_1 \mid B) + P(A_2 \mid B)$ if $A_1 \cap A_2 = \varnothing$ (1.98)

(*a*)   From definition (1.55),

$$P(A|B) = \frac{P(A \cap B)}{P(B)} \qquad P(B) > 0$$

By axiom 1, $P(A \cap B) \geq 0$. Thus,

$$P(A \mid B) \geq 0$$

(*b*)   By Eq. (1.5), $S \cap B = B$. Then

$$P(S|B) = \frac{P(S \cap B)}{P(B)} = \frac{P(B)}{P(B)} = 1$$

(*c*)   By definition (1.55),

$$P(A_1 \cup A_2 | B) = \frac{P[(A_1 \cup A_2) \cap B]}{P(B)}$$

Now by Eqs. (1.14) and (1.17), we have

$$(A_1 \cup A_2) \cap B = (A_1 \cap B) \cup (A_2 \cap B)$$

and $A_1 \cap A_2 = \varnothing$ implies that $(A_1 \cap B) \cap (A_2 \cap B) = \varnothing$. Thus, by axiom 3 we get

$$P(A_1 \cup A_2 | B) = \frac{P(A_1 \cap B) + P(A_2 \cap B)}{P(B)} = \frac{P(A_1 \cap B)}{P(B)} + \frac{P(A_2 \cap B)}{P(B)}$$

$$= P(A_1 | B) + P(A_2 | B) \qquad \text{if } A_1 \cap A_2 = \varnothing$$

1.44.   Find $P(A \mid B)$ if (*a*) $A \cap B = \varnothing$, (*b*) $A \subset B$, and (*c*) $B \subset A$.

(*a*)   If $A \cap B = \varnothing$, then $P(A \cap B) = P(\varnothing) = 0$. Thus,

$$P(A|B) = \frac{P(A \cap B)}{P(B)} = \frac{P(\varnothing)}{P(B)} = 0$$

(*b*)   If $A \subset B$, then $A \cap B = A$ and

$$P(A|B) = \frac{P(A \cap B)}{P(B)} = \frac{P(A)}{P(B)}$$

(*c*)   If $B \subset A$, then $A \cap B = B$ and

$$P(A|B) = \frac{P(A \cap B)}{P(B)} = \frac{P(B)}{P(B)} = 1$$

1.45.   Show that if $P(A \mid B) > P(A)$, then $P(B \mid A) > P(B)$.

If $P(A|B) = \dfrac{P(A \cap B)}{P(B)} > P(A)$, then $P(A \cap B) > P(A)P(B)$. Thus,

$$P(B|A) = \frac{P(A \cap B)}{P(A)} > \frac{P(A)P(B)}{P(A)} = P(B) \qquad \text{or} \qquad P(B|A) > P(B)$$

1.46. Show that

$$P(\overline{A}|B) = 1 - P(A|B) \tag{1.99}$$

By Eqs. (1.6) and (1.7), we have

$$A \cup \overline{A} = S, \qquad A \cap \overline{A} = \varnothing$$

Then by Eqs. (1.97) and (1.98), we get

$$P(A \cup \overline{A}|B) = P(S|B) = 1 = P(A|B) + P(\overline{A}|B)$$

Thus we obtain

$$P(\overline{A}|B) = 1 - P(A|B)$$

note that Eq. (1.99) is similar to property 1 (Eq. (1.39)).

1.47. Show that

$$P(A_1 \cup A_2|B) = P(A_1|B) + P(A_2|B) - P(A_1 \cap A_2|B) \tag{1.100}$$

Using a Venn diagram, we see that

$$A_1 \cup A_2 = A_1 \cup (A_2 \setminus A_1) = A_1 \cup (A_2 \cap \overline{A}_1)$$

and

$$A_1 \cup (A_2 \cap \overline{A}_1) = \varnothing$$

By Eq. (1.98) we have

$$P(A_1 \cup A_2|B) = P\big[A_1 \cup \big(A_2 \cap \overline{A}_1\big)|B\big] = P(A_1|B) + P\big(A_2 \cap \overline{A}_1|B\big) \tag{1.100}$$

Again, using a Venn diagram, we see that

$$A_2 = (A_1 \cap A_2) \cup (A_2 \setminus A_1) = (A_1 \cap A_2) \cup (A_2 \cap \overline{A}_1)$$

and

$$(A_1 \cap A_2) \cup (A_2 \cap \overline{A}_1) = \varnothing$$

By Eq. (1.98) we have

$$P(A_2|B) = P\big[(A_1 \cap A_2) \cup \big(A_2 \cap \overline{A}_1\big)|B\big] = P\big(A_1 \cap A_2|B\big) + P\big(A_2 \cap \overline{A}_1|B\big)$$

Thus,

$$P\left(A_2 \cap \bar{A}_1 \mid B\right) = P\left(A_2 \mid B\right) - P\left(A_1 \cap A_2 \mid B\right) \tag{1.101}$$

Substituting Eq. (1.101) into Eq. (1.100), we obtain

$$P\left(A_1 \cup A_2 \mid B\right) = P\left(A_1 \mid B\right) + P\left(A_2 \mid B\right) - P\left(A_1 \cap A_2 \mid B\right)$$

Note that Eq. (1.100) is similar to property 5 (Eq. (1.43)).

1.48. Consider the experiment of throwing the two fair dice of Prob. 1.37 behind you; you are then informed that the sum is not greater than 3.

   (*a*) Find the probability of the event that two faces are the same without the information given.

   (*b*) Find the probability of the same event with the information given.

   (*a*) Let A be the event that two faces are the same. Then from Fig. 1-3 (Prob. 1.5) and by Eq. (1.54), we have

$$A = \{(i, i): i = 1, 2, \ldots, 6\}$$

and

$$P(A) = \frac{n(A)}{n(S)} = \frac{6}{36} = \frac{1}{6}$$

   (*b*) Let B be the event that the sum is not greater than 3. Again from Fig. 1-3, we see that

$$B = \{(i, j): i + j \le 3\} = \{(1, 1), (1, 2), (2, 1)\}$$

and

$$P(B) = \frac{n(B)}{n(S)} = \frac{3}{36} = \frac{1}{12}$$

Now $A \cap B$ is the event that two faces are the same and also that their sum is not greater than 3. Thus,

$$P(A \cap B) = \frac{n(A \cap B)}{n(S)} = \frac{1}{36}$$

Then by definition (1.55), we obtain

$$P(A \mid B) = \frac{P(A \cap B)}{P(B)} = \frac{\frac{1}{36}}{\frac{1}{12}} = \frac{1}{3}$$

Note that the probability of the event that two faces are the same doubled from $\frac{1}{6}$ to $\frac{1}{3}$ with the information given.

Alternative Solution:

There are 3 elements in B, and 1 of them belongs to A. Thus, the probability of the same event with the information given is $\frac{1}{3}$.

1.49. Two manufacturing plants produce similar parts. Plant 1 produces 1,000 parts, 100 of which are defective. Plant 2 produces 2,000 parts, 150 of which are defective. A part is selected at random and found to be defective. What is the probability that it came from plant 1?

Let B be the event that "the part selected is defective," and let A be the event that "the part selected came from plant 1." Then $A \cap B$ is the event that the item selected is defective and came from plant 1. Since a part is selected at random, we assume equally likely events, and using Eq. (1.54), we have

$$P(A \cap B) = \frac{100}{3000} = \frac{1}{30}$$

Similarly, since there are 3000 parts and 250 of them are defective, we have

$$P(B) = \frac{250}{3000} = \frac{1}{12}$$

By Eq. (1.55), the probability that the part came from plant 1 is

$$P(A|B) = \frac{P(A \cap B)}{P(B)} = \frac{\frac{1}{30}}{\frac{1}{12}} = \frac{2}{5} = 0.4$$

Alternative Solution:

There are 250 defective parts, and 100 of these are from plant 1. Thus, the probability that the defective part came from plant 1 is $\frac{100}{250} = 0.4$.

1.50. Mary has two children. One child is a boy. What is the probability that the other child is a girl?

Let S be the sample space of all possible events $S = \{B\,B, B\,G, G\,B, G\,G\}$ where $B\,B$ denotes the events the first child is a boy and the second is also a boy, and BG denotes the event the first child is a boy and the second is a girl, and so on.

Now we have

$$P[\{BB\}] = P[\{BG\}] = P[\{GB\}] = P[\{GG\}] = 1/4$$

Let B be the event that there is at least one boy; $B = \{B\,B, B\,G, G\,B\}$, and A be the event that there is at least one girl; $A = \{B\,G, G\,B, G\,G\}$

Then

$$A \cap B = \{BG, GB\} \quad \text{and} \quad P(A \cap B) = 1/2$$
$$P(B) = 3/4$$

Now, by Eq. (1.55) we have

$$P(A|B) = \frac{P(A \cap B)}{P(B)} = \frac{(1/2)}{(3/4)} = \frac{2}{3}$$

Note that one would intuitively think the answer is 1/2 because the second event looks independent of the first. This problem illustrates that the initial intuition can be misleading.

1.51. A lot of 100 semiconductor chips contains 20 that are defective. Two chips are selected at random, without replacement, from the lot.

(*a*) What is the probability that the first one selected is defective?

(*b*) What is the probability that the second one selected is defective given that the first one was defective?

(*c*) What is the probability that both are defective?

(a) Let A denote the event that the first one selected is defective. Then, by Eq. (1.54),

$$P(A) = \frac{20}{100} = 0.2$$

(b) Let B denote the event that the second one selected is defective. After the first one selected is defective, there are 99 chips left in the lot, with 19 chips that are defective. Thus, the probability that the second one selected is defective given that the first one was defective is

$$P(B \mid A) = \frac{19}{99} = 0.192$$

(c) By Eq. (1.57), the probability that both are defective is

$$P(A \cap B) = P(B \mid A)P(A) = \left(\frac{19}{99}\right)(0.2) = 0.0384$$

1.52. A number is selected at random from $\{1, 2, \ldots, 100\}$. Given that the number selected is divisible by 2, find the probability that it is divisible by 3 or 5.

Let

$$A_2 = \text{event that the number is divisible by 2}$$
$$A_3 = \text{event that the number is divisible by 3}$$
$$A_5 = \text{event that the number is divisible by 5}$$

Then the desired probability is

$$P(A_3 \cup A_5 \mid A_2) = \frac{P[(A_3 \cup A_5) \cap A_2]}{P(A_2)} \qquad \text{[Eq. (1.55)]}$$

$$= \frac{P[(A_3 \cap A_2) \cup (A_5 \cap A_2)]}{P(A_2)} \qquad \text{[Eq. (1.18)]}$$

$$= \frac{P(A_3 \cap A_2) + P(A_5 \cap A_2) - P(A_3 \cap A_5 \cap A_2)}{P(A_2)} \qquad \text{[Eq. (1.44)]}$$

Now

$$A_3 \cap A_2 = \text{event that the number is divisible by 6}$$
$$A_5 \cap A_2 = \text{event that the number is divisible by 10}$$
$$A_3 \cap A_5 \cap A_2 = \text{event that the number is divisible by 30}$$

and

$$P(A_3 \cap A_2) = \frac{16}{100} \qquad P(A_5 \cap A_2) = \frac{10}{100} \qquad P(A_3 \cap A_5 \cap A_2) = \frac{3}{100}$$

Thus,

$$P(A_3 \cup A_5 \mid A_2) = \frac{\frac{16}{100} + \frac{10}{100} - \frac{3}{100}}{\frac{50}{100}} = \frac{23}{50} = 0.46$$

1.53. Show that

$$P(A \cap B \cap C) = P(A)P(B \mid A)P(C \mid A \cap B) \qquad (1.102)$$

By definition Eq. (1.55) we have

$$P(A)P(B|A)P(C|A \cap B) = P(A)\frac{P(B \cap A)}{P(A)}\frac{P(C \cap A \cap B)}{P(A \cap B)}$$
$$= P(A \cap B \cap C)$$

since $P(B \cap A) = P(A \cap B)$ and $P(C \cap A \cap B) = P(A \cap B \cap C)$.

1.54. Let $A_1, A_2, \ldots, A_n$ be events in a sample space S. Show that

$$P(A_1 \cap A_2 \cap \cdots \cap A_n) = P(A_1)P(A_2 \mid A_1)P(A_3 \mid A_1 \cap A_2) \cdots P(A_n \mid A_1 \cap A_2 \cap \cdots \cap A_{n-1}) \qquad (1.103)$$

We prove Eq. (1.103) by induction. Suppose Eq. (1.103) is true for $n = k$:

$$P(A_1 \cap A_2 \cap \cdots \cap A_k) = P(A_1)P(A_2 \mid A_1)P(A_3 \mid A_1 \cap A_2) \cdots P(A_k \mid A_1 \cap A_2 \cap \cdots \cap A_{k-1})$$

Multiplying both sides by $P(A_{k+1} \mid A_1 \cap A_2 \cap \cdots \cap A_k)$, we have

$$P(A_1 \cap A_2 \cap \cdots \cap A_k)P(A_{k+1} \mid A_1 \cap A_2 \cap \cdots \cap A_k) = P(A_1 \cap A_2 \cap \cdots \cap A_{k+1})$$

and

$$P(A_1 \cap A_2 \cap \cdots \cap A_{k+1}) = P(A_1)P(A_2 \mid A_1)P(A_3 \mid A_1 \cap A_2) \cdots P(A_{k+1} \mid A_1 \cap A_2 \cap \cdots \cap A_k)$$

Thus, Eq. (1.103) is also true for $n = k + 1$. By Eq. (1.57), Eq. (1.103) is true for $n = 2$. Thus, Eq. (1.103) is true for $n \geq 2$.

1.55. Two cards are drawn at random from a deck. Find the probability that both are aces.

Let A be the event that the first card is an ace, and let B be the event that the second card is an ace. The desired probability is $P(B \cap A)$. Since a card is drawn at random, $P(A) = \frac{4}{52}$. Now if the first card is an ace, then there will be 3 aces left in the deck of 51 cards. Thus, $P(B \mid A) = \frac{3}{51}$. By Eq. (1.57),

$$P(B \cap A) = P(B|A)P(A) = \left(\frac{3}{51}\right)\left(\frac{4}{52}\right) = \frac{1}{221}$$

Check:

By counting technique, we have

$$P(B \cap A) = \frac{\binom{4}{2}}{\binom{52}{2}} = \frac{(4)(3)}{(52)(51)} = \frac{1}{221}$$

1.56. There are two identical decks of cards, each possessing a distinct symbol so that the cards from each deck can be identified. One deck of cards is laid out in a fixed order, and the other deck is shuffled and the cards laid out one by one on top of the fixed deck. Whenever two cards with the same symbol occur in the same position, we say that a *match* has occurred. Let the number of cards in the deck be 10. Find the probability of getting a match at the first four positions.

Let A_i, $i = 1, 2, 3, 4$, be the events that a match occurs at the ith position. The required probability is

$$P(A_1 \cap A_2 \cap A_3 \cap A_4)$$

By Eq. (1.103),

$$P(A_1 \cap A_2 \cap A_3 \cap A_4) = P(A_1)P(A_2 \mid A_1)P(A_3 \mid A_1 \cap A_2)P(A_4 \mid A_1 \cap A_2 \cap A_3)$$

There are 10 cards that can go into position 1, only one of which matches. Thus, $P(A_1) = \frac{1}{10}$. $P(A_2 \mid A_1)$ is the conditional probability of a match at position 2 given a match at position 1. Now there are 9 cards left to go into position 2, only one of which matches. Thus, $P(A_2 \mid A_1) = \frac{1}{9}$. In a similar fashion, we obtain $P(A_3 \mid A_1 \cap A_2) = \frac{1}{8}$ and $P(A_4 \mid A_1 \cap A_2 \cap A_3) = \frac{1}{7}$. Thus,

$$P(A_1 \cap A_2 \cap A_3 \cap A_4) = \left(\frac{1}{10}\right)\left(\frac{1}{9}\right)\left(\frac{1}{8}\right)\left(\frac{1}{7}\right) = \frac{1}{5040}$$

Total Probability

1.57. Verify Eq. (1.60).

Since $B \cap S = B$ [and using Eq. (1.59)], we have

$$\begin{aligned} B = B \cap S &= B \cap (A_1 \cup A_2 \cup \cdots \cup A_n) \\ &= (B \cap A_1) \cup (B \cap A_2) \cup \cdots \cup (B \cap A_n) \end{aligned} \tag{1.104}$$

Now the events $B \cap A_i$, $i = 1, 2, \ldots, n$, are mutually exclusive, as seen from the Venn diagram of Fig. 1-14. Then by axiom 3 of probability and Eq. (1.57), we obtain

$$P(B) = P(B \cap S) = \sum_{i=1}^{n} P(B \cap A_i) = \sum_{i=1}^{n} P(B \mid A_i)P(A_i)$$

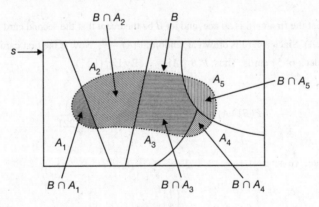

Fig. 1-14

1.58. Show that for any events A and B in S,

$$P(B) = P(B \mid A)P(A) + P(B \mid \bar{A})P(\bar{A}) \tag{1.105}$$

From Eq. (1.78) (Prob. 1.29), we have

$$P(B) = P(B \cap A) + P(B \cap \bar{A})$$

Using Eq. (1.55), we obtain

$$P(B) = P(B \mid A)P(A) + P(B \mid \bar{A})P(\bar{A})$$

Note that Eq. (1.105) is the special case of Eq. (1.60).

1.59. Suppose that a laboratory test to detect a certain disease has the following statistics. Let

A = event that the tested person has the disease

B = event that the test result is positive

It is known that

$$P(B \mid A) = 0.99 \quad \text{and} \quad P(B \mid \bar{A}) = 0.005$$

and 0.1 percent of the population actually has the disease. What is the probability that a person has the disease given that the test result is positive?

From the given statistics, we have

$$P(A) = 0.001 \quad \text{then} \quad P(\bar{A}) = 0.999$$

The desired probability is $P(A \mid B)$. Thus, using Eqs. (1.58) and (1.105), we obtain

$$P(A|B) = \frac{P(B|A)P(A)}{P(B|A)P(A) + P(B|\bar{A})P(\bar{A})}$$

$$= \frac{(0.99)(0.001)}{(0.99)(0.001) + (0.005)(0.999)} = 0.165$$

Note that in only 16.5 percent of the cases where the tests are positive will the person actually have the disease even though the test is 99 percent effective in detecting the disease when it is, in fact, present.

1.60. A company producing electric relays has three manufacturing plants producing 50, 30, and 20 percent, respectively, of its product. Suppose that the probabilities that a relay manufactured by these plants is defective are 0.02, 0.05, and 0.01, respectively.

(a) If a relay is selected at random from the output of the company, what is the probability that it is defective?

(b) If a relay selected at random is found to be defective, what is the probability that it was manufactured by plant 2?

(a) Let B be the event that the relay is defective, and let A_i be the event that the relay is manufactured by plant i ($i = 1, 2, 3$). The desired probability is $P(B)$. Using Eq. (1.60), we have

$$P(B) = \sum_{i=1}^{3} P(B|A_i)P(A_i)$$

$$= (0.02)(0.5) + (0.05)(0.3) + (0.01)(0.2) = 0.027$$

(b) The desired probability is $P(A_2 \mid B)$. Using Eq. (1.58) and the result from part (a), we obtain

$$P(A_2|B) = \frac{P(B|A_2)P(A_2)}{P(B)} = \frac{(0.05)(0.3)}{0.027} = 0.556$$

1.61. Two numbers are chosen at random from among the numbers 1 to 10 without replacement. Find the probability that the second number chosen is 5.

Let A_i, $i = 1, 2, \ldots, 10$ denote the event that the first number chosen is i. Let B be the event that the second number chosen is 5. Then by Eq. (1.60),

$$P(B) = \sum_{i=1}^{10} P(B|A_i)P(A_i)$$

Now $P(A_i) = \frac{1}{10}$. $P(B \mid A_i)$ is the probability that the second number chosen is 5, given that the first is i. If $i = 5$, then $P(B \mid A_i) = 0$. If $i \neq 5$, then $P(B \mid A_i) = \frac{1}{9}$. Hence,

$$P(B) = \sum_{i=1}^{10} P(B|A_i)P(A_i) = 9\left(\frac{1}{9}\right)\left(\frac{1}{10}\right) = \frac{1}{10}$$

1.62. Consider the binary communication channel shown in Fig. 1-15. The channel input symbol X may assume the state 0 or the state 1, and, similarly, the channel output symbol Y may assume either the state 0 or the state 1. Because of the channel noise, an input 0 may convert to an output 1 and vice versa. The channel is characterized by the channel transition probabilities $p_0, q_0, p_1,$ and q_1, defined by

$$p_0 = P(y_1|x_0) \qquad \text{and} \qquad p_1 = P(y_0|x_1)$$
$$q_0 = P(y_0|x_0) \qquad \text{and} \qquad q_1 = P(y_1|x_1)$$

where x_0 and x_1 denote the events $(X = 0)$ and $(X = 1)$, respectively, and y_0 and y_1 denote the events $(Y = 0)$ and $(Y = 1)$, respectively. Note that $p_0 + q_0 = 1 = p_1 + q_1$. Let $P(x_0) = 0.5, p_0 = 0.1,$ and $p_1 = 0.2$.

(a) Find $P(y_0)$ and $P(y_1)$.

(b) If a 0 was observed at the output, what is the probability that a 0 was the input state?

(c) If a 1 was observed at the output, what is the probability that a 1 was the input state?

(d) Calculate the probability of error P_e.

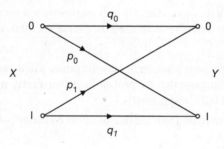

Fig. 1-15

(a) We note that

$$P(x_1) = 1 - P(x_0) = 1 - 0.5 = 0.5$$
$$P(y_0|x_0) = q_0 = 1 - p_0 = 1 - 0.1 = 0.9$$
$$P(y_1|x_1) = q_1 = 1 - p_1 = 1 - 0.2 = 0.8$$

Using Eq. (1.60), we obtain

$$P(y_0) = P(y_0|x_0)P(x_0) + P(y_0|x_1)P(x_1) = 0.9(0.5) + 0.2(0.5) = 0.55$$
$$P(y_1) = P(y_1|x_0)P(x_0) + P(y_1|x_1)P(x_1) = 0.1(0.5) + 0.8(0.5) = 0.45$$

(b) Using Bayes' rule (1.58), we have

$$P(x_0|y_0) = \frac{P(x_0)P(y_0|x_0)}{P(y_0)} = \frac{(0.5)(0.9)}{0.55} = 0.818$$

(c) Similarly,

$$P(x_1|y_1) = \frac{P(x_1)P(y_1|x_1)}{P(y_1)} = \frac{(0.5)(0.8)}{0.45} = 0.889$$

(d) The probability of error is

$$P_e = P(y_1|x_0)P(x_0) + P(y_0|x_1)P(x_1) = 0.1(0.5) + 0.2(0.5) = 0.15.$$

Independent Events

1.63. Let A and B be events in an event space F. Show that if A and B are independent, then so are (a) A and $\bar{B}$, (b) $\bar{A}$ and B, and (c) $\bar{A}$ and $\bar{B}$.

(a) From Eq. (1.79) (Prob. 1.29), we have

$$P(A) = P(A \cap B) + P(A \cap \bar{B})$$

Since A and B are independent, using Eqs. (1.62) and (1.39), we obtain

$$P(A \cap \bar{B}) = P(A) - P(A \cap B) = P(A) - P(A)P(B)$$
$$= P(A)[1 - P(B)] = P(A)P(\bar{B}) \tag{1.106}$$

Thus, by definition (1.62), A and $\bar{B}$ are independent.

(b) Interchanging A and B in Eq. (1.106), we obtain

$$P(B \cap \bar{A}) = P(B)\,P(\bar{A})$$

which indicates that $\bar{A}$ and B are independent.

(c) We have

$$
\begin{aligned}
P(\bar{A} \cap \bar{B}) &= P[\overline{(A \cup B)}] && [\text{Eq. (1.20)}]\\
&= 1 - P(A \cup B) && [\text{Eq. (1.39)}]\\
&= 1 - P(A) - P(B) + P(A \cap B) && [\text{Eq. (1.44)}]\\
&= 1 - P(A) - P(B) + P(A)P(B) && [\text{Eq. (1.62)}]\\
&= 1 - P(A) - P(B)[1 - P(A)]\\
&= [1 - P(A)][1 - P(B)]\\
&= P(\bar{A})P(\bar{B}) && [\text{Eq. (1.39)}]
\end{aligned}
$$

Hence, $\bar{A}$ and $\bar{B}$ are independent.

1.64. Let A and B be events defined in an event space F. Show that if both $P(A)$ and $P(B)$ are nonzero, then events A and B cannot be both mutually exclusive and independent.

Let A and B be mutually exclusive events and $P(A) \neq 0$, $P(B) \neq 0$. Then $P(A \cap B) = P(\varnothing) = 0$ but $P(A)P(B) \neq 0$. Since

$$P(A \cap B) \neq P(A)P(B)$$

A and B cannot be independent.

1.65. Show that if three events A, B, and C are independent, then A and $(B \cup C)$ are independent.

We have

$$
\begin{aligned}
P[A \cap (B \cup C)] &= P[(A \cap B) \cup (A \cap C)] && [\text{Eq. (1.18)}]\\
&= P(A \cap B) + P(A \cap C) - P(A \cap B \cap C) && [\text{Eq. (1.44)}]\\
&= P(A)P(B) + P(A)P(C) - P(A)P(B)P(C) && [\text{Eq. (1.66)}]\\
&= P(A)P(B) + P(A)P(C) - P(A)P(B \cap C) && [\text{Eq. (1.66)}]\\
&= P(A)[P(B) + P(C) - P(B \cap C)]\\
&= P(A)P(B \cup C) && [\text{Eq. (1.44)}]
\end{aligned}
$$

Thus, A and $(B \cup C)$ are independent.

1.66. Consider the experiment of throwing two fair dice (Prob. 1.37). Let A be the event that the sum of the dice is 7, B be the event that the sum of the dice is 6, and C be the event that the first die is 4. Show that events A and C are independent, but events B and C are not independent.

From Fig. 1-3 (Prob. 1.5), we see that

$$A = \{\zeta_{16}, \zeta_{25}, \zeta_{34}, \zeta_{43}, \zeta_{52}, \zeta_{61}\}$$
$$B = \{\zeta_{15}, \zeta_{24}, \zeta_{33}, \zeta_{42}, \zeta_{51}\}$$
$$C = \{\zeta_{41}, \zeta_{42}, \zeta_{43}, \zeta_{44}, \zeta_{45}, \zeta_{46}\}$$

and

$$A \cap C = \{\zeta_{43}\} \qquad B \cap C = \{\zeta_{42}\}$$

Now

$$P(A) = \frac{6}{36} = \frac{1}{6} \qquad P(B) = \frac{5}{56} \qquad P(C) = \frac{6}{36} = \frac{1}{6}$$

and

$$P(A \cap C) = \frac{1}{36} = P(A)P(C)$$

Thus, events A and C are independent. But

$$P(B \cap C) = \frac{1}{36} \neq P(B)P(C)$$

Thus, events B and C are not independent.

1.67. In the experiment of throwing two fair dice, let A be the event that the first die is odd, B be the event that the second die is odd, and C be the event that the sum is odd. Show that events A, B, and C are pairwise independent, but A, B, and C are not independent.

From Fig. 1-3 (Prob. 1.5), we see that

$$P(A) = P(B) = P(C) = \frac{18}{36} = \frac{1}{2}$$

$$P(A \cap B) = P(A \cap C) = P(B \cap C) = \frac{9}{36} = \frac{1}{4}$$

Thus

$$P(A \cap B) = \frac{1}{4} = P(A)P(B)$$

$$P(A \cap C) = \frac{1}{4} = P(A)P(C)$$

$$P(B \cap C) = \frac{1}{4} = P(B)P(C)$$

which indicates that A, B, and C are pairwise independent. However, since the sum of two odd numbers is even, $\{A \cap B \cap C\} = \varnothing$ and

$$P(A \cap B \cap C) = 0 \neq \frac{1}{8} = P(A)P(B)P(C)$$

which shows that A, B, and C are not independent.

1.68. A system consisting of n separate components is said to be a *series* system if it functions when all n components function (Fig. 1-16). Assume that the components fail independently and that the probability of failure of component i is $p_i, i = 1, 2, \ldots, n$. Find the probability that the system functions.

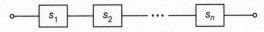

Fig. 1-16 Series system.

Let A_i be the event that component s_i functions. Then

$$P(A_i) = 1 - P(\overline{A}_i) = 1 - p_i$$

Let A be the event that the system functions. Then, since A_i's are independent, we obtain

$$P(A) = P\left(\bigcap_{i=1}^{n} A_i\right) = \prod_{i=1}^{n} P(A_i) = \prod_{i=1}^{n}(1 - p_i) \qquad (1.107)$$

1.69. A system consisting of n separate components is said to be a *parallel* system if it functions when at least one of the components functions (Fig. 1-17). Assume that the components fail independently and that the probability of failure of component i is p_i, $i = 1, 2, \ldots, n$. Find the probability that the system functions.

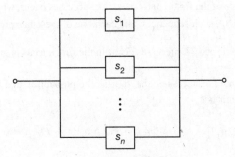

Fig. 1-17 Parallel system.

Let A be the event that component s_i functions. Then

$$P(\overline{A}_i) = p_i$$

Let A be the event that the system functions. Then, since $\overline{A}_i$'s are independent, we obtain

$$P(A) = 1 - P(\overline{A}) = 1 - P\left(\bigcap_{i=1}^{n} \overline{A}_i\right) = 1 - \prod_{i=1}^{n} p_i \qquad (1.108)$$

1.70. Using Eqs. (1.107) and (1.108), redo Prob. 1.40.

From Prob. 1.40, $p_i = \frac{1}{2}$, $i = 1, 2, 3, 4$, where p_i is the probability of failure of switch s_i. Let A be the event that there exists a closed path between a and b. Using Eq. (1.108), the probability of failure for the parallel combination of switches 3 and 4 is

$$p_{34} = p_3 p_4 = \left(\frac{1}{2}\right)\left(\frac{1}{2}\right) = \frac{1}{4}$$

Using Eq. (1.107), the probability of failure for the combination of switches 2, 3, and 4 is

$$p_{234} = 1 - \left(1 - \frac{1}{2}\right)\left(1 - \frac{1}{4}\right) = 1 - \frac{3}{8} = \frac{5}{8}$$

Again, using Eq. (1.108), we obtain

$$P(A) = 1 - p_1 p_{234} = 1 - \left(\frac{1}{2}\right)\left(\frac{5}{8}\right) = 1 - \frac{5}{16} = \frac{11}{16}$$

1.71. A *Bernoulli* experiment is a random experiment, the outcome of which can be classified in but one of two mutually exclusive and exhaustive ways, say success or failure. A sequence of Bernoulli trials occurs when a Bernoulli experiment is performed several independent times so that the probability of success, say p, remains the same from trial to trial. Now an infinite sequence of Bernoulli trials is performed. Find the probability that (*a*) at least 1 success occurs in the first n trials; (*b*) exactly k successes occur in the first n trials; (*c*) all trials result in successes.

(*a*) In order to find the probability of at least 1 success in the first n trials, it is easier to first compute the probability of the complementary event, that of no successes in the first n trials. Let A_i denote the event of a failure on the ith trial. Then the probability of no successes is, by independence,

$$P(A_1 \cap A_2 \cap \cdots \cap A_n) = P(A_1)P(A_2) \cdots P(A_n) = (1-p)^n \tag{1.109}$$

Hence, the probability that at least 1 success occurs in the first n trials is $1 - (1-p)^n$.

(*b*) In any particular sequence of the first n outcomes, if k successes occur, where $k = 0, 1, 2, \ldots, n$, then $n - k$ failures occur. There are $\binom{n}{k}$ such sequences, and each one of these has probability $p^k(1-p)^{n-k}$.

Thus, the probability that exactly k successes occur in the first n trials is given by $\binom{n}{k}p^k(1-p)^{n-k}$.

(*c*) Since $\bar{A}_i$ denotes the event of a success on the ith trial, the probability that all trials resulted in successes in the first n trials is, by independence,

$$P(\bar{A}_1 \cap \bar{A}_2 \cap \cdots \cap \bar{A}_n) = P(\bar{A}_1)P(\bar{A}_2) \cdots P(\bar{A}_n) = p^n \tag{1.110}$$

Hence, using the continuity theorem of probability (1.89) (Prob. 1.34), the probability that all trials result in successes is given by

$$P\left(\bigcap_{i=1}^{\infty} \bar{A}_i\right) = P\left(\lim_{n\to\infty} \bigcap_{i=1}^{n} \bar{A}_i\right) = \lim_{n\to\infty} P\left(\bigcap_{i=1}^{n} \bar{A}_i\right) = \lim_{n\to\infty} p^n = \begin{cases} 0 & p < 1 \\ 1 & p = 1 \end{cases}$$

1.72. Let S be the sample space of an experiment and $S = \{A, B, C\}$, where $P(A) = p, P(B) = q$, and $P(C) = r$. The experiment is repeated infinitely, and it is assumed that the successive experiments are independent. Find the probability of the event that A occurs before B.

Suppose that A occurs for the first time at the nth trial of the experiment. If A is to have occurred before B, then C must have occurred on the first $(n-1)$ trials. Let D be the event that A occurs before B.

Then

$$D = \bigcup_{n=1}^{\infty} D_n$$

where D_n is the event that C occurs on the first $(n-1)$ trials and A occurs on the nth trial. Since D_n's are mutually exclusive, we have

$$P(D) = \sum_{n=1}^{\infty} P(D_n)$$

Since the trials are independent, we have

$$P(D_n) = [P(C)]^{n-1} P(A) = r^{n-1}p$$

Thus,

$$P(D) = \sum_{n=1}^{\infty} r^{n-1}p = p\sum_{k=0}^{\infty} r^k = \frac{p}{1-r} = \frac{p}{p+q}$$

or

$$P(D) = \frac{P(A)}{P(A) + P(B)} \qquad (1.111)$$

since $p + q + r = 1$.

1.73. In a gambling game, craps, a pair of dice is rolled and the outcome of the experiment is the sum of the dice. The player wins on the first roll if the sum is 7 or 11 and loses if the sum is 2, 3, or 12. If the sum is 4, 5, 6, 8, 9, or 10, that number is called the player's "point." Once the point is established, the rule is: If the player rolls a 7 before the point, the player loses; but if the point is rolled before a 7, the player wins. Compute the probability of winning in the game of craps.

Let A, B, and C be the events that the player wins, the player wins on the first roll, and the player gains point, respectively. Then $P(A) = P(B) + P(C)$. Now from Fig. 1-3 (Prob. 1.5),

$$P(B) = P(\text{sum} = 7) + P(\text{sum} = 11) = \frac{6}{36} + \frac{2}{36} = \frac{2}{9}$$

Let A_k be the event that point of k occurs before 7. Then

$$P(C) = \sum_{k \in \{4,5,6,8,9,10\}} P(A_k)P(\text{point} = k)$$

By Eq. (1.111) (Prob. 1.72),

$$P(A_k) = \frac{P(\text{sum} = k)}{P(\text{sum} = k) + P(\text{sum} = 7)} \qquad (1.112)$$

Again from Fig. 1-3,

$$P(\text{sum} = 4) = \frac{3}{36} \qquad P(\text{sum} = 5) = \frac{4}{36} \qquad P(\text{sum} = 6) = \frac{5}{36}$$

$$P(\text{sum} = 8) = \frac{5}{36} \qquad P(\text{sum} = 9) = \frac{4}{36} \qquad P(\text{sum} = 10) = \frac{3}{36}$$

Now by Eq. (1.112),

$$P(A_4) = \frac{1}{3} \qquad P(A_5) = \frac{2}{5} \qquad P(A_6) = \frac{5}{11}$$

$$P(A_8) = \frac{5}{11} \qquad P(A_9) = \frac{2}{5} \qquad P(A_{10}) = \frac{1}{3}$$

Using these values, we obtain

$$P(A) = P(B) + P(C) = \frac{2}{9} + \frac{134}{495} = 0.49293$$

SUPPLEMENTARY PROBLEMS

1.74. Consider the experiment of selecting items from a group consisting of three items $\{a, b, c\}$.

  (*a*) Find the sample space S_1 of the experiment in which two items are selected without replacement.

  (*b*) Find the sample space S_2 of the experiment in which two items are selected with replacement.

1.75. Let A and B be arbitrary events. Then show that $A \subset B$ if and only if $A \cup B = B$.

1.76. Let A and B be events in the sample space S. Show that if $A \subset B$, then $\bar{B} \subset \bar{A}$.

1.77. Verify Eq. (1.19).

1.78. Show that

$$(A \cap B) \setminus C = (A \setminus C) \cap (B \setminus C)$$

1.79. Let A and B be any two events in S. Express the following events in terms of A and B.

  (*a*) At least one of the events occurs.

  (*b*) Exactly one of two events occurs.

1.80. Show that A and B are disjoint if and only if

$$A \cup B = A \,\Delta\, B$$

1.81. Let $A, B,$ and C be any three events in S. Express the following events in terms of these events.

  (*a*) Either B or C occurs, but not A.

  (*b*) Exactly one of the events occurs.

  (*c*) Exactly two of the events occur.

1.82. Show that $F = \{S, \varnothing\}$ is an event space.

1.83. Let $S = \{1, 2, 3, 4\}$ and $F_1 = \{S, \varnothing, \{1, 3\}, \{2, 4\}\}$, $F_2 = \{S, \varnothing, \{1, 3\}\}$. Show that F_1 is an event space, and F_2 is not an event space.

1.84. In an experiment one card is selected from an ordinary 52-card deck. Define the events: A = select a King, B = select a Jack or a Queen, C = select a Heart.

  Find $P(A)$, $P(B)$, and $P(C)$.

1.85. A random experiment has sample space $S = \{a, b, c\}$. Suppose that $P(\{a, c\}) = 0.75$ and $P(\{b, c\}) = 0.6$. Find the probabilities of the elementary events.

1.86. Show that

  (*a*) $P(\bar{A} \cup \bar{B}) = 1 - P(A \cap B)$

  (*b*) $P(A \cap B) \geq 1 - P(\bar{A}) - P(\bar{B})$

  (*c*) $P(A \,\Delta\, B) = P(A \cup B) - P(A \cap B)$

1.87. Let A, B, and C be three events in S. If $P(A) = P(B) = \frac{1}{4}, P(C) = \frac{1}{3}, P(A \cap B) = \frac{1}{8}, P(A \cap C) = \frac{1}{6}$, and $P(B \cap C) = 0$, find $P(A \cup B \cup C)$.

1.88. Verify Eq. (1.45).

1.89. Show that

$$P(A_1 \cap A_2 \cap \cdots \cap A_n) \geq P(A_1) + P(A_2) + \cdots + P(A_n) - (n-1)$$

1.90. In an experiment consisting of 10 throws of a pair of fair dice, find the probability of the event that at least one double 6 occurs.

1.91. Show that if $P(A) > P(B)$, then $P(A \mid B) > P(B \mid A)$.

1.92. Show that

(a) $P(A \mid A) = 1$

(b) $P(A \cap B \mid C) = P(A \mid C) P(B \mid A \cap C)$

1.93. Show that

$$P(A \cap B \cap C) = P(A \mid B \cap C) P(B \mid C) P(C)$$

1.94. An urn contains 8 white balls and 4 red balls. The experiment consists of drawing 2 balls from the urn without replacement. Find the probability that both balls drawn are white.

1.95. There are 100 patients in a hospital with a certain disease. Of these, 10 are selected to undergo a drug treatment that increases the percentage cured rate from 50 percent to 75 percent. What is the probability that the patient received a drug treatment if the patient is known to be cured?

1.96. Two boys and two girls enter a music hall and take four seats at random in a row. What is the probability that the girls take the two end seats?

1.97. Let A and B be two independent events in S. It is known that $P(A \cap B) = 0.16$ and $P(A \cup B) = 0.64$. Find $P(A)$ and $P(B)$.

1.98. Consider the random experiment of Example 1.7 of rolling a die. Let A be the event that the outcome is an odd number and B the event that the outcome is less than 3. Show that events A and B are independent.

1.99. The relay network shown in Fig. 1-18 operates if and only if there is a closed path of relays from left to right. Assume that relays fail independently and that the probability of failure of each relay is as shown. What is the probability that the relay network operates?

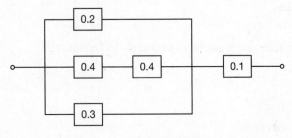

Fig. 1-18

ANSWERS TO SUPPLEMENTARY PROBLEMS

1.74. (a) $S_1 = \{ab, ac, ba, bc, ca, cb\}$

(b) $S_2 = \{aa, ab, ac, ba, bb, bc, ca, cb, cc\}$

1.75. *Hint*: Draw a Venn diagram.

1.76. *Hint*: Draw a Venn diagram.

1.77. *Hint*: Draw a Venn diagram.

1.78. *Hint*: Use Eqs. (1.10) and (1.17).

1.79. (a) $A \cup B$; (b) $A \Delta B$

1.80. *Hint*: Follow Prob. 1.19.

1.81. (a) $\bar{A} \cap (B \cup C)$

(b) $\{A \cap \overline{(B \cup C)}\} \cup \{B \cap \overline{(A \cup C)}\} \cup \{C \cap \overline{(A \cup B)}\}$

(c) $\{(A \cap B) \cap \bar{C}\} \cup \{(A \cap C) \cap \bar{B}\} \cup \{(B \cap C) \cap \bar{A}\}$

1.82. *Hint*: Follow Prob. 1.23.

1.83. *Hint*: Follow Prob. 1.23.

1.84. $P(A) = 1/13, P(B) = 2/13, P(C) = 13/52$

1.85. $P(a) = 0.4, P(b) = 0.25, P(c) = 0.35$

1.86. *Hint*: (a) Use Eqs. (1.21) and (1.39).

(b) Use Eqs. (1.43), (1.39), and (1.42).

(c) Use a Venn diagram.

1.87. $\dfrac{13}{24}$

1.88. *Hint*: Prove by induction.

1.89. *Hint*: Use induction to generalize Bonferroni's inequality (1.77) (Prob. 1.28).

1.90. 0.246

1.91. *Hint*: Use Eqs. (1.55) and (1.56).

1.92. *Hint:*　Use definition Eq.(1.55).

1.93. *Hint:*　Follow Prob. 1.53.

1.94. 0.424

1.95. 0.143

1.96. $\dfrac{1}{6}$

1.97. $P(A) = P(B) = 0.4$

1.98. *Hint:*　Show that $P(A \cap B) = P(A)P(B) = 1/6$.

1.99. 0.865

CHAPTER 2

Random Variables

2.1 Introduction

In this chapter, the concept of a random variable is introduced. The main purpose of using a random variable is so that we can define certain probability functions that make it both convenient and easy to compute the probabilities of various events.

2.2 Random Variables

A. Definitions:

Consider a random experiment with sample space S. A *random variable* $X(\zeta)$ is a single-valued real function that assigns a real number called the *value* of $X(\zeta)$ to each sample point ζ of S. Often, we use a single letter X for this function in place of $X(\zeta)$ and use r.v. to denote the random variable.

Note that the terminology used here is traditional. Clearly a random variable is not a variable at all in the usual sense, and it is a function.

The sample space S is termed the *domain* of the r.v. X, and the collection of all numbers [values of $X(\zeta)$] is termed the *range* of the r.v. X. Thus, the range of X is a certain subset of the set of all real numbers (Fig. 2-1).

Note that two or more different sample points might give the same value of $X(\zeta)$, but two different numbers in the range cannot be assigned to the same sample point.

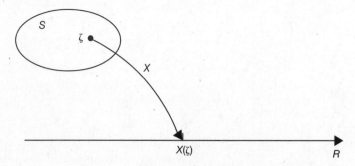

Fig. 2-1 Random variable X as a function.

EXAMPLE 2.1 In the experiment of tossing a coin once (Example 1.1), we might define the r.v. X as (Fig. 2-2)

$$X(H) = 1 \qquad X(T) = 0$$

Note that we could also define another r.v., say Y or Z, with

$$Y(H) = 0, Y(T) = 1 \quad \text{or} \quad Z(H) = 0, Z(T) = 0$$

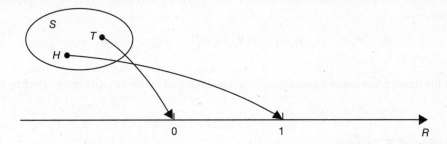

Fig. 2-2 One random variable associated with coin tossing.

B. Events Defined by Random Variables:

If X is a r.v. and x is a fixed real number, we can define the event $(X = x)$ as

$$(X = x) = \{\zeta: X(\zeta) = x\} \tag{2.1}$$

Similarly, for fixed numbers x, x_1, and x_2, we can define the following events:

$$(X \leq x) = \{\zeta: X(\zeta) \leq x\}$$
$$(X > x) = \{\zeta: X(\zeta) > x\} \tag{2.2}$$
$$(x_1 < X \leq x_2) = \{\zeta: x_1 < X(\zeta) \leq x_2\}$$

These events have probabilities that are denoted by

$$P(X = x) = P\{\zeta: X(\zeta) = x\}$$
$$P(X \leq x) = P\{\zeta: X(\zeta) \leq x\} \tag{2.3}$$
$$P(X > x) = P\{\zeta: X(\zeta) > x\}$$
$$P(x_1 < X \leq x_2) = P\{\zeta: x_1 < X(\zeta) \leq x_2\}$$

EXAMPLE 2.2 In the experiment of tossing a fair coin three times (Prob. 1.1), the sample space S_1 consists of eight equally likely sample points $S_1 = \{HHH, \ldots, TTT\}$. If X is the r.v. giving the number of heads obtained, find (a) $P(X = 2)$; (b) $P(X < 2)$.

(a) Let $A \subset S_1$ be the event defined by $X = 2$. Then, from Prob. 1.1, we have

$$A = (X = 2) = \{\zeta: X(\zeta) = 2\} = \{HHT, HTH, THH\}$$

Since the sample points are equally likely, we have

$$P(X = 2) = P(A) = \frac{3}{8}$$

(b) Let $B \subset S_1$ be the event defined by $X < 2$. Then

$$B = (X < 2) = \{\zeta: X(\zeta) < 2\} = \{HTT, THT, TTH, TTT\}$$

and $$P(X < 2) = P(B) = \frac{4}{8} = \frac{1}{2}$$

2.3 Distribution Functions

A. Definition:

The *distribution function* [or *cumulative distribution function* (cdf)] of X is the function defined by

$$F_X(x) = P(X \le x) \qquad -\infty < x < \infty \tag{2.4}$$

Most of the information about a random experiment described by the r.v. X is determined by the behavior of $F_X(x)$.

B. Properties of $F_X(x)$:

Several properties of $F_X(x)$ follow directly from its definition (2.4).

1. $0 \le F_X(x) \le 1$ (2.5)
2. $F_X(x_1) \le F_X(x_2)$ if $x_1 < x_2$ (2.6)
3. $\lim_{x \to \infty} F_X(x) = F_X(\infty) = 1$ (2.7)
4. $\lim_{x \to -\infty} F_X(x) = F_X(-\infty) = 0$ (2.8)
5. $\lim_{x \to a^+} F_X(x) = F_X(a^+) = F_X(a)$ $a^+ = \lim_{0 < \varepsilon \to 0} a + \varepsilon$ (2.9)

Property 1 follows because $F_X(x)$ is a probability. Property 2 shows that $F_X(x)$ is a nondecreasing function (Prob. 2.5). Properties 3 and 4 follow from Eqs. (1.36) and (1.40):

$$\lim_{x \to \infty} P(X \le x) = P(X \le \infty) = P(S) = 1$$

$$\lim_{x \to -\infty} P(X \le x) = P(X \le -\infty) = P(\varnothing) = 0$$

Property 5 indicates that $F_X(x)$ is *continuous on the right*. This is the consequence of the definition (2.4).

TABLE 2-1

x	$(X \le x)$	$F_X(x)$
-1	$\varnothing$	0
0	$\{TTT\}$	$\frac{1}{8}$
1	$\{TTT, TTH, THT, HTT\}$	$\frac{4}{8} = \frac{1}{2}$
2	$\{TTT, TTH, THT, HTT, HHT, HTH, THH\}$	$\frac{7}{8}$
3	S	1
4	S	1

EXAMPLE 2.3 Consider the r.v. X defined in Example 2.2. Find and sketch the cdf $F_X(x)$ of X.

Table 2-1 gives $F_X(x) = P(X \le x)$ for $x = -1, 0, 1, 2, 3, 4$. Since the value of X must be an integer, the value of $F_X(x)$ for noninteger values of x must be the same as the value of $F_X(x)$ for the nearest smaller integer value of x. The $F_X(x)$ is sketched in Fig. 2-3. Note that $F_X(x)$ has jumps at $x = 0, 1, 2, 3$, and that at each jump the upper value is the correct value for $F_X(x)$.

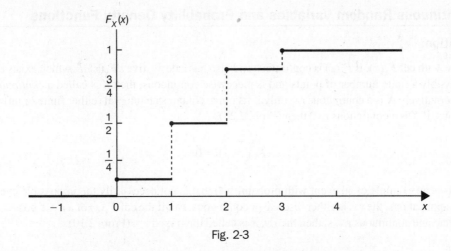

Fig. 2-3

C. Determination of Probabilities from the Distribution Function:

From definition (2.4), we can compute other probabilities, such as $P(a < X \le b)$, $P(X > a)$, and $P(X < b)$ (Prob. 2.6):

$$P(a < X \le b) = F_X(b) - F_X(a) \tag{2.10}$$

$$P(X > a) = 1 - F_X(a) \tag{2.11}$$

$$P(X < b) = F_X(b^-) \qquad b^- = \lim_{0 < \varepsilon \to 0} b - \varepsilon \tag{2.12}$$

2.4 Discrete Random Variables and Probability Mass Functions

A. Definition:

Let X be a r.v. with cdf $F_X(x)$. If $F_X(x)$ changes values only in jumps (at most a countable number of them) and is constant between jumps—that is, $F_X(x)$ is a staircase function (see Fig. 2-3)—then X is called a *discrete* random variable. Alternatively, X is a discrete r.v. only if its range contains a finite or countably infinite number of points. The r.v. X in Example 2.3 is an example of a discrete r.v.

B. Probability Mass Functions:

Suppose that the jumps in $F_X(x)$ of a discrete r.v. X occur at the points $x_1, x_2, \ldots$, where the sequence may be either finite or countably infinite, and we assume $x_i < x_j$ if $i < j$.

Then

$$F_X(x_i) - F_X(x_{i-1}) = P(X \le x_i) - P(X \le x_{i-1}) = P(X = x_i) \tag{2.13}$$

Let

$$p_X(x) = P(X = x) \tag{2.14}$$

The function $p_X(x)$ is called the *probability mass function* (pmf) of the discrete r.v. X.

Properties of $p_X(x)$:

1. $0 \le p_X(x_k) \le 1 \qquad k = 1, 2, \ldots$ (2.15)
2. $p_X(x) = 0 \qquad$ if $x \ne x_k \ (k = 1, 2, \ldots)$ (2.16)
3. $\sum_k p_X(x_k) = 1$ (2.17)

The cdf $F_X(x)$ of a discrete r.v. X can be obtained by

$$F_X(x) = P(X \le x) = \sum_{x_k \le x} p_X(x_k) \tag{2.18}$$

2.5   Continuous Random Variables and Probability Density Functions

A.   Definition:

Let X be a r.v. with cdf $F_X(x)$. If $F_X(x)$ is continuous and also has a derivative $dF_X(x)/dx$ which exists everywhere except at possibly a finite number of points and is piecewise continuous, then X is called a *continuous* random variable. Alternatively, X is a continuous r.v. only if its range contains an interval (either finite or infinite) of real numbers. Thus, if X is a continuous r.v., then (Prob. 2.20)

$$P(X = x) = 0 \tag{2.19}$$

Note that this is an example of an event with probability 0 that is not necessarily the impossible event $\varnothing$.

In most applications, the r.v. is either discrete or continuous. But if the cdf $F_X(x)$ of a r.v. X possesses features of both discrete and continuous r.v.'s, then the r.v. X is called the mixed r.v. (Prob. 2.10).

B.   Probability Density Functions:

Let
$$f_X(x) = \frac{dF_X(x)}{dx} \tag{2.20}$$

The function $f_X(x)$ is called the *probability density function* (pdf) of the continuous r.v. X.

Properties of $f_X(x)$:

1. $f_X(x) \geq 0$ $\hspace{6cm}$ (2.21)
2. $\int_{-\infty}^{\infty} f_X(x)\, dx = 1$ $\hspace{4.5cm}$ (2.22)
3. $f_X(x)$ is piecewise continuous.
4. $P(a < X \leq b) = \int_a^b f_X(x)\, dx$ $\hspace{3.5cm}$ (2.23)

The cdf $F_X(x)$ of a continuous r.v. X can be obtained by

$$F_X(x) = P(X \leq x) = \int_{-\infty}^{x} f_X(\xi)\, d\xi \tag{2.24}$$

By Eq. (2.19), if X is a continuous r.v., then

$$P(a < X \leq b) = P(a \leq X \leq b) = P(a \leq X < b) = P(a < X < b)$$
$$= \int_a^b f_X(x)\, dx = F_X(b) - F_X(a) \tag{2.25}$$

2.6   Mean and Variance

A.   Mean:

The *mean* (or *expected value*) of a r.v. X, denoted by μ_X or $E(X)$, is defined by

$$\mu_X = E(X) = \begin{cases} \displaystyle\sum_k x_k p_X(x_k) & X\text{: discrete} \\[2ex] \displaystyle\int_{-\infty}^{\infty} x f_X(x)\, dx & X\text{: continuous} \end{cases} \tag{2.26}$$

B. Moment:

The *nth moment* of a r.v. X is defined by

$$E(X^n) = \begin{cases} \displaystyle\sum_k x_k^{\ n} p_X(x_k) & X: \text{discrete} \\[2ex] \displaystyle\int_{-\infty}^{\infty} x^n f_X(x)\,dx & X: \text{continuous} \end{cases} \tag{2.27}$$

Note that the mean of X is the first moment of X.

C. Variance:

The *variance* of a r.v. X, denoted by σ_X^2 or $\text{Var}(X)$, is defined by

$$\sigma_X^2 = \text{Var}(X) = E\{[X - E(X)]^2\} \tag{2.28}$$

Thus,

$$\sigma_X^2 = \begin{cases} \displaystyle\sum_k (x_k - \mu_X)^2 p_X(x_k) & X: \text{discrete} \\[2ex] \displaystyle\int_{-\infty}^{\infty} (x - \mu_X)^2 f_X(x)\,dx & X: \text{continuous} \end{cases} \tag{2.29}$$

Note from definition (2.28) that

$$\text{Var}(X) \geq 0 \tag{2.30}$$

The *standard deviation* of a r.v. X, denoted by σ_X, is the positive square root of $\text{Var}(X)$. Expanding the right-hand side of Eq. (2.28), we can obtain the following relation:

$$\text{Var}(X) = E(X^2) - [E(X)]^2 \tag{2.31}$$

which is a useful formula for determining the variance.

2.7 Some Special Distributions

In this section we present some important special distributions.

A. Bernoulli Distribution:

A r.v. X is called a *Bernoulli* r.v. with parameter p if its pmf is given by

$$p_X(k) = P(X = k) = p^k(1 - p)^{1-k} \qquad k = 0, 1 \tag{2.32}$$

where $0 \leq p \leq 1$. By Eq. (2.18), the cdf $F_X(x)$ of the Bernoulli r.v. X is given by

$$F_X(x) = \begin{cases} 0 & x < 0 \\ 1 - p & 0 \leq x < 1 \\ 1 & x \geq 1 \end{cases} \tag{2.33}$$

Fig. 2-4 illustrates a Bernoulli distribution.

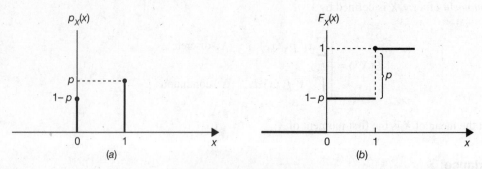

Fig. 2-4 Bernoulli distribution.

The mean and variance of the Bernoulli r.v. X are

$$\mu_{X} = E(X) = p \tag{2.34}$$

$$\sigma_X^2 = \mathrm{Var}(X) = p(1 - p) \tag{2.35}$$

A Bernoulli r.v. X is associated with some experiment where an outcome can be classified as either a "success" or a "failure," and the probability of a success is p and the probability of a failure is $1 - p$. Such experiments are often called *Bernoulli trials* (Prob. 1.71).

B. Binomial Distribution:

A r.v. X is called a *binomial* r.v. with parameters (n, p) if its pmf is given by

$$p_X(k) = P(X = k) = \binom{n}{k} p^k (1 - p)^{n-k} \qquad k = 0, 1, \ldots, n \tag{2.36}$$

where $0 \le p \le 1$ and

$$\binom{n}{k} = \frac{n!}{k!(n-k)!}$$

which is known as the binomial coefficient. The corresponding cdf of X is

$$F_X(x) = \sum_{k=0}^{n} \binom{n}{k} p^k (1 - p)^{n-k} \qquad n \le x < n + 1 \tag{2.37}$$

Fig. 2-5 illustrates the binomial distribution for $n = 6$ and $p = 0.6$.

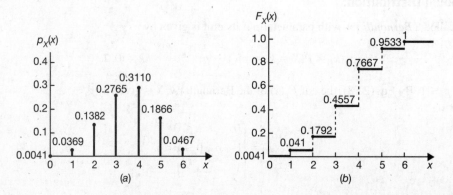

Fig. 2-5 Binomial distribution with $n = 6$, $p = 0.6$.

The mean and variance of the binomial r.v. X are (Prob. 2.30)

$$\mu_X = E(X) = np \tag{2.38}$$
$$\sigma_X^2 = \text{Var}(X) = np(1 - p) \tag{2.39}$$

A binomial r.v. X is associated with some experiments in which n independent Bernoulli trials are performed and X represents the number of successes that occur in the n trials. Note that a Bernoulli r.v. is just a binomial r.v. with parameters $(1, p)$.

C. Geometric Distribution:

A r.v. X is called a *geometric* r.v. with parameter p if its pmf is given by

$$p_X(X) = P(X = x) = (1 - p)^{x-1}p \qquad x = 1, 2, \dots \tag{2.40}$$

where $0 \le p \le 1$. The cdf $F_X(x)$ of the geometric r.v. X is given by

$$F_X(X) = P(X \le x) = 1 - (1 - p)^x \qquad x = 1, 2, \dots \tag{2.41}$$

Fig. 2-6 illustrates the geometric distribution with $p = 0.25$.

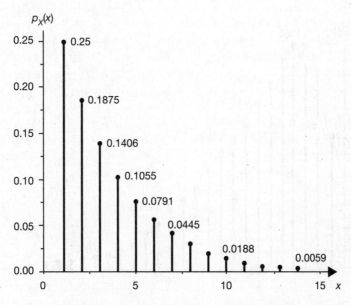

Fig. 2-6 Geometric distribution with $p = 0.25$.

The mean and variance of the geometric r.v. X are (Probs. 2.29 and 4.55)

$$\mu_X = E(X) = \frac{1}{p} \tag{2.42}$$

$$\sigma_X^2 = \text{Var}(X) = \frac{1 - p}{p^2} \tag{2.43}$$

A geometric r.v. X is associated with some experiments in which a sequence of Bernoulli trials with probability p of success. The sequence is observed until the first success occurs. The r.v. X denotes the trial number on which the first success occurs.

Memoryless property of the geometric distribution:

If X is a geometric r.v., then it has the following significant property (Probs. 2.17, 2.57).

$$P(X > i + j \mid X > i) = P(X > j) \qquad i, j \geq 1 \qquad (2.44)$$

Equation (2.44) indicates that suppose after i flips of a coin, no "head" has turned up yet, then the probability for no "head" to turn up for the next j flips of the coin is exactly the same as the probability for no "head" to turn up for the first i flips of the coin.

Equation (2.44) is known as the *memoryless property*. Note that memoryless property Eq. (2.44) is only valid when i, j are integers. The geometric distribution is the only discrete distribution that possesses this property.

D. Negative Binomial Distribution:

A r.v. X is called a *negative binomial* r.v. with parameters p and k if its pmf is given by

$$P_X(x) = P(X = x) = \binom{x-1}{k-1} p^k (1-p)^{x-k} \qquad x = k, k+1, \dots \qquad (2.45)$$

where $0 \leq p \leq 1$.

Fig. 2-7 illustrates the negative binomial distribution for $k = 2$ and $p = 0.25$.

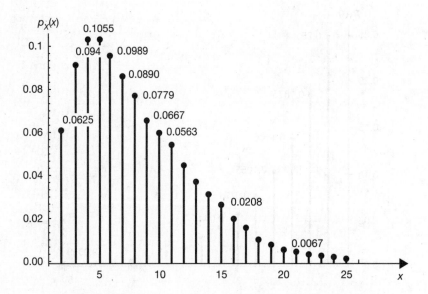

Fig. 2-7 Negative binomial distribution with $k = 2$ and $p = 0.25$.

The mean and variance of the negative binomial r.v. X are (Probs. 2.80, 4.56).

$$\mu_X = E(X) = \frac{k}{p} \qquad (2.46)$$

$$\sigma_X^2 = \mathrm{Var}(X) = \frac{k(1-p)}{p^2} \qquad (2.47)$$

A negative binomial r.v. X is associated with sequence of independent Bernoulli trials with the probability of success p, and X represents the number of trials until the kth success is obtained. In the experiment of flipping a coin, if $X = x$, then it must be true that there were exactly $k - 1$ heads thrown in the first $x - 1$ flippings, and a head must have been thrown on the xth flipping. There are $\begin{pmatrix} x-1 \\ k-1 \end{pmatrix}$ sequences of length x with these properties, and each of them is assigned the same probability of $p^{k-1}(1-p)^{x-k}$.

Note that when $k = 1$, X is a geometrical r.v. A negative binomial r.v. is sometimes called a *Pascal* r.v.

E. Poisson Distribution:

A r.v. X is called a *Poisson* r.v. with parameter $\lambda \, (> 0)$ if its pmf is given by

$$p_X(k) = P(X = k) = e^{-\lambda} \frac{\lambda^k}{k!} \qquad k = 0, 1, \dots \tag{2.48}$$

The corresponding cdf of X is

$$F_X(x) = e^{-\lambda} \sum_{k=0}^{n} \frac{\lambda^k}{k!} \qquad n \le x < n+1 \tag{2.49}$$

Fig. 2-8 illustrates the Poisson distribution for $\lambda = 3$.

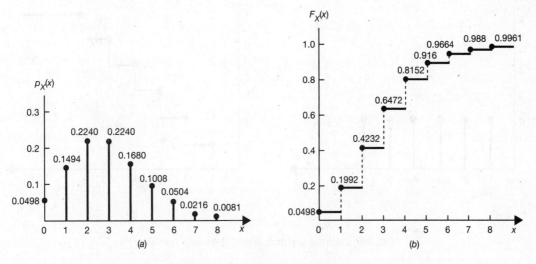

Fig. 2-8 Poisson distribution with $\lambda = 3$.

The mean and variance of the Poisson r.v. X are (Prob. 2.31)

$$\mu_X = E(X) = \lambda \tag{2.50}$$
$$\sigma_X^2 = \text{Var}(X) = \lambda \tag{2.51}$$

The Poisson r.v. has a tremendous range of applications in diverse areas because it may be used as an approximation for a binomial r.v. with parameters (n, p) when n is large and p is small enough so that np is of a moderate size (Prob. 2.43).

Some examples of Poisson r.v.'s include

1. The number of telephone calls arriving at a switching center during various intervals of time
2. The number of misprints on a page of a book
3. The number of customers entering a bank during various intervals of time

F. Discrete Uniform Distribution:

A r.v. X is called a *discrete uniform* r.v. if its pmf is given by

$$p_X(x) = P(X = x) = \frac{1}{n} \qquad 1 \le x \le n \tag{2.52}$$

The cdf $F_X(x)$ of the discrete uniform r.v. X is given by

$$F_X(x) = P(X \le x) = \begin{cases} 0 & 0 < x < 1 \\ \dfrac{\lfloor x \rfloor}{n} & 1 < x < n \\ 1 & n \le x \end{cases} \tag{2.53}$$

where $\lfloor x \rfloor$ denotes the integer less than or equal to x.

Fig. 2-9 illustrates the discrete uniform distribution for $n = 6$.

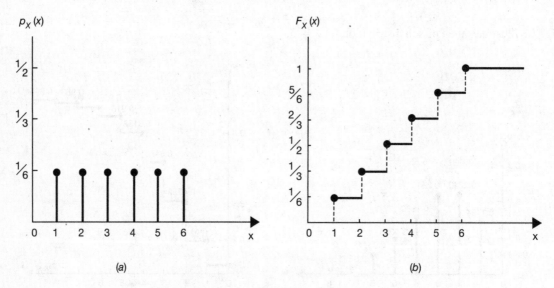

Fig. 2-9 Discrete uniform distribution with $n = 6$.

The mean and variance of the discrete uniform r.v. X are (Prob. 2.32)

$$\mu_X = E(X) = \frac{1}{2}(n + 1) \tag{2.54}$$

$$\sigma_X^2 = \text{Var}(X) = \frac{1}{12}(n^2 - 1) \tag{2.55}$$

The discrete uniform r.v. X is associated with cases where all finite outcomes of an experiment are equally likely. If the sample space is a countably infinite set, such as the set of positive integers, then it is not possible to define a discrete uniform r.v. X. If the sample space is an uncountable set with finite length such as the interval (a, b), then a continuous uniform r.v. X will be utilized.

G. Continuous Uniform Distribution:

A r.v. X is called a *continuous uniform* r.v. over (a, b) if its pdf is given by

$$f_X(x) = \begin{cases} \dfrac{1}{b-a} & a < x < b \\ 0 & \text{otherwise} \end{cases} \tag{2.56}$$

The corresponding cdf of X is

$$F_X(x) = \begin{cases} 0 & x \leq a \\ \dfrac{x-a}{b-a} & a < x < b \\ 1 & x \geq b \end{cases} \tag{2.57}$$

Fig. 2-10 illustrates a *continuous* uniform distribution.
The mean and variance of the uniform r.v. X are (Prob. 2.34)

$$\mu_X = E(X) = \frac{a+b}{2} \tag{2.58}$$

$$\sigma_X^2 = \text{Var}(X) = \frac{(b-a)^2}{12} \tag{2.59}$$

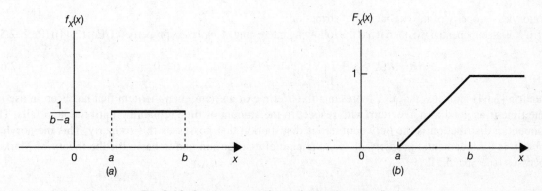

Fig. 2-10 Continuous uniform distribution over (a, b).

A uniform r.v. X is often used where we have no prior knowledge of the actual pdf and all continuous values in some range seem equally likely (Prob. 2.75).

H. Exponential Distribution:

A r.v. X is called an *exponential* r.v. with parameter $\lambda \, (> 0)$ if its pdf is given by

$$f_X(x) = \begin{cases} \lambda e^{-\lambda x} & x > 0 \\ 0 & x < 0 \end{cases} \tag{2.60}$$

which is sketched in Fig. 2-11(a). The corresponding cdf of X is

$$F_X(x) = \begin{cases} 1 - e^{-\lambda x} & x \geq 0 \\ 0 & x < 0 \end{cases} \tag{2.61}$$

which is sketched in Fig. 2-11(b).

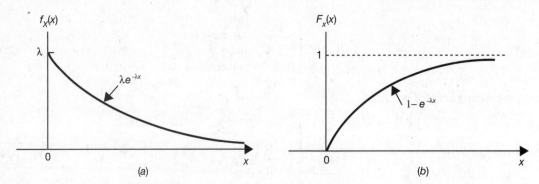

Fig. 2-11 Exponential distribution.

The mean and variance of the exponential r.v. X are (Prob. 2.35)

$$\mu_X = E(X) = \frac{1}{\lambda} \tag{2.62}$$

$$\sigma_X^2 = \text{Var}(X) = \frac{1}{\lambda^2} \tag{2.63}$$

Memoryless property of the exponential distribution:

If X is an exponential r.v., then it has the following interesting *memoryless* property (cf. Eq. (2.44)) (Prob. 2.58)

$$P(X > s + t \mid X > s) = P(X > t) \qquad s, t \ge 0 \tag{2.64}$$

Equation (2.64) indicates that if X represents the lifetime of an item, then the item that has been in use for some time is as good as a new item with respect to the amount of time remaining until the item fails. The exponential distribution is the only continuous distribution that possesses this property. This memoryless property is a fundamental property of the exponential distribution and is basic for the theory of Markov processes (see Sec. 5.5).

I. Gamma Distribution:

A r.v. X is called a gamma r.v. with parameter (α, λ) ($\alpha > 0$ and $\lambda > 0$) if its pdf is given by

$$f_X(x) = \begin{cases} \dfrac{\lambda e^{-\lambda x}(\lambda x)^{\alpha-1}}{\Gamma(\alpha)} & x > 0 \\ 0 & x < 0 \end{cases} \tag{2.65}$$

where $\Gamma(\alpha)$ is the gamma function defined by

$$\Gamma(\alpha) = \int_0^\infty e^{-x} x^{\alpha-1}\, dx \qquad \alpha > 0 \tag{2.66}$$

and it satisfies the following recursive formula (Prob. 2.26)

$$\Gamma(\alpha + 1) = \alpha\, \Gamma(\alpha) \qquad \alpha > 0 \tag{2.67}$$

The pdf $f_X(x)$ with $(\alpha, \lambda) = (1, 1), (2, 1),$ and $(5, 2)$ are plotted in Fig. 2-12.

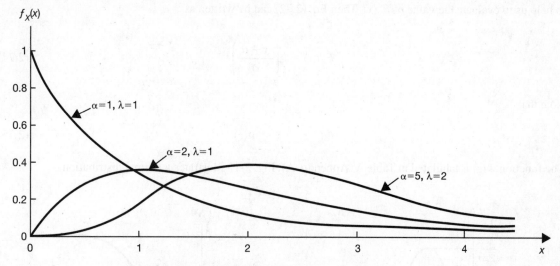

Fig. 2-12 Gamma distributions for selected values of α and λ

The mean and variance of the gamma r.v. are (Prob. 4.65)

$$\mu_X = E(X) = \frac{\alpha}{\lambda} \tag{2.68}$$

$$\sigma_X^2 = \text{Var}(X) = \frac{\alpha}{\lambda^2} \tag{2.69}$$

Note that when $\alpha = 1$, the gamma r.v. becomes an exponential r.v. with parameter λ [Eq. (2.60)], and when $\alpha = n/2$, $\lambda = 1/2$, the gamma pdf becomes

$$f_X(x) = \frac{(1/2)e^{-x/2}\,(x/2)^{n/2-1}}{\Gamma(n/2)} \tag{2.70}$$

which is the *chi-squared* r.v. pdf with n degrees of freedom (Prob. 4.40). When $\alpha = n$ (integer), the gamma distribution is sometimes known as the Erlang distribution.

J. Normal (or Gaussian) Distribution:

A r.v. X is called a *normal* (or *Gaussian*) r.v. if its pdf is given by

$$f_X(x) = \frac{1}{\sqrt{2\pi}\sigma}\, e^{-(x-\mu)^2/(2\sigma^2)} \tag{2.71}$$

The corresponding cdf of X is

$$F_X(x) = \frac{1}{\sqrt{2\pi}\sigma} \int_{-\infty}^{x} e^{-(\xi-\mu)^2/(2\sigma^2)}\, d\xi = \frac{1}{\sqrt{2\pi}} \int_{-\infty}^{(x-\mu)/\sigma} e^{-\xi^2/2}\, d\xi \tag{2.72}$$

This integral cannot be evaluated in a closed form and must be evaluated numerically. It is convenient to use the function $\Phi(z)$, defined as

$$\Phi(z) = \frac{1}{\sqrt{2\pi}} \int_{-\infty}^{z} e^{-\xi^2/2}\, d\xi \tag{2.73}$$

to help us to evaluate the value of $F_X(x)$. Then Eq. (2.72) can be written as

$$F_X(x) = \Phi\left(\frac{x - \mu}{\sigma}\right) \tag{2.74}$$

Note that

$$\Phi(-z) = 1 - \Phi(z) \tag{2.75}$$

The function $\Phi(z)$ is tabulated in Table A (Appendix A). Fig. 2-13 illustrates a normal distribution.

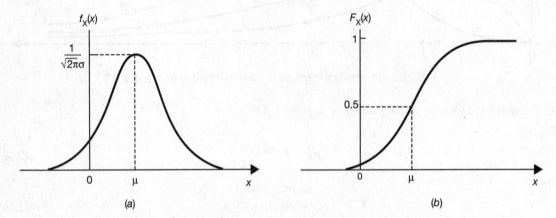

Fig. 2-13 Normal distribution.

The mean and variance of the normal r.v. X are (Prob. 2.36)

$$\mu_X = E(X) = \mu \tag{2.76}$$

$$\sigma_X^2 = \text{Var}(X) = \sigma^2 \tag{2.77}$$

We shall use the notation $N(\mu; \sigma^2)$ to denote that X is normal with mean μ and variance σ^2. A normal r.v. Z with zero mean and unit variance—that is, $Z = N(0; 1)$—is called a *standard normal* r.v. Note that the cdf of the standard normal r.v. is given by Eq. (2.73). The normal r.v. is probably the most important type of continuous r.v. It has played a significant role in the study of random phenomena in nature. Many naturally occurring random phenomena are approximately normal. Another reason for the importance of the normal r.v. is a remarkable theorem called the *central limit theorem*. This theorem states that the sum of a large number of independent r.v.'s, under certain conditions, can be approximated by a normal r.v. (see Sec. 4.9C).

2.8   Conditional Distributions

In Sec. 1.6 the conditional probability of an event A given event B is defined as

$$P(A|B) = \frac{P(A \cap B)}{P(B)} \quad P(B) > 0$$

The conditional cdf $F_X(x \mid B)$ of a r.v. X given event B is defined by

$$F_X(x|B) = P(X \leq x|B) = \frac{P\{(X \leq x) \cap B\}}{P(B)} \tag{2.78}$$

The conditional cdf $F_X(x \mid B)$ has the same properties as $F_X(x)$. (See Prob. 1.43 and Sec. 2.3.) In particular,

$$F_X(-\infty \mid B) = 0 \qquad F_X(\infty \mid B) = 1 \tag{2.79}$$
$$P(a < X \le b \mid B) = F_X(b \mid B) - F_X(a \mid B) \tag{2.80}$$

If X is a discrete r.v., then the conditional pmf $p_X(x_k \mid B)$ is defined by

$$p_X(x_k \mid B) = P(X = x_k \mid B) = \frac{P\{(X = x_k) \cap B\}}{P(B)} \tag{2.81}$$

If X is a continuous r.v., then the conditional pdf $f_X(x \mid B)$ is defined by

$$f_X(x \mid B) = \frac{dF_X(x \mid B)}{dx} \tag{2.82}$$

SOLVED PROBLEMS

Random Variables

2.1. Consider the experiment of throwing a fair die. Let X be the r.v. which assigns 1 if the number that appears is even and 0 if the number that appears is odd.

(a) What is the range of X?

(b) Find $P(X = 1)$ and $P(X = 0)$.

The sample space S on which X is defined consists of 6 points which are equally likely:

$$S = \{1, 2, 3, 4, 5, 6\}$$

(a) The range of X is $R_X = \{0, 1\}$.

(b) $(X = 1) = \{2, 4, 6\}$. Thus, $P(X = 1) = \frac{3}{6} = \frac{1}{2}$. Similarly, $(X = 0) = \{1, 3, 5\}$, and $P(X = 0) = \frac{1}{2}$.

2.2. Consider the experiment of tossing a coin three times (Prob. 1.1). Let X be the r.v. giving the number of heads obtained. We assume that the tosses are independent and the probability of a head is p.

(a) What is the range of X?

(b) Find the probabilities $P(X = 0)$, $P(X = 1)$, $P(X = 2)$, and $P(X = 3)$.

The sample space S on which X is defined consists of eight sample points (Prob. 1.1):

$$S = \{HHH, HHT, \ldots, TTT\}$$

(a) The range of X is $R_X = \{0, 1, 2, 3\}$.

(b) If $P(H) = p$, then $P(T) = 1 - p$. Since the tosses are independent, we have

$$P(X = 0) = P[\{TTT\}] = (1 - p)^3$$
$$P(X = 1) = P[\{HTT\}] + P[\{THT\}] + P[\{TTH\}] = 3(1 - p)^2 p$$
$$P(X = 2) = P[\{HHT\}] + P[\{HTH\}] + P[\{THH\}] = 3(1 - p)p^2$$
$$P(X = 3) = P[\{HHH\}] = p^3$$

2.3. An information source generates symbols at random from a four-letter alphabet $\{a, b, c, d\}$ with probabilities $P(a) = \frac{1}{2}$, $P(b) = \frac{1}{4}$, and $P(c) = P(d) = \frac{1}{8}$. A coding scheme encodes these symbols into binary codes as follows:

a	0
b	10
c	110
d	111

Let X be the r.v. denoting the length of the code—that is, the number of binary symbols (bits).

  (*a*)   What is the range of X?

  (*b*)   Assuming that the generations of symbols are independent, find the probabilities $P(X = 1)$, $P(X = 2)$, $P(X = 3)$, and $P(X > 3)$.

  (*a*)   The range of X is $R_X = \{1, 2, 3\}$.

  (*b*) $P(X = 1) = P[\{a\}] = P(a) = \frac{1}{2}$

        $P(X = 2) = P[\{b\}] = P(b) = \frac{1}{4}$

        $P(X = 3) = P[\{c, d\}] = P(c) + P(d) = \frac{1}{4}$

        $P(X > 3) = P(\varnothing) = 0$

2.4. Consider the experiment of throwing a dart onto a circular plate with unit radius. Let X be the r.v. representing the distance of the point where the dart lands from the origin of the plate. Assume that the dart always lands on the plate and that the dart is equally likely to land anywhere on the plate.

  (*a*)   What is the range of X?

  (*b*)   Find (i) $P(X < a)$ and (ii) $P(a < X < b)$, where $a < b \leq 1$.

  (*a*)   The range of X is $R_X = \{x: 0 \leq x < 1\}$.

  (*b*) (i) $(X < a)$ denotes that the point is inside the circle of radius a. Since the dart is equally likely to fall anywhere on the plate, we have (Fig. 2-14)

$$P(X < a) = \frac{\pi a^2}{\pi 1^2} = a^2$$

     (ii) $(a < X < b)$ denotes the event that the point is inside the annular ring with inner radius a and outer radius b. Thus, from Fig. 2-14, we have

$$P(a < X < b) = \frac{\pi(b^2 - a^2)}{\pi 1^2} = b^2 - a^2$$

Distribution Function

2.5. Verify Eq. (2.6).

Let $x_1 < x_2$. Then $(X \leq x_1)$ is a subset of $(X \leq x_2)$; that is, $(X \leq x_1) \subset (X \leq x_2)$. Then, by Eq. (1.41), we have

$$P(X \leq x_1) \leq P(X \leq x_2) \qquad \text{or} \qquad F_X(x_1) \leq F_X(x_2)$$

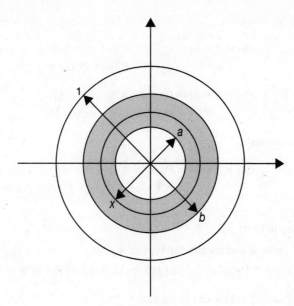

Fig. 2-14

2.6. Verify (*a*) Eq. (2.10); (*b*) Eq. (2.11); (*c*) Eq. (2.12).

(*a*) Since $(X \le b) = (X \le a) \cup (a < X \le b)$ and $(X \le a) \cap (a < X \le b) = \emptyset$, we have

$$P(X \le b) = P(X \le a) + P(a < X \le b)$$

or $\qquad\qquad F_X(b) = F_X(a) + P(a < X \le b)$

Thus, $\qquad\qquad P(a < X \le b) = F_X(b) - F_X(a)$

(*b*) Since $(X \le a) \cup (X > a) = S$ and $(X \le a) \cap (X > a) = \emptyset$, we have

$$P(X \le a) + P(X > a) = P(S) = 1$$

Thus, $\qquad\qquad P(X > a) = 1 - P(X \le a) = 1 - F_X(a)$

(*c*) Now $\qquad P(X < b) = P[\lim_{\substack{\varepsilon \to 0 \\ \varepsilon > 0}} X \le b - \varepsilon] = \lim_{\substack{\varepsilon \to 0 \\ \varepsilon > 0}} P(X \le b - \varepsilon)$

$$= \lim_{\substack{\varepsilon \to 0 \\ \varepsilon > 0}} F_X(b - \varepsilon) = F_X(b^-)$$

2.7. Show that

(*a*) $P(a \le X \le b) = P(X = a) + F_X(b) - F_X(a)$ $\qquad\qquad$ (2.83)

(*b*) $P(a < X < b) = F_X(b) - F_X(a) - P(X = b)$ $\qquad\qquad$ (2.84)

(*c*) $P(a \le X < b) = P(X = a) + F_X(b) - F_X(a) - P(X = b)$ $\qquad\qquad$ (2.85)

(*a*) Using Eqs. (1.37) and (2.10), we have

$$P(a \le X \le b) = P[(X = a) \cup (a < X \le b)]$$
$$= P(X = a) + P(a < X \le b)$$
$$= P(X = a) + F_X(b) - F_X(a)$$

(*b*) We have

$$P(a < X \le b) = P[(a < X < b) \cup (X = b)]$$
$$= P(a < X < b) + P(X = b)$$

Again using Eq. (2.10), we obtain

$$P(a < X < b) = P(a < X \le b) - P(X = b)$$
$$= F_X(b) - F_X(a) - P(X = b)$$

(c)   Similarly, $P(a \le X \le b) = P[(a \le X < b) \cup (X = b)]$
$$= P(a \le X < b) + P(X = b)$$

Using Eq. (2.83), we obtain

$$P(a \le X < b) = P(a \le X \le b) - P(X = b)$$
$$= P(X = a) + F_X(b) - F_X(a) - P(X = b)$$

2.8.   Let X be the r.v. defined in Prob. 2.3.

(a)   Sketch the cdf $F_X(x)$ of X and specify the type of X.

(b)   Find (i) $P(X \le 1)$, (ii) $P(1 < X \le 2)$, (iii) $P(X > 1)$, and (iv) $P(1 \le X \le 2)$.

(a)   From the result of Prob. 2.3 and Eq. (2.18), we have

$$F_X(x) = P(X \le x) = \begin{cases} 0 & x < 1 \\ \dfrac{1}{2} & 1 \le x < 2 \\ \dfrac{3}{4} & 2 \le x < 3 \\ 1 & x \ge 3 \end{cases}$$

which is sketched in Fig. 2-15. The r.v. X is a discrete r.v.

(b)   (i)   We see that

$$P(X \le 1) = F_X(1) = \frac{1}{2}$$

(ii)   By Eq. (2.10),

$$P(1 < X \le 2) = F_X(2) - F_X(1) = \frac{3}{4} - \frac{1}{2} = \frac{1}{4}$$

(iii)   By Eq. (2.11),

$$P(X > 1) = 1 - F_X(1) = 1 - \frac{1}{2} = \frac{1}{2}$$

(iv)   By Eq. (2.83),

$$P(1 \le X \le 2) = P(X = 1) + F_X(2) - F_X(1) = \frac{1}{2} + \frac{3}{4} - \frac{1}{2} = \frac{3}{4}$$

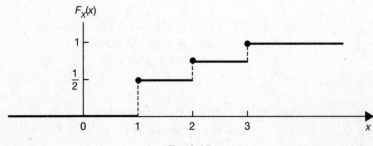

Fig. 2-15

2.9. Sketch the cdf $F_X(x)$ of the r.v. X defined in Prob. 2.4 and specify the type of X.

From the result of Prob. 2.4, we have

$$F_X(x) = P(X \le x) = \begin{cases} 0 & x < 0 \\ x^2 & 0 \le x < 1 \\ 1 & 1 \le x \end{cases}$$

which is sketched in Fig. 2-16. The r.v. X is a continuous r.v.

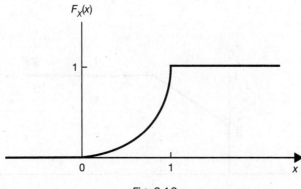

Fig. 2-16

2.10. Consider the function given by

$$F(x) = \begin{cases} 0 & x < 0 \\ x + \dfrac{1}{2} & 0 \le x < \dfrac{1}{2} \\ 1 & x \ge \dfrac{1}{2} \end{cases}$$

(a) Sketch $F(x)$ and show that $F(x)$ has the properties of a cdf discussed in Sec. 2.3B.

(b) If X is the r.v. whose cdf is given by $F(x)$, find (i) $P(X \le \frac{1}{4})$, (ii) $P(0 < X \le \frac{1}{4})$, (iii) $P(X = 0)$, and (iv) $P(0 \le X \le \frac{1}{4})$.

(c) Specify the type of X.

(a) The function $F(x)$ is sketched in Fig. 2-17. From Fig. 2-17, we see that $0 \le F(x) \le 1$ and $F(x)$ is a nondecreasing function, $F(-\infty) = 0$, $F(\infty) = 1$, $F(0) = \frac{1}{2}$, and $F(x)$ is continuous on the right. Thus, $F(x)$ satisfies all the properties [Eqs. (2.5) to (2.9)] required of a cdf.

(b) (i) We have

$$P\left(X \le \frac{1}{4}\right) = F\left(\frac{1}{4}\right) = \frac{1}{4} + \frac{1}{2} = \frac{3}{4}$$

(ii) By Eq. (2.10),

$$P\left(0 < X \le \frac{1}{4}\right) = F\left(\frac{1}{4}\right) - F(0) = \frac{3}{4} - \frac{1}{2} = \frac{1}{4}$$

(iii) By Eq. (2.12),

$$P(X=0) = P(X \le 0) - P(X < 0) = F(0) - F(0^-) = \frac{1}{2} - 0 = \frac{1}{2}$$

(iv) By Eq. (2.83),

$$P\left(0 \le X \le \frac{1}{4}\right) = P(X=0) + F\left(\frac{1}{4}\right) - F(0) = \frac{1}{2} + \frac{3}{4} - \frac{1}{2} = \frac{3}{4}$$

(c)　The r.v. X is a mixed r.v.

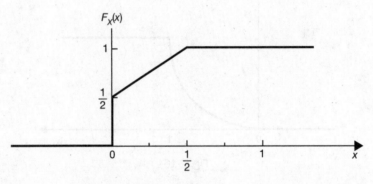

Fig. 2-17

2.11. Find the values of constants a and b such that

$$F(x) = \begin{cases} 1 - ae^{-x/b} & x \ge 0 \\ 0 & x < 0 \end{cases}$$

is a valid cdf.

To satisfy property 1 of $F_X(x)$ $[0 \le F_X(x) \le 1]$, we must have $0 \le a \le 1$ and $b > 0$. Since $b > 0$, property 3 of $F_X(x)$ $[F_X(\infty) = 1]$ is satisfied. It is seen that property 4 of $F_X(x)$ $[F_X(-\infty) = 0]$ is also satisfied. For $0 \le a \le 1$ and $b > 0$, $F(x)$ is sketched in Fig. 2-18. From Fig. 2-18, we see that $F(x)$ is a nondecreasing function and continuous on the right, and properties 2 and 5 of $F_X(x)$ are satisfied. Hence, we conclude that $F(x)$ given is a valid cdf if $0 \le a \le 1$ and $b > 0$. Note that if $a = 0$, then the r.v. X is a discrete r.v.; if $a = 1$, then X is a continuous r.v.; and if $0 < a < 1$, then X is a mixed r.v.

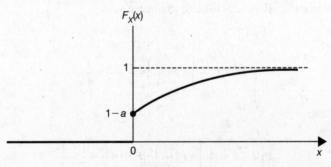

Fig. 2-18

Discrete Random Variables and Pmf's

2.12. Suppose a discrete r.v. X has the following pmfs:

$$p_X(1) = \tfrac{1}{2} \qquad p_X(2) = \tfrac{1}{4} \qquad p_X(3) = \tfrac{1}{8} \qquad p_X(4) = \tfrac{1}{8}$$

(a) Find and sketch the cdf $F_X(x)$ of the r.v. X.

(b) Find (i) $P(X \le 1)$, (ii) $P(1 < X \le 3)$, (iii) $P(1 \le X \le 3)$.

(a) By Eq. (2.18), we obtain

$$F_X(x) = P(X \le x) = \begin{cases} 0 & x < 1 \\ \dfrac{1}{2} & 1 \le x < 2 \\ \dfrac{3}{4} & 2 \le x < 3 \\ \dfrac{7}{8} & 3 \le x < 4 \\ 1 & x \ge 4 \end{cases}$$

which is sketched in Fig. 2-19.

(b) (i) By Eq. (2.12), we see that

$$P(X < 1) = F_X(1^-) = 0$$

(ii) By Eq. (2.10),

$$P(1 < X \le 3) = F_X(3) - F_X(1) = \frac{7}{8} - \frac{1}{2} = \frac{3}{8}$$

(iii) By Eq. (2.83),

$$P(1 \le X \le 3) = P(X = 1) + F_X(3) - F_X(1) = \frac{1}{2} + \frac{7}{8} - \frac{1}{2} = \frac{7}{8}$$

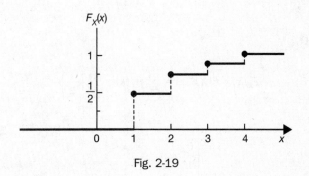

Fig. 2-19

2.13. (a) Verify that the function $p(x)$ defined by

$$p(x) = \begin{cases} \dfrac{3}{4}\left(\dfrac{1}{4}\right)^x & x = 0, 1, 2, \dots \\ 0 & \text{otherwise} \end{cases}$$

is a pmf of a discrete r.v. X.

(b) Find (i) $P(X = 2)$, (ii) $P(X \le 2)$, (iii) $P(X \ge 1)$.

(a) It is clear that $0 \le p(x) \le 1$ and

$$\sum_{i=0}^{\infty} p(i) = \frac{3}{4} \sum_{i=0}^{\infty} \left(\frac{1}{4}\right)^i = \frac{3}{4} \frac{1}{1 - \frac{1}{4}} = 1$$

Thus, $p(x)$ satisfies all properties of the pmf [Eqs. (2.15) to (2.17)] of a discrete r.v. X.

(b) (i) By definition (2.14),

$$P(X = 2) = p(2) = \frac{3}{4}\left(\frac{1}{4}\right)^2 = \frac{3}{64}$$

(ii) By Eq. (2.18),

$$P(X \le 2) = \sum_{i=0}^{2} p(i) = \frac{3}{4}\left(1 + \frac{1}{4} + \frac{1}{16}\right) = \frac{63}{64}$$

(iii) By Eq. (1.39),

$$P(X \ge 1) = 1 - P(X = 0) = 1 - p(0) = 1 - \frac{3}{4} = \frac{1}{4}$$

2.14. Consider the experiment of tossing an honest coin repeatedly (Prob. 1.41). Let the r.v. X denote the number of tosses required until the first head appears.

(a) Find and sketch the pmf $p_X(x)$ and the cdf $F_X(x)$ of X.

(b) Find (i) $P(1 < X \le 4)$, (ii) $P(X > 4)$.

(a) From the result of Prob. 1.41, the pmf of X is given by

$$p_X(x) = p_X(k) = P(X = k) = \left(\frac{1}{2}\right)^k \qquad k = 1, 2, \ldots$$

Then by Eq. (2.18),

$$F_X(x) = P(X \le x) = \sum_{k=1}^{m \le x} p_X(k) = \sum_{k=1}^{m \le x} \left(\frac{1}{2}\right)^k$$

or

$$F_X(x) = \begin{cases} 0 & x < 1 \\ \dfrac{1}{2} & 1 \le x < 2 \\ \dfrac{3}{4} & 2 \le x < 3 \\ \vdots & \vdots \\ 1 - \left(\dfrac{1}{2}\right)^n & n \le x < n + 1 \\ \vdots & \vdots \end{cases}$$

These functions are sketched in Fig. 2-20.

(b)   (i)   By Eq. (2.10),

$$P(1 < X \leq 4) = F_X(4) - F_X(1) = \frac{15}{16} - \frac{1}{2} = \frac{7}{16}$$

(ii)   By Eq. (1.39),

$$P(X > 4) = 1 - P(X \leq 4) = 1 - F_X(4) = 1 - \frac{15}{16} = \frac{1}{16}$$

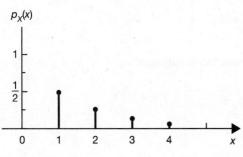 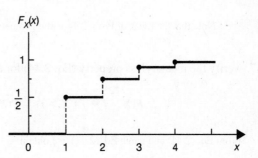

Fig. 2-20

2.15 Let X be a binomial r.v. with parameters (n, p).

(a)   Show that $p_X(x)$ given by Eq. (2.36) satisfies Eq. (2.17).

(b)   Find $P(X > 1)$ if $n = 6$ and $p = 0.1$.

(a)   Recall that the binomial expansion formula is given by

$$(a + b)^n = \sum_{k=0}^{n} \binom{n}{k} a^k b^{n-k} \tag{2.86}$$

Thus, by Eq. (2.36),

$$\sum_{k=0}^{n} p_X(k) = \sum_{k=0}^{n} \binom{n}{k} p^k (1 - p)^{n-k} = (p + 1 - p)^n = 1^n = 1$$

(b)   Now

$$P(X > 1) = 1 - P(X = 0) - P(X = 1)$$

$$= 1 - \binom{6}{0}(0.1)^0(0.9)^6 - \binom{6}{1}(0.1)^1(0.9)^5$$

$$= 1 - (0.9)^6 - 6(0.1)(0.9)^5 \approx 0.114$$

2.16. Let X be a geometric r.v. with parameter p.

(a)   Show that $p_X(x)$ given by Eq. (2.40) satisfies Eq. (2.17).

(b)   Find the cdf $F_X(x)$ of X.

(a)   Recall that for a geometric series, the sum is given by

$$\sum_{n=0}^{\infty} ar^n = \sum_{n=1}^{\infty} ar^{n-1} = \frac{a}{1 - r} \qquad |r| < 1 \tag{2.87}$$

Thus,

$$\sum_x p_X(x) = \sum_{i=1}^{\infty} p_X(i) = \sum_{i=1}^{\infty} (1 - p)^{i-1} p = \frac{p}{1 - (1 - p)} = \frac{p}{p} = 1$$

(*b*) Using Eq. (2.87), we obtain

$$P(X > k) = \sum_{i=k+1}^{\infty} (1-p)^{i-1}p = \frac{(1-p)^k p}{1-(1-p)} = (1-p)^k \qquad (2.88)$$

Thus,

$$P(X \le k) = 1 - P(X > k) = 1 - (1-p)^k \qquad (2.89)$$

and

$$F_X(x) = P(X \le x) = 1 - (1-p)^x \qquad x = 1, 2, \dots \qquad (2.90)$$

Note that the r.v. X of Prob. 2.14 is the geometric r.v. with $p = \frac{1}{2}$.

2.17. Verify the memoryless property (Eq. 2.44) for the geometric distribution.

$$P(X > i + j \mid X > i) = P(X > j) \qquad i, j \ge 1$$

From Eq. (2.41) and using Eq. (2.11), we obtain

$$P(X > i) = 1 - P(X \le 1) = (1-p)^i \qquad \text{for } i \ge 1 \qquad (2.91)$$

Now, by definition of conditional probability, Eq. (1.55), we obtain

$$P(X > i + j \mid X > i) = \frac{P[\{X > i + j\} \cap \{X > i\}]}{P(X > i)}$$

$$= \frac{P(X > i + j)}{P(X > i)}$$

$$= \frac{(1-p)^{i+j}}{(1-p)^i} = (1-p)^j = P(X > j) \qquad i, j > 1$$

2.18. Let X be a negative binomial r.v. with parameters p and k. Show that $p_X(x)$ given by Eq. (2.45) satisfies Eq. (2.17) for $k = 2$.

From Eq. (2.45)

$$p_X(x) = \binom{x-1}{k-1} p^k (1-p)^{x-k} \qquad x = k, k+1, \dots$$

Let $k = 2$ and $1 - p = q$. Then

$$p_X(x) = \binom{x-1}{1} p^2 q^{x-2} = (x-1)p^2 q^{x-2} \qquad x = 2, 3, \dots \qquad (2.92)$$

$$\sum_{x=2}^{\infty} p_X(x) = \sum_{x=2}^{\infty} (x-1) p^2 q^{x-2} = p^2 + 2p^2 q + 3p^2 q^2 + 4p^2 q^3 + \cdots \qquad (2.93)$$

Now let

$$S = p^2 + 2p^2 q + 3p^2 q^2 + 4p^2 q^3 + \cdots$$

Then

$$qS = p^2 q + 2p^2 q^2 + 3p^2 q^3 + 4p^2 q^4 + \cdots$$

Subtracting the second series from the first series, we obtain

$$(1-q)S = p^2 + p^2 q + p^2 q^2 + p^2 q^3 + \cdots$$
$$= p^2(1 + q + q^2 + q^3 + \cdots) = p^2 \frac{1}{1-q}$$

and we have

$$S = p^2 \frac{1}{(1-q)^2} = p^2 \frac{1}{p^2} = 1$$

Thus,

$$\sum_{x=2}^{\infty} p_X(x) = 1$$

2.19. Let X be a Poisson r.v. with parameter λ.

(a)   Show that $p_X(x)$ given by Eq. (2.48) satisfies Eq. (2.17).

(b)   Find $P(X > 2)$ with $\lambda = 4$.

(a)   By Eq. (2.48),

$$\sum_{k=0}^{\infty} p_X(k) = e^{-\lambda} \sum_{k=0}^{\infty} \frac{\lambda^k}{k!} = e^{-\lambda} e^{\lambda} = 1$$

(b)   With $\lambda = 4$, we have

$$p_X(k) = e^{-4} \frac{4^k}{k!}$$

and

$$P(X \le 2) = \sum_{k=0}^{2} p_X(k) = e^{-4}(1 + 4 + 8) \approx 0.238$$

Thus,

$$P(X > 2) = 1 - P(X \le 2) \approx 1 - 0.238 = 0.762$$

Continuous Random Variables and Pdf's

2.20. Verify Eq. (2.19).

From Eqs. (1.41) and (2.10), we have

$$P(X = x) \le P(x - \varepsilon < X \le x) = F_X(x) - F_X(x - \varepsilon)$$

for any $\varepsilon \ge 0$. As $F_X(x)$ is continuous, the right-hand side of the above expression approaches 0 as $\varepsilon \to 0$. Thus, $P(X = x) = 0$.

2.21. The pdf of a continuous r.v. X is given by

$$f_X(x) = \begin{cases} \dfrac{1}{3} & 0 < x < 1 \\[2mm] \dfrac{2}{3} & 1 < x < 2 \\[2mm] 0 & \text{otherwise} \end{cases}$$

Find the corresponding cdf $F_X(x)$ and sketch $f_X(x)$ and $F_X(x)$.

By Eq. (2.24), the cdf of X is given by

$$F_X(x) = \begin{cases} 0 & x < 0 \\ \int_0^x \frac{1}{3}\,d\xi = \frac{x}{3} & 0 \le x < 1 \\ \int_0^1 \frac{1}{3}\,d\xi + \int_1^x \frac{2}{3}\,d\xi = \frac{2}{3}x - \frac{1}{3} & 1 \le x < 2 \\ \int_0^1 \frac{1}{3}\,d\xi + \int_1^2 \frac{2}{3}\,d\xi = 1 & 2 \le x \end{cases}$$

The functions $f_X(x)$ and $F_X(x)$ are sketched in Fig. 2-21.

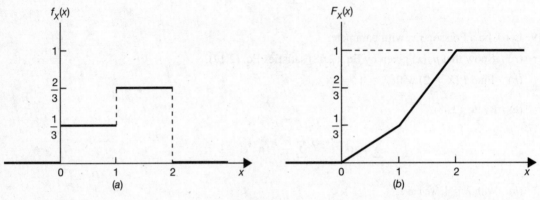

Fig. 2-21

2.22. Let X be a continuous r.v. X with pdf

$$f_X(x) = \begin{cases} kx & 0 < x < 1 \\ 0 & \text{otherwise} \end{cases}$$

where k is a constant.

(*a*)   Determine the value of k and sketch $f_X(x)$.

(*b*)   Find and sketch the corresponding cdf $F_X(x)$.

(*c*)   Find $P(\frac{1}{4} < X \le 2)$.

(*a*)   By Eq. (2.21), we must have $k > 0$, and by Eq. (2.22),

$$\int_0^1 kx\,dx = \frac{k}{2} = 1$$

Thus, $k = 2$ and

$$f_X(x) = \begin{cases} 2x & 0 < x < 1 \\ 0 & \text{otherwise} \end{cases}$$

which is sketched in Fig. 2-22(*a*).

(b)   By Eq. (2.24), the cdf of X is given by

$$F_X(x) = \begin{cases} 0 & x < 0 \\ \int_0^x 2\xi \, d\xi = x^2 & 0 \le x < 1 \\ \int_0^1 2\xi \, d\xi = 1 & 1 \le x \end{cases}$$

which is sketched in Fig. 2-22(b).

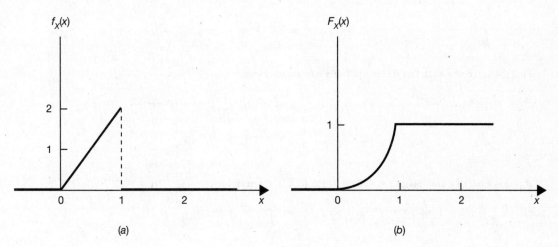

Fig. 2-22

(c)   By Eq. (2.25),

$$P\left(\frac{1}{4} < X \le 2\right) = F_X(2) - F_X\left(\frac{1}{4}\right) = 1 - \left(\frac{1}{4}\right)^2 = \frac{15}{16}$$

2.23. Show that the pdf of a normal r.v. X given by Eq. (2.71) satisfies Eq. (2.22).

From Eq. (2.71),

$$\int_{-\infty}^{\infty} f_x(x) \, dx = \frac{1}{\sqrt{2\pi}\sigma} \int_{-\infty}^{\infty} e^{-(x-\mu)^2/(2\sigma^2)} \, dx$$

Let $y = (x - \mu)/(\sqrt{2}\sigma)$. Then $dx = \sqrt{2}\sigma \, dy$ and

$$\frac{1}{\sqrt{2\pi}\sigma} \int_{-\infty}^{\infty} e^{-(x-\mu)^2/(2\sigma^2)} \, dx = \frac{1}{\sqrt{\pi}} \int_{-\infty}^{\infty} e^{-y^2} \, dy$$

Let

$$\int_{-\infty}^{\infty} e^{-y^2} \, dy = I$$

Then

$$I^2 = \left[\int_{-\infty}^{\infty} e^{-x^2} \, dx\right]\left[\int_{-\infty}^{\infty} e^{-y^2} \, dy\right] = \int_{-\infty}^{\infty}\int_{-\infty}^{\infty} e^{-(x^2+y^2)} \, dx \, dy$$

Letting $x = r \cos \theta$ and $y = r \sin \theta$ (that is, using polar coordinates), we have

$$I^2 = \int_0^{2\pi} \int_0^\infty e^{-r^2} r \, dr \, d\theta = 2\pi \int_0^\infty e^{-r^2} r \, dr = \pi$$

Thus,

$$I = \int_{-\infty}^\infty e^{-y^2} \, dy = \sqrt{\pi} \tag{2.94}$$

and

$$\int_{-\infty}^\infty f_X(x) \, dx = \frac{1}{\sqrt{\pi}} \int_{-\infty}^\infty e^{-y^2} \, dy = \frac{1}{\sqrt{\pi}} \sqrt{\pi} = 1$$

2.24.   Consider a function

$$f(x) = \frac{1}{\sqrt{\pi}} e^{(-x^2 + x - a)} \qquad -\infty < x < \infty$$

Find the value of a such that $f(x)$ is a pdf of a continuous r.v. X.

$$f(x) = \frac{1}{\sqrt{\pi}} e^{(-x^2 + x - a)} = \frac{1}{\sqrt{\pi}} e^{-(x^2 - x + 1/4 + a - 1/4)}$$

$$= \left[\frac{1}{\sqrt{\pi}} e^{-(x-1/2)^2} \right] e^{-(a-1/4)}$$

If $f(x)$ is a pdf of a continuous r.v. X, then by Eq. (2.22), we must have

$$\int_{-\infty}^\infty f(x) \, dx = e^{-(a-1/4)} \int_{-\infty}^\infty \frac{1}{\sqrt{\pi}} e^{-(x-1/2)^2} \, dx = 1$$

Now by Eq. (2.52), the pdf of $N\left(\dfrac{1}{2}; \dfrac{1}{2}\right)$ is $\dfrac{1}{\sqrt{\pi}} e^{-(x-1/2)^2}$. Thus,

$$\int_{-\infty}^\infty \frac{1}{\sqrt{\pi}} e^{-(x-1/2)^2} \, dx = 1 \quad \text{and} \quad \int_{-\infty}^\infty f(x) \, dx = e^{-(a-1/4)} = 1$$

from which we obtain $a = \dfrac{1}{4}$.

2.25. A r.v. X is called a *Rayleigh* r.v. if its pdf is given by

$$f_X(x) = \begin{cases} \dfrac{x}{\sigma^2} e^{-x^2/(2\sigma^2)} & x > 0 \\ 0 & x < 0 \end{cases} \tag{2.95}$$

(a)   Determine the corresponding cdf $F_X(x)$.

(b)   Sketch $f_X(x)$ and $F_X(x)$ for $\sigma = 1$.

(a)   By Eq. (2.24), the cdf of X is

$$F_X(x) = \int_0^x \frac{\xi}{\sigma^2} e^{-\xi^2/(2\sigma^2)} \, d\xi \qquad x \geq 0$$

Let $y = \xi^2/(2\sigma^2)$. Then $dy = (1/\sigma^2)\xi \, d\xi$, and

$$F_X(x) = \int_0^{x^2/(2\sigma^2)} e^{-y} \, dy = 1 - e^{-x^2/(2\sigma^2)} \tag{2.96}$$

(b) With $\sigma = 1$, we have

$$f_X(x) = \begin{cases} xe^{-x^2/2} & x > 0 \\ 0 & x < 0 \end{cases}$$

and

$$F_X(x) = \begin{cases} 1 - e^{-x^2/2} & x \geq 0 \\ 0 & x < 0 \end{cases}$$

These functions are sketched in Fig. 2-23.

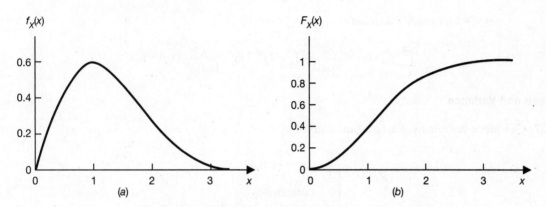

Fig. 2-23 Rayleigh distribution with $\sigma = 1$.

2.26. Consider a gamma r.v. X with parameter (α, λ) $(\alpha > 0$ and $\lambda > 0)$ defined by Eq. (2.65).

(a) Show that the gamma function has the following properties:

1. $\Gamma(\alpha + 1) = \alpha\Gamma(\alpha)$ $\alpha > 0$ (2.97)

2. $\Gamma(k + 1) = k!$ $k\ (\geq 0)$: integer   (2.98)

3. $\Gamma\left(\frac{1}{2}\right) = \sqrt{\pi}$ (2.99)

(b) Show that the pdf given by Eq. (2.65) satisfies Eq. (2.22).

(a) Integrating Eq. (2.66) by parts ($u = x^{\alpha-1}, dv = e^{-x}\,dx$), we obtain

$$\begin{aligned} \Gamma(\alpha) &= -e^{-x}x^{\alpha-1}\Big|_0^\infty + \int_0^\infty e^{-x}(\alpha-1)x^{\alpha-2}\,dx \\ &= (\alpha-1)\int_0^\infty e^{-x}x^{\alpha-2}\,dx = (\alpha-1)\Gamma(\alpha-1) \end{aligned}$$

$$\text{(2.100)}$$

Replacing α by $\alpha + 1$ in Eq. (2.100), we get Eq. (2.97).

Next, by applying Eq. (2.97) repeatedly using an integral value of α, say $\alpha = k$, we obtain

$$\Gamma(k + 1) = k\Gamma(k) = k(k - 1)\Gamma(k - 1) = k(k - 1) \cdots (2)\Gamma(1)$$

Since

$$\Gamma(1) = \int_0^\infty e^{-x}\,dx = 1$$

it follows that $\Gamma(k + 1) = k!$. Finally, by Eq. (2.66),

$$\Gamma\left(\frac{1}{2}\right) = \int_0^\infty e^{-x}x^{-1/2}\,dx$$

Let $y = x^{1/2}$. Then $dy = \frac{1}{2} x^{-1/2} dx$, and

$$\Gamma\left(\frac{1}{2}\right) = 2 \int_0^\infty e^{-y^2} dy = \int_{-\infty}^\infty e^{-y^2} dy = \sqrt{\pi}.$$

in view of Eq. (2.94).

(b) Now

$$\int_{-\infty}^\infty f_X(x) \, dx = \int_0^\infty \frac{\lambda e^{-\lambda x}(\lambda x)^{\alpha-1}}{\Gamma(\alpha)} \, dx = \frac{\lambda^\alpha}{\Gamma(\alpha)} \int_0^\infty e^{-\lambda x} x^{\alpha-1} \, dx$$

Let $y = \lambda x$. Then $dy = \lambda \, dx$ and

$$\int_{-\infty}^\infty f_X(x) \, dx = \frac{\lambda^\alpha}{\Gamma(\alpha)\lambda^\alpha} \int_0^\infty e^{-y} y^{\alpha-1} \, dy = \frac{\lambda^\alpha}{\Gamma(\alpha)\lambda^\alpha} \Gamma(\alpha) = 1$$

Mean and Variance

2.27. Consider a discrete r.v. X whose pmf is given by

$$p_X(x) = \begin{cases} \dfrac{1}{3} & x = -1, 0, 1 \\ 0 & \text{otherwise} \end{cases}$$

(a) Plot $p_X(x)$ and find the mean and variance of X.

(b) Repeat (a) if the pmf is given by

$$p_X(x) = \begin{cases} \dfrac{1}{3} & x = -2, 0, 2 \\ 0 & \text{otherwise} \end{cases}$$

(a) The pmf $p_X(x)$ is plotted in Fig. 2-24(a). By Eq. (2.26), the mean of X is

$$\mu_X = E(X) = \frac{1}{3}(-1 + 0 + 1) = 0$$

By Eq. (2.29), the variance of X is

$$\sigma_X^2 = \text{Var}(X) = E[(X - \mu_X)^2] = E(X^2) = \frac{1}{3}[(-1)^2 + (0)^2 + (1)^2] = \frac{2}{3}$$

(b) The pmf $p_X(x)$ is plotted in Fig. 2-24(b). Again by Eqs. (2.26) and (2.29), we obtain

$$\mu_X = E(X) = \frac{1}{3}(-2 + 0 + 2) = 0$$

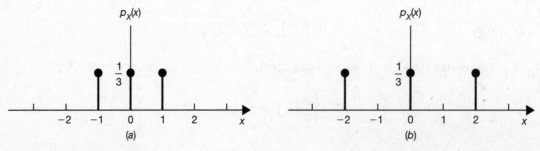

Fig. 2-24

$$\sigma_X^2 = \mathrm{Var}(X) = \frac{1}{3}[(-2)^2 + (0)^2 + (2)^2] = \frac{8}{3}$$

Note that the variance of X is a measure of the spread of a distribution about its mean.

2.28. Let a r.v. X denote the outcome of throwing a fair die. Find the mean and variance of X.

Since the die is fair, the pmf of X is

$$p_X(x) = p_X(k) = \frac{1}{6} \qquad k = 1, 2, \ldots, 6$$

By Eqs. (2.26) and (2.29), the mean and variance of X are

$$\mu_X = E(X) = \frac{1}{6}(1 + 2 + 3 + 4 + 5 + 6) = \frac{7}{2} = 3.5$$

$$\sigma_X^2 = \frac{1}{6}\left[\left(1 - \frac{7}{2}\right)^2 + \left(2 - \frac{7}{2}\right)^2 + \left(3 - \frac{7}{2}\right)^2 + \left(4 - \frac{7}{2}\right)^2 + \left(5 - \frac{7}{2}\right)^2 + \left(6 - \frac{7}{2}\right)^2\right] = \frac{35}{12}$$

Alternatively, the variance of X can be found as follows:

$$E(X^2) = \frac{1}{6}(1^2 + 2^2 + 3^2 + 4^2 + 5^2 + 6^2) = \frac{91}{6}$$

Hence, by Eq. (2.31),

$$\sigma_X^2 = E(X^2) - [E(X)]^2 = \frac{91}{6} - \left(\frac{7}{2}\right)^2 = \frac{35}{12}$$

2.29. Find the mean and variance of the geometric r.v. X defined by Eq. (2.40).

To find the mean and variance of a geometric r.v. X, we need the following results about the sum of a geometric series and its first and second derivatives. Let

$$g(r) = \sum_{n=0}^{\infty} ar^n = \frac{a}{1-r} \qquad |r| < 1 \tag{2.101}$$

Then

$$g'(r) = \frac{dg(r)}{dr} = \sum_{n=1}^{\infty} anr^{n-1} = \frac{a}{(1-r)^2} \tag{2.102}$$

$$g''(r) = \frac{d^2g(r)}{dr^2} = \sum_{n=2}^{\infty} an(n-1)r^{n-2} = \frac{2a}{(1-r)^3} \tag{2.103}$$

By Eqs. (2.26) and (2.40), and letting $q = 1 - p$, the mean of X is given by

$$\mu_X = E(X) = \sum_{x=1}^{\infty} xq^{x-1}p = \frac{p}{(1-q)^2} = \frac{p}{p^2} = \frac{1}{p} \tag{2.104}$$

where Eq. (2.102) is used with $a = p$ and $r = q$.

To find the variance of X, we first find $E[X(X-1)]$. Now,

$$E[X(X-1)] = \sum_{x=1}^{\infty} x(x-1)q^{x-1}p = \sum_{x=2}^{\infty} pqx(x-1)q^{x-2}$$

$$= \frac{2pq}{(1-q)^3} = \frac{2pq}{p^3} = \frac{2q}{p^2} = \frac{2(1-p)}{p^2} \tag{2.105}$$

where Eq. (2.103) is used with $a = pq$ and $r = q$.

Since $E[X(X-1)] = E(X^2 - X) = E(X^2) - E(X)$, we have

$$E(X^2) = E[X(X-1)] + E(X) = \frac{2(1-p)}{p^2} + \frac{1}{p} = \frac{2-p}{p^2} \tag{2.106}$$

Then by Eq. (2.31), the variance of X is

$$\sigma_X^2 = \text{Var}(X) = E(X^2) - E[(X)]^2 = \frac{2-p}{p^2} - \frac{1}{p^2} = \frac{1-p}{p^2} \tag{2.107}$$

2.30. Let X be a binomial r.v. with parameters (n, p). Verify Eqs. (2.38) and (2.39).

By Eqs. (2.26) and (2.36), and letting $q = 1 - p$, we have

$$E(X) = \sum_{k=0}^{n} k p_X(k) = \sum_{k=0}^{n} k \binom{n}{k} p^k q^{n-k}$$

$$= \sum_{k=0}^{n} k \frac{n!}{(n-k)! k!} p^k q^{n-k}$$

$$= np \sum_{k=1}^{n} \frac{(n-1)!}{(n-k)!(k-1)!} p^{k-1} q^{n-k}$$

Letting $i = k - 1$ and using Eq. (2.86), we obtain

$$E(X) = np \sum_{i=0}^{n-1} \frac{(n-1)!}{(n-1-i)! i!} p^i q^{n-1-i}$$

$$= np \sum_{i=0}^{n-1} \binom{n-1}{i} p^i q^{n-1-i}$$

$$= np(p+q)^{n-1} = np(1)^{n-1} = np$$

Next,

$$E[X(X-1)] = \sum_{k=0}^{n} k(k-1) p_X(k) = \sum_{k=0}^{n} k(k-1) \binom{n}{k} p^k q^{n-k}$$

$$= \sum_{k=0}^{n} k(k-1) \frac{n!}{(n-k)! k!} p^k q^{n-k}$$

$$= n(n-1)p^2 \sum_{k=2}^{n} \frac{(n-2)!}{(n-k)!(k-2)!} p^{k-2} q^{n-k}$$

Similarly, letting $i = k - 2$ and using Eq. (2.86), we obtain

$$E[X(X-1)] = n(n-1)p^2 \sum_{i=0}^{n-2} \frac{(n-2)!}{(n-2-i)! i!} p^i q^{n-2-i}$$

$$= n(n-1)p^2 \sum_{i=0}^{n-2} \binom{n-2}{i} p^i q^{n-2-i}$$

$$= n(n-1)p^2(p+q)^{n-2} = n(n-1)p^2$$

Thus,

$$E(X)^2 = E[X(X-1)] + E(X) = n(n-1)p^2 + np \tag{2.108}$$

and by Eq. (2.31),

$$\sigma_X^2 = \text{Var}(x) = n(n-1)p^2 + np - (np)^2 = np(1-p)$$

2.31. Let X be a Poisson r.v. with parameter λ. Verify Eqs. (2.50) and (2.51).

By Eqs. (2.26) and (2.48),

$$E(X) = \sum_{k=0}^{\infty} k p_X(k) = \sum_{k=0}^{\infty} k e^{-\lambda} \frac{\lambda^k}{k!} = 0 + \sum_{k=1}^{\infty} e^{-\lambda} \frac{\lambda^k}{(k-1)!}$$

$$= \lambda e^{-\lambda} \sum_{k=1}^{\infty} \frac{\lambda^{k-1}}{(k-1)!} = \lambda e^{-\lambda} \sum_{i=0}^{\infty} \frac{\lambda^i}{i!} = \lambda e^{-\lambda} e^{\lambda} = \lambda$$

Next,
$$E[X(X-1)] = \sum_{k=0}^{\infty} k(k-1) e^{-\lambda} \frac{\lambda^k}{k!} = \lambda^2 e^{-\lambda} \sum_{k=2}^{\infty} \frac{\lambda^{k-2}}{(k-2)!}$$

$$= \lambda^2 e^{-\lambda} \sum_{i=0}^{\infty} \frac{\lambda^i}{i!} = \lambda^2 e^{-\lambda} e^{\lambda} = \lambda^2$$

Thus,
$$E(X^2) = E[X(X-1)] + E(X) = \lambda^2 + \lambda \qquad (2.109)$$

and by Eq. (2.31),

$$\sigma_X^2 = \text{Var}(X) = E(X^2) - [E(X)]^2 = \lambda^2 + \lambda - \lambda^2 = \lambda$$

2.32. Let X be a discrete uniform r.v. defined by Eq. (2.52). Verify Eqs. (2.54) and (2.55), i.e.,

$$\mu_X = E(X) = \frac{1}{2}(n+1) \qquad \sigma_X^2 = \text{Var}(X) = \frac{1}{12}(n^2 - 1)$$

By Eqs. (2.52) and (2.26), we have

$$E(X) = \sum_{x=1}^{n} x\, p_X(x) = \frac{1}{n} \sum_{x=1}^{n} x = \frac{1}{n} \frac{1}{2} n(n+1) = \frac{1}{2}(n+1)$$

$$E(X^2) = \sum_{x=1}^{n} x^2 p_X(x) = \frac{1}{n} \sum_{x=1}^{n} x^2 = \frac{1}{n} \frac{1}{3} n(n+1)\left(n + \frac{1}{2}\right) = \frac{1}{3}(n+1)\left(n + \frac{1}{2}\right)$$

Now, using Eq. (2.31), we obtain

$$\text{Var}(X) = E(X^2) - [E(X)]^2$$

$$= \frac{1}{3}(n+1)\left(n + \frac{1}{2}\right) - \frac{1}{4}(n+1)^2$$

$$= \frac{1}{12}(n+1)(n-1) = \frac{1}{12}(n^2 - 1)$$

2.33. Find the mean and variance of the r.v. X of Prob. 2.22.

From Prob. 2.22, the pdf of X is

$$f_X(x) = \begin{cases} 2x & 0 < x < 1 \\ 0 & \text{otherwise} \end{cases}$$

By Eq. (2.26), the mean of X is

$$\mu_x = E(X) = \int_0^1 x(2x)\, dx = 2 \left. \frac{x^3}{3} \right|_0^1 = \frac{2}{3}$$

By Eq. (2.27), we have

$$E(X^2) = \int_0^1 x^2(2x)\,dx = 2\left.\frac{x^4}{4}\right|_0^1 = \frac{1}{2}$$

Thus, by Eq. (2.31), the variance of X is

$$\sigma_X^2 = \mathrm{Var}(X) = E(X^2) - [E(X)]^2 = \frac{1}{2} - \left(\frac{2}{3}\right)^2 = \frac{1}{18}$$

2.34. Let X be a uniform r.v. over (a, b). Verify Eqs. (2.58) and (2.59).

By Eqs. (2.56) and (2.26), the mean of X is

$$\mu_X = E(X) = \int_a^b x\frac{1}{b-a}\,dx = \frac{1}{b-a}\left.\frac{x^2}{2}\right|_a^b = \frac{1}{2}\frac{b^2-a^2}{b-a} = \frac{1}{2}(b+a)$$

By Eq. (2.27), we have

$$E(X^2) = \int_a^b x^2\frac{1}{b-a}\,dx = \frac{1}{b-a}\left.\frac{x^3}{3}\right|_a^b = \frac{1}{3}(b^2 + ab + a^2) \qquad (2.110)$$

Thus, by Eq. (2.31), the variance of X is

$$\sigma_X^2 = \mathrm{Var}(X) = E(X^2) - [E(X)]^2$$
$$= \frac{1}{3}(b^2 + ab + a^2) - \frac{1}{4}(b+a)^2 = \frac{1}{12}(b-a)^2$$

2.35. Let X be an exponential r.v. X with parameter λ. Verify Eqs. (2.62) and (2.63).

By Eqs. (2.60) and (2.26), the mean of X is

$$\mu_X = E(X) = \int_0^\infty x\lambda e^{-\lambda x}\,dx$$

Integrating by parts ($u = x, dv = \lambda e^{-\lambda x}dx$) yields

$$E(X) = -xe^{-\lambda x}\Big|_0^\infty + \int_0^\infty e^{-\lambda x}\,dx = \frac{1}{\lambda}$$

Next, by Eq. (2.27),

$$E(X^2) = \int_0^\infty x^2\lambda e^{-\lambda x}\,dx$$

Again integrating by parts ($u = x^2, dv = \lambda e^{-\lambda x}\,dx$), we obtain

$$E(X^2) = -x^2 e^{-\lambda x}\Big|_0^\infty + 2\int_0^\infty xe^{-\lambda x}\,dx = \frac{2}{\lambda^2} \qquad (2.111)$$

Thus, by Eq. (2.31), the variance of X is

$$\sigma_X^2 = \mathrm{E}(X^2) - [E(X)]^2 = \frac{2}{\lambda^2} - \left(\frac{1}{\lambda}\right)^2 = \frac{1}{\lambda^2}$$

2.36. Let $X = N(\mu; \sigma^2)$. Verify Eqs. (2.76) and (2.77).

Using Eqs. (2.71) and (2.26), we have

$$\mu_X = E(X) = \frac{1}{\sqrt{2\pi}\sigma} \int_{-\infty}^{\infty} x e^{-(x-\mu)^2/(2\sigma^2)} \, dx$$

Writing x as $(x - \mu) + \mu$, we have

$$E(X) = \frac{1}{\sqrt{2\pi}\sigma} \int_{-\infty}^{\infty} (x-\mu) \, e^{-(x-\mu)^2/(2\sigma^2)} \, dx + \mu \frac{1}{\sqrt{2\pi}\sigma} \int_{-\infty}^{\infty} e^{-(x-\mu)^2/(2\sigma^2)} \, dx$$

Letting $y = x - \mu$ in the first integral, we obtain

$$E(X) = \frac{1}{\sqrt{2\pi}\sigma} \int_{-\infty}^{\infty} y e^{-y^2/(2\sigma^2)} \, dy + \mu \int_{-\infty}^{\infty} f_X(x) \, dx$$

The first integral is zero, since its integrand is an odd function. Thus, by the property of pdf Eq. (2.22), we get

$$\mu_X = E(X) = \mu$$

Next, by Eq. (2.29),

$$\sigma_X^2 = E[(X - \mu)^2] = \frac{1}{\sqrt{2\pi}\sigma} \int_{-\infty}^{\infty} (x-\mu)^2 \, e^{-(x-\mu)^2/(2\sigma^2)} \, dx$$

From Eqs. (2.22) and (2.71), we have

$$\int_{-\infty}^{\infty} e^{-(x-\mu)^2/(2\sigma^2)} \, dx = \sigma\sqrt{2\pi}$$

Differentiating with respect to σ, we obtain

$$\int_{-\infty}^{\infty} \frac{(x-\mu)^2}{\sigma^3} e^{-(x-\mu)^2/(2\sigma^2)} \, dx = \sqrt{2\pi}$$

Multiplying both sides by $\sigma^2/\sqrt{2\pi}$, we have

$$\frac{1}{\sqrt{2\pi}\sigma} \int_{-\infty}^{\infty} (x-\mu)^2 \, e^{-(x-\mu)^2/(2\sigma^2)} \, dx = \sigma^2$$

Thus,
$$\sigma_X^2 = \text{Var}(X) = \sigma^2$$

2.37. Find the mean and variance of a Rayleigh r.v. defined by Eq. (2.95) (Prob. 2.25).

Using Eqs. (2.95) and (2.26), we have

$$\mu_X = E(X) = \int_0^{\infty} x \frac{x}{\sigma^2} e^{-x^2/(2\sigma^2)} \, dx = \frac{1}{\sigma^2} \int_0^{\infty} x^2 e^{-x^2/(2\sigma^2)} \, dx$$

Now the variance of $N(0; \sigma^2)$ is given by

$$\frac{1}{\sqrt{2\pi}\sigma} \int_{-\infty}^{\infty} x^2 e^{-x^2/(2\sigma^2)} \, dx = \sigma^2$$

Since the integrand is an even function, we have

$$\frac{1}{\sqrt{2\pi}\sigma} \int_0^\infty x^2 e^{-x^2/(2\sigma^2)}\,dx = \frac{1}{2}\sigma^2$$

or

$$\int_0^\infty x^2 e^{-x^2/(2\sigma^2)}\,dx = \frac{1}{2}\sqrt{2\pi}\sigma^3 = \sqrt{\frac{\pi}{2}}\,\sigma^3$$

Then

$$\mu_X = E(X) = \frac{1}{\sigma^2}\sqrt{\frac{\pi}{2}}\,\sigma^3 = \sqrt{\frac{\pi}{2}}\,\sigma \tag{2.112}$$

Next,

$$E(X^2) = \int_0^\infty x^2 \frac{x}{\sigma^2} e^{-x^2/(2\sigma^2)}\,dx = \frac{1}{\sigma^2}\int_0^\infty x^3 e^{-x^2/(2\sigma^2)}\,dx$$

Let $y = x^2/(2\sigma^2)$. Then $dy = x\,dx/\sigma^2$, and so

$$E(X^2) = 2\sigma^2 \int_0^\infty y e^{-y}\,dy = 2\sigma^2 \tag{2.113}$$

Hence, by Eq. (2.31),

$$\sigma_X^2 = E(X^2) - [E(X)]^2 = \left(2 - \frac{\pi}{2}\right)\sigma^2 \approx 0.429\sigma^2 \tag{2.114}$$

2.38. Consider a continuous r.v. X with pdf $f_X(x)$. If $f_X(x) = 0$ for $x < 0$, then show that, for any $a > 0$,

$$P(X \geq a) \leq \frac{\mu_X}{a} \tag{2.115}$$

where $\mu_X = E(X)$. This is known as the *Markov inequality*.

From Eq. (2.23),

$$P(X \geq a) = \int_a^\infty f_X(x)\,dx$$

Since $f_X(x) = 0$ for $x < 0$,

$$\mu_X = E(X) = \int_0^\infty x f_X(x)\,dx \geq \int_a^\infty x f_X(x)\,dx \geq a \int_a^\infty f_X(x)\,dx$$

Hence,

$$\int_a^\infty f_X(x)\,dx = P(X \geq a) \leq \frac{\mu_X}{a}$$

2.39. For any $a > 0$, show that

$$P(|X - \mu_X| \geq a) \leq \frac{\sigma_X^2}{a^2} \tag{2.116}$$

where μ_X and σ_X^2 are the mean and variance of X, respectively. This is known as the *Chebyshev inequality*.

From Eq. (2.23),

$$P(|X - \mu_X| \geq a) = \int_{-\infty}^{\mu_X - a} f_X(x)\,dx + \int_{\mu_X + a}^\infty f_X(x)\,dx = \int_{|x - \mu_X| \geq a} f_X(x)\,dx$$

By Eq. (2.29),

$$\sigma_X^2 = \int_{-\infty}^{\infty} (x - \mu_X)^2 f_X(x)\, dx \geq \int_{|x - \mu_X| \geq a} (x - \mu_X)^2 f_X(x)\, dx \geq a^2 \int_{|x - \mu_X| \geq a} f_X(x)\, dx$$

Hence,
$$\int_{|x - \mu_X| \geq a} f_X(x)\, dx \leq \frac{\sigma_X^2}{a^2} \qquad \text{or} \qquad P\left(|X - \mu_X| \geq a\right) \leq \frac{\sigma_X^2}{a^2}$$

Note that by setting $a = k\sigma_X$ in Eq. (2.116), we obtain

$$P\left(|X - \mu_X| \geq k\sigma_x\right) \leq \frac{1}{k^2} \tag{2.117}$$

Equation (2.117) says that the probability that a r.v. will fall k or more standard deviations from its mean is $\leq 1/k^2$. Notice that nothing at all is said about the distribution function of X. The Chebyshev inequality is therefore quite a generalized statement. However, when applied to a particular case, it may be quite weak.

Special Distributions

2.40. A binary source generates digits 1 and 0 randomly with probabilities 0.6 and 0.4, respectively.

   (*a*)   What is the probability that two 1s and three 0s will occur in a five-digit sequence?

   (*b*)   What is the probability that at least three 1s will occur in a five-digit sequence?

   (*a*)   Let X be the r.v. denoting the number of 1s generated in a five-digit sequence. Since there are only two possible outcomes (1 or 0), the probability of generating 1 is constant, and there are five digits, it is clear that X is a binomial r.v. with parameters $(n, p) = (5, 0.6)$. Hence, by Eq. (2.36), the probability that two 1s and three 0s will occur in a five-digit sequence is

$$P(X = 2) = \binom{5}{2} (0.6)^2 (0.4)^3 = 0.23$$

   (*b*)   The probability that at least three 1s will occur in a five-digit sequence is

$$P(X \geq 3) = 1 - P(X \leq 2)$$

where
$$P(X \leq 2) = \sum_{k=0}^{2} \binom{5}{k} (0.6)^k (0.4)^{5-k} = 0.317$$

Hence,
$$P(X \geq 3) = 1 - 0.317 = 0.683$$

2.41. A fair coin is flipped 10 times. Find the probability of the occurrence of 5 or 6 heads.

Let the r.v. X denote the number of heads occurring when a fair coin is flipped 10 times. Then X is a binomial r.v. with parameters $(n, p) = (10, \frac{1}{2})$. Thus, by Eq. (2.36),

$$P(5 \leq X \leq 6) = \sum_{k=5}^{6} \binom{10}{k} \left(\frac{1}{2}\right)^k \left(\frac{1}{2}\right)^{10-k} = 0.451$$

2.42. Let X be a binomial r.v. with parameters (n, p), where $0 < p < 1$. Show that as k goes from 0 to n, the pmf $p_X(k)$ of X first increases monotonically and then decreases monotonically, reaching its largest value when k is the largest integer less than or equal to $(n + 1)p$.

By Eq. (2.36), we have

$$\frac{p_X(k)}{p_X(k-1)} = \frac{\binom{n}{k} p^k (1-p)^{n-k}}{\binom{n}{k-1} p^{k-1}(1-p)^{n-k+1}} = \frac{(n-k+1)p}{k(1-p)} \tag{2.118}$$

Hence, $p_X(k) \geq p_X(k-1)$ if and only if $(n-k+1)p \geq k(1-p)$ or $k \leq (n+1)p$. Thus, we see that $p_X(k)$ increases monotonically and reaches its maximum when k is the largest integer less than or equal to $(n+1)p$ and then decreases monotonically.

2.43. Show that the Poisson distribution can be used as a convenient approximation to the binomial distribution for large n and small p.

From Eq. (2.36), the pmf of the binomial r.v. with parameters (n,p) is

$$p_X(k) = \binom{n}{k} p^k (1-p)^{n-k} = \frac{n(n-1)(n-2)\cdots(n-k+1)}{k!} p^k (1-p)^{n-k}$$

Multiplying and dividing the right-hand side by n^k, we have

$$\binom{n}{k} p^k (1-p)^{n-k} = \frac{\left(1-\dfrac{1}{n}\right)\left(1-\dfrac{2}{n}\right)\cdots\left(1-\dfrac{k-1}{n}\right)}{k!}(np)^k \left(1-\dfrac{np}{n}\right)^{n-k}$$

If we let $n \to \infty$ in such a way that $np = \lambda$ remains constant, then

$$\left(1-\frac{1}{n}\right)\left(1-\frac{2}{n}\right)\cdots\left(1-\frac{k-1}{n}\right) \xrightarrow[n\to\infty]{} 1$$

$$\left(1-\frac{np}{n}\right)^{n-k} = \left(1-\frac{\lambda}{n}\right)^n \left(1-\frac{\lambda}{n}\right)^{-k} \xrightarrow[n\to\infty]{} e^{-\lambda}(1) = e^{-\lambda}$$

where we used the fact that

$$\lim_{n\to\infty}\left(1-\frac{\lambda}{n}\right)^n = e^{-\lambda}$$

Hence, in the limit as $n \to \infty$ with $np = \lambda$ (and as $p = \lambda/n \to 0$),

$$\binom{h}{k} p^k (1-p)^{n-k} \xrightarrow[n\to\infty]{} e^{-\lambda}\frac{\lambda^k}{k!} \qquad np = \lambda$$

Thus, in the case of large n and small p,

$$\binom{n}{k} p^k (1-p)^{n-k} \approx e^{-\lambda}\frac{\lambda^k}{k!} \qquad np = \lambda \tag{2.119}$$

which indicates that the binomial distribution can be approximated by the Poisson distribution.

2.44. A noisy transmission channel has a per-digit error probability $p = 0.01$.

(a)   Calculate the probability of more than one error in 10 received digits.

(b)   Repeat (a), using the Poisson approximation Eq. (2.119).

(*a*) It is clear that the number of errors in 10 received digits is a binomial r.v. X with parameters $(n, p) = (10, 0.01)$. Then, using Eq. (2.36), we obtain

$$P(X > 1) = 1 - P(X = 0) - P(X = 1)$$

$$= 1 - \binom{10}{0}(0.01)^0\,(0.99)^{10} - \binom{10}{1}(0.01)^1\,(0.99)^9$$

$$= 0.0042$$

(*b*) Using Eq. (2.119) with $\lambda = np = 10(0.01) = 0.1$, we have

$$P(X > 1) = 1 - P(X = 0) - P(X = 1)$$

$$= 1 - e^{-0.1}\frac{(0.1)^0}{0!} - e^{-0.1}\frac{(0.1)^1}{1!}$$

$$= 0.0047$$

2.45. The number of telephone calls arriving at a switchboard during any 10-minute period is known to be a Poisson r.v. X with $\lambda = 2$.

(*a*) Find the probability that more than three calls will arrive during any 10-minute period.

(*b*) Find the probability that no calls will arrive during any 10-minute period.

(*a*) From Eq. (2.48), the pmf of X is

$$p_X(k) = P(X = k) = e^{-2}\frac{2^k}{k!} \qquad k = 0, 1, \ldots$$

Thus, $$P(X > 3) = - P(X \le 3) = 1 - \sum_{k=0}^{3} e^{-2}\frac{2^k}{k!}$$

$$= 1 - e^{-2}\left(1 + 2 + \frac{4}{2} + \frac{8}{6}\right) \approx 0.143$$

(*b*) $P(X = 0) = p_X(0) = e^{-2} \approx 0.135$

2.46. Consider the experiment of throwing a pair of fair dice.

(*a*) Find the probability that it will take less than six tosses to throw a 7.

(*b*) Find the probability that it will take more than six tosses to throw a 7.

(*a*) From Prob. 1.37(*a*), we see that the probability of throwing a 7 on any toss is $\frac{1}{6}$. Let X denote the number of tosses required for the first success of throwing a 7. Then, it is clear that X is a geometric r.v. with parameter $p = \frac{1}{6}$. Thus, using Eq. (2.90) of Prob. 2.16, we obtain

$$P(X < 6) = P(X \le 5) = F_X(5) = 1 - \left(\frac{5}{6}\right)^5 \approx 0.598$$

(*b*) Similarly, we get

$$P(X > 6) = 1 - P(X \le 6) = 1 - F_X(6)$$

$$= 1 - \left[1 - \left(\frac{5}{6}\right)^6\right] = \left(\frac{5}{6}\right)^6 \approx 0.335$$

2.47. Consider the experiment of rolling a fair die. Find the average number of rolls required in order to obtain a 6.

Let X denote the number of trials (rolls) required until the number 6 first appears. Then X is given by geometrical r.v. with parameter $p = \frac{1}{6}$. From Eq. (2.104) of Prob. 2.29, the mean of X is given by

$$\mu_X = E(X) = \frac{1}{p} = \frac{1}{\frac{1}{6}} = 6$$

Thus, the average number of rolls required in order to obtain a 6 is 6.

2.48. Consider an experiment of tossing a fair coin repeatedly until the coin lands heads the sixth time. Find the probability that it takes exactly 10 tosses.

The number of tosses of a fair coin it takes to get 6 heads is a negative binomial r.v. X with parameters $p = 0.5$ and $k = 6$. Thus, by Eq. (2.45), the probability that it takes exactly 10 tosses is

$$P(X = 10) = \binom{10-1}{6-1}\left(\frac{1}{2}\right)^6\left(1 - \frac{1}{2}\right)^4$$

$$= \binom{9}{5}\left(\frac{1}{2}\right)^6\left(\frac{1}{2}\right)^4 = \frac{9!}{5!4!}\left(\frac{1}{2}\right)^{10} = 0.123$$

2.49. Assume that the length of a phone call in minutes is an exponential r.v. X with parameter $\lambda = \frac{1}{10}$. If someone arrives at a phone booth just before you arrive, find the probability that you will have to wait (*a*) less than 5 minutes, and (*b*) between 5 and 10 minutes.

(*a*)   From Eq. (2.60), the pdf of X is

$$f_X(x) = \begin{cases} \dfrac{1}{10}e^{-x/10} & x > 0 \\ 0 & x < 0 \end{cases}$$

Then

$$P(X < 5) = \int_0^5 \frac{1}{10}e^{-x/10}\,dx = -e^{-x/10}\Big|_0^5 = 1 - e^{-0.5} \approx 0.393$$

(*b*)   Similarly,

$$P(5 < X < 10) = \int_5^{10} \frac{1}{10}e^{-x/10}\,dx = e^{-0.5} - e^{-1} \approx 0.239$$

2.50. All manufactured devices and machines fail to work sooner or later. Suppose that the failure rate is constant and the time to failure (in hours) is an exponential r.v. X with parameter λ.

(*a*)   Measurements show that the probability that the time to failure for computer memory chips in a given class exceeds 10^4 hours is e^{-1} (≈ 0.368). Calculate the value of the parameter λ.

(*b*)   Using the value of the parameter λ determined in part (*a*), calculate the time x_0 such that the probability that the time to failure is less than x_0 is 0.05.

(*a*)   From Eq. (2.61), the cdf of X is given by

$$F_X(x) = \begin{cases} 1 - e^{-\lambda x} & x > 0 \\ 0 & x < 0 \end{cases}$$

Now

$$P(X > 10^4) = 1 - P(X \leq 10^4) = 1 - F_X(10^4)$$
$$= 1 - (1 - e^{-\lambda 10^4}) = e^{-\lambda 10^4} = e^{-1}$$

from which we obtain $\lambda = 10^{-4}$.

(*b*)   We want

$$F_X(x_0) = P(X \leq x_0) = 0.05$$

Hence,

$$1 - e^{-\lambda x_0} = 1 - e^{-10^{-4}x_0} = 0.05$$

or

$$e^{-10^{-4}x_0} = 0.95$$

from which we obtain

$$x_0 = -10^4 \ln(0.95) = 513 \text{ hours}$$

2.51. A production line manufactures 1000-ohm (Ω) resistors that have 10 percent tolerance. Let X denote the resistance of a resistor. Assuming that X is a normal r.v. with mean 1000 and variance 2500, find the probability that a resistor picked at random will be rejected.

Let A be the event that a resistor is rejected. Then $A = \{X < 900\} \cup \{X > 1100\}$. Since $\{X < 900\} \cap \{X > 1100\} = \varnothing$, we have

$$P(A) = P(X < 900) + P(X > 1100) = F_X(900) + [1 - F_X(1100)]$$

Since X is a normal r.v. with $\mu = 1000$ and $\sigma^2 = 2500$ ($\sigma = 50$), by Eq. (2.74) and Table A (Appendix A),

$$F_X(900) = \Phi\left(\frac{900 - 1000}{50}\right) = \Phi(-2) = 1 - \Phi(2)$$

$$F_X(1100) = \Phi\left(\frac{1100 - 1000}{50}\right) = \Phi(2)$$

Thus,

$$P(A) = 2[1 - \Phi(2)] \approx 0.045$$

2.52. The radial miss distance [in meters (m)] of the landing point of a parachuting sky diver from the center of the target area is known to be a Rayleigh r.v. X with parameter $\sigma^2 = 100$.

(*a*)   Find the probability that the sky diver will land within a radius of 10 m from the center of the target area.

(*b*)   Find the radius r such that the probability that $X > r$ is e^{-1} (≈ 0.368).

(*a*)   Using Eq. (2.96) of Prob. 2.25, we obtain

$$P(X \leq 10) = F_X(10) = 1 - e^{-100/200} = 1 - e^{-0.5} \approx 0.393$$

(*b*)   Now

$$P(X > r) = 1 - P(X \leq r) = 1 - F_X(r)$$
$$= 1 - (1 - e^{-r^2/200}) = e^{-r^2/200} = e^{-1}$$

from which we obtain $r^2 = 200$ and $r = \sqrt{200} = 14.142$ m.

Conditional Distributions

2.53. Let X be a Poisson r.v. with parameter λ. Find the conditional pmf of X given $B = (X$ is even).

From Eq. (2.48), the pdf of X is

$$p_X(k) = e^{-\lambda} \frac{\lambda^k}{k!} \qquad k = 0, 1, \ldots$$

Then the probability of event B is

$$P(B) = P(X = 0, 2, 4, \ldots) = \sum_{k = \text{even}}^{\infty} e^{-\lambda} \frac{\lambda^k}{k!}$$

Let $A = \{X$ is odd$\}$. Then the probability of event A is

$$P(A) = P(X = 1, 3, 5, \ldots) = \sum_{k = \text{odd}}^{\infty} e^{-\lambda} \frac{\lambda^k}{k!}$$

Now

$$\sum_{k = \text{even}}^{\infty} e^{-\lambda} \frac{\lambda^k}{k!} + \sum_{k = \text{odd}}^{\infty} e^{-\lambda} \frac{\lambda^k}{k!} = e^{-\lambda} \sum_{k=0}^{\infty} \frac{\lambda^k}{k!} = e^{-\lambda} e^{\lambda} = 1 \tag{2.120}$$

$$\sum_{k = \text{even}}^{\infty} e^{-\lambda} \frac{\lambda^k}{k!} - \sum_{k = \text{odd}}^{\infty} e^{-\lambda} \frac{\lambda^k}{k!} = e^{-\lambda} \sum_{k=0}^{\infty} \frac{(-\lambda)^k}{k!} = e^{-\lambda} e^{-\lambda} = e^{-2\lambda} \tag{2.121}$$

Hence, adding Eqs. (2.120) and (2.121), we obtain

$$P(B) = \sum_{k = \text{even}}^{\infty} e^{-\lambda} \frac{\lambda^k}{k!} = \frac{1}{2}(1 + e^{-2\lambda}) \tag{2.122}$$

Now, by Eq. (2.81), the pmf of X given B is

$$p_X(k|B) = \frac{P\{(X = k) \cap B\}}{P(B)}$$

If k is even, $(X = k) \subset B$ and $(X = k) \cap B = (X = k)$. If k is odd, $(X = k) \cap B = \varnothing$. Hence,

$$p_X(k \mid B) = \begin{cases} \dfrac{P(X = k)}{P(B)} = \dfrac{2e^{-\lambda} \lambda^k}{(1 + e^{-2\lambda})k!} & k \text{ even} \\[2ex] \dfrac{P(\varnothing)}{P(B)} = 0 & k \text{ odd} \end{cases}$$

2.54. Show that the conditional cdf and pdf of X given the event $B = (a < X \leq b)$ are as follows:

$$F_X(x \mid a < X \leq b) = \begin{cases} 0 & x \leq a \\[1ex] \dfrac{F_X(x) - F_X(a)}{F_X(b) - F_X(a)} & a < x \leq b \\[1ex] 1 & x > b \end{cases} \tag{2.123}$$

$$f_X(x \mid a < X \leq b) = \begin{cases} 0 & x \leq a \\[1ex] \dfrac{f_X(x)}{\int_a^b f_X(\xi)\, d\xi} & a < x \leq b \\[1ex] 0 & x > b \end{cases} \tag{2.124}$$

Substituting $B = (a < X \le b)$ in Eq. (2.78), we have

$$F_X(x \mid a < X \le b) = P(X \le x \mid a < X \le b) = \frac{P\{(X \le x) \cap (a < X \le b)\}}{P(a < X \le b)}$$

Now $\qquad (X \le x) \cap (a < X \le b) = \begin{cases} \varnothing & x \le a \\ (a < X \le x) & a < x \le b \\ (a < X \le b) & x > b \end{cases}$

Hence, $\qquad F_X(x \mid a < X \le b) = \dfrac{P(\varnothing)}{P(a < X \le b)} = 0 \qquad\qquad x \le a$

$$F_X(x \mid a < X \le b) = \frac{P(a < X \le x)}{P(a < X \le b)} = \frac{F_X(x) - F_X(a)}{F_X(b) - F_X(a)} \qquad a < x \le b$$

$$F_X(x \mid a < X \le b) = \frac{P(a < X \le b)}{P(a < X \le b)} = 1 \qquad\qquad x > b$$

By Eq. (2.82), the conditional pdf of X given $a < X \le b$ is obtained by differentiating Eq. (2.123) with respect to x. Thus,

$$F_X(x \mid a < X \le b) = \begin{cases} 0 & X \le a \\ \dfrac{f_X(x)}{F_X(b) - F_X(a)} = \dfrac{f_X(x)}{\int_a^b f_X(\xi)\,d\xi} & a < x \le b \\ 0 & x > b \end{cases}$$

2.55. Recall the parachuting sky diver problem (Prob. 2.52). Find the probability of the sky diver landing within a 10-m radius from the center of the target area given that the landing is within 50 m from the center of the target area.

From Eq. (2.96) (Prob. 2.25) with $\sigma^2 = 100$, we have

$$F_X(x) = 1 - e^{-x^2/200}$$

Setting $x = 10$ and $b = 50$ and $a = -\infty$ in Eq. (2.123), we obtain

$$P(X \le 10 \mid X \le 50) = F_X(10 \mid X \le 50) = \frac{F_X(10)}{F_X(50)}$$

$$= \frac{1 - e^{-100/200}}{1 - e^{-2500/200}} \approx 0.393$$

2.56. Let $X = N(0; \sigma^2)$. Find $E(X \mid X > 0)$ and $\mathrm{Var}(X \mid X > 0)$.

From Eq. (2.71), the pdf of $X = N(0; \sigma^2)$ is

$$f_X(x) = \frac{1}{\sqrt{2\pi}\,\sigma}\, e^{-x^2/(2\sigma^2)}$$

Then by Eq. (2.124),

$$f_X(x \mid X > 0) = \begin{cases} 0 & x < 0 \\ \dfrac{f_X(x)}{\int_0^\infty f_X(\xi)\,d\xi} = 2\,\dfrac{1}{\sqrt{2\pi}\,\sigma}\, e^{-x^2/(2\sigma^2)} & x \ge 0 \end{cases} \qquad (2.125)$$

Hence, $\qquad E(X \mid X > 0) = 2\,\dfrac{1}{\sqrt{2\pi}\,\sigma} \int_0^\infty x\, e^{-x^2/(2\sigma^2)}\, dx$

Let $y = x^2/(2\sigma^2)$. Then $dy = x \, dx/\sigma^2$, and we get

$$E(X \mid X > 0) = \frac{2\sigma}{\sqrt{2\pi}} \int_0^\infty e^{-y} \, dy = \sigma \sqrt{\frac{2}{\pi}} \tag{2.126}$$

Next,

$$E(X^2 \mid X > 0) = 2 \frac{1}{\sqrt{2\pi}\sigma} \int_0^\infty x^2 e^{-x^2/(2\sigma^2)} \, dx$$

$$= \frac{1}{\sqrt{2\pi}\sigma} \int_{-\infty}^\infty x^2 e^{-x^2/(2\sigma^2)} \, dx = \text{Var}(X) = \sigma^2 \tag{2.127}$$

Then by Eq. (2.31), we obtain

$$\text{Var}(X \mid X > 0) = E(X^2 \mid X > 0) - [E(X \mid X > 0)]^2$$

$$= \sigma^2 \left(1 - \frac{2}{\pi}\right) \approx 0.363 \, \sigma^2 \tag{2.128}$$

2.57. If X is a nonnegative integer valued r.v. and it satisfies the following memoryless property [see Eq. (2.44)]

$$P(X > i + j \mid X > i) = P(X > j) \qquad i, j \geq 1 \tag{2.129}$$

then show that X must be a geometric r.v.

Let $p_k = P(X = k), \qquad k = 1, 2, 3, \ldots$, then

$$P(X > j) = \sum_{k=j+1}^\infty p_k = S_j \tag{2.130}$$

By Eq. (1.55) and using Eqs. (2.129) and (2.130), we have

$$P(X > i + j \mid X > i) = \frac{P(X > i + j)}{P(X > i)} = \frac{S_{i+j}}{S_i} = P(X > j) = S_j$$

Hence,

$$S_{i+j} = S_i S_j \tag{2.131}$$

Setting $i = j = 1$, we have

$$S_2 = S_1 S_1 = S_1^2, S_3 = S_1 S_2 = S_1^3, \ldots, \text{ and } S_{i+1} = S_1^{i+1}$$

Now

$$S_1 = P(X > 1) = 1 - P(X = 1) = 1 - p$$

Thus,

$$S_x = P(X > x) = S_1^x = (1 - p)^x$$

Finally, by Eq. (2.130), we get

$$P(X = x) = P(X > x - 1) - P(X > x)$$
$$= (1 - p)^{x-1} - (1 - p)^x$$
$$= (1 - p)^{x-1} [1 - (1 - p)] = (1 - p)^{x-1} p \qquad x = 1, 2, \ldots$$

Comparing with Eq. (2.40) we conclude that X is a geometric r.v. with parameter p.

2.58. If X is nonnegative continuous r.v. and it satisfies the following memoryless property (see Eq. (2.64))

$$P(X > s + t \mid X > s) = P(X > t) \qquad s, t \geq 0 \qquad (2.132)$$

then show that X must be an exponential r.v.

By Eq. (1.55), Eq. (2.132) reduces to

$$P(X > s + t \mid X > s) = \frac{P(\{X > s + t\} \cap \{X > s\})}{P(X > s)} = \frac{P(X > s + t)}{P(X > s)} = P(X > t)$$

Hence,

$$P(X > s + t) = P(X > s)P(X > t) \qquad (2.133)$$

Let

$$P(X > t) = g(t) \qquad t \geq 0$$

Then Eq. (2.133) becomes

$$g(s + t) = g(s)g(t) \qquad s, t \geq 0$$

which is satisfied only by exponential function, that is,

$$g(t) = e^{\alpha t}$$

Since $P(X > \infty) = 0$ we let $\alpha = -\lambda \ (\lambda > 0)$, then

$$P(X > x) = g(x) = e^{-\lambda x} \qquad x \geq 0 \qquad (2.134)$$

Now if X is a continuous r.v.

$$F_X(x) = P(X \leq x) = 1 - P(X > x) = 1 - e^{-\lambda x} \qquad x \geq 0$$

Comparing with Eq. (2.61), we conclude that X is an exponential r.v. with parameter λ.

SUPPLEMENTARY PROBLEMS

2.59. Consider the experiment of tossing a coin. Heads appear about once out of every three tosses. If this experiment is repeated, what is the probability of the event that heads appear exactly twice during the first five tosses?

2.60. Consider the experiment of tossing a fair coin three times (Prob. 1.1). Let X be the r.v. that counts the number of heads in each sample point. Find the following probabilities:

(a) $P(X \leq 1)$; (b) $P(X > 1)$; and (c) $P(0 < X < 3)$.

2.61. Consider the experiment of throwing two fair dice (Prob. 1.37). Let X be the r.v. indicating the sum of the numbers that appear.

(a) What is the range of X?

(b) Find (i) $P(X = 3)$; (ii) $P(X \leq 4)$; and (iii) $P(3 < X \leq 7)$.

2.62. Let X denote the number of heads obtained in the flipping of a fair coin twice.

    (*a*)   Find the pmf of X.

    (*b*)   Compute the mean and the variance of X.

2.63. Consider the discrete r.v. X that has the pmf

$$p_X(x_k) = (\tfrac{1}{2})^{x_k} \qquad x_k = 1, 2, 3, \ldots$$

Let $A = \{\zeta: X(\zeta) = 1, 3, 5, 7, \ldots\}$. Find $P(A)$.

2.64. Consider the function given by

$$p(x) = \begin{cases} \dfrac{k}{x^2} & x = 1, 2, 3, \ldots \\ 0 & \text{otherwise} \end{cases}$$

where k is a constant. Find the value of k such that $p(x)$ can be the pmf of a discrete r.v. X.

2.65. It is known that the floppy disks produced by company A will be defective with probability 0.01. The company sells the disks in packages of 10 and offers a guarantee of replacement that at most 1 of the 10 disks is defective. Find the probability that a package purchased will have to be replaced.

2.66. Consider an experiment of tossing a fair coin sequentially until "head" appears. What is the probability that the number of tossing is less than 5?

2.67. Given that X is a Poisson r.v. and $p_X(0) = 0.0498$, compute $E(X)$ and $P(X \geq 3)$.

2.68. A digital transmission system has an error probability of 10^{-6} per digit. Find the probability of three or more errors in 10^6 digits by using the Poisson distribution approximation.

2.69. Show that the pmf $p_X(x)$ of a Poisson r.v. X with parameter λ satifsies the following recursion formula:

$$p_X(k+1) = \frac{\lambda}{k+1}\, p_X(k) \qquad p_X(k-1) = \frac{k}{\lambda}\, p_X(k)$$

2.70. The continuous r.v. X has the pdf

$$f_X(x) = \begin{cases} k(x - x^2) & 0 < x < 1 \\ 0 & \text{otherwise} \end{cases}$$

where k is a constant. Find the value of k and the cdf of X.

2.71. The continuous r.v. X has the pdf

$$f_X(x) = \begin{cases} k(2x - x^2) & 0 < x < 2 \\ 0 & \text{otherwise} \end{cases}$$

where k is a constant. Find the value of k and $P(X > 1)$.

2.72. A r.v. X is defined by the cdf

$$F_X(x) = \begin{cases} 0 & x < 0 \\ \dfrac{1}{2}x & 0 \le x < 1 \\ k & 1 \le x \end{cases}$$

(a) Find the value of k.

(b) Find the type of X.

(c) Find (i) $P(\frac{1}{2} < X \le 1)$; (ii) $P(\frac{1}{2} < X < 1)$; and (iii) $P(X > 2)$.

2.73. It is known that the time (in hours) between consecutive traffic accidents can be described by the exponential r.v. X with parameter $\lambda = \frac{1}{60}$. Find (i) $P(X \le 60)$; (ii) $P(X > 120)$; and (iii) $P(10 < X \le 100)$.

2.74. Binary data are transmitted over a noisy communication channel in a block of 16 binary digits. The probability that a received digit is in error as a result of channel noise is 0.01. Assume that the errors occurring in various digit positions within a block are independent.

(a) Find the mean and the variance of the number of errors per block.

(b) Find the probability that the number of errors per block is greater than or equal to 4.

2.75. Let the continuous r.v. X denote the weight (in pounds) of a package. The range of weight of packages is between 45 and 60 pounds.

(a) Determine the probability that a package weighs more than 50 pounds.

(b) Find the mean and the variance of the weight of packages.

2.76. In the manufacturing of computer memory chips, company A produces one defective chip for every nine good chips. Let X be time to failure (in months) of chips. It is known that X is an exponential r.v. with parameter $\lambda = \frac{1}{2}$ for a defective chip and $\lambda = \frac{1}{10}$ with a good chip. Find the probability that a chip purchased randomly will fail before (a) six months of use; and (b) one year of use.

2.77. The *median* of a continuous r.v. X is the value of $x = x_0$ such that $P(X \ge x_0) = P(X \le x_0)$. The *mode* of X is the value of $x = x_m$ at which the pdf of X achieves its maximum value.

(a) Find the median and mode of an exponential r.v. X with parameter λ.

(b) Find the median and mode of a normal r.v. $X = N(\mu, \sigma^2)$.

2.78. Let the r.v. X denote the number of defective components in a random sample of n components, chosen without replacement from a total of N components, r of which are defective. The r.v. X is known as the *hypergeometric* r.v. with parameters (N, r, n).

(a) Find the pmf of X.

(b) Find the mean and variance of X.

2.79. A lot consisting of 100 fuses is inspected by the following procedure: Five fuses are selected randomly, and if all five "blow" at the specified amperage, the lot is accepted. Suppose that the lot contains 10 defective fuses. Find the probability of accepting the lot.

2.80. Let X be the negative binomial r.v. with parameters p and k. Verify Eqs. (2.46) and (2.47), that is,

$$\mu_X = E(X) = \frac{k}{p} \qquad \sigma_X^2 = \text{Var}(X) = \frac{k(1-p)}{p^2}$$

2.81. Suppose the probability that a bit transmitted through a digital communication channel and received in error is 0.1. Assuming that the transmissions are independent events, find the probability that the third error occurs at the 10th bit.

2.82. A r.v. X is called a *Laplace* r.v. if its pdf is given by

$$f_X(x) = ke^{-\lambda|x|} \qquad \lambda > 0, \quad -\infty < x < \infty$$

where k is a constant.

(*a*) Find the value of k.

(*b*) Find the cdf of X.

(*c*) Find the mean and the variance of X.

2.83. A r.v. X is called a *Cauchy* r.v. if its pdf is given by

$$f_X(x) = \frac{k}{a^2 + x^2} \qquad -\infty < x < \infty$$

where $a \, (> 0)$ and k are constants.

(*a*) Find the value of k.

(*b*) Find the cdf of X.

(*c*) Find the mean and the variance of X.

ANSWERS TO SUPPLEMENTARY PROBLEMS

2.59. 0.329

2.60. (*a*) $\dfrac{1}{2}$, (*b*) $\dfrac{1}{2}$, (*c*) $\dfrac{3}{4}$

2.61. (*a*) $R_X = \{2, 3, 4, \ldots, 12\}$

(*b*) (i) $\dfrac{1}{18}$; (ii) $\dfrac{1}{6}$; (iii) $\dfrac{1}{2}$

2.62. (*a*) $p_X(0) = \dfrac{1}{4}$, $p_X(1) = \dfrac{1}{2}$, $p_X(2) = \dfrac{1}{4}$

(*b*) $E(X) = 1, \text{Var}(X) = \dfrac{1}{2}$

2.63. $\dfrac{2}{3}$

2.64. $k = 6/\pi^2$

2.65. 0.004

2.66. 0.9375

2.67. $E(X) = 3, P(X \geq 3) = 0.5767$

2.68. 0.08

2.69. *Hint:* Use Eq. (2.48).

2.70. $k = 6$; $F_X(x) = \begin{cases} 0 & x \le 0 \\ 3x^2 - 2x^3 & 0 < x \le 1 \\ 1 & x > 1 \end{cases}$

2.71. $k = \dfrac{3}{4}$; $P(X > 1) = \dfrac{1}{2}$

2.72. (a) $k = 1$.

   (b) Mixed r.v.

   (c) (i) $\dfrac{3}{4}$; (ii) $\dfrac{1}{4}$; (iii) 0

2.73. (i) 0.632; (ii) 0.135; (iii) 0.658

2.74. (a) $E(X) = 0.16$, $\text{Var}(X) = 0.158$

   (b) 0.165×10^{-4}

2.75. *Hint:* Assume that X is uniformly distributed over $(45, 60)$.

   (a) $\dfrac{2}{3}$;   (b) $E(X) = 52.5$, $\text{Var}(X) = 18.75$

2.76. (a) 0.501;   (b) 0.729

2.77. (a) $x_0 = (\ln 2)/\lambda = 0.693/\lambda$, $x_m = 0$

   (b) $x_0 = x_m = \mu$

2.78. *Hint:* To find $E(X)$, note that

$$\binom{r}{x} = \frac{r}{x}\binom{r-1}{x-1} \quad \text{and} \quad \binom{N}{n} = \sum_{x=0}^{n}\binom{r}{x}\binom{N-r}{n-x}$$

To find $\text{Var}(X)$, first find $E[X(X-1)]$.

   (a) $p_X(x) = \dfrac{\dbinom{r}{x}\dbinom{N-r}{n-x}}{\dbinom{N}{n}}$ $x = 0, 1, 2, \ldots, \min\{r, n\}$

   (b) $E(X) = n\left(\dfrac{r}{N}\right)$, $\text{Var}(X) = n\left(\dfrac{r}{N}\right)\left(1 - \dfrac{r}{N}\right)\left(\dfrac{N-n}{N-1}\right)$

2.79. *Hint:* Let X be a r.v. equal to the number of defective fuses in the sample of 5 and use the result of Prob. 2.78.

   0.584

2.80. *Hint:*　To find $E(X)$, use Maclaurin's series expansions of the negative binomial $h(q) = (1 - q)^{-r}$ and its derivatives $h'(q)$ and $h''(q)$, and note that

$$h(q) = (1 - q)^{-r} = \sum_{k=0}^{\infty} \binom{r + k - 1}{r - 1} q^k = \sum_{x=r}^{\infty} \binom{x - 1}{r - 1} q^{x-r}$$

To find $\text{Var}(X)$, first find $E[(X - r)(X - r - 1)]$ using $h''(q)$.

2.81.　0.017

2.82.　(*a*)　$k = \lambda/2$ 　　　　　　(*b*)　$F_X(x) = \begin{cases} \dfrac{1}{2} e^{\lambda x} & x < 0 \\[2mm] 1 - \dfrac{1}{2} e^{-\lambda x} & x \geq 0 \end{cases}$

　　　　　(*c*)　$E(X) = 0$, $\text{Var}(X) = 2/\lambda^2$

2.83.　(*a*)　$k = a/\pi$ 　　　　　　(*b*)　$F_X(x) = \dfrac{1}{2} + \dfrac{1}{\pi} \tan^{-1}\left(\dfrac{x}{a}\right)$

　　　　　(*c*)　$E(X) = 0$, $\text{Var}(X)$ does not exist.

CHAPTER 3

Multiple Random Variables

3.1 Introduction

In many applications it is important to study two or more r.v.'s defined on the same sample space. In this chapter, we first consider the case of two r.v.'s, their associated distribution, and some properties, such as independence of the r.v.'s. These concepts are then extended to the case of many r.v.'s defined on the same sample space.

3.2 Bivariate Random Variables

A. Definition:

Let S be the sample space of a random experiment. Let X and Y be two r.v.'s. Then the pair (X, Y) is called a *bivariate* r.v. (or *two-dimensional random vector*) if each of X and Y associates a real number with every element of S. Thus, the bivariate r.v. (X, Y) can be considered as a function that to each point ζ in S assigns a point (x, y) in the plane (Fig. 3-1). The range space of the bivariate r.v. (X, Y) is denoted by R_{xy} and defined by

$$R_{xy} = \{(x, y); \zeta \in S \text{ and } X(\zeta) = x, Y(\zeta) = y\}$$

If the r.v.'s X and Y are each, by themselves, discrete r.v.'s, then (X, Y) is called a *discrete* bivariate r.v. Similarly, if X and Y are each, by themselves, continuous r.v.'s, then (X, Y) is called a *continuous* bivariate r.v. If one of X and Y is discrete while the other is continuous, then (X, Y) is called a *mixed* bivariate r.v.

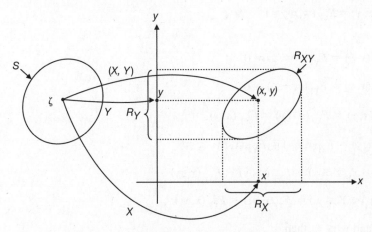

Fig. 3-1 (X, Y) as a function from S to the plane.

3.3 Joint Distribution Functions

A. Definition:

The *joint cumulative distribution function* (or joint cdf) of X and Y, denoted by $F_{XY}(x, y)$, is the function defined by

$$F_{XY}(x, y) = P(X \leq x, Y \leq y) \tag{3.1}$$

The event $(X \leq x, Y \leq y)$ in Eq. (3.1) is equivalent to the event $A \cap B$, where A and B are events of S defined by

$$A = \{\zeta \in S; X(\zeta) \leq x\} \qquad \text{and} \qquad B = \{\zeta \in S; Y(\zeta) \leq y\} \tag{3.2}$$

and $\qquad\qquad\qquad\qquad\qquad P(A) = F_X(x) \qquad\qquad P(B) = F_Y(y)$

Thus, $\qquad\qquad\qquad\qquad\qquad\quad F_{XY}(x, y) = P(A \cap B) \tag{3.3}$

If, for particular values of x and y, A and B were independent events of S, then by Eq. (1.62),

$$F_{XY}(x, y) = P(A \cap B) = P(A)P(B) = F_X(x)F_Y(y)$$

B. Independent Random Variables:

Two r.v.'s X and Y will be called *independent* if

$$F_{XY}(x, y) = F_X(x)F_Y(y) \tag{3.4}$$

for every value of x and y.

C. Properties of $F_{XY}(x, y)$:

The joint cdf of two r.v.'s has many properties analogous to those of the cdf of a single r.v.

1. $0 \leq F_{XY}(x, y) \leq 1 \tag{3.5}$

2. If $x_1 \leq x_2$, and $y_1 \leq y_2$, then

$$F_{XY}(x_1, y_1) \leq F_{XY}(x_2, y_1) \leq F_{XY}(x_2, y_2) \tag{3.6a}$$

$$F_{XY}(x_1, y_1) \leq F_{XY}(x_1, y_2) \leq F_{XY}(x_2, y_2) \tag{3.6b}$$

3. $\displaystyle\lim_{\substack{x \to \infty \\ y \to \infty}} F_{XY}(x, y) = F_{XY}(\infty, \infty) = 1 \tag{3.7}$

4. $\displaystyle\lim_{x \to -\infty} F_{XY}(x, y) = F_{XY}(-\infty, y) = 0 \tag{3.8a}$

$\displaystyle\lim_{y \to -\infty} F_{XY}(x, y) = F_{XY}(x, -\infty) = 0 \tag{3.8b}$

5. $\displaystyle\lim_{x \to a^+} F_{XY}(x, y) = F_{XY}(a^+, y) = F_{XY}(a, y) \tag{3.9a}$

$\displaystyle\lim_{y \to b^+} F_{XY}(x, y) = F_{XY}(x, b^+) = F_{XY}(x, b) \tag{3.9b}$

6. $P(x_1 < X \leq x_2, Y \leq y) = F_{XY}(x_2, y) - F_{XY}(x_1, y) \tag{3.10}$

$P(X \leq x, y_1 < Y \leq y_2) = F_{XY}(x, y_2) - F_{XY}(x, y_1) \tag{3.11}$

7. If $x_1 \leq x_2$ and $y_1 \leq y_2$, then

$$F_{XY}(x_2, y_2) - F_{XY}(x_1, y_2) - F_{XY}(x_2, y_1) + F_{XY}(x_1, y_1) \geq 0 \tag{3.12}$$

Note that the left-hand side of Eq. (3.12) is equal to $P(x_1 < X \leq x_2, y_1 < Y \leq y_2)$ (Prob. 3.5).

D. Marginal Distribution Functions:

Now

$$\lim_{y \to \infty} (X \le x, Y \le y) = (X \le x, Y \le \infty) = (X \le x)$$

since the condition $y \le \infty$ is always satisfied. Then

$$\lim_{y \to \infty} F_{XY}(x, y) = F_{XY}(x, \infty) = F_X(x) \qquad (3.13)$$

Similarly,

$$\lim_{y \to \infty} F_{XY}(x, y) = F_{XY}(\infty, y) = F_Y(y) \qquad (3.14)$$

The cdf's $F_X(x)$ and $F_Y(y)$, when obtained by Eqs. (3.13) and (3.14), are referred to as the *marginal cdf's* of X and Y, respectively.

3.4 Discrete Random Variables—Joint Probability Mass Functions

A. Joint Probability Mass Functions:

Let (X, Y) be a discrete bivariate r.v., and let (X, Y) take on the values (x_i, y_j) for a certain allowable set of integers i and j. Let

$$p_{XY}(x_i, y_j) = P(X = x_i, Y = y_j) \qquad (3.15)$$

The function $p_{XY}(x_i, y_j)$ is called the *joint probability mass function* (joint pmf) of (X, Y).

B. Properties of $p_{XY}(x_i, y_j)$:

1. $0 \le p_{XY}(x_i, y_j) \le 1$ (3.16)
2. $\sum_{x_i} \sum_{y_j} p_{XY}(x_i, y_j) = 1$ (3.17)
3. $P[(X, Y) \in A] = \sum_{(x_i, y_j) \in R_A} p_{XY}(x_i, y_j)$ (3.18)

where the summation is over the points (x_i, y_j) in the range space R_A corresponding to the event A. The joint cdf of a discrete bivariate r.v. (X, Y) is given by

$$F_{XY}(x, y) = \sum_{x_i \le x} \sum_{y_j \le y} p_{XY}(x_i, y_j) \qquad (3.19)$$

C. Marginal Probability Mass Functions:

Suppose that for a fixed value $X = x_i$, the r.v. Y can take on only the possible values y_j ($j = 1, 2, ..., n$). Then

$$P(X = x_i) = p_X(x_i) = \sum_{y_j} p_{XY}(x_i, y_j) \qquad (3.20)$$

where the summation is taken over all possible pairs (x_i, y_j) with x_i fixed. Similarly,

$$P(Y = y_j) = p_Y(y_j) = \sum_{x_i} p_{XY}(x_i, y_j) \qquad (3.21)$$

where the summation is taken over all possible pairs (x_i, y_j) with y_j fixed. The pmf's $p_X(x_i)$ and $p_Y(y_j)$, when obtained by Eqs. (3.20) and (3.21), are referred to as the *marginal* pmf's of X and Y, respectively.

D. Independent Random Variables:

If X and Y are independent r.v.'s, then (Prob. 3.10)

$$p_{XY}(x_i, y_j) = p_X(x_i)p_Y(y_j) \qquad (3.22)$$

3.5   Continuous Random Variables—Joint Probability Density Functions

A.   Joint Probability Density Functions:

Let (X, Y) be a continuous bivariate r.v. with cdf $F_{XY}(x, y)$ and let

$$f_{XY}(x, y) = \frac{\partial^2 F_{XY}(x, y)}{\partial x \, \partial y} \tag{3.23}$$

The function $f_{XY}(x, y)$ is called the *joint probability density function* (joint pdf) of (X, Y). By integrating Eq. (3.23), we have

$$F_{XY}(x, y) = \int_{-\infty}^{x} \int_{-\infty}^{y} f_{XY}(\xi, \eta) \, d\eta \, d\xi \tag{3.24}$$

B.   Properties of $f_{XY}(x, y)$:

1. $f_{XY}(x, y) \geq 0$ (3.25)

2. $\int_{-\infty}^{\infty}\int_{-\infty}^{\infty} f_{XY}(x, y) \, dx \, dy = 1$ (3.26)

3. $f_{XY}(x, y)$ is continuous for all values of x or y except possibly a finite set.

4. $P[(X, Y) \in A] = \iint\limits_{R_A} f_{XY}(x, y) \, dx \, dy$ (3.27)

5. $P(a < X \leq b, c < Y \leq d) = \int_{c}^{d}\int_{a}^{b} f_{XY}(x, y) \, dx \, dy$ (3.28)

Since $P(X = a) = 0 = P(Y = c)$ [by Eq. (2.19)], it follows that

$$P(a < X \leq b, c < Y \leq d) = P(a \leq X \leq b, c \leq Y \leq d) = P(a \leq X < b, c \leq Y < d)$$

$$= P(a < X < b, c < Y < d) = \int_{c}^{d}\int_{a}^{b} f_{XY}(x, y) \, dx \, dy \tag{3.29}$$

C.   Marginal Probability Density Functions:

By Eq. (3.13),

$$F_X(x) = F_{XY}(x, \infty) = \int_{-\infty}^{x}\int_{-\infty}^{\infty} f_{XY}(\xi, \eta) \, d\eta \, d\xi$$

Hence, $$f_X(x) = \frac{dF_X(x)}{dx} = \int_{-\infty}^{\infty} f_{XY}(x, \eta) \, d\eta$$

or $$f_X(x) = \int_{-\infty}^{\infty} f_{XY}(x, y) \, dy \tag{3.30}$$

Similarly, $$f_Y(y) = \int_{-\infty}^{\infty} f_{XY}(x, y) \, dx \tag{3.31}$$

The pdf's $f_X(x)$ and $f_Y(y)$, when obtained by Eqs. (3.30) and (3.31), are referred to as the *marginal* pdf's of X and Y, respectively.

D. Independent Random Variables:

If X and Y are independent r.v.'s, by Eq. (3.4),

$$F_{XY}(x, y) = F_X(x)F_Y(y)$$

Then

$$\frac{\partial^2 F_{XY}(x, y)}{\partial x\, \partial y} = \frac{\partial}{\partial x} F_X(x) \frac{\partial}{\partial y} F_Y(y)$$

or

$$f_{XY}(x, y) = f_X(x)f_Y(y) \qquad (3.32)$$

analogous with Eq. (3.22) for the discrete case. Thus, we say that the continuous r.v.'s X and Y are independent r.v.'s if and only if Eq. (3.32) is satisfied.

3.6 Conditional Distributions

A. Conditional Probability Mass Functions:

If (X, Y) is a discrete bivariate r.v. with joint pmf $p_{XY}(x_i, y_j)$, then the *conditional* pmf of Y, given that $X = x_i$, is defined by

$$p_{Y|X}(y_j|x_i) = \frac{p_{XY}(x_i, y_j)}{p_X(x_i)} \qquad p_X(x_i) > 0 \qquad (3.33)$$

Similarly, we can define $p_{X|Y}(x_i|y_j)$ as

$$p_{X|Y}(x_i|y_j) = \frac{p_{XY}(x_i, y_j)}{p_Y(y_j)} \qquad p_Y(y_j) > 0 \qquad (3.34)$$

B. Properties of $p_{Y|X}(y_j|x_i)$:

1. $0 \le p_{Y|X}(y_j|x_i) \le 1$ (3.35)
2. $\sum_{y_j} p_{Y|X}(y_j|x_i) = 1$ (3.36)

Notice that if X and Y are independent, then by Eq. (3.22),

$$p_{Y|X}(y_j|x_i) = p_Y(y_j) \qquad \text{and} \qquad p_{X|Y}(x_i|y_j) = p_X(x_i) \qquad (3.37)$$

C. Conditional Probability Density Functions:

If (X, Y) is a continuous bivariate r.v. with joint pdf $f_{XY}(x, y)$, then the *conditional* pdf of Y, given that $X = x$, is defined by

$$f_{Y|X}(y|x) = \frac{f_{XY}(x, y)}{f_X(x)} \qquad f_X(x) > 0 \qquad (3.38)$$

Similarly, we can define $f_{X|Y}(x|y)$ as

$$f_{X|Y}(x|y) = \frac{f_{XY}(x, y)}{f_Y(y)} \qquad f_Y(y) > 0 \qquad (3.39)$$

D.   Properties of $f_{Y|X}(y|x)$:

1. $f_{Y|X}(y|x) \geq 0$ (3.40)
2. $\int_{-\infty}^{\infty} f_{Y|X}(y|x)\, dy = 1$ (3.41)

As in the discrete case, if X and Y are independent, then by Eq. (3.32),

$$f_{Y|X}(y|x) = f_Y(y) \qquad\text{and}\qquad f_{X|Y}(x|y) = f_X(x) \tag{3.42}$$

3.7   Covariance and Correlation Coefficient

The (k, n)th moment of a bivariate r.v. (X, Y) is defined by

$$m_{kn} = E(X^k Y^n) = \begin{cases} \displaystyle\sum_{y_j}\sum_{x_i} x_i^k y_j^n p_{XY}(x_i, y_j) & \text{(discrete case)} \\[2ex] \displaystyle\int_{-\infty}^{\infty}\int_{-\infty}^{\infty} x^k y^n f_{XY}(x, y)\, dx\, dy & \text{(continuous case)} \end{cases} \tag{3.43}$$

If $n = 0$, we obtain the kth moment of X, and if $k = 0$, we obtain the nth moment of Y. Thus,

$$m_{10} = E(X) = \mu_X \qquad\text{and}\qquad m_{01} = E(Y) = \mu_Y \tag{3.44}$$

If (X, Y) is a discrete bivariate r.v., then using Eqs. (3.43), (3.20), and (3.21), we obtain

$$\mu_X = E(X) = \sum_{y_j}\sum_{x_i} x_i p_{XY}(x_i, y_j)$$

$$= \sum_{x_i} x_i \left[\sum_{y_j} p_{XY}(x_i, y_j) \right] = \sum_{x_i} x_i p_X(x_i) \tag{3.45a}$$

$$\mu_Y = E(Y) = \sum_{x_i}\sum_{y_j} y_j p_{XY}(x_i, y_j)$$

$$= \sum_{y_j} y_j \left[\sum_{x_i} p_{XY}(x_i, y_j) \right] = \sum_{y_j} y_j p_Y(y_j) \tag{3.45b}$$

Similarly, we have

$$E(X^2) = \sum_{y_j}\sum_{x_i} x_i^2 p_{XY}(x_i, y_j) = \sum_{x_i} x_i^2 p_X(x_i) \tag{3.46a}$$

$$E(Y^2) = \sum_{y_j}\sum_{x_i} y_j^2 p_{XY}(x_i, y_j) = \sum_{y_j} y_j^2 p_Y(y_j) \tag{3.46b}$$

If (X, Y) is a continuous bivariate r.v., then using Eqs. (3.43), (3.30), and (3.31), we obtain

$$\mu_X = E(X) = \int_{-\infty}^{\infty}\int_{-\infty}^{\infty} x f_{XY}(x, y)\, dx\, dy$$

$$= \int_{-\infty}^{\infty} x \left[\int_{-\infty}^{\infty} f_{XY}(x, y)\, dy \right] dx = \int_{-\infty}^{\infty} x f_X(x)\, dx \tag{3.47a}$$

$$\mu_Y = E(Y) = \int_{-\infty}^{\infty}\int_{-\infty}^{\infty} y f_{XY}(x, y)\, dx\, dy$$

$$= \int_{-\infty}^{\infty} y \left[\int_{-\infty}^{\infty} f_{XY}(x, y)\, dx \right] dy = \int_{-\infty}^{\infty} y f_Y(y)\, dy \tag{3.47b}$$

Similarly, we have

$$E(X^2) = \int_{-\infty}^{\infty} \int_{-\infty}^{\infty} x^2 f_{XY}(x, y)\, dx\, dy = \int_{-\infty}^{\infty} x^2 f_X(x)\, dx \tag{3.48a}$$

$$E(Y^2) = \int_{-\infty}^{\infty} \int_{-\infty}^{\infty} y^2 f_{XY}(x, y)\, dx\, dy = \int_{-\infty}^{\infty} y^2 f_Y(y)\, dy \tag{3.48b}$$

The variances of X and Y are easily obtained by using Eq. (2.31). The $(1, 1)$th joint moment of (X, Y),

$$m_{11} = E(XY) \tag{3.49}$$

is called the *correlation* of X and Y. If $E(XY) = 0$, then we say that X and Y are *orthogonal*. The *covariance* of X and Y, denoted by $\text{Cov}(X, Y)$ or σ_{XY}, is defined by

$$\text{Cov}(X, Y) = \sigma_{XY} = E[(X - \mu_X)(Y - \mu_Y)] \tag{3.50}$$

Expanding Eq. (3.50), we obtain

$$\text{Cov}(X, Y) = E(XY) - E(X)E(Y) \tag{3.51}$$

If $\text{Cov}(X, Y) = 0$, then we say that X and Y are *uncorrelated*. From Eq. (3.51), we see that X and Y are uncorrelated if

$$E(XY) = E(X)E(Y) \tag{3.52}$$

Note that if X and Y are independent, then it can be shown that they are uncorrelated (Prob. 3.32), but the converse is not true in general; that is, the fact that X and Y are uncorrelated does not, in general, imply that they are independent (Probs. 3.33, 3.34, and 3.38). The *correlation coefficient*, denoted by $\rho(X, Y)$ or ρ_{XY}, is defined by

$$\rho(X, Y) = \rho_{XY} = \frac{\text{Cov}(X, Y)}{\sigma_X \sigma_Y} = \frac{\sigma_{XY}}{\sigma_X \sigma_Y} \tag{3.53}$$

It can be shown that (Prob. 3.36)

$$|\rho_{XY}| \le 1 \qquad \text{or} \qquad -1 \le \rho_{XY} \le 1 \tag{3.54}$$

Note that the correlation coefficient of X and Y is a measure of linear dependence between X and Y (see Prob. 4.46).

3.8   Conditional Means and Conditional Variances

If (X, Y) is a discrete bivariate r.v. with joint pmf $p_{XY}(x_i, y_j)$, then the *conditional mean* (or *conditional expectation*) of Y, given that $X = x_i$, is defined by

$$\mu_{Y|x_i} = E(Y|x_i) = \sum_{y_j} y_j p_{Y|X}(y_j|x_i) \tag{3.55}$$

The *conditional variance* of Y, given that $X = x_i$, is defined by

$$\sigma_{Y|x_i}^2 = \text{Var}(Y|x_i) = E[(Y - \mu_{Y|x_i})^2 | x_i] = \sum_{y_j} (y_j - \mu_{Y|x_i})^2 p_{Y|X}(y_j|x_i) \tag{3.56}$$

which can be reduced to

$$\text{Var}(Y \mid x_i) = E(Y^2 \mid x_i) - [E(Y \mid x_i)]^2 \tag{3.57}$$

The conditional mean of X, given that $Y = y_j$, and the conditional variance of X, given that $Y = y_j$, are given by similar expressions. Note that the conditional mean of Y, given that $X = x_i$, is a function of x_i alone. Similarly, the conditional mean of X, given that $Y = y_j$, is a function of y_j alone.

If (X, Y) is a continuous bivariate r.v. with joint pdf $f_{XY}(x, y)$, the conditional mean of Y, given that $X = x$, is defined by

$$\mu_{Y|x} = E(Y|x) = \int_{-\infty}^{\infty} y f_{Y|X}(y|x)\,dy \tag{3.58}$$

The conditional variance of Y, given that $X = x$, is defined by

$$\sigma_{Y|x}^2 = \text{Var}(Y|x) = E[(Y - \mu_{Y|x})^2 \mid x] = \int_{-\infty}^{\infty}(y - \mu_{Y|x})^2 f_{Y|X}(y|x)\,dy \tag{3.59}$$

which can be reduced to

$$\text{Var}(Y \mid x) = E(Y^2 \mid x) - [E(Y \mid x)]^2 \tag{3.60}$$

The conditional mean of X, given that $Y = y$, and the conditional variance of X, given that $Y = y$, are given by similar expressions. Note that the conditional mean of Y, given that $X = x$, is a function of x alone. Similarly, the conditional mean of X, given that $Y = y$, is a function of y alone (Prob. 3.40).

3.9 *N*-Variate Random Variables

In previous sections, the extension from one r.v. to two r.v.'s has been made. The concepts can be extended easily to any number of r.v.'s defined on the same sample space. In this section we briefly describe some of the extensions.

A. Definitions:

Given an experiment, the n-tuple of r.v.'s $(X_1, X_2, \ldots, X_n)$ is called an *n-variate* r.v. (or *n-dimensional random vector*) if each $X_i, i = 1, 2, \ldots, n$, associates a real number with every sample point $\zeta \in S$. Thus, an n-variate r.v. is simply a rule associating an n-tuple of real numbers with every $\zeta \in S$.

Let $(X_1, \ldots, X_n)$ be an n-variate r.v. on S. Then its joint cdf is defined as

$$F_{X_1 \ldots X_n}(x_1, \ldots, x_n) = P(X_1 \le x_1, \ldots, X_n \le x_n) \tag{3.61}$$

Note that

$$F_{X_1 \ldots X_n}(\infty, \cdots, \infty) = 1 \tag{3.62}$$

The marginal joint cdf's are obtained by setting the appropriate X_i's to $+\infty$ in Eq. (3.61). For example,

$$F_{X_1 \ldots X_{n-1}}(x_1, \ldots, x_{n-1}) = F_{X_1 \ldots X_{n-1} X_n}(x_1, \ldots, x_{n-1}, \infty) \tag{3.63}$$

$$F_{X_1 X_2}(x_1, x_2) = F_{X_1 X_2 X_3 \ldots X_n}(x_1, x_2, \infty, \cdots, \infty) \tag{3.64}$$

A discrete n-variate r.v. will be described by a joint pmf defined by

$$p_{X_1 \ldots X_n}(x_1, \ldots, x_n) = P(X_1 = x_1, \ldots, X_n = x_n) \tag{3.65}$$

The probability of any *n*-dimensional event *A* is found by summing Eq. (3.65) over the points in the *n*-dimensional range space R_A corresponding to the event *A*:

$$P[(X_1, \ldots, X_n) \in A] = \sum_{(x_1, \ldots, x_n) \in R_A} \cdots \sum p_{X_1 \cdots X_n}(x_1, \ldots, x_n) \tag{3.66}$$

Properties of $p_{X_1 \ldots X_n}(x_1, \ldots, x_n)$:

1. $0 \le p_{X_1 \cdots X_n}(x_1, \ldots, x_n) \le 1 \tag{3.67}$

2. $\displaystyle\sum_{x_1} \cdots \sum_{x_n} p_{X_1 \cdots X_n}(x_1, \ldots, x_n) = 1 \tag{3.68}$

The marginal pmf's of one or more of the r.v.'s are obtained by summing Eq. (3.65) appropriately. For example,

$$p_{X_1 \cdots X_{n-1}}(x_1, \ldots, x_{n-1}) = \sum_{x_n} p_{X_1 \cdots X_n}(x_1, \ldots, x_n) \tag{3.69}$$

$$p_{X_1}(x_1) = \sum_{x_2} \cdots \sum_{x_n} p_{X_1 \cdots X_n}(x_1, \ldots, x_n) \tag{3.70}$$

Conditional pmf's are defined similarly. For example,

$$p_{X_n | X_1 \cdots X_{n-1}}(x_n | x_1, \ldots, x_{n-1}) = \frac{p_{X_1 \cdots X_n}(x_1, \ldots, x_n)}{p_{X_1 \cdots X_{n-1}}(x_1, \ldots, x_{n-1})} \tag{3.71}$$

A continuous *n*-variate r.v. will be described by a joint pdf defined by

$$f_{X_1 \cdots X_n}(x_1, \ldots, x_n) = \frac{\partial^n F_{X_1 \cdots X_n}(x_1, \ldots, x_n)}{\partial x_1 \cdots \partial x_n} \tag{3.72}$$

Then

$$F_{X_1 \cdots X_n}(x_1, \ldots, x_n) = \int_{-\infty}^{x_n} \cdots \int_{-\infty}^{x_1} f_{X_1 \cdots X_n}(\xi_1, \ldots, \xi_n) \, d\xi_1 \cdots d\xi_n \tag{3.73}$$

and

$$P[(X_1, \ldots, X_n) \in A] = \int_{(x_1, \ldots, x_n) \in R_A} \cdots \int f_{X_1 \cdots X_n}(\xi_1, \ldots, \xi_n) \, d\xi_1 \cdots d\xi_n \tag{3.74}$$

Properties of $f_{X_1 \ldots X_n}(x_1, \ldots, x_n)$:

1. $f_{X1 \ldots X_n}(x_1, \ldots, x_n) \ge 0 \tag{3.75}$

2. $\displaystyle\int_{-\infty}^{\infty} \cdots \int_{-\infty}^{\infty} f_{X_1 \cdots X_n}(x_1, \cdots, x_n) \, dx_1 \cdots dx_n = 1 \tag{3.76}$

The marginal pdf's of one or more of the r.v.'s are obtained by integrating Eq. (3.72) appropriately. For example,

$$f_{X_1 \cdots X_{n-1}}(x_1, \ldots, x_{n-1}) = \int_{-\infty}^{\infty} f_{X_1 \cdots X_n}(x_1, \ldots, x_n) \, dx_n \tag{3.77}$$

$$f_{X_1}(x_1) = \int_{-\infty}^{\infty} \cdots \int_{-\infty}^{\infty} f_{X_1 \cdots X_n}(x_1, \ldots, x_n) \, dx_2 \cdots dx_n \tag{3.78}$$

Conditional pdf's are defined similarly. For example,

$$f_{X_n|X_1\cdots X_{n-1}}(x_n|x_1,\ldots,x_{n-1}) = \frac{f_{X_1\cdots X_n}(x_1,\ldots,x_n)}{f_{X_1\cdots X_{n-1}}(x_1,\ldots,x_{n-1})} \tag{3.79}$$

The r.v.'s $X_1,\ldots,X_n$ are said to be mutually independent if

$$p_{X_1\cdots X_n}(x_1,\ldots,x_n) = \prod_{i=1}^{n} p_{X_i}(x_i) \tag{3.80}$$

for the discrete case, and

$$f_{X_1\cdots X_n}(x_1,\ldots,x_n) = \prod_{i=1}^{n} f_{X_i}(x_i) \tag{3.81}$$

for the continuous case.

The mean (or expectation) of X_i in $(X_1,\ldots,X_n)$ is defined as

$$\mu_i = E(X_i) = \begin{cases} \sum_{x_n}\cdots\sum_{x_1} x_i p_{X_1\cdots X_n}(x_1,\ldots,x_n) & \text{(discrete case)} \\ \int_{-\infty}^{\infty}\cdots\int_{-\infty}^{\infty} x_i f_{X_1\cdots X_n}(x_1,\ldots,x_n)\,dx_1\cdots dx_n & \text{(continuous case)} \end{cases} \tag{3.82}$$

The variance of X_i is defined as

$$\sigma_i^2 = \text{Var}(X_i) = E[(X_i - \mu_i)^2] \tag{3.83}$$

The covariance of X_i and X_j is defined as

$$\sigma_{ij} = \text{Cov}(X_i, X_j) = E[(X_i - \mu_i)(X_j - \mu_j)] \tag{3.84}$$

The correlation coefficient of X_i and X_j is defined as

$$\rho_{ij} = \frac{\text{Cov}(X_i, X_j)}{\sigma_i \sigma_j} = \frac{\sigma_{ij}}{\sigma_i \sigma_j} \tag{3.85}$$

3.10 Special Distributions

A. Multinomial Distribution:

The multinomial distribution is an extension of the binomial distribution. An experiment is termed a multinomial trial with parameters $p_1, p_2, \ldots, p_k$, if it has the following conditions:

1. The experiment has k possible outcomes that are mutually exclusive and exhaustive, say $A_1, A_2, \ldots, A_k$.

2. $P(A_i) = p_i$ $i = 1,\ldots,k$ and $\sum_{i=1}^{k} p_i = 1$ \hfill (3.86)

Consider an experiment which consists of n repeated, independent, multinomial trials with parameters $p_1, p_2, \ldots, p_k$. Let X_i be the r.v. denoting the number of trials which result in A_i. Then $(X_1, X_2, \ldots, X_k)$ is called the *multinomial* r.v. with parameters $(n, p_1, p_2, \ldots, p_k)$ and its pmf is given by (Prob. 3.46)

$$p_{X_1 X_2\cdots X_k}(x_1, x_2, \ldots, x_k) = \frac{n!}{x_1! x_2!\cdots x_k!} p_1^{x_1} p_2^{x_2}\cdots p_k^{x_k} \tag{3.87}$$

for $x_i = 0, 1, \ldots, n, i = 1, \ldots, k$, such that $\sum_{i=1}^{k} x_i = n$.

Note that when $k = 2$, the multinomial distribution reduces to the binomial distribution.

B. Bivariate Normal Distribution:

A bivariate r.v. (X, Y) is said to be a bivariate *normal* (or *Gaussian*) r.v. if its joint pdf is given by

$$f_{XY}(x, y) = \frac{1}{2\pi\sigma_x\sigma_y(1 - \rho^2)^{1/2}} \exp\left[-\frac{1}{2}q(x, y)\right] \tag{3.88}$$

where

$$q(x, y) = \frac{1}{1 - \rho^2}\left[\left(\frac{x - \mu_x}{\sigma_x}\right)^2 - 2\rho\left(\frac{x - \mu_x}{\sigma_x}\right)\left(\frac{y - \mu_y}{\sigma_y}\right) + \left(\frac{y - \mu_y}{\sigma_y}\right)^2\right] \tag{3.89}$$

and $\mu_x, \mu_y, \sigma_x^2, \sigma_y^2$ are the means and variances of X and Y, respectively. It can be shown that ρ is the correlation coefficient of X and Y (Prob. 3.50) and that X and Y are independent when $\rho = 0$ (Prob. 3.49).

C. N-variate Normal Distribution:

Let $(X_1, \ldots, X_n)$ be an n-variate r.v. defined on a sample space S. Let **X** be an n-dimensional *random vector* expressed as an $n \times 1$ matrix:

$$\mathbf{X} = \begin{bmatrix} X_1 \\ \vdots \\ X_n \end{bmatrix} \tag{3.90}$$

Let **x** be an n-dimensional vector ($n \times 1$ matrix) defined by

$$\mathbf{x} = \begin{bmatrix} x_1 \\ \vdots \\ x_n \end{bmatrix} \tag{3.91}$$

The n-variate r.v. $(X_1, \ldots, X_n)$ is called an n-variate normal r.v. if its joint pdf is given by

$$f_{\mathbf{X}}(\mathbf{x}) = \frac{1}{(2\pi)^{n/2}|\det K|^{1/2}} \exp\left[-\frac{1}{2}(\mathbf{x} - \boldsymbol{\mu})^T K^{-1}(\mathbf{x} - \boldsymbol{\mu})\right] \tag{3.92}$$

where T denotes the "transpose," $\boldsymbol{\mu}$ is the *vector mean*, K is the *covariance matrix* given by

$$\boldsymbol{\mu} = E[X] = \begin{bmatrix} \mu_1 \\ \vdots \\ \mu_n \end{bmatrix} = \begin{bmatrix} E(X_1) \\ \vdots \\ E(X_n) \end{bmatrix} \tag{3.93}$$

$$K = \begin{bmatrix} \sigma_{11} & \cdots & \sigma_{1n} \\ \vdots & \ddots & \vdots \\ \sigma_{n1} & \cdots & \sigma_{nn} \end{bmatrix} \qquad \sigma_{ij} = \text{Cov}(X_i, X_j) \tag{3.94}$$

and det K is the determinant of the matrix K. Note that $f_{\mathbf{X}}(\mathbf{x})$ stands for $f_{X_1 \cdots X_n}(x_1, \ldots, x_n)$.

SOLVED PROBLEMS

Bivariate Random Variables and Joint Distribution Functions

3.1. Consider an experiment of tossing a fair coin twice. Let (X, Y) be a bivariate r.v., where X is the number of heads that occurs in the two tosses and Y is the number of tails that occurs in the two tosses.

    (*a*) What is the range R_X of X?

    (*b*) What is the range R_Y of Y?

    (*c*) Find and sketch the range R_{XY} of (X, Y).

    (*d*) Find $P(X = 2, Y = 0), P(X = 0, Y = 2)$, and $P(X = 1, Y = 1)$.

The sample space S of the experiment is

$$S = \{HH, HT, TH, TT\}$$

    (*a*) $R_X = \{0, 1, 2\}$

    (*b*) $R_Y = \{0, 1, 2\}$

    (*c*) $R_{XY} = \{(2, 0), (1, 1), (0, 2)\}$ which is sketched in Fig. 3-2.

    (*d*) Since the coin is fair, we have

$$P(X = 2, Y = 0) = P\{HH\} = \frac{1}{4}$$
$$P(X = 0, Y = 2) = P\{TT\} = \frac{1}{4}$$
$$P(X = 1, Y = 1) = P\{HT, TH\} = \frac{1}{2}$$

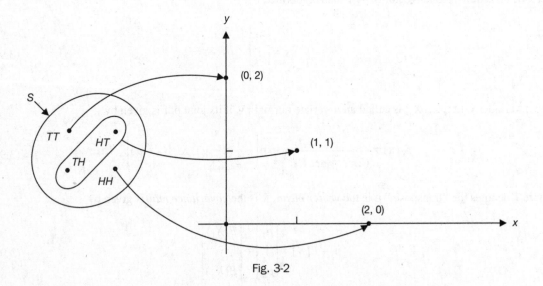

Fig. 3-2

3.2. Consider a bivariate r.v. (X, Y). Find the region of the xy plane corresponding to the events

$$A = \{X + Y \le 2\} \qquad B = \{X^2 + Y^2 < 4\}$$
$$C = \{\min(X, Y) \le 2\} \qquad D = \{\max(X, Y) \le 2\}$$

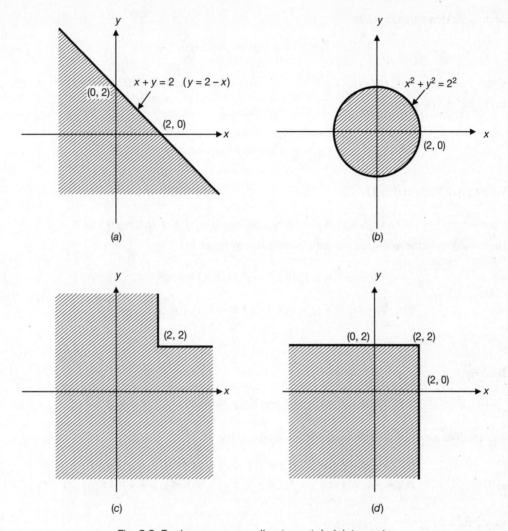

Fig. 3-3 Regions corresponding to certain joint events.

The region corresponding to event A is expressed by $x + y \leq 2$, which is shown in Fig. 3-3(a), that is, the region below and including the straight line $x + y = 2$.

The region corresponding to event B is expressed by $x^2 + y^2 < 2^2$, which is shown in Fig. 3-3(b), that is, the region within the circle with its center at the origin and radius 2.

The region corresponding to event C is shown in Fig. 3-3(c), which is found by noting that

$$\{\min(X, Y) \leq 2\} = (X \leq 2) \cup (Y \leq 2)$$

The region corresponding to event D is shown in Fig. 3-3(d), which is found by noting that

$$\{\max(X, Y) \leq 2\} = (X \leq 2) \cap (Y \leq 2)$$

3.3. Verify Eqs. (3.7), (3.8a), and (3.8b).

Since $\{X \leq \infty, Y \leq \infty\} = S$ and by Eq. (1.36),

$$P(X \leq \infty, Y \leq \infty) = F_{XY}(\infty, \infty) = P(S) = 1$$

Next, as we know, from Eq. (2.8),

$$P(X = -\infty) = P(Y = -\infty) = 0$$

Since $\quad (X = -\infty, Y \le y) \subset (X = -\infty) \quad$ and $\quad (X \le x, Y \le -\infty) \subset (Y = -\infty)$

and by Eq. (1.41), we have

$$P(X = -\infty, Y \le y) = F_{XY}(-\infty, y) = 0$$
$$P(X \le x, Y = -\infty) = F_{XY}(x, -\infty) = 0$$

3.4. Verify Eqs. (3.10) and (3.11).

Clearly $\qquad (X \le x_2, Y \le y) = (X \le x_1, Y \le y) \cup (x_1 < X \le x_2, Y \le y)$

The two events on the right-hand side are disjoint; hence by Eq. (1.37),

$$P(X \le x_2, Y \le y) = P(X \le x_1, Y \le y) + P(x_1 < X \le x_2, Y \le y)$$

or $\qquad P(x_1 < X \le x_2, Y \le y) = P(X \le x_2, Y \le y) - P(X \le x_1, Y \le y)$

$$= F_{XY}(x_2, y) - F_{XY}(x_1, y)$$

Similarly,

$$(X \le x, Y \le y_2) = (X \le x, Y \le y_1) \cup (X \le x, y_1 < Y \le y_2)$$

Again the two events on the right-hand side are disjoint, hence

$$P(X \le x, Y \le y_2) = P(X \le x, Y \le y_1) + P(X \le x, y_1 < Y \le y_2)$$

or $\qquad P(X \le x, y_1 < Y \le y_2) = P(X \le x, Y \le y_2) - P(X \le x, Y \le y_1) = F_{XY}(x, y_2) - F_{XY}(x, y_1)$

3.5. Verify Eq. (3.12).

Clearly

$$(x_1 < X \le x_2, Y \le y_2) = (x_1 < X \le x_2, Y \le y_1) \cup (x_1 < X \le x_2, y_1 < Y \le y_2)$$

The two events on the right-hand side are disjoint; hence

$$P(x_1 < X \le x_2, Y \le y_2) = P(x_1 < X \le x_2, Y \le y_1) + P(x_1 < X \le x_2, y_1 < Y \le y_2)$$

Then using Eq. (3.10), we obtain

$$\begin{aligned}
P(x_1 < X \le x_2, y_1 < Y \le y_2) &= P(x_1 < X \le x_2, Y \le y_2) - P(x_1 < X \le x_2, Y \le y_1) \\
&= F_{XY}(x_2, y_2) - F_{XY}(x_1, y_2) - [F_{XY}(x_2, y_1) - F_{XY}(x_1, y_1)] \\
&= F_{XY}(x_2, y_2) - F_{XY}(x_1, y_2) - F_{XY}(x_2, y_1) + F_{XY}(x_1, y_1)
\end{aligned} \qquad (3.95)$$

Since the probability must be nonnegative, we conclude that

$$F_{XY}(x_2, y_2) - F_{XY}(x_1, y_2) - F_{XY}(x_2, y_1) + F_{XY}(x_1, y_1) \ge 0$$

if $x_2 \ge x_1$ and $y_2 \ge y_1$.

3.6. Consider a function

$$F(x, y) = \begin{cases} 1 - e^{-(x+y)} & 0 \le x < \infty, 0 \le y < \infty \\ 0 & \text{otherwise} \end{cases}$$

Can this function be a joint cdf of a bivariate r.v. (X, Y)?

It is clear that $F(x, y)$ satisfies properties 1 to 5 of a cdf [Eqs. (3.5) to (3.9)]. But substituting $F(x, y)$ in Eq. (3.12) and setting $x_2 = y_2 = 2$ and $x_1 = y_1 = 1$, we get

$$F(2, 2) - F(1, 2) - F(2, 1) + F(1, 1) = (1 - e^{-4}) - (1 - e^{-3}) - (1 - e^{-3}) + (1 - e^{-2})$$
$$= -e^{-4} + 2e^{-3} - e^{-2} = -(e^{-2} - e^{-1})^2 < 0$$

Thus, property 7 [Eq. (3.12)] is not satisfied. Hence, $F(x, y)$ cannot be a joint cdf.

3.7. Consider a bivariate r.v. (X, Y). Show that if X and Y are independent, then every event of the form $(a < X \le b)$ is independent of every event of the form $(c < Y \le d)$.

By definition (3.4), if X and Y are independent, we have

$$F_{XY}(x, y) = F_X(x)F_Y(y)$$

Setting $x_1 = a, x_2 = b, y_1 = c$, and $y_2 = d$ in Eq. (3.95) (Prob. 3.5), we obtain

$$\begin{aligned} P(a < X \le b, c < Y \le d) &= F_{XY}(b, d) - F_{XY}(a, d) - F_{XY}(b, c) + F_{XY}(a, c) \\ &= F_X(b)F_Y(d) - F_X(a)F_Y(d) - F_X(b)F_Y(c) + F_X(a)F_Y(c) \\ &= [F_X(b) - F_X(a)][F_Y(d) - F_Y(c)] \\ &= P(a < X \le b)P(c < Y \le d) \end{aligned}$$

which indicates that event $(a < X \le b)$ and event $(c < Y \le d)$ are independent [Eq. (1.62)].

3.8. The joint cdf of a bivariate r.v. (X, Y) is given by

$$F_{XY}(x, y) = \begin{cases} (1 - e^{-\alpha x})(1 - e^{-\beta y}) & x \ge 0, y \ge 0, \alpha, \beta > 0 \\ 0 & \text{otherwise} \end{cases}$$

(a) Find the marginal cdf's of X and Y.

(b) Show that X and Y are independent.

(c) Find $P(X \le 1, Y \le 1), P(X \le 1), P(Y > 1)$, and $P(X > x, Y > y)$.

(a) By Eqs. (3.13) and (3.14), the marginal cdf's of X and Y are

$$F_X(x) = F_{XY}(x, \infty) = \begin{cases} 1 - e^{-\alpha x} & x \ge 0 \\ 0 & x < 0 \end{cases}$$

$$F_Y(y) = F_{XY}(\infty, y) = \begin{cases} 1 - e^{-\beta y} & y \ge 0 \\ 0 & y < 0 \end{cases}$$

(b) Since $F_{XY}(x, y) = F_X(x)F_Y(y)$, X and Y are independent.

(c)　$P(X \le 1, Y \le 1) = F_{XY}(1, 1) = (1 - e^{-\alpha})(1 - e^{-\beta})$

$P(X \le 1) = F_X(1) = (1 - e^{-\alpha})$

$P(Y > 1) = 1 - P(Y \le 1) = 1 - F_Y(1) = e^{-\beta}$

By De Morgan's law (1.21), we have

$$\overline{(X > x) \cap (Y > y)} = \overline{(X > x)} \cup \overline{(Y > y)} = (X \le x) \cup (Y \le y)$$

Then by Eq. (1.43),

$$
\begin{aligned}
P[\overline{(X > x) \cap (Y > y)}] &= P(X \le x) + P(Y \le y) - P(X \le x, Y \le y) \\
&= F_X(x) + F_Y(y) - F_{XY}(x, y) \\
&= (1 - e^{-\alpha x}) + (1 - e^{-\beta y}) - (1 - e^{-\alpha x})(1 - e^{-\beta y}) \\
&= 1 - e^{-\alpha x} e^{-\beta y}
\end{aligned}
$$

Finally, by Eq. (1.39), we obtain

$$P(X > x, Y > y) = 1 - P[\overline{(X > x) \cap (Y > y)}] = e^{-\alpha x} e^{-\beta y}$$

3.9. The joint cdf of a bivariate r.v. (X, Y) is given by

$$
F_{XY}(x, y) = \begin{cases}
0 & x < 0 \quad \text{or} \quad y < 0 \\
p_1 & 0 \le x < a, \quad 0 \le y < b \\
p_2 & x \ge a, \quad 0 \le y < b \\
p_3 & 0 \le x < a, \quad y \ge b \\
1 & x \ge a, \quad y \ge b
\end{cases}
$$

(a)　Find the marginal cdf's of X and Y.

(b)　Find the conditions on p_1, p_2, and p_3 for which X and Y are independent.

(a)　By Eq. (3.13), the marginal cdf of X is given by

$$
F_X(x) = F_{XY}(x, \infty) = \begin{cases}
0 & x < 0 \\
p_3 & 0 \le x < a \\
1 & x \ge a
\end{cases}
$$

By Eq. (3.14), the marginal cdf of Y is given by

$$
F_Y(y) = F_{XY}(\infty, y) = \begin{cases}
0 & y < 0 \\
p_2 & 0 \le y < b \\
1 & y \ge b
\end{cases}
$$

(b)　For X and Y to be independent, by Eq. (3.4), we must have $F_{XY}(x, y) = F_X(x)F_Y(y)$. Thus, for $0 \le x < a, 0 \le y < b$, we must have $p_1 = p_2 p_3$ for X and Y to be independent.

Discrete Bivariate Random Variables—Joint Probability Mass Functions

3.10. Verify Eq. (3.22).

If X and Y are independent, then by Eq. (1.62),

$$p_{XY}(x_i, y_j) = P(X = x_i, Y = y_j) = P(X = x_i)P(Y = y_j) = p_X(x_i)p_Y(y_j)$$

3.11. Two fair dice are thrown. Consider a bivariate r.v. (X, Y). Let $X = 0$ or 1 according to whether the first die shows an even number or an odd number of dots. Similarly, let $Y = 0$ or 1 according to the second die.

(a) Find the range R_{XY} of (X, Y).

(b) Find the joint pmf's of (X, Y).

(a) The range of (X, Y) is

$$R_{XY} = \{(0,0), (0,1), (1,0), (1,1)\}$$

(b) It is clear that X and Y are independent and

$$P(X = 0) = P(X = 1) = \frac{3}{6} = \frac{1}{2}$$
$$P(Y = 0) = P(Y = 1) = \frac{3}{6} = \frac{1}{2}$$

Thus $\quad p_{XY}(i,j) = P(X = i, Y = j) = P(X = i)P(Y = j) = \frac{1}{4} \qquad i,j = 0,1$

3.12. Consider the binary communication channel shown in Fig. 3-4 (Prob. 1.62). Let (X, Y) be a bivariate r.v., where X is the input to the channel and Y is the output of the channel. Let $P(X = 0) = 0.5$, $P(Y = 1 \mid X = 0) = 0.1$, and $P(Y = 0 \mid X = 1) = 0.2$.

(a) Find the joint pmf's of (X, Y).

(b) Find the marginal pmf's of X and Y.

(c) Are X and Y independent?

(a) From the results of Prob. 1.62, we found that

$$P(X = 1) = 1 - P(X = 0) = 0.5$$
$$P(Y = 0 \mid X = 0) = 0.9 \qquad P(Y = 1 \mid X = 1) = 0.8$$

Then by Eq. (1.57), we obtain

$$P(X = 0, Y = 0) = P(Y = 0 \mid X = 0)P(X = 0) = 0.9(0.5) = 0.45$$
$$P(X = 0, Y = 1) = P(Y = 1 \mid X = 0)P(X = 0) = 0.1(0.5) = 0.05$$
$$P(X = 1, Y = 0) = P(Y = 0 \mid X = 1)P(X = 1) = 0.2(0.5) = 0.1$$
$$P(X = 1, Y = 1) = P(Y = 1 \mid X = 1) P(X = 1) = 0.8(0.5) = 0.4$$

Hence, the joint pmf's of (X, Y) are

$$p_{XY}(0,0) = 0.45 \qquad p_{XY}(0,1) = 0.05$$
$$p_{XY}(1,0) = 0.1 \qquad p_{XY}(1,1) = 0.4$$

(b) By Eq. (3.20), the marginal pmf's of X are

$$p_X(0) = \sum_{y_j} p_{XY}(0, y_j) = 0.45 + 0.05 = 0.5$$

$$p_X(1) = \sum_{y_j} p_{XY}(1, y_j) = 0.1 + 0.4 = 0.5$$

By Eq. (3.21), the marginal pmf's of Y are

$$p_Y(0) = \sum_{x_i} p_{XY}(x_i, 0) = 0.45 + 0.1 = 0.55$$

$$p_Y(1) = \sum_{x_i} p_{XY}(x_i, 1) = 0.05 + 0.4 = 0.45$$

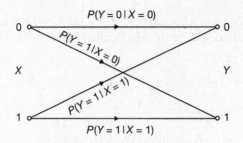

Fig. 3-4 Binary communication channel.

(*c*)   Now

$$p_X(0)p_Y(0) = 0.5(0.55) = 0.275 \neq p_{XY}(0,0) = 0.45$$

Hence, X and Y are not independent.

3.13. Consider an experiment of drawing randomly three balls from an urn containing two red, three white, and four blue balls. Let (X, Y) be a bivariate r.v. where X and Y denote, respectively, the number of red and white balls chosen.

(*a*)   Find the range of (X, Y).

(*b*)   Find the joint pmf's of (X, Y).

(*c*)   Find the marginal pmf's of X and Y.

(*d*)   Are X and Y independent?

(*a*)   The range of (X, Y) is given by

$$R_{XY} = \{(0,0),(0,1),(0,2),(0,3),(1,0),(1,1),(1,2),(2,0),(2,1)\}$$

(*b*)   The joint pmf's of (X, Y)

$$P_{XY}(i,j) = P(X = i, Y = j) \qquad i = 0,1,2 \qquad j = 0,1,2,3$$

are given as follows:

$$p_{XY}(0,0) = \binom{4}{3} \Big/ \binom{9}{3} = \frac{4}{84} \qquad p_{XY}(0,1) = \binom{3}{1}\binom{4}{2} \Big/ \binom{9}{3} = \frac{18}{84}$$

$$p_{XY}(0,2) = \binom{3}{2}\binom{4}{1} \Big/ \binom{9}{3} = \frac{12}{84} \qquad p_{XY}(0,3) = \binom{3}{3} \Big/ \binom{9}{3} = \frac{1}{84}$$

$$p_{XY}(1,0) = \binom{2}{1}\binom{4}{2} \Big/ \binom{9}{3} = \frac{12}{84} \qquad p_{XY}(1,1) = \binom{2}{1}\binom{3}{1}\binom{4}{1} \Big/ \binom{9}{3} = \frac{24}{84}$$

$$p_{XY}(1,2) = \binom{2}{1}\binom{3}{2} \Big/ \binom{9}{3} = \frac{6}{84}$$

$$p_{XY}(2,0) = \binom{2}{2}\binom{4}{1} \Big/ \binom{9}{3} = \frac{4}{84} \qquad p_{XY}(2,1) = \binom{2}{2}\binom{3}{1} \Big/ \binom{9}{3} = \frac{3}{84}$$

which are expressed in tabular form as in Table 3-1.

(c) The marginal pmf's of X are obtained from Table 3-1 by computing the row sums, and the marginal pmf's of Y are obtained by computing the column sums. Thus,

$$p_X(0) = \frac{35}{84} \qquad p_X(1) = \frac{42}{84} \qquad p_X(2) = \frac{7}{84}$$

$$p_Y(0) = \frac{20}{84} \qquad p_Y(1) = \frac{45}{84} \qquad p_Y(2) = \frac{18}{84} \qquad p_Y(3) = \frac{1}{84}$$

TABLE 3-1 $p_{XY}(i, j)$

i	j 0	1	2	3
0	$\frac{4}{84}$	$\frac{18}{84}$	$\frac{12}{84}$	$\frac{1}{84}$
1	$\frac{12}{84}$	$\frac{24}{84}$	$\frac{6}{84}$	0
2	$\frac{4}{84}$	$\frac{3}{84}$	0	0

(d) Since

$$p_{XY}(0,0) = \frac{4}{84} \neq p_X(0)p_Y(0) = \frac{35}{84}\left(\frac{20}{84}\right)$$

X and Y are not independent.

3.14. The joint pmf of a bivariate r.v. (X, Y) is given by

$$p_{XY}(x_i, y_j) = \begin{cases} k(2x_i + y_j) & x_i = 1, 2; \ y_j = 1, 2 \\ 0 & \text{otherwise} \end{cases}$$

where k is a constant.

(a) Find the value of k.

(b) Find the marginal pmf's of X and Y.

(c) Are X and Y independent?

(a) By Eq. (3.17),

$$\sum_{x_i}\sum_{y_j} p_{XY}(x_i, y_j) = \sum_{x_i=1}^{2}\sum_{y_j=1}^{2} k(2x_i + y_j)$$

$$= k[(2+1) + (2+2) + (4+1) + (4+2)] = k(18) = 1$$

Thus, $k = \frac{1}{18}$.

(b) By Eq. (3.20), the marginal pmf's of X are

$$p_X(x_i) = \sum_{y_j} p_{XY}(x_i, y_j) = \sum_{y_j=1}^{2} \frac{1}{18}(2x_i + y_j)$$

$$= \frac{1}{18}(2x_i + 1) + \frac{1}{18}(2x_i + 2) = \frac{1}{18}(4x_i + 3) \qquad x_i = 1, 2$$

CHAPTER 3 *Multiple Random Variables*

By Eq. (3.21), the marginal pmf's of Y are

$$p_Y(y_j) = \sum_{x_i} p_{XY}(x_i, y_j) = \sum_{x_i=1}^{2} \frac{1}{18}(2x_i + y_j)$$

$$= \frac{1}{18}(2 + y_j) + \frac{1}{18}(4 + y_j) = \frac{1}{18}(2y_j + 6) \qquad y_j = 1, 2$$

(c) Now $p_X(x_i)p_Y(y_j) \neq p_{XY}(x_i, y_j)$; hence X and Y are not independent.

3.15. The joint pmf of a bivariate r.v. (X, Y) is given by

$$p_{XY}(x_i, y_j) = \begin{cases} kx_i^2 y_j & x_i = 1, 2; \ y_j = 1, 2, 3 \\ 0 & \text{otherwise} \end{cases}$$

where k is a constant.

(a) Find the value of k.

(b) Find the marginal pmf's of X and Y.

(c) Are X and Y independent?

(a) By Eq. (3.17),

$$\sum_{x_i}\sum_{y_j} p_{XY}(x_i, y_j) = \sum_{x_i=1}^{2}\sum_{y_j=1}^{3} kx_i^2 y_j$$

$$= k(1 + 2 + 3 + 4 + 8 + 12) = k(30) = 1$$

Thus, $k = \frac{1}{30}$.

(b) By Eq. (3.20), the marginal pmf's of X are

$$p_X(x_i) = \sum_{y_j} p_{XY}(x_i, y_j) = \sum_{y_j=1}^{3} \frac{1}{30}x_i^2 y_j = \frac{1}{5}x_i^2 \qquad x_i = 1, 2$$

By Eq. (3.21), the marginal pmf's of Y are

$$p_Y(y_j) = \sum_{x_i} p_{XY}(x_i, y_j) = \sum_{x_i=1}^{2} \frac{1}{30}x_i^2 y_j = \frac{1}{6}y_j \qquad y_j = 1, 2, 3$$

(c) Now

$$p_X(x_i)p_Y(y_j) = \frac{1}{30}x_i^2 y_j = p_{XY}(x_i \ y_j)$$

Hence, X and Y are independent.

3.16. Consider an experiment of tossing two coins three times. Coin A is fair, but coin B is not fair, with $P(H) = \frac{1}{4}$ and $P(T) = \frac{3}{4}$. Consider a bivariate r.v. (X, Y), where X denotes the number of heads resulting from coin A and Y denotes the number of heads resulting from coin B.

(a) Find the range of (X, Y).

(b) Find the joint pmf's of (X, Y).

(c) Find $P(X = Y), P(X > Y)$, and $P(X + Y \leq 4)$.

(a) The range of (X, Y) is given by

$$R_{XY} = \{(i, j): i, j = 0, 1, 2, 3\}$$

(b) It is clear that the r.v.'s X and Y are independent, and they are both binomial r.v.'s with parameters $(n, p) = (3, \frac{1}{2})$ and $(n, p) = (3, \frac{1}{4})$, respectively. Thus, by Eq. (2.36), we have

$$p_X(0) = P(X = 0) = \binom{3}{0}\left(\frac{1}{2}\right)^3 = \frac{1}{8} \qquad p_X(1) = P(X = 1) = \binom{3}{1}\left(\frac{1}{2}\right)^3 = \frac{3}{8}$$

$$p_X(2) = P(X = 2) = \binom{3}{2}\left(\frac{1}{2}\right)^3 = \frac{3}{8} \qquad p_X(3) = P(X = 3) = \binom{3}{3}\left(\frac{1}{2}\right)^3 = \frac{1}{8}$$

$$p_Y(0) = P(Y = 0) = \binom{3}{0}\left(\frac{1}{4}\right)^0\left(\frac{3}{4}\right)^3 = \frac{27}{64} \qquad p_Y(1) = P(Y = 1) = \binom{3}{1}\left(\frac{1}{4}\right)\left(\frac{3}{4}\right)^2 = \frac{27}{64}$$

$$p_Y(2) = P(Y = 2) = \binom{3}{2}\left(\frac{1}{4}\right)^2\left(\frac{3}{4}\right) = \frac{9}{64} \qquad p_Y(3) = P(Y = 3) = \binom{3}{3}\left(\frac{1}{4}\right)^3\left(\frac{3}{4}\right)^0 = \frac{1}{64}$$

Since X and Y are independent, the joint pmf's of (X, Y) are

$$p_{XY}(i, j) = p_X(i)\, p_Y(j) \qquad i, j = 0, 1, 2, 3$$

which are tabulated in Table 3-2.

(c) From Table 3-2, we have

$$P(X = Y) = \sum_{i=0}^{3} p_{XY}(i, i) = \frac{1}{512}(27 + 81 + 27 + 1) = \frac{136}{512}$$

$$p(X > Y) = \sum_{i>j}^{3} p_{XY}(i, j) = P_{XY}(1, 0) + P_{XY}(2, 0) + P_{XY}(3, 0) + p_{XY}(2, 1) + p_{XY}(3, 1) + p_{XY}(3, 2)$$

$$= \frac{1}{512}(81 + 81 + 27 + 81 + 27 + 9) = \frac{306}{512}$$

Table 3-2 $p_{XY}(i, j)$

			j	
i	0	1	2	3
0	$\frac{27}{512}$	$\frac{27}{512}$	$\frac{9}{512}$	$\frac{1}{512}$
1	$\frac{81}{512}$	$\frac{81}{512}$	$\frac{27}{512}$	$\frac{3}{512}$
2	$\frac{81}{512}$	$\frac{81}{512}$	$\frac{27}{512}$	$\frac{3}{512}$
3	$\frac{27}{512}$	$\frac{27}{512}$	$\frac{9}{512}$	$\frac{1}{512}$

$$P(X + Y > 4) = \sum_{i+j>4} p_{XY}(i, j) = P_{XY}(2, 3) + P_{XY}(3, 2) + P_{XY}(3, 3)$$

$$= \frac{1}{512}(3 + 9 + 1) = \frac{13}{512}$$

Thus,
$$P(X + y \le 4) = 1 - P(X + Y > 4) = 1 - \frac{13}{512} = \frac{499}{512}$$

Continuous Bivariate Random Variables—Probability Density Functions

3.17. The joint pdf of a bivariate r.v. (X, Y) is given by

$$f_{XY}(x, y) = \begin{cases} k(x + y) & 0 < x < 2, 0 < y < 2 \\ 0 & \text{otherwise} \end{cases}$$

where k is a constant.

(a) Find the value of k.

(b) Find the marginal pdf's of X and Y.

(c) Are X and Y independent?

(a) By Eq. (3.26),

$$\int_{-\infty}^{\infty}\int_{-\infty}^{\infty} f_{XY}(x, y)\, dx\, dy = k \int_0^2\int_0^2 (x + y)\, dx\, dy$$

$$= k \int_0^2 \left(\frac{x^2}{2} + xy \right)\bigg|_{x=0}^{x=2} dy$$

$$= k \int_0^2 (2 + 2y)\, dy = 8k = 1$$

Thus, $k = \frac{1}{8}$.

(b) By Eq. (3.30), the marginal pdf of X is

$$f_X(x) = \int_{-\infty}^{\infty} f_{XY}(x, y)\, dy = \frac{1}{8}\int_0^2 (x + y)\, dy$$

$$= \frac{1}{8}\left(xy + \frac{y^2}{2} \right)\bigg|_{y=0}^{y=2} = \begin{cases} \frac{1}{4}(x + 1) & 0 < x < 2 \\ 0 & \text{otherwise} \end{cases}$$

Since $f_{XY}(x, y)$ is symmetric with respect to x and y, the marginal pdf of Y is

$$f_Y(y) = \begin{cases} \frac{1}{4}(y + 1) & 0 < y < 2 \\ 0 & \text{otherwise} \end{cases}$$

(c) Since $f_{XY}(x, y) \neq f_X(x)f_Y(y)$, X and Y are not independent.

3.18. The joint pdf of a bivariate r.v. (X, Y) is given by

$$f_{XY}(x, y) = \begin{cases} kxy & 0 < x < 1, 0 < y < 1 \\ 0 & \text{otherwise} \end{cases}$$

where k is a constant.

(a) Find the value of k.

(b) Are X and Y independent?

(c) Find $P(X + Y < 1)$.

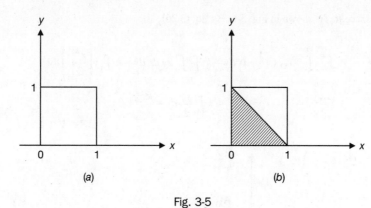

Fig. 3-5

(a) The range space R_{XY} is shown in Fig. 3-5(a). By Eq. (3.26),

$$\int_{-\infty}^{\infty}\int_{-\infty}^{\infty}f_{XY}(x,y)\,dx\,dy = k\int_0^1\int_0^1 xy\,dx\,dy = k\int_0^1 y\left(\frac{x^2}{2}\Big|_0^1\right)dy$$

$$= k\int_0^1\frac{y}{2}\,dy = \frac{k}{4} = 1$$

Thus $k = 4$.

(b) To determine whether X and Y are independent, we must find the marginal pdf's of X and Y. By Eq. (3.30),

$$f_X(x) = \begin{cases}\int_0^1 4xy\,dy = 2x & 0 < x < 1 \\ 0 & \text{otherwise}\end{cases}$$

By symmetry,

$$f_Y(y) = \begin{cases}2y & 0 < y < 1 \\ 0 & \text{otherwise}\end{cases}$$

Since $f_{XY}(x,y) = f_X(x)f_Y(y)$, X and Y are independent.

(c) The region in the xy plane corresponding to the event $(X + Y < 1)$ is shown in Fig. 3-5(b) as a shaded area. Then

$$P(X+Y<1) = \int_0^1\int_0^{1-y}4xy\,dx\,dy = \int_0^1 4y\left(\frac{x^2}{2}\Big|_0^{1-y}\right)dy$$

$$= \int_0^1 4y\left[\frac{1}{2}(1-y)^2\right]dy = \frac{1}{6}$$

3.19. The joint pdf of a bivariate r.v. (X, Y) is given by

$$f_{XY}(x,y) = \begin{cases}kxy & 0 < x < y < 1 \\ 0 & \text{otherwise}\end{cases}$$

where k is a constant.

(a) Find the value of k.

(b) Are X and Y independent?

(*a*)   The range space R_{XY} is shown in Fig. 3-6. By Eq. (3.26),

$$\int_{-\infty}^{\infty}\int_{-\infty}^{\infty} f_{XY}(x, y)\, dx\, dy = k \int_0^1\int_0^y xy\, dx\, dy = k \int_0^1 y\left(\frac{x^2}{2}\Big|_0^y\right) dy$$

$$= k \int_0^1 \frac{y^3}{2}\, dy = \frac{k}{8} = 1$$

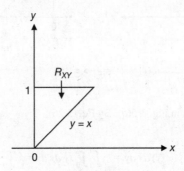

Fig. 3-6

Thus $k = 8$.

(*b*)   By Eq. (3.30), the marginal pdf of X is

$$f_X(x) = \begin{cases} \int_x^1 8xy\, dy = 4x(1 - x^2) & 0 < x < 1 \\ 0 & \text{otherwise} \end{cases}$$

By Eq. (3.31), the marginal pdf of Y is

$$f_Y(y) = \begin{cases} \int_0^y 8xy\, dx = 4y^3 & 0 < y < 1 \\ 0 & \text{otherwise} \end{cases}$$

Since $f_{XY}(x,y) \neq f_X(x)\, f_Y(y)$, X and Y are not independent.

Note that if the range space R_{XY} depends functionally on x or y, then X and Y cannot be independent r.v.'s.

3.20. The joint pdf of a bivariate r.v. (X, Y) is given by

$$f_{XY}(x, y) = \begin{cases} k & 0 < y \leq x < 1 \\ 0 & \text{otherwise} \end{cases}$$

where k is a constant.

(*a*)   Determine the value of k.

(*b*)   Find the marginal pdf's of X and Y.

(*c*)   Find $P(0 < X < \frac{1}{2}, 0 < Y < \frac{1}{2})$.

(*a*)   The range space R_{XY} is shown in Fig. 3-7. By Eq. (3.26),

$$\int_{-\infty}^{\infty}\int_{-\infty}^{\infty} f_{XY}(x, y)\, dx\, dy = k \iint_{R_{XY}} dx\, dy = k \times \text{area}(R_{XY}) = k\left(\frac{1}{2}\right) = 1$$

Thus $k = 2$.

(b) By Eq. (3.30), the marginal pdf of X is

$$f_X(x) = \begin{cases} \int_0^x 2\,dy = 2x & 0 < x < 1 \\ 0 & \text{otherwise} \end{cases}$$

By Eq. (3.31), the marginal pdf of Y is

$$f_Y(y) = \begin{cases} \int_y^1 2\,dx = 2(1-y) & 0 < y < 1 \\ 0 & \text{otherwise} \end{cases}$$

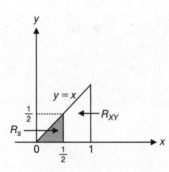

Fig. 3-7

(c) The region in the xy plane corresponding to the event $(0 < X < \frac{1}{2}, 0 < Y < \frac{1}{2})$ is shown in Fig. 3-7 as the shaded area R_s. Then

$$P\left(0 < X < \frac{1}{2}, 0 < Y < \frac{1}{2}\right) = P\left(0 < X < \frac{1}{2}, 0 < Y < X\right)$$

$$= \iint_{R_s} f_{XY}(x,y)\,dx\,dy = 2\iint_{R_s} dx\,dy = 2 \times \text{area}(R_s) = 2\left(\frac{1}{8}\right) = \frac{1}{4}$$

Note that the bivariate r.v. (X, Y) is said to be uniformly distributed over the region R_{XY} if its pdf is

$$f_{XY}(x, y) = \begin{cases} k & (x,y) \in R_{XY} \\ 0 & \text{otherwise} \end{cases} \tag{3.96}$$

where k is a constant. Then by Eq. (3.26), the constant k must be $k = 1/(\text{area of } R_{XY})$.

3.21. Suppose we select one point at random from within the circle with radius R. If we let the center of the circle denote the origin and define X and Y to be the coordinates of the point chosen (Fig. 3-8), then (X, Y) is a uniform bivariate r.v. with joint pdf given by

$$f_{XY}(x,y) = \begin{cases} k & x^2 + y^2 \le R^2 \\ 0 & x^2 + y^2 > R^2 \end{cases}$$

where k is a constant.

(a) Determine the value of k.

(b) Find the marginal pdf's of X and Y.

(c) Find the probability that the distance from the origin of the point selected is not greater than a.

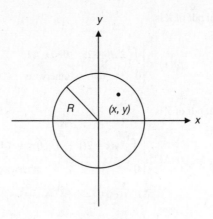

Fig. 3-8

(*a*)　By Eq. (3.26),

$$\int_{-\infty}^{\infty}\int_{-\infty}^{\infty} f_{XY}(x,y)\,dx\,dy = k \iint_{x^2+y^2\le R^2} dx\,dy = k(\pi R^2) = 1$$

Thus, $k = 1/\pi R^2$.

(*b*)　By Eq. (3.30), the marginal pdf of X is

$$f_X(x) = \frac{1}{\pi R^2}\int_{-\sqrt{(R^2-x^2)}}^{\sqrt{(R^2-x^2)}} dy = \frac{2}{\pi R^2}\sqrt{R^2-x^2}\qquad x^2\le R^2$$

Hence,

$$f_X(x) = \begin{cases} \dfrac{2}{\pi R^2}\sqrt{R^2-x^2} & |x|\le R \\ 0 & |x|>R \end{cases}$$

By symmetry, the marginal pdf of Y is

$$f_Y(y) = \begin{cases} \dfrac{2}{\pi R^2}\sqrt{R^2-y^2} & |y|\le R \\ 0 & |y|>R \end{cases}$$

(*c*)　For $0\le a\le R$,

$$P(X^2+Y^2\le a) = \iint_{x^2+y^2\le a^2} f_{XY}(x,y)\,dx\,dy$$

$$= \frac{1}{\pi R^2}\iint_{x^2+y^2\le a^2} dx\,dy = \frac{\pi a^2}{\pi R^2} = \frac{a^2}{R^2}$$

3.22. The joint pdf of a bivariate r.v. (X, Y) is given by

$$f_{XY}(x,y) = \begin{cases} ke^{-(ax+by)} & x>0,\ y>0 \\ 0 & \text{otherwise} \end{cases}$$

where a and b are positive constants and k is a constant.

(*a*)　Determine the value of k.

(*b*)　Are X and Y independent?

(a) By Eq. (3.26),

$$\int_{-\infty}^{\infty}\int_{-\infty}^{\infty} f_{XY}(x,y)\,dx\,dy = k\int_{0}^{\infty}\int_{0}^{\infty} e^{-(ax+by)}dx\,dy$$

$$= k\int_{0}^{\infty} e^{-ax}dx\int_{0}^{\infty} e^{-by}dy = \frac{k}{ab} = 1$$

Thus, $k = ab$.

(b) By Eq. (3.30), the marginal pdf of X is

$$f_X(x) = abe^{-ax}\int_{0}^{\infty} e^{-by}\,dy = ae^{-ax} \qquad x>0$$

By Eq. (3.31), the marginal pdf of Y is

$$f_Y(y) = abe^{-by}\int_{0}^{\infty} e^{-ax}\,dx = be^{-by} \qquad y>0$$

Since $f_{XY}(x,y) = f_X(x)f_Y(y)$, X and Y are independent.

3.23. A manufacturer has been using two different manufacturing processes to make computer memory chips. Let (X, Y) be a bivariate r.v., where X denotes the time to failure of chips made by process A and Y denotes the time to failure of chips made by process B. Assuming that the joint pdf of (X, Y) is

$$f_{XY}(x,y) = \begin{cases} abe^{-(ax+by)} & x>0,\, y>0 \\ 0 & \text{otherwise} \end{cases}$$

where $a = 10^{-4}$ and $b = 1.2(10^{-4})$, determine $P(X > Y)$.

The region in the xy plane corresponding to the event $(X > Y)$ is shown in Fig. 3-9 as the shaded area. Then

$$P(X>Y) = ab\int_{0}^{\infty}\int_{0}^{x} e^{-(ax+by)}dy\,dx$$

$$= ab\int_{0}^{\infty} e^{-ax}\left[\int_{0}^{x} e^{-by}dy\right]dx = a\int_{0}^{\infty} e^{-ax}(1-e^{-bx})\,dx$$

$$= \frac{b}{a+b} = \frac{1.2(10^{-4})}{(1+1.2)(10^{-4})} = 0.545$$

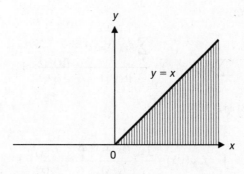

Fig. 3-9

3.24. A smooth-surface table is ruled with equidistant parallel lines a distance D apart. A needle of length L, where $L \leq D$, is randomly dropped onto this table. What is the probability that the needle will intersect one of the lines? (This is known as *Buffon's* needle problem.)

We can determine the needle's position by specifying a bivariate r.v. (X, Θ), where X is the distance from the middle point of the needle to the nearest parallel line and Θ is the angle from the vertical to the needle (Fig. 3-10). We interpret the statement "the needle is randomly dropped" to mean that both X and Θ have uniform distributions and that X and Θ are independent. The possible values of X are between 0 and $D/2$, and the possible values of Θ are between 0 and $\pi/2$. Thus, the joint pdf of (X, Θ) is

$$f_{X\Theta}(x, \theta) = f_X(x)f_\Theta(\theta) = \begin{cases} \dfrac{4}{\pi D} & 0 \leq x \leq \dfrac{D}{2}, 0 \leq \theta \leq \dfrac{\pi}{2} \\ 0 & \text{otherwise} \end{cases}$$

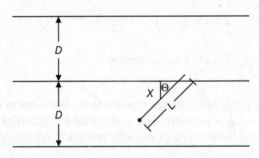

Fig. 3-10 Buffon's needle problem.

From Fig. 3-10, we see that the condition for the needle to intersect a line is $X < L/2 \cos \theta$. Thus, the probability that the needle will intersect a line is

$$P\left(X < \frac{L}{2}\cos\theta\right) = \int_0^{\pi/2} \int_0^{(L/2)\cos\theta} f_{X\Theta}(x, \theta)\, dx\, d\theta$$

$$= \frac{4}{\pi D} \int_0^{\pi/2} \left[\int_0^{(L/2)\cos\theta} dx\right] d\theta$$

$$= \frac{4}{\pi D} \int_0^{\pi/2} \frac{L}{2}\cos\theta\, d\theta = \frac{2L}{\pi D}$$

Conditional Distributions

3.25. Verify Eqs. (3.36) and (3.41).

(*a*) By Eqs. (3.33) and (3.20),

$$\sum_{y_j} p_{Y|X}(y_j|x_i) = \frac{\sum_{y_j} p_{YX}(x_i, y_j)}{p_X(x_i)} = \frac{p_X(x_i)}{p_X(x_i)} = 1$$

(*b*) Similarly, by Eqs. (3.38) and (3.30),

$$\int_{-\infty}^{\infty} f_{Y|X}(y|x)\, = \frac{\int_{-\infty}^{\infty} f_{YX}(x, y)\, dy}{f_X(x)} = \frac{f_X(x)}{f_X(x)} = 1$$

3.26. Consider the bivariate r.v. (X, Y) of Prob. 3.14.

    (*a*) Find the conditional pmf's $p_{Y|X}(y_j \mid x_i)$ and $p_{X|Y}(x_i \mid y_j)$.

    (*b*) Find $P(Y = 2 \mid X = 2)$ and $P(X = 2 \mid Y = 2)$.

    (*a*) From the results of Prob. 3.14, we have

$$p_{XY}(x_i, x_j) = \begin{cases} \dfrac{1}{18}(2x_i + y_i) & x_i = 1, 2; \ y_j = 1, 2 \\ 0 & \text{otherwise} \end{cases}$$

$$p_X(x_i) = \frac{1}{18}(4x_i + 3) \qquad x_i = 1, 2$$

$$p_Y(y_i) = \frac{1}{18}(2y_i + 6) \qquad y_i = 1, 2$$

Thus, by Eqs. (3.33) and (3.34),

$$p_{Y|X}(y_j \mid x_i) = \frac{2x_i + y_j}{4x_i + 3} \qquad y_j = 1, 2; \ x_i = 1, 2$$

$$p_{X|Y}(x_i \mid y_j) = \frac{2x_i + y_j}{2y_j + 6} \qquad x_i = 1, 2; \ y_j = 1, 2$$

    (*b*) Using the results of part (*a*), we obtain

$$P(Y = 2 \mid X = 2) = p_{Y|X}(2|2) = \frac{2(2) + 2}{4(2) + 3} = \frac{6}{11}$$

$$P(X = 2 \mid Y = 2) = p_{X|Y}(2|2) = \frac{2(2) + 2}{2(2) + 6} = \frac{3}{5}$$

3.27. Find the conditional pmf's $p_{Y|X}(y_j \mid x_i)$ and $p_{X|Y}(x_i \mid y_j)$ for the bivariate r.v. (X, Y) of Prob. 3.15.

From the results of Prob. 3.15, we have

$$p_{XY}(x_i, y_j) = \begin{cases} \dfrac{1}{30}x_i^2 y_j & x_i = 1, 2; \ y_j = 1, 2, 3 \\ 0 & \text{otherwise} \end{cases}$$

$$p_X(x_i) = \frac{1}{5}x_i^2 \qquad x_i = 1, 2$$

$$p_Y(y_j) = \frac{1}{6}y_j \qquad y_j = 1, 2, 3$$

Thus, by Eqs. (3.33) and (3.34),

$$p_{Y|X}(y_j \mid x_i) = \frac{\dfrac{1}{30}x_i^2 y_j}{\dfrac{1}{5}x_i^2} = \frac{1}{6}y_j \qquad y_j = 1, 2, 3; \ x_i = 1, 2$$

$$p_{X|Y}(x_i \mid y_j) = \frac{\dfrac{1}{30}x_i^2 y_j}{\dfrac{1}{6}y_j} = \frac{1}{5}x_i^2 \qquad x_i = 1, 2; \ y_j = 1, 2, 3$$

Note that $P_{Y|X}(y_j \mid x_i) = p_Y(y_j)$ and $p_{X|Y}(x_i \mid y_j) = p_X(x_i)$, as must be the case since X and Y are independent, as shown in Prob. 3.15.

3.28. Consider the bivariate r.v. (X, Y) of Prob. 3.17.

    (a) Find the conditional pdf's $f_{Y|X}(y\,|\,x)$ and $f_{X|Y}(x\,|\,y)$.

    (b) Find $P(0 < Y < \frac{1}{2}\,|\,X = 1)$.

    (a) From the results of Prob. 3.17, we have

$$f_{XY}(x,y) = \begin{cases} \dfrac{1}{8}(x+y) & 0<x<2, 0<y<2 \\ 0 & \text{otherwise} \end{cases}$$

$$f_X(x) = \frac{1}{4}(x+1) \qquad 0<x<2$$

$$f_Y(y) = \frac{1}{4}(y+1) \qquad 0<y<2$$

Thus, by Eqs. (3.38) and (3.39),

$$f_{Y|X}(y|x) = \frac{\frac{1}{8}(x+y)}{\frac{1}{4}(x+1)} = \frac{1}{2}\frac{x+y}{x+1} \qquad 0<x<2, 0<y<2$$

$$f_{X|Y}(x|y) = \frac{\frac{1}{8}(x+y)}{\frac{1}{4}(y+1)} = \frac{1}{2}\frac{x+y}{y+1} \qquad 0<x<2, 0<y<2$$

    (b) Using the results of part (a), we obtain

$$P(0<Y<\frac{1}{2}|X=1) = \int_0^{1/2} f_{Y|X}(y|x=1) = \frac{1}{2}\int_0^{1/2}\left(\frac{1+y}{2}\right)dy = \frac{5}{32}$$

3.29. Find the conditional pdf's $f_{Y|X}(y|x)$ and $f_{X|Y}(x|y)$ for the bivariate r.v. (X, Y) of Prob. 3.18.

From the results of Prob. 3.18, we have

$$f_{XY}(x, y) = \begin{cases} 4xy & 0<x<1, 0<y<1 \\ 0 & \text{otherwise} \end{cases}$$

$$f_X(x) = 2x \qquad 0<x<1$$

$$f_Y(y) = 2y \qquad 0<y<1$$

Thus, by Eqs. (3.38) and (3.39),

$$f_{Y|X}(y|x) = \frac{4xy}{2x} = 2y \qquad 0<y<1, 0<x<1$$

$$f_{X|Y}(x|y) = \frac{4xy}{2y} = 2x \qquad 0<x<1, 0<y<1$$

Again note that $f_{Y|X}(y\,|\,x) = f_Y(y)$ and $f_{X|Y}(x\,|\,y) = f_X(x)$, as must be the case since X and Y are independent, as shown in Prob. 3.18.

3.30. Find the conditional pdf's $f_{Y|X}(y\,|\,x)$ and $f_{X|Y}(x\,|\,y)$ for the bivariate r.v. (X, Y) of Prob. 3.20.

From the results of Prob. 3.20, we have

$$f_{XY}(x, y) = \begin{cases} 2 & 0<y\leq x<1 \\ 0 & \text{otherwise} \end{cases}$$

$$f_X(x) = 2x \qquad 0<x<1$$

$$f_Y(y) = 2(1-y) \qquad 0<y<1$$

Thus, by Eqs. (3.38) and (3.39),

$$f_{Y|X}(y|x) = \frac{1}{x} \qquad y \le x < 1, 0 < x < 1$$

$$f_{X|Y}(x|y) = \frac{1}{1-y} \qquad y \le x < 1, 0 < x < 1$$

3.31. The joint pdf of a bivariate r.v. (X, Y) is given by

$$f_{XY}(x, y) = \begin{cases} \dfrac{1}{y} e^{-x/y} e^{-y} & x > 0, y > 0 \\ 0 & \text{otherwise} \end{cases}$$

(a) Show that $f_{XY}(x, y)$ satisfies Eq. (3.26).

(b) Find $P(X > 1 \mid Y = y)$.

(a) We have

$$\int_{-\infty}^{\infty} \int_{-\infty}^{\infty} f_{XY}(x, y) \, dx \, dy = \int_0^{\infty} \int_0^{\infty} \frac{1}{y} e^{-x/y} e^{-y} \, dx \, dy$$

$$= \int_0^{\infty} \frac{1}{y} e^{-y} \left(\int_0^{\infty} e^{-x/y} dx \right) dy$$

$$= \int_0^{\infty} \frac{1}{y} e^{-y} \left[-y e^{-x/y} \Big|_{x=0}^{x=\infty} \right] dy$$

$$= \int_0^{\infty} e^{-y} \, dy = 1$$

(b) First we must find the marginal pdf on Y. By Eq. (3.31),

$$f_Y(y) = \int_{-\infty}^{\infty} f_{XY}(x, y) \, dx = \frac{1}{y} e^{-y} \int_0^{\infty} e^{-x/y} dx$$

$$= \frac{1}{y} e^{-y} \left[-y e^{-x/y} \Big|_{x=0}^{x=\infty} \right] = e^{-y}$$

By Eq. (3.39), the conditional pdf of X is

$$f_{X|Y}(x|y) = \frac{f_{XY}(x, y)}{f_Y(y)} \begin{cases} \dfrac{1}{y} e^{-x/y} & x > 0, y > 0 \\ 0 & \text{otherwise} \end{cases}$$

Then

$$P(X > 1 | Y = y) = \int_1^{\infty} f_{X|Y}(x, y) \, dx = \int_1^{\infty} \frac{1}{y} e^{-x/y} \, dx$$

$$= -e^{-x/y} \Big|_{x=1}^{x=\infty} = e^{-1/y}$$

Covariance and Correlation Coefficients

3.32. Let (X, Y) be a bivariate r.v. If X and Y are independent, show that X and Y are uncorrelated.

If (X, Y) is a discrete bivariate r.v., then by Eqs. (3.43) and (3.22),

$$E(XY) = \sum_{y_j} \sum_{x_i} x_i y_j p_{XY}(x_i, y_j) = \sum_{y_j} \sum_{x_i} x_i y_j p_X(x_i) p_Y(y_j)$$

$$= \left[\sum_{x_i} x_i p_X(x_i) \right] \left[\sum_{y_j} y_j p_Y(y_j) \right] = E(X)E(Y)$$

If (X, Y) is a continuous bivariate r.v., then by Eqs. (3.43) and (3.32),

$$E(XY) = \int_{-\infty}^{\infty}\int_{-\infty}^{\infty} xy\, f_{XY}(x, y)\, dx\, dy = \int_{-\infty}^{\infty}\int_{-\infty}^{\infty} xy\, f_X(x)f_Y(y)\, dx\, dy$$
$$= \int_{-\infty}^{\infty} xf_X(x)\, dx \int_{-\infty}^{\infty} yf_Y(y)\, dy = E(X)E(Y)$$

Thus, X and Y are uncorrelated by Eq. (3.52).

3.33. Suppose the joint pmf of a bivariate r.v. (X, Y) is given by

$$p_{XY}(x_i, y_j) = \begin{cases} \dfrac{1}{3} & (0, 1), (1, 0), (2, 1) \\ 0 & \text{otherwise} \end{cases}$$

(*a*) Are X and Y independent?

(*b*) Are X and Y uncorrelated ?

(*a*) By Eq. (3.20), the marginal pmf's of X are

$$p_X(0) = \sum_{y_j} p_{XY}(0, y_j) = p_{XY}(0, 1) = \frac{1}{3}$$

$$p_X(1) = \sum_{y_j} p_{XY}(1, y_j) = p_{XY}(1, 0) = \frac{1}{3}$$

$$p_X(2) = \sum_{y_j} p_{XY}(2, y_j) = p_{XY}(2, 1) = \frac{1}{3}$$

By Eq. (3.21), the marginal pmf's of Y are

$$p_Y(0) = \sum_{x_i} p_{XY}(x_i, 0) = p_{XY}(1, 0) = \frac{1}{3}$$

$$p_Y(1) = \sum_{x_i} p_{XY}(x_i, 1) = p_{XY}(0, 1) + p_{XY}(2, 1) = \frac{2}{3}$$

and
$$p_{XY}(0, 1) = \frac{1}{3} \neq p_X(0)p_Y(1) = \frac{2}{9}$$

Thus, X and Y are not independent.

(*b*) By Eqs. (3.45*a*), (3.45*b*), and (3.43), we have

$$E(X) = \sum_{x_i} x_i p_X(x_i) = (0)\left(\frac{1}{3}\right) + (1)\left(\frac{1}{3}\right) + (2)\left(\frac{1}{3}\right) = 1$$

$$E(Y) = \sum_{y_j} y_j p_y(y_j) = (0)\left(\frac{1}{3}\right) + (1)\left(\frac{2}{3}\right) = \frac{2}{3}$$

$$E(XY) = \sum_{y_j}\sum_{x_i} x_i y_j p_{XY}(x_i, y_j)$$

$$= (0)(1)\left(\frac{1}{3}\right) + (1)(0)\left(\frac{1}{3}\right) + (2)(1)\left(\frac{1}{3}\right) = \frac{2}{3}$$

Now by Eq. (3.51),

$$\text{Cov}(X, Y) = E(XY) - E(X)E(Y) = \frac{2}{3} - (1)\left(\frac{2}{3}\right) = 0$$

Thus, X and Y are uncorrelated.

3.34. Let (X, Y) be a bivariate r.v. with the joint pdf

$$f_{XY}(x, y) = \frac{x^2 + y^2}{4\pi} e^{-(x^2 + y^2)/2} \qquad -\infty < x < \infty, -\infty < y < \infty$$

Show that X and Y are not independent but are uncorrelated.

By Eq. (3.30), the marginal pdf of X is

$$f_X(x) = \frac{1}{4\pi} \int_{-\infty}^{\infty} (x^2 + y^2) e^{-(x^2 + y^2)/2} dy$$

$$= \frac{e^{-x^2/2}}{2\sqrt{2\pi}} \left(x^2 \int_{-\infty}^{\infty} \frac{1}{\sqrt{2\pi}} e^{-y^2/2} dy + \int_{-\infty}^{\infty} \frac{1}{\sqrt{2\pi}} y^2 e^{-y^2/2} dy \right)$$

Noting that the integrand of the first integral in the above expression is the pdf of $N(0; 1)$ and the second integral in the above expression is the variance of $N(0; 1)$, we have

$$f_X(x) = \frac{1}{2\sqrt{2\pi}} (x^2 + 1) e^{-x^2/2} \qquad -\infty < x < \infty$$

Since $f_{XY}(x, y)$ is symmetric in x and y, we have

$$f_Y(y) = \frac{1}{2\sqrt{2\pi}} (y^2 + 1) e^{-y^2/2} \qquad -\infty < y < \infty$$

Now $f_{XY}(x, y) \neq f_X(x) f_Y(y)$, and hence X and Y are not independent. Next, by Eqs. (3.47a) and (3.47b),

$$E(X) = \int_{-\infty}^{\infty} x f_X(x) \, dx = 0$$

$$E(Y) = \int_{-\infty}^{\infty} y f_Y(y) \, dy = 0$$

since for each integral the integrand is an odd function. By Eq. (3.43),

$$E(XY) = \int_{-\infty}^{\infty} \int_{-\infty}^{\infty} xy f_{XY}(x, y) \, dx \, dy = 0$$

The integral vanishes because the contributions of the second and the fourth quadrants cancel those of the first and the third. Thus, $E(XY) = E(X)E(Y)$, and so X and Y are uncorrelated.

3.35. Let (X, Y) be a bivariate r.v. Show that

$$[E(XY)]^2 \leq E(X^2)E(Y^2) \tag{3.97}$$

This is known as the *Cauchy-Schwarz inequality*.

Consider the expression $E[(X - \alpha Y)^2]$ for any two r.v.'s X and Y and a real variable α. This expression, when viewed as a quadratic in α, is greater than or equal to zero; that is,

$$E[(X - \alpha Y)^2] \geq 0$$

for any value of α. Expanding this, we obtain

$$E(X^2) - 2\alpha E(XY) + \alpha^2 E(Y^2) \geq 0$$

Choose a value of α for which the left-hand side of this inequality is minimum,

$$\alpha = \frac{E(XY)}{E(Y^2)}$$

which results in the inequality

$$E(X^2) - \frac{[E(XY)]^2}{E(Y^2)} \geq 0 \qquad \text{or} \qquad [E(XY)]^2 \leq E(X^2)E(Y^2)$$

3.36. Verify Eq. (3.54).

From the Cauchy-Schwarz inequality [Eq. (3.97)], we have

$$\{E[(X - \mu_X)(Y - \mu_Y)]\}^2 \leq E[(X - \mu_X)^2]E[(Y - \mu_Y)^2]$$

or

$$\sigma_{XY}^2 \leq \sigma_X^2 \sigma_Y^2$$

Then

$$\rho_{XY}^2 = \frac{\sigma_{XY}^2}{\sigma_X^2 \sigma_Y^2} \leq 1$$

Since ρ_{XY} is a real number, this implies

$$|\rho_{XY}| \leq 1 \qquad \text{or} \qquad -1 \leq \rho_{XY} \leq 1$$

3.37. Let (X, Y) be the bivariate r.v. of Prob. 3.12.

   (a) Find the mean and the variance of X.

   (b) Find the mean and the variance of Y.

   (c) Find the covariance of X and Y.

   (d) Find the correlation coefficient of X and Y.

   (a) From the results of Prob. 3.12, the mean and the variance of X are evaluated as follows:

$$E(X) = \sum_{x_i} x_i p_X(x_i) = (0)(0.5) + (1)(0.5) = 0.5$$

$$E(X^2) = \sum_{x_i} x_i^2 p_X(x_i) = (0)^2(0.5) + (1)^2(0.5) = 0.5$$

$$\sigma_X^2 = E(X^2) - [E(X)]^2 = 0.5 - (0.5)^2 = 0.25$$

   (b) Similarly, the mean and the variance of Y are

$$E(Y) = \sum_{y_j} y_j p_Y(y_j) = (0)(0.55) + (1)(0.45) = 0.45$$

$$E(Y^2) = \sum_{y_j} y_j^2 p_Y(y_j) = (0)^2(0.55) + (1)^2(0.45) = 0.45$$

$$\sigma_Y^2 = E(Y^2) - [E(Y)]^2 = 0.45 - (0.45)^2 = 0.2475$$

(c)   By Eq. (3.43),

$$E(XY) = \sum_{y_j} \sum_{x_i} x_i y_j p_{XY}(x_i, y_j)$$

$$= (0)(0)(0.45) + (0)(1)(0.05) + (1)(0)(0.1) + (1)(1)(0.4)$$

$$= 0.4$$

By Eq. (3.51), the covariance of X and Y is

$$\text{Cov}(X, Y) = E(XY) - E(X)\,E(Y) = 0.4 - (0.5)(0.45) = 0.175$$

(d)   By Eq. (3.53), the correlation coefficient of X and Y is

$$\rho_{XY} = \frac{\text{Cov}(X, Y)}{\sigma_X \sigma_Y} = \frac{0.175}{\sqrt{(0.25)(0.2475)}} = 0.704$$

3.38. Suppose that a bivariate r.v. (X, Y) is uniformly distributed over a unit circle (Prob. 3.21).

(a)   Are X and Y independent?

(b)   Are X and Y correlated?

(a)   Setting $R = 1$ in the results of Prob. 3.21, we obtain

$$f_{XY}(x, y) = \begin{cases} \dfrac{1}{\pi} & x^2 + y^2 < 1 \\[2mm] 0 & x^2 + y^2 > 1 \end{cases}$$

$$f_X(x) = \frac{2}{\pi}\sqrt{1 - x^2} \qquad |x| < 1$$

$$f_Y(y) = \frac{2}{\pi}\sqrt{1 - y^2} \qquad |y| < 1$$

Since $f_{XY}(x, y) \neq f_X(x)\,f_Y(y)$, X and Y are not independent.

(b)   By Eqs. (3.47a) and (3.47b), the means of X and Y are

$$E(X) = \frac{2}{\pi}\int_{-1}^{1} x\sqrt{1 - x^2}\; dx = 0$$

$$E(Y) = \frac{2}{\pi}\int_{-1}^{1} y\sqrt{1 - y^2}\; dy = 0$$

since each integrand is an odd function.

Next, by Eq. (3.43),

$$E(XY) = \frac{1}{\pi}\iint\limits_{x^2 + y^2 < 1} xy\, dx\, dy = 0$$

The integral vanishes because the contributions of the second and the fourth quadrants cancel those of the first and the third. Hence, $E(XY) = E(X)E(Y) = 0$ and X and Y are uncorrelated.

Conditional Means and Conditional Variances

3.39. Consider the bivariate r.v. (X, Y) of Prob. 3.14 (or Prob. 3.26). Compute the conditional mean and the conditional variance of Y given $x_i = 2$.

From Prob. 3.26, the conditional pmf $p_{Y|X}(y_j|x_i)$ is

$$p_{Y|X}(y_j|x_i) = \frac{2x_i + y_j}{4x_i + 3} \qquad y_j = 1, 2; x_i = 1, 2$$

Thus,
$$p_{Y|X}(y_j|2) = \frac{4 + y_j}{11} \qquad y_j = 1, 2$$

and by Eqs. (3.55) and (3.56), the conditional mean and the conditional variance of Y given $x_i = 2$ are

$$\mu_{Y|2} = E(Y|x_i = 2) = \sum_{y_j} y_j p_{Y|X}(y_j|2) = \sum_{y_j} y_j \left(\frac{4 + y_j}{11}\right)$$

$$= (1)\left(\frac{5}{11}\right) + (2)\left(\frac{6}{11}\right) = \frac{17}{11} \approx 1.545$$

$$\sigma_{Y|2}^2 = E\left[\left(Y - \frac{17}{11}\right)^2 \bigg| x_i = 2\right] = \sum_{y_j}\left(y_j - \frac{17}{11}\right)^2\left(\frac{4 + y_j}{11}\right)$$

$$= \left(\frac{-6}{11}\right)^2\left(\frac{5}{11}\right) + \left(\frac{5}{11}\right)^2\left(\frac{6}{11}\right) = \frac{330}{1331} \approx 0.248$$

3.40. Let (X, Y) be the bivariate r.v. of Prob. 3.20 (or Prob. 3.30). Compute the conditional means $E(Y|x)$ and $E(X|y)$.

From Prob. 3.30,

$$f_{Y|X}(y|x) = \frac{1}{x} \qquad y \le x < 1, 0 < x < 1$$

$$f_{X|Y}(x|y) = \frac{1}{1 - y} \qquad y \le x < 1, 0 < x < 1$$

By Eq. (3.58), the conditional mean of Y, given $X = x$, is

$$E(Y|x) = \int_{-\infty}^{\infty} y f_{Y|X}(y|x)\, dy = \int_0^x y\left(\frac{1}{x}\right) dy = \frac{y^2}{2x}\bigg|_{y=0}^{y=x} = \frac{x}{2} \qquad 0 < x < 1$$

Similarly, the conditional mean of X, given $Y = y$, is

$$E(X|y) = \int_{-\infty}^{\infty} x f_{X|Y}(x|y)\, dx = \int_y^1 x\left(\frac{1}{1 - y}\right) dx = \frac{x^2}{2(1 - y)}\bigg|_{x=y}^{x=1} = \frac{1 + y}{2} \qquad 0 < y < 1$$

Note that $E(Y|x)$ is a function of x only and $E(X|y)$ is a function of y only.

3.41. Let (X, Y) be the bivariate r.v. of Prob. 3.20 (or Prob. 3.30). Compute the conditional variances $\mathrm{Var}(Y|x)$ and $\mathrm{Var}(X|y)$.

Using the results of Prob. 3.40 and Eq. (3.59), the conditional variance of Y, given $X = x$, is

$$\mathrm{Var}(Y|x) = E\{[Y - E(Y|x)]^2 x\} = \int_{-\infty}^{\infty}\left(y - \frac{x}{2}\right)^2 f_{Y|X}(y|x)\, dy$$

$$= \int_0^x \left(y - \frac{x}{2}\right)^2 \left(\frac{1}{x}\right) dy = \frac{1}{3x}\left(y - \frac{x}{2}\right)^3\bigg|_{y=0}^{y=x} = \frac{x^2}{12}$$

Similarly, the conditional variance of X, given $Y = y$, is

$$\text{Var}(X \mid y) = E\{[X - E(X \mid y)]^2 \, y \mid\} = \int_{-\infty}^{\infty} \left(x - \frac{1+y}{2} \right)^2 f_{X \mid Y}(x \mid y) \, dx$$

$$= \int_{y}^{1} \left(x - \frac{1+y}{2} \right)^2 \left(\frac{1}{1-y} \right) dx = \frac{1}{3(1-y)} \left(x - \frac{1+y}{2} \right)^3 \Bigg|_{x=y}^{x=1} = \frac{(1-y)^2}{12}$$

N-Dimensional Random Vectors

3.42. Let (X_1, X_2, X_3, X_4) be a four-dimensional random vector, where X_k ($k = 1, 2, 3, 4$) are independent Poisson r.v.'s with parameter 2.

(a) Find $P(X_1 = 1, X_2 = 3, X_3 = 2, X_4 = 1)$.

(b) Find the probability that exactly one of the X_k's equals zero.

(a) By Eq. (2.48), the pmf of X_k is

$$p_{X_k}(i) = P(X_k = i) = e^{-2} \frac{2^i}{i!} \qquad i = 0, 1, \dots \tag{3.98}$$

Since the X_k's are independent, by Eq. (3.80),

$$P(X_1 = 1, X_2 = 3, X_3 = 2, X_4 = 1) = p_{X_1}(1) p_{X_2}(3) p_{X_3}(2) p_{X_4}(1)$$

$$= \left(\frac{e^{-2} 2}{1!} \right) \left(\frac{e^{-2} 2^3}{3!} \right) \left(\frac{e^{-2} 2^2}{2!} \right) \left(\frac{e^{-2} 2}{1!} \right) = \frac{e^{-8} 2^7}{12} \approx 3.58(10^{-3})$$

(b) First, we find the probability that $X_k = 0$, $k = 1, 2, 3, 4$. From Eq. (3.98),

$$P(X_k = 0) = e^{-2} \qquad k = 1, 2, 3, 4$$

Next, we treat zero as "success." If Y denotes the number of successes, then Y is a binomial r.v. with parameters $(n, p) = (4, e^{-2})$. Thus, the probability that exactly one of the X_k's equals zero is given by [Eq. (2.36)]

$$P(Y = 1) = \binom{4}{1} (e^{-2})(1 - e^{-2})^3 \approx 0.35$$

3.43. Let (X, Y, Z) be a trivariate r.v., where X, Y, and Z are independent uniform r.v.'s over $(0, 1)$. Compute $P(Z \geq XY)$.

Since X, Y, Z are independent and uniformly distributed over $(0, 1)$, we have

$$f_{XYZ}(x, y, z) = f_X(x) f_Y(y) f_Z(z) = 1 \qquad 0 < x < 1, 0 < y < 1, 0 < z < 1$$

Then

$$P(Z \geq XY) = \iiint_{z > xy} f_{XYZ}(x, y, z) \, dx \, dy \, dz = \int_0^1 \int_0^1 \int_{xy}^1 dz \, dy \, dx$$

$$= \int_0^1 \int_0^1 (1 - xy) \, dy \, dx = \int_0^1 \left(1 - \frac{x}{2} \right) dx = \frac{3}{4}$$

3.44. Let (X, Y, Z) be a trivariate r.v. with joint pdf

$$f_{XYZ}(x, y, z) = \begin{cases} ke^{-(ax+by+cz)} & x > 0, y > 0, z > 0 \\ 0 & \text{otherwise} \end{cases}$$

where $a, b, c > 0$ and k are constants.

(a) Determine the value of k.

(b) Find the marginal joint pdf of X and Y.

(c) Find the marginal pdf of X.

(d) Are X, Y, and Z independent?

(a) By Eq. (3.76),

$$\int_{-\infty}^{\infty}\int_{-\infty}^{\infty}\int_{-\infty}^{\infty} f_{XYZ}(x, y, z)\, dx\, dy\, dz = k\int_0^\infty\int_0^\infty\int_0^\infty e^{-(ax+by+cz)}\, dx\, dy\, dz$$

$$= k\int_0^\infty e^{-ax}dx\int_0^\infty e^{-by}dy\int_0^\infty e^{-cz}dz = \frac{k}{abc} = 1$$

Thus, $k = abc$.

(b) By Eq. (3.77), the marginal joint pdf of X and Y is

$$f_{XY}(x, y) = \int_{-\infty}^{\infty} f_{XYZ}(x, y, z)\, dz = abc\int_0^\infty e^{-(ax+by+cz)}\, dz$$

$$= abce^{-(ax+by)}\int_0^\infty e^{-cz}dz = abe^{-(ax+by)} \qquad x > 0, y > 0$$

(c) By Eq. (3.78), the marginal pdf of X is

$$f_X(x) = \int_{-\infty}^{\infty}\int_{-\infty}^{\infty} f_{XYZ}(x, y, z)\, dy\, dz = abc\int_0^\infty\int_0^\infty e^{-(ax+by+cz)}\, dy\, dz$$

$$= abce^{-ax}\int_0^\infty e^{-by}dy\int_0^\infty e^{-cz}dz = ae^{-ax} \qquad x > 0$$

(d) Similarly, we obtain

$$f_Y(y) = \int_{-\infty}^{\infty}\int_{-\infty}^{\infty} f_{XYZ}(x, y, z)\, dx\, dz = be^{-by} \qquad y > 0$$

$$f_Z(z) = \int_{-\infty}^{\infty}\int_{-\infty}^{\infty} f_{XYZ}(x, y, z)\, dx\, dy = ce^{-cz} \qquad z > 0$$

Since $f_{XYZ}(x, y, z) = f_X(x)\, f_Y(y)\, f_Z(z)$, X, Y, and Z are independent.

3.45. Show that

$$f_{XYZ}(x, y, z) = f_{Z|X,Y}(z\,|\,x, y)f_{Y|X}(y\,|\,x)f_X(x) \tag{3.99}$$

By definition (3.79),

$$f_{Z|X,Y}(z|x, y) = \frac{f_{XYZ}(x, y, z)}{f_{XY}(x, y)}$$

Hence

$$f_{XYZ}(x, y, z) = f_{Z|X,Y}(z|x, y)f_{XY}(x, y) \tag{3.100}$$

Now, by Eq. (3.38),

$$f_{XY}(x, y) = f_{Y|X}(y \mid x) f_X(x)$$

Substituting this expression into Eq. (3.100), we obtain

$$f_{XYZ}(x, y, z) = f_{Z|X,Y}(z \mid x, y) f_{Y|X}(y \mid x) f_X(x)$$

Special Distributions

3.46. Derive Eq. (3.87).

Consider a sequence of n independent multinomial trials. Let A_i ($i = 1, 2, \ldots, k$) be the outcome of a single trial. The r.v. X_i is equal to the number of times A_i occurs in the n trials. If $x_1, x_2, \ldots, x_k$ are nonnegative integers such that their sum equals n, then for such a sequence the probability that A_i occurs x_i times, $i = 1, 2, \ldots, k$—that is, $P(X_1 = x_1, X_2 = x_2, \ldots, X_k = x_k)$—can be obtained by counting the number of sequences containing exactly x_1 A_1's, x_2 A_2's, $\ldots, x_k$ A_k's and multiplying by $p_1^{x_1} p_2^{x_2} \cdots p_k^{x_k}$. The total number of such sequences is given by the number of ways we could lay out in a row n things, of which x_1 are of one kind, x_2 are of a second kind, $\ldots, x_k$ are of a kth kind. The number of ways we could choose x_1 positions for the A_1's is $\binom{n}{x_1}$; after having put the A_1's in their position, the number of ways we could choose positions for the A_2's is $\binom{n - x_1}{x_2}$, and so on. Thus, the total number of sequences with x_1 A_1's, x_2 A_2's, $\ldots, x_k$ A_k's is given by

$$\binom{n}{x_1}\binom{n - x_1}{x_2}\binom{n - x_1 - x_2}{x_3}\cdots\binom{n - x_1 - x_2 - \cdots - x_{k-1}}{x_k}$$

$$= \frac{n!}{x_1!(n - x_1)!}\frac{(n - x_1)!}{x_2!(n - x_1 - x_2)!}\cdots\frac{(n - x_1 - x_2 - \cdots - x_{k-1})!}{x_k!0!}$$

$$= \frac{n!}{x_1! x_2! \cdots x_k!}$$

Thus, we obtain

$$p_{X_1 X_2 \cdots X_k}(x_1, x_2, \ldots, x_k) = \frac{n!}{x_1! x_2! \cdots x_k!} p_1^{x_1} p_2^{x_2} \cdots p_k^{x_k}$$

3.47. Suppose that a fair die is rolled seven times. Find the probability that 1 and 2 dots appear twice each; 3, 4, and 5 dots once each; and 6 dots not at all.

Let $(X_1, X_2, \ldots, X_6)$ be a six-dimensional random vector, where X_i denotes the number of times i dots appear in seven rolls of a fair die. Then $(X_1, X_2, \ldots, X_6)$ is a multinomial r.v. with parameters $(7, p_1, p_2, \ldots, p_6)$ where $p_i = \frac{1}{6}$ ($i = 1, 2, \ldots 6$). Hence, by Eq. (3.87),

$$P(X_1 = 2, X_2 = 2, X_3 = 1, X_4 = 1, X_5 = 1, X_6 = 0) = \frac{7!}{2!2!1!1!1!0!}\left(\frac{1}{6}\right)^2\left(\frac{1}{6}\right)^2\left(\frac{1}{6}\right)^1\left(\frac{1}{6}\right)^1\left(\frac{1}{6}\right)^1\left(\frac{1}{6}\right)^0$$

$$= \frac{7!}{2!2!}\left(\frac{1}{6}\right)^7 = \frac{35}{6^5} \approx 0.0045$$

3.48. Show that the pmf of a multinomial r.v. given by Eq. (3.87) satisfies the condition (3.68); that is,

$$\sum \sum \cdots \sum p_{X_1 X_2 \cdots X_k}(x_1, x_2, \ldots, x_k) = 1 \tag{3.101}$$

where the summation is over the set of all nonnegative integers $x_1, x_2, \ldots, x_k$ whose sum is n.

The *multinomial theorem* (which is an extension of the binomial theorem) states that

$$(a_1 + a_2 + \cdots + a_k)^n = \sum \binom{n}{x_1 \, x_2 \cdots x_k} a_1{}^{x_1} a_2{}^{x_2} \cdots a_k{}^{x_k} \tag{3.102}$$

where $x_1 + x_2 + \cdots + x_k = n$ and

$$\binom{n}{x_1 \, x_2 \cdots x_k} = \frac{n!}{x_1! \, x_2! \cdots x_k!}$$

is called the *multinomial coefficient*, and the summation is over the set of all nonnegative integers $x_1, x_2, \ldots, x_k$ whose sum is n.

Thus, setting $a_i = p_i$ in Eq. (3.102), we obtain

$$\Sigma \, \Sigma \cdots \Sigma \, p_{X_1 X_2 \cdots X_k}(x_1, x_2, \ldots, x_k) = (p_1 + p_2 + \cdots + p_k)^n = (1)^n = 1$$

3.49. Let (X, Y) be a bivariate normal r.v. with its pdf given by Eq. (3.88).

(a) Find the marginal pdf's of X and Y.

(b) Show that X and Y are independent when $\rho = 0$.

(a) By Eq. (3.30), the marginal pdf of X is

$$f_X(x) = \int_{-\infty}^{\infty} f_{XY}(x, y) \, dy$$

From Eqs. (3.88) and (3.89), we have

$$f_{XY}(x, y) = \frac{1}{2\pi \sigma_X \sigma_Y (1 - \rho^2)^{1/2}} \exp\left[-\frac{1}{2} q(x, y)\right]$$

$$q(x, y) = \frac{1}{1 - \rho^2}\left[\left(\frac{x - \mu_X}{\sigma_X}\right)^2 - 2\rho\left(\frac{x - \mu_X}{\sigma_X}\right)\left(\frac{y - \mu_Y}{\sigma_Y}\right) + \left(\frac{y - \mu_Y}{\sigma_Y}\right)^2\right]$$

Rewriting $q(x, y)$,

$$q(x, y) = \frac{1}{1 - \rho^2}\left[\left(\frac{y - \mu_Y}{\sigma_Y}\right) - \rho\left(\frac{x - \mu_X}{\sigma_X}\right)\right]^2 + \left(\frac{x - \mu_X}{\sigma_X}\right)^2$$

$$= \frac{1}{(1 - \rho^2)\sigma_Y{}^2}\left[y - \mu_Y - \rho\frac{\sigma_Y}{\sigma_X}(x - \mu_X)\right]^2 + \left(\frac{x - \mu_X}{\sigma_X}\right)^2$$

Then

$$f_X(x) = \frac{\exp\left[-\dfrac{1}{2}\left(\dfrac{x - \mu_X}{\sigma_X}\right)^2\right]}{\sqrt{2\pi}\,\sigma_X} \int_{-\infty}^{\infty} \frac{1}{\sqrt{2\pi}\,\sigma_Y(1 - \rho^2)^{1/2}} \exp\left[-\frac{1}{2} q_1(x, y)\right] dy$$

where

$$q_1(x, y) = \frac{1}{(1 - \rho^2)\sigma_Y{}^2}\left[y - \mu_Y - \rho\frac{\sigma_Y}{\sigma_X}(x - \mu_X)\right]^2$$

Comparing the integrand with Eq. (2.71), we see that the integrand is a normal pdf with mean $\mu_Y + \rho(\sigma_Y/\sigma_X)(x - \mu_X)$ and variance $(1 - \rho^2)\sigma_Y{}^2$. Thus, the integral must be unity and we obtain

$$f_X(x) = \frac{1}{\sqrt{2\pi}\,\sigma_X} \exp\left[\frac{-(x - \mu_X)^2}{2\sigma_X^2}\right] \tag{3.103}$$

In a similar manner, the marginal pdf of Y is

$$f_Y(y) = \frac{1}{\sqrt{2\pi}\sigma_Y} \exp\left[\frac{-(y - \mu_Y)^2}{2\sigma_Y{}^2}\right] \tag{3.104}$$

(b)　When $\rho = 0$, Eq. (3.88) reduces to

$$f_{XY}(x, y) = \frac{1}{2\pi\sigma_X\sigma_Y} \exp\left\{-\frac{1}{2}\left[\left(\frac{x - \mu_X}{\sigma_X}\right)^2 + \left(\frac{y - \mu_Y}{\sigma_Y}\right)^2\right]\right\}$$

$$= \frac{1}{\sqrt{2\pi}\sigma_X} \exp\left[-\frac{1}{2}\left(\frac{x - \mu_X}{\sigma_X}\right)^2\right] \frac{1}{\sqrt{2\pi}\sigma_Y} \exp\left[-\frac{1}{2}\left(\frac{y - \mu_Y}{\sigma_Y}\right)^2\right]$$

$$= f_X(x)f_Y(y)$$

Hence, X and Y are independent.

3.50. Show that ρ in Eq. (3.88) is the correlation coefficient of X and Y.

By Eqs. (3.50) and (3.53), the correlation coefficient of X and Y is

$$\rho_{XY} = E\left[\left(\frac{X - \mu_X}{\sigma_X}\right)\left(\frac{Y - \mu_Y}{\sigma_Y}\right)\right]$$

$$= \int_{-\infty}^{\infty}\int_{-\infty}^{\infty}\left[\left(\frac{x - \mu_X}{\sigma_X}\right)\left(\frac{y - \mu_Y}{\sigma_Y}\right)\right] f_{XY}(x, y)\, dx\, dy \tag{3.105}$$

where $f_{XY}(x, y)$ is given by Eq. (3.88). By making a change in variables $v = (x - \mu_X)/\sigma_X$ and $w = (y - \mu_Y)/\sigma_Y$, we can write Eq. (3.105) as

$$\rho_{XY} = \int_{-\infty}^{\infty}\int_{-\infty}^{\infty} vw\, \frac{1}{2\pi(1 - \rho^2)^{1/2}} \exp\left[-\frac{1}{2(1 - \rho^2)}(v^2 - 2\rho vw + w^2)\right] dv\, dw$$

$$= \int_{-\infty}^{\infty} \frac{w}{\sqrt{2\pi}}\left\{\int_{-\infty}^{\infty} \frac{v}{\sqrt{2\pi}(1 - \rho^2)^{1/2}} \exp\left[-\frac{(v - \rho w)^2}{2(1 - \rho^2)}\right] dv\right\} e^{-w^2/2} dw$$

The term in the curly braces is identified as the mean of $V = N(\rho w\,;\, 1 - \sigma^2)$, and so

$$\rho_{XY} = \int_{-\infty}^{\infty} \frac{w}{\sqrt{2\pi}}(\rho w)\, e^{-w^2/2} dw = \rho\int_{-\infty}^{\infty} w^2 \frac{1}{\sqrt{2\pi}}\, e^{-w^2/2} dw$$

The last integral is the variance of $W = N(0;\, 1)$, and so it is equal to 1 and we obtain $\rho_{XY} = \rho$.

3.51. Let (X, Y) be a bivariate normal r.v. with its pdf given by Eq. (3.88). Determine $E(Y \mid x)$.

By Eq. (3.58),

$$E(Y \mid x) = \int_{-\infty}^{\infty} y f_{Y \mid X}(y \mid x)\, dy \tag{3.106}$$

where

$$f_{Y \mid X}(y \mid x) = \frac{f_{XY}(x, y)}{f_X(x)} \tag{3.107}$$

Substituting Eqs. (3.88) and (3.103) into Eq. (3.107), and after some cancellation and rearranging, we obtain

$$f_{Y|X}(y|x) = \frac{1}{\sqrt{2\pi}\sigma_Y(1-\rho^2)^{1/2}} \exp\left\{-\frac{1}{2\sigma_Y^2(1-\rho^2)}\left[y - \rho\frac{\sigma_Y}{\sigma_X}(x-\mu_X) - \mu_Y\right]^2\right\}$$

which is equal to the pdf of a normal r.v. with mean $\mu_Y + \rho(\sigma_Y/\sigma_X)(x-\mu_x)$ and variance $(1-\rho^2)\sigma_Y^2$. Thus, we get

$$E(Y|x) = \mu_Y + \rho\frac{\sigma_Y}{\sigma_X}(x-\mu_X) \tag{3.108}$$

Note that when X and Y are independent, then $\rho = 0$ and $E(Y|x) = \mu_Y = E(Y)$.

3.52. The joint pdf of a bivariate r.v. (X, Y) is given by

$$f_{XY}(x, y) = \frac{1}{2\sqrt{3}\pi}\exp\left[-\frac{1}{3}(x^2 - xy + y^2 + x - 2y + 1)\right] \qquad -\infty < x, y < \infty$$

(a) Find the means of X and Y.

(b) Find the variances of X and Y.

(c) Find the correlation coefficient of X and Y.

We note that the term in the bracket of the exponential is a quadratic function of x and y, and hence $f_{XY}(x, y)$ could be a pdf of a bivariate normal r.v. If so, then it is simpler to solve equations for the various parameters. Now, the given joint pdf of (X, Y) can be expressed as

$$f_{XY}(x, y) = \frac{1}{2\sqrt{3}\pi}\exp\left[-\frac{1}{2}q(x, y)\right]$$

where

$$q(x, y) = \frac{2}{3}(x^2 - xy + y^2 + x - 2y + 1)$$

$$= \frac{2}{3}[x^2 - x(y-1) + (y-1)^2]$$

Comparing the above expressions with Eqs. (3.88) and (3.89), we see that $f_{XY}(x, y)$ is the pdf of a bivariate normal r.v. with $\mu_X = 0$, $\mu_Y = 1$, and the following equations:

$$2\pi\sigma_X\sigma_Y\sqrt{1-\rho^2} = 2\sqrt{3}\,\pi$$

$$(1-\rho^2)\sigma_X^2 = (1-\rho^2)\sigma_Y^2 = \frac{3}{2}$$

$$\frac{2\rho}{\sigma_X\sigma_Y(1-\rho^2)} = \frac{2}{3}$$

Solving for σ_X^2, σ_Y^2, and ρ, we get

$$\sigma_X^2 = \sigma_Y^2 = 2 \qquad \text{and} \qquad \rho = \frac{1}{2}$$

Hence,

(a) The mean of X is zero, and the mean of Y is 1.

(b) The variance of both X and Y is 2.

(c) The correlation coefficient of X and Y is $\frac{1}{2}$.

3.53. Consider a bivariate r.v. (X, Y), where X and Y denote the horizontal and vertical miss distances, respectively, from a target when a bullet is fired. Assume that X and Y are independent and that the probability of the bullet landing on any point of the xy plane depends only on the distance of the point from the target. Show that (X, Y) is a bivariate normal r.v.

From the assumption, we have

$$f_{XY}(x, y) = f_X(x)\, f_Y(y) = g(x^2 + y^2) \tag{3.109}$$

for some function g. Differentiating Eq. (3.109) with respect to x, we have

$$f_X'(x) f_Y(y) = 2x g'(x^2 + y^2) \tag{3.110}$$

Dividing Eq. (3.110) by Eq. (3.109) and rearranging, we get

$$\frac{f_X'(x)}{2x f_X(x)} = \frac{g'(x^2 + y^2)}{g(x^2 + y^2)} \tag{3.111}$$

Note that the left-hand side of Eq. (3.111) depends only on x, whereas the right-hand side depends only on $x^2 + y^2$; thus,

$$\frac{f_X'(x)}{x f_X(x)} = c \tag{3.112}$$

where c is a constant. Rewriting Eq.(3.112) as

$$\frac{f_X'(x)}{f_X(x)} = cx \qquad \text{or} \qquad \frac{d}{dx}[\ln f_X(x)] = cx \tag{3.113}$$

and integrating both sides, we get

$$\ln f_X(x) = \frac{c}{2} x^2 + a \qquad \text{or} \qquad f_X(x) = k e^{cx^2/2}$$

where a and k are constants. By the properties of a pdf, the constant c must be negative, and setting $c = -1/\sigma^2$, we have

$$f_X(x) = k e^{-x^2/(2\sigma^2)}$$

Thus, by Eq. (2.71), $X = N(0; \sigma^2)$ and

$$f_X(x) = \frac{1}{\sqrt{2\pi}\,\sigma} e^{-x^2/2(2\sigma^2)}$$

In a similar way, we can obtain the pdf of Y as

$$f_Y(y) = \frac{1}{\sqrt{2\pi}\,\sigma} e^{-y^2/(2\sigma^2)}$$

Since X and Y are independent, the joint pdf of (X, Y) is

$$f_{XY}(x, y) = f_X(x) f_Y(y) = \frac{1}{2\pi\sigma^2} e^{-(x^2+y^2)/(2\sigma^2)}$$

which indicates that (X, Y) is a bivariate normal r.v.

3.54. Let $(X_1, X_2, \ldots, X_n)$ be an n-variate normal r.v. with its joint pdf given by Eq. (3.92). Show that if the covariance of X_i and X_j is zero for $i \neq j$, that is,

$$\text{Cov}(X_i, X_j) = \sigma_{ij} = \begin{cases} \sigma_i^2 & i = j \\ 0 & i \neq j \end{cases} \tag{3.114}$$

then $X_1, X_2, \ldots, X_n$ are independent.

From Eq. (3.94) with Eq. (3.114), the covariance matrix K becomes

$$K = \begin{bmatrix} \sigma_i^2 & 0 & \cdots & 0 \\ 0 & \sigma_2^2 & \cdots & 0 \\ \vdots & \vdots & \ddots & \vdots \\ 0 & 0 & \cdots & \sigma_n^2 \end{bmatrix} \tag{3.115}$$

It therefore follows that

$$\left| \det K \right|^{1/2} = \sigma_1 \sigma_2 \cdots \sigma_n = \prod_{i=1}^{n} \sigma_i \tag{3.116}$$

and

$$K^{-1} = \begin{bmatrix} \dfrac{1}{\sigma_i^2} & 0 & \cdots & 0 \\ 0 & \dfrac{1}{\sigma_2^2} & \cdots & 0 \\ \vdots & \vdots & \ddots & \vdots \\ 0 & 0 & \cdots & \dfrac{1}{\sigma_n^2} \end{bmatrix} \tag{3.117}$$

Then we can write

$$(\mathbf{x} - \boldsymbol{\mu})^T K^{-1}(\mathbf{x} - \boldsymbol{\mu}) = \sum_{i=1}^{n} \left(\frac{x_i - \mu_i}{\sigma_i} \right)^2 \tag{3.118}$$

Substituting Eqs. (3.116) and (3.118) into Eq. (3.92), we obtain

$$f_{X_1 \cdots X_n}(x_1, \ldots, x_n) = \frac{1}{(2\pi)^{n/2} \left(\displaystyle\prod_{i=1}^{n} \sigma_i \right)} \exp\left[-\frac{1}{2} \sum_{i=1}^{n} \left(\frac{x_i - \mu_i}{\sigma_i} \right)^2 \right] \tag{3.119}$$

Now Eq. (3.119) can be rewritten as

$$f_{X_1 \cdots X_n}(x_1, \ldots, x_n) = \prod_{i=1}^{n} f_{X_i}(x_i) \tag{3.120}$$

where

$$f_{X_i}(x_i) = \frac{1}{\sqrt{2\pi} \, \sigma_i} e^{-(x_i - \mu_i)^2 / (2\sigma_i^2)}$$

Thus we conclude that $X_1, X_2, \ldots, X_n$ are independent.

SUPPLEMENTARY PROBLEMS

3.55. Consider an experiment of tossing a fair coin three times. Let (X, Y) be a bivariate r.v., where X denotes the number of heads on the first two tosses and Y denotes the number of heads on the third toss.

(a) Find the range of X.

(b) Find the range of Y.

(c) Find the range of (X, Y).

(d) Find (i) $P(X \le 2, Y \le 1)$; (ii) $P(X \le 1, Y \le 1)$; and (iii) $P(X \le 0, Y \le 0)$.

3.56. Let $F_{XY}(x, y)$ be a joint cdf of a bivariate r.v. (X, Y). Show that

$$P(X > a, Y > c) = 1 - F_X(a) - F_Y(c) + F_{XY}(a, c)$$

where $F_X(x)$ and $F_Y(y)$ are marginal cdf's of X and Y, respectively.

3.57. Let the joint pmf of (X, Y) be given by

$$p_{XY}(x_i, y_j) = \begin{cases} k(x_i + y_j) & x_i = 1, 2, 3; \, y_j = 1, 2 \\ 0 & \text{otherwise} \end{cases}$$

where k is a constant.

(a) Find the value of k.

(b) Find the marginal pmf's of X and Y.

3.58. The joint pdf of (X, Y) is given by

$$f_{XY}(x, y) = \begin{cases} ke^{-(x+2y)} & x > 0, y > 0 \\ 0 & \text{otherwise} \end{cases}$$

where k is a constant.

(a) Find the value of k.

(b) Find $P(X > 1, Y < 1), P(X < Y)$, and $P(X \le 2)$.

3.59. Let (X, Y) be a bivariate r.v., where X is a uniform r.v. over $(0, 0.2)$ and Y is an exponential r.v. with parameter 5, and X and Y are independent.

(a) Find the joint pdf of (X, Y).

(b) Find $P(Y \le X)$.

3.60. Let the joint pdf of (X, Y) be given by

$$f_{XY}(x, y) = \begin{cases} xe^{-x(y+1)} & x > 0, y > 0 \\ 0 & \text{otherwise} \end{cases}$$

(a) Show that $f_{XY}(x, y)$ satisfies Eq.(3.26).

(b) Find the marginal pdf's of X and Y.

3.61. The joint pdf of (X, Y) is given by

$$f_{XY}(x, y) = \begin{cases} kx^2(4 - y) & x < y < 2x, 0 < x < 2 \\ 0 & \text{otherwise} \end{cases}$$

where k is a constant.

(a) Find the value of k.

(b) Find the marginal pdf's of X and Y.

3.62. The joint pdf of (X, Y) is given by

$$f_{XY}(x, y) = \begin{cases} xye^{-(x^2+y^2)/2} & x > 0, y > 0 \\ 0 & \text{otherwise} \end{cases}$$

(a) Find the marginal pdf's of X and Y.

(b) Are X and Y independent?

3.63. The joint pdf of (X, Y) is given by

$$f_{XY}(x, y) = \begin{cases} e^{-(x+y)} & x > 0, y > 0 \\ 0 & \text{otherwise} \end{cases}$$

(a) Are X and Y independent ?

(b) Find the conditional pdf's of X and Y.

3.64. The joint pdf of (X, Y) is given by

$$f_{XY}(x, y) = \begin{cases} e^{-y} & 0 < x \le y \\ 0 & \text{otherwise} \end{cases}$$

(a) Find the conditional pdf's of Y, given that $X = x$.

(b) Find the conditional cdf's of Y, given that $X = x$.

3.65. Consider the bivariate r.v. (X, Y) of Prob. 3.14.

(a) Find the mean and the variance of X.

(b) Find the mean and the variance of Y.

(c) Find the covariance of X and Y.

(d) Find the correlation coefficient of X and Y.

3.66. Consider a bivariate r.v. (X, Y) with joint pdf

$$f_{XY}(x, y) = \frac{1}{2\pi\sigma^2} e^{-(x^2+y^2)/(2\sigma^2)} \qquad -\infty < x, y < \infty$$

Find $P[(X, Y) \,|\, x^2 + y^2 \le a^2]$.

3.67. Let (X, Y) be a bivariate normal r.v., where X and Y each have zero mean and variance σ^2, and the correlation coefficient of X and Y is ρ. Find the joint pdf of (X, Y).

3.68. The joint pdf of a bivariate r.v. (X, Y) is given by

$$f_{XY}(x, y) = \frac{1}{\sqrt{3}\pi} \exp\left[-\frac{2}{3}(x^2 - xy + y^2)\right]$$

(a) Find the means and variances of X and Y.

(b) Find the correlation coefficient of X and Y.

3.69. Let (X, Y, Z) be a trivariate r.v., where X, Y, and Z are independent and each has a uniform distribution over $(0, 1)$. Compute $P(X \geq Y \geq Z)$.

ANSWERS TO SUPPLEMENTARY PROBLEMS

3.55. (a) $R_X = \{0, 1, 2\}$

(b) $R_Y = \{0, 1\}$

(c) $R_{XY} = \{(0,0), (0,1), (1,0), (1,1), (2,0), (2,1)\}$

(d) (i) $P(X \leq 2, Y \leq 1) = 1$; (ii) $P(X \leq 1, Y \leq 1) = \dfrac{3}{4}$; and (iii) $P(X \leq 0, Y \leq 0) = \dfrac{1}{8}$

3.56. *Hint:* Set $x_1 = a, y_1 = c$, and $x_2 = y_2 = \infty$ in Eq. (3.95) and use Eqs. (3.13) and (3.14).

3.57. (a) $k = \dfrac{1}{21}$

(b) $p_X(x_i) = \dfrac{1}{21}(2x_i + 3) \qquad x_i = 1, 2, 3$

$P_Y(y_j) = \dfrac{1}{21}(6 + 3y_j) \qquad y_j = 1, 2$

3.58. (a) $k = 2$

(b) $P(X > 1, Y < 1) = e^{-1} - e^{-3} \approx 0.318; \ P(X < Y) = \dfrac{1}{3}; \ P(X \leq 2) = 1 - e^{-2} \approx 0.865$

3.59. (a) $f_{XY}(x, y) = \begin{cases} 25e^{-(5y)} & 0 < x < 0.2, y > 0 \\ 0 & \text{otherwise} \end{cases}$

(b) $P(Y \leq X) = e^{-1} \approx 0.368$

3.60. (b) $f_X(x) = e^{-x} \qquad\qquad x > 0$

$f_Y(y) = \dfrac{1}{(y+1)^2} \qquad y > 0$

3.61. (a) $k = \dfrac{5}{32}$

(b) $f_X(x) = \dfrac{5}{32} x^3 \left(4 - \dfrac{3}{2}x\right) \qquad\qquad 0 < x < 2$

$f_Y(y) = \begin{cases} \left(\dfrac{5}{32}\right)\dfrac{7}{24}(4 - y)y^3 & 0 < y < 2 \\ \left(\dfrac{5}{32}\right)\dfrac{1}{3}(4 - y)\left(8 - \dfrac{1}{8}y^3\right) & 2 < y < 4 \\ 0 & \text{otherwise} \end{cases}$

3.62. (a) $f_X(x) = xe^{-x^2/2} \qquad x > 0$

$f_Y(y) = ye^{-y^2/2} \qquad y > 0$

(b) Yes

3.63. (a) Yes

(b) $f_{X|Y}(x|y) = e^{-x}$ $x > 0$

$f_{Y|X}(y|x) = e^{-y}$ $y > 0$

3.64. (a) $f_{Y|X}(y|x) = e^{x-y}$ $y \geq x$

(b) $F_{Y|X}(y|x) = \begin{cases} 0 & y \leq x \\ 1 - e^{x-y} & y \geq x \end{cases}$

3.65. (a) $E(X) = \dfrac{29}{18}, \operatorname{Var}(X) = \dfrac{77}{324}$

(b) $E(Y) = \dfrac{14}{9}, \operatorname{Var}(Y) = \dfrac{20}{81}$

(c) $\operatorname{Cov}(X, Y) = -\dfrac{1}{162}$

(d) $\rho = -0.025$

3.66. $1 - e^{-a^2/(2\sigma^2)}$

3.67. $f_{XY}(x, y) = \dfrac{1}{2\pi\sigma^2(1-\rho^2)^{1/2}} \exp\left[-\dfrac{1}{2}\dfrac{x^2 - 2\rho xy + y^2}{\sigma^2(1-p^2)}\right]$

3.68. (a) $\mu_X = \mu_Y = 0$ $\sigma_X^2 = \sigma_Y^2 = 1$

(b) $\rho = \dfrac{1}{2}$

3.69. $\dfrac{1}{6}$

CHAPTER 4

Functions of Random Variables, Expectation, Limit Theorems

4.1 Introduction

In this chapter we study a few basic concepts of functions of random variables and investigate the expected value of a certain function of a random variable. The techniques of moment generating functions and characteristic functions, which are very useful in some applications, are presented. Finally, the laws of large numbers and the central limit theorem, which is one of the most remarkable results in probability theory, are discussed.

4.2 Functions of One Random Variable

A. Random Variable $g(X)$:

Given a r.v. X and a function $g(x)$, the expression

$$Y = g(X) \tag{4.1}$$

defines a new r.v. Y. With y a given number, we denote D_Y the subset of R_x (range of X) such that $g(x) \le y$. Then

$$(Y \le y) = [g(X) \le y] = (X \in D_Y) \tag{4.2}$$

where $(X \in D_Y)$ is the event consisting of all outcomes ζ such that the point $X(\zeta) \in D_Y$. Hence,

$$F_Y(y) = P(Y \le y) = P[g(X) \le y] = P(X \in D_Y) \tag{4.3}$$

If X is a continuous r.v. with pdf $f_X(x)$, then

$$F_Y(y) = \int_{D_Y} f_X(x)\,dx \tag{4.4}$$

B. Determination of $f_Y(y)$ from $f_X(x)$:

Let X be a continuous r.v. with pdf $f_X(x)$. If the transformation $y = g(x)$ is one-to-one and has the inverse transformation

$$x = g^{-1}(y) = h(y) \tag{4.5}$$

then the pdf of Y is given by (Prob. 4.2)

$$f_Y(y) = f_X(x) \left| \frac{dx}{dy} \right| = f_X[h(y)] \left| \frac{dh(y)}{dy} \right| \tag{4.6}$$

Note that if $g(x)$ is a continuous monotonic increasing or decreasing function, then the transformation $y = g(x)$ is one-to-one. If the transformation $y = g(x)$ is not one-to-one, $f_Y(Y)$ is obtained as follows: Denoting the real roots of $y = g(x)$ by x_k, that is,

$$y = g(x_1) = \cdots = g(x_k) = \cdots \tag{4.7}$$

then

$$f_Y(y) = \sum_k \frac{f_X(x_k)}{|g'(x_k)|} \tag{4.8}$$

where $g'(x)$ is the derivative of $g(x)$.

4.3　Functions of Two Random Variables

A.　One Function of Two Random Variables:

Given two r.v.'s X and Y and a function $g(x, y)$, the expression

$$Z = g(X, Y) \tag{4.9}$$

defines a new r.v. Z. With z a given number, we denote D_z the subset of R_{XY} [range of (X, Y)] such that $g(x, y) \le z$. Then

$$(Z \le z) = [g(X, Y) \le z] = \{(X, Y) \in D_z\} \tag{4.10}$$

where $\{(X, Y) \in D_z\}$ is the event consisting of all outcomes ζ such that point $\{X(\zeta), Y(\zeta)\} \in D_z$. Hence,

$$F_Z(z) = P(Z \le z) = P[g(X, Y) \le z] = P\{(X, Y) \in D_z\} \tag{4.11}$$

If X and Y are continuous r.v.'s with joint pdf $f_{XY}(x, y)$, then

$$F_Z(z) = \iint_{D_z} f_{XY}(x, y)\, dx\, dy \tag{4.12}$$

B.　Two Functions of Two Random Variables:

Given two r.v.'s X and Y and two functions $g(x, y)$ and $h(x, y)$, the expression

$$Z = g(X, Y) \qquad W = h(X, Y) \tag{4.13}$$

defines two new r.v.'s Z and W. With z and w two given numbers, we denote D_{ZW} the subset of R_{XY} [range of (X, Y)] such that $g(x, y) \le z$ and $h(x, y) \le w$. Then

$$(Z \le z, W \le w) = [g(X, Y) \le z, h(X, Y) \le w] = \{(X, Y) \in D_{ZW}\} \tag{4.14}$$

where $\{(X, Y) \in D_{ZW}\}$ is the event consisting of all outcomes ζ such that point $\{X(\zeta), Y(\zeta)\} \in D_{ZW}$. Hence,

$$\begin{aligned} F_{ZW}(z, w) = P(Z \le z, W \le w) &= P[g(X, Y) \le z, h(X, Y) \le w] \\ &= P\{(X, Y) \in D_{ZW}\} \end{aligned} \tag{4.15}$$

In the continuous case, we have

$$F_{ZW}(z, w) = \iint\limits_{D_{ZW}} f_{XY}(x, y)\, dx\, dy \tag{4.16}$$

Determination of $f_{ZW}(z, w)$ from $f_{XY}(x, y)$:
Let X and Y be two continuous r.v.'s with joint pdf $f_{XY}(x, y)$. If the transformation

$$z = g(x, y) \qquad w = h(x, y) \tag{4.17}$$

is one-to-one and has the inverse transformation

$$x = q(z, w) \qquad y = r(z, w) \tag{4.18}$$

then the joint pdf of Z and W is given by

$$f_{ZW}(z, w) = f_{XY}(x, y)\,|J(x, y)|^{-1} \tag{4.19}$$

where $x = q(z, w)$, $y = r(z, w)$, and

$$J(x, y) = \begin{vmatrix} \dfrac{\partial g}{\partial x} & \dfrac{\partial g}{\partial y} \\[2mm] \dfrac{\partial h}{\partial x} & \dfrac{\partial h}{\partial y} \end{vmatrix} = \begin{vmatrix} \dfrac{\partial z}{\partial x} & \dfrac{\partial z}{\partial y} \\[2mm] \dfrac{\partial w}{\partial x} & \dfrac{\partial w}{\partial y} \end{vmatrix} \tag{4.20}$$

which is the Jacobian of the transformation (4.17). If we define

$$\bar{J}(z, w) = \begin{vmatrix} \dfrac{\partial q}{\partial z} & \dfrac{\partial q}{\partial w} \\[2mm] \dfrac{\partial r}{\partial z} & \dfrac{\partial r}{\partial w} \end{vmatrix} = \begin{vmatrix} \dfrac{\partial x}{\partial z} & \dfrac{\partial x}{\partial w} \\[2mm] \dfrac{\partial y}{\partial z} & \dfrac{\partial y}{\partial w} \end{vmatrix} \tag{4.21}$$

then

$$\left| \bar{J}(z, w) \right| = |J(x, y)|^{-1} \tag{4.22}$$

and Eq. (4.19) can be expressed as

$$f_{ZW}(z, w) = f_{XY}[q(z, w), r(z, w)]\,|\bar{J}(z, w)| \tag{4.23}$$

4.4 Functions of n Random Variables

A. One Function of n Random Variables:

Given n r.v.'s $X_1, \ldots, X_n$ and a function $g(x_1, \ldots, x_n)$, the expression

$$Y = g(X_1, \ldots, X_n) \tag{4.24}$$

defines a new r.v. Y. Then

$$(Y \le y) = [g(X_1, \ldots, X_n) \le y] = [(X_1, \ldots, X_n) \in D_Y] \tag{4.25}$$

and

$$F_Y(y) = P[g(X_1, \ldots, X_n) \le y] = P[(X_1, \ldots, X_n) \in D_Y] \tag{4.26}$$

where D_Y is the subset of the range of $(X_1, \ldots, X_n)$ such that $g(x_1, \ldots, x_n) \le y$. If $X_1, \ldots, X_n$ are continuous r.v.'s with joint pdf $f_{X_1, \ldots, X_n}(x_1, \ldots, x_n)$, then

$$F_Y(y) = \int_{D_Y} \cdots \int f_{X_1 \cdots X_n}(x_1, \ldots, x_n)\, dx_1 \cdots dx_n \tag{4.27}$$

B.   *n* Functions of *n* Random Variables:

When the joint pdf of n r.v.'s $X_1, \ldots, X_n$ is given and we want to determine the joint pdf of n r.v.'s $Y_1, \ldots, Y_n$, where

$$
\begin{aligned}
Y_1 &= g_1(X_1, \ldots, X_n) \\
&\;\;\vdots \\
Y_n &= g_n(X_1, \ldots, X_n)
\end{aligned}
\tag{4.28}
$$

the approach is the same as for two r.v.'s. We shall assume that the transformation

$$
\begin{aligned}
y_1 &= g_1(x_1, \ldots, x_n) \\
&\;\;\vdots \\
y_n &= g_n(x_1, \ldots, x_n)
\end{aligned}
\tag{4.29}
$$

is one-to-one and has the inverse transformation

$$
\begin{aligned}
x_1 &= h_1(y_1, \ldots, y_n) \\
&\;\;\vdots \\
x_n &= h_n(y_1, \ldots, y_n)
\end{aligned}
\tag{4.30}
$$

Then the joint pdf of $Y_1, \ldots, Y_n$, is given by

$$f_{Y_1 \ldots Y_n}(y_1, \ldots, y_n) = f_{X_1 \ldots X_n}(x_1, \ldots, x_n) \,|\, J(x_1, \ldots, x_n)|^{-1} \tag{4.31}$$

where

$$J(x_1, \ldots, x_n) = \begin{vmatrix} \dfrac{\partial g_1}{\partial x_1} & \cdots & \dfrac{\partial g_1}{\partial x_n} \\ \vdots & \ddots & \vdots \\ \dfrac{\partial g_n}{\partial x_1} & \cdots & \dfrac{\partial g_n}{\partial x_n} \end{vmatrix} \tag{4.32}$$

which is the Jacobian of the transformation (4.29).

4.5   Expectation

A.   Expectation of a Function of One Random Variable:

The expectation of $Y = g(X)$ is given by

$$E(Y) = E[g(X)] = \begin{cases} \displaystyle\sum_i g(x_i) p_X(x_i) & \text{(discrete case)} \\[2mm] \displaystyle\int_{-\infty}^{\infty} g(x) f_X(x)\, dx & \text{(continuous case)} \end{cases} \tag{4.33}$$

B. Expectation of a Function of More Than One Random Variable:

Let $X_1, \ldots, X_n$ be n r.v.'s, and let $Y = g(X_1, \ldots, X_n)$. Then

$$
E(Y) = E[g(X)] =
\begin{cases}
\displaystyle\sum_{x_i} \cdots \sum_{x_n} g(x_1, \ldots, x_n)\, p_{X_1 \cdots X_n}(x_1, \ldots, x_n) & \text{(discrete case)} \\[2ex]
\displaystyle\int_{-\infty}^{\infty} \cdots \int_{-\infty}^{\infty} g(x_1, \ldots, x_n)\, f_{X_1 \cdots X_n}(x_1, \ldots, x_n)\, dx_1 \cdots dx_n & \text{(continuous case)}
\end{cases}
\tag{4.34}
$$

C. Linearity Property of Expectation:

Note that the expectation operation is linear (Prob. 4.45), and we have

$$
E\!\left(\sum_{i=1}^{n} a_i\, X_i \right) = \sum_{i=1}^{n} a_i\, E(X_i)
\tag{4.35}
$$

where a_i's are constants. If r.v.'s X and Y are independent, then we have (Prob. 4.47)

$$
E[g(X)h(Y)] = E[g(X)]E[h(Y)]
\tag{4.36}
$$

The relation (4.36) can be generalized to a mutually independent set of n r.v.'s $X_1, \ldots, X_n$:

$$
E\!\left[\prod_{i=1}^{n} g_i(X_i) \right] = \prod_{i=1}^{n} E[g_i(X_i)]
\tag{4.37}
$$

D. Conditional Expectation as a Random Variable:

In Sec. 3.8 we defined the conditional expectation of Y given $X = x$, $E(Y \mid x)$ [Eq. (3.58)], which is, in general, a function of x, say $H(x)$. Now $H(X)$ is a function of the r.v. X; that is,

$$
H(X) = E(Y \mid X)
\tag{4.38}
$$

Thus, $E(Y \mid X)$ is a function of the r.v. X. Note that $E(Y \mid X)$ has the following property (Prob. 4.44):

$$
E[E(Y \mid X)] = E(Y)
\tag{4.39}
$$

E. Jensen's Inequality:

A twice differentiable real-valued function $g(x)$ is said to be *convex* if $g''(x) \geq 0$ for all x; similarly, it is said to be *concave* if $g''(x) \leq 0$.

Examples of convex functions include x^2, $|x|$, e^x, $x \log x$ $(x \geq 0)$, and so on. Examples of concave functions include $\log x$ and $\sqrt{x}$ $(x \geq 0)$. If $g(x)$ is convex, then $h(x) = -g(x)$ is concave and vice versa.

Jensen's Inequality:

If $g(x)$ is a convex function, then

$$
E[g(x)] \geq g(E[X])
\tag{4.40}
$$

provided that the expectations exist and are finite.

Equation (4.40) is known as *Jensen's inequality* (for proof see Prob. 4.50).

F.　Cauchy-Schwarz Inequality:

Assume that $E(X^2), E(Y^2) < \infty$, then

$$E\left(|XY|\right) \le \sqrt{E(X^2)\,E(Y^2)} \tag{4.41}$$

Equation (4.41) is known as Cauchy-Schwarz inequality (for proof see Prob. 4.51).

4.6　Probability Generating Functions

A.　Definition:

Let X be a nonnegative integer-valued discrete r.v. with pmf $p_X(x)$. The *probability generating function* (or *z-transform*) of X is defined by

$$G_X(z) = E(z^X) = \sum_{x=0}^{\infty} p_X(x)\,z^x \tag{4.42}$$

where z is a variable.
Note that

$$\left|G_X(z)\right| \le \sum_{x=0}^{\infty} \left|p_X(x)\right|\left|z\right|^x \le \sum_{x=0}^{\infty} p_X(x) = 1 \quad \text{for } |z| < 1 \tag{4.43}$$

B.　Properties of $G_X(z)$:

Differentiating Eq. (4.42) repeatedly, we have

$$G_X'(z) = \sum_{x=1}^{\infty} x\, p_X(x)\, z^{x-1} = p_X(1) + 2\, p_X(2)\, z + 3\, p_X(3)z^2 + \cdots \tag{4.44}$$

$$G_X''(z) = \sum_{x=2}^{\infty} x\,(x-1)\, p_X(x)\, z^{x-2} = 2p_X(2) + 3 \cdot 2\, p_X(3)z + \cdots \tag{4.45}$$

In general,

$$G_X^{(n)}(z) = \sum_{x=n}^{\infty} x\,(x-1)\cdots(x-n+1)\, p_X(x)\, z^{x-n} = \sum_{x=n}^{\infty} \binom{x}{n} n!\, p_X(x)\, z^{x-n} \tag{4.46}$$

Then, we have

(1)　$p_X(0) = P(X = 0) = G_X(0)$ 　　　　　　　　　　　　　　　　　　　　　 (4.47)

(2)　$p_X(n) = P(X = n) = \dfrac{1}{n!} G_X^{(n)}(0)$ 　　　　　　　　　　　　　　　 (4.48)

(3)　$E(X) = G'(1)$ 　　　　　　　　　　　　　　　　　　　　　　　　　　　 (4.49)

(4)　$E[X(X-1)(X-2)\cdots(X-n+1)] = G_X^{(n)}(1)$ 　　　　　　　　　　 (4.50)

One of the useful properties of the probability generating function is that it turns a sum into product.

$$E\left(z^{(X_1 + X_2)}\right) = E\left(z^{X_1} z^{X_2}\right) = E\left(z^{X_1}\right) E\left(z^{X_2}\right) \tag{4.51}$$

Suppose that $X_1, X_2, \ldots, X_n$ are independent nonnegative integer-valued r.v.'s, and let $Y = X_1 + X_2 + \cdots + X_n$. Then

$$(5) \quad G_Y(z) = \prod_{i=1}^{n} G_{X_i}(z) \tag{4.52}$$

Note that property (2) indicates that the probability generating function determines the distribution. Property (4) is known as the nth *factorial moment*.

Setting $n = 2$ in Eq. (4.50) and by Eq. (4.35), we have

$$E[X(X-1)] = E(X^2 - X) = E(X^2) - E(X) = G_X''(1) \tag{4.53}$$

Thus, using Eq. (4.49)

$$E(X^2) = G_X'(1) + G_X''(1) \tag{4.54}$$

Using Eq. (2.31), we obtain

$$\mathrm{Var}(X) = G_X'(1) + G_X''(1) - [G_X'(1)]^2 \tag{4.55}$$

C. Lemma for Probability Generating Function:

Lemma 4.1:   If two nonnegative integer-valued discrete r.v.'s have the same probability generating functions, then they must have the same distribution.

4.7   Moment Generating Functions

A. Definition:

The *moment generating function* of a r.v. X is defined by

$$M_X(t) = E(e^{tX}) = \begin{cases} \displaystyle\sum_i e^{tx_i} p_X(x_i) & \text{(discrete case)} \\ \displaystyle\int_{-\infty}^{\infty} e^{tx} f_X(x)\, dx & \text{(continuous case)} \end{cases} \tag{4.56}$$

where t is a real variable. Note that $M_X(t)$ may not exist for all r.v.'s X. In general, $M_X(t)$ will exist only for those values of t for which the sum or integral of Eq. (4.56) converges absolutely. Suppose that $M_X(t)$ exists. If we express e^{tX} formally and take expectation, then

$$\begin{aligned} M_X(t) = E(e^{tX}) &= E\left[1 + tX + \frac{1}{2!}(tX)^2 + \cdots + \frac{1}{k!}(tX)^k + \cdots\right] \\ &= 1 + tE(X) + \frac{t^2}{2!}E(X^2) + \cdots + \frac{t^k}{k!}E(X^k) + \cdots \end{aligned} \tag{4.57}$$

and the kth moment of X is given by

$$m_k = E(X^k) = M_X^{(k)}(0) \qquad k = 1, 2, \ldots \tag{4.58}$$

where

$$M_X^{(k)}(0) = \frac{d^k}{dt^k} M_X(t)\bigg|_{t=0} \tag{4.59}$$

Note that by substituting z by e^t, we can obtain the moment generating function for a nonnegative integer-valued discrete r.v. from the probability generating function.

B. Joint Moment Generating Function:

The joint moment generating function $M_{XY}(t_1, t_2)$ of two r.v.'s X and Y is defined by

$$M_{XY}(t_1, t_2) = E[e^{(t_1 X + t_2 Y)}] \tag{4.60}$$

where t_1 and t_2 are real variables. Proceeding as we did in Eq. (4.57), we can establish that

$$M_{XY}(t_1, t_2) = E[e^{(t_1 X + t_2 Y)}] = \sum_{k=0}^{\infty} \sum_{n=0}^{\infty} \frac{t_1^k t_2^n}{k! \, n!} E(X^k Y^n) \tag{4.61}$$

and the (k, n) joint moment of X and Y is given by

$$m_{kn} = E(X^k Y^n) = M_{XY}^{(kn)}(0, 0) \tag{4.62}$$

where

$$M_{XY}^{(kn)}(0, 0) = \frac{\partial^{k+n}}{\partial^k t_1 \partial^n t_2} M_{XY}(t_1, t_2) \bigg|_{t_1 = t_2 = 0} \tag{4.63}$$

In a similar fashion, we can define the joint moment generating function of n r.v.'s $X_1, \ldots, X_n$ by

$$M_{X_1 \cdots X_n}(t_1, \ldots, t_n) = [e^{(t_1 X_1 + \cdots + t_n X_n)}] \tag{4.64}$$

from which the various moments can be computed. If $X_1, \ldots, X_n$ are independent, then

$$\begin{aligned} M_{X_1 \cdots X_n}(t_1, \ldots, t_n) &= E\big[e^{(t_1 X_1 + \cdots + t_n X_n)}\big] = E\big(e^{t_1 X_1} \cdots e^{t_n X_n}\big) \\ &= E(e^{t_1 X_1}) \cdots E(e^{t_n X_n}) = M_{X_1}(t_1) \cdots M_{X_n}(t_n) \end{aligned} \tag{4.65}$$

C. Lemmas for Moment Generating Functions:

Two important lemmas concerning moment generating functions are stated in the following:

Lemma 4.2: If two r.v.'s have the same moment generating functions, then they must have the same distribution.

Lemma 4.3: Given cdf's $F(x), F_1(x), F_2(x), \ldots$ with corresponding moment generating functions $M(t), M_1(t), M_2(t), \ldots$, then $F_n(x) \to F(x)$ if $M_n(t) \to M(t)$.

4.8 Characteristic Functions

A. Definition:

The *characteristic function* of a r.v. X is defined by

$$\Psi_X(\omega) = E(e^{j\omega X}) = \begin{cases} \sum_i e^{j\omega x_i} p_X(x_i) & \text{(discrete case)} \\[2ex] \int_{-\infty}^{\infty} e^{j\omega x} f_X(x) \, dx & \text{(continuous case)} \end{cases} \tag{4.66}$$

where ω is a real variable and $j = \sqrt{-1}$. Note that $\Psi_X(\omega)$ is obtained by replacing t in $M_X(t)$ by $j\omega$ if $M_X(t)$ exists. Thus, the characteristic function has all the properties of the moment generating function. Now

$$|\Psi_X(\omega)| = \left| \sum_i e^{j\omega x_i} p_X(x_i) \right| \le \sum_i \left| e^{j\omega x_i} p_X(x_i) \right| = \sum_i p_X(x_i) = 1 < \infty$$

for the discrete case and

$$|\Psi_X(\omega)| = \left| \int_{-\infty}^{\infty} e^{j\omega x} f_X(x)\, dx \right| \le \int_{-\infty}^{\infty} \left| e^{j\omega x} f_X(x)\, dx \right| = \int_{-\infty}^{\infty} f_X(x)\, dx = 1 < \infty$$

for the continuous case. Thus, the characteristic function $\Psi_X(\omega)$ is always defined even if the moment generating function $M_X(t)$ is not (Prob. 4.76). Note that $\Psi_X(\omega)$ of Eq. (4.66) for the continuous case is the Fourier transform (with the sign of j reversed) of $f_X(x)$. Because of this fact, if $\Psi_X(\omega)$ is known, $f_X(x)$ can be found from the inverse Fourier transform; that is,

$$f_X(x) = \frac{1}{2\pi} \int_{-\infty}^{\infty} \Psi_X(\omega) e^{-j\omega x}\, d\omega \tag{4.67}$$

B. Joint Characteristic Functions:

The joint characteristic function $\Psi_{XY}(\omega_1, \omega_2)$ of two r.v.'s X and Y is defined by

$$\Psi_{XY}(\omega_1, \omega_2) = E\left[e^{j(\omega_1 X + \omega_2 Y)} \right]$$
$$= \begin{cases} \sum_i \sum_k e^{j(\omega_1 x_i + \omega_2 y_k)}\, p_{XY}(x_i, y_k) & \text{(discrete case)} \\ \int_{-\infty}^{\infty} \int_{-\infty}^{\infty} e^{j(\omega_1 x + \omega_2 y)}\, f_{XY}(x, y)\, dx\, dy & \text{(continuous case)} \end{cases} \tag{4.68}$$

where ω_1 and ω_2 are real variables.

The expression of Eq. (4.68) for the continuous case is recognized as the two-dimensional Fourier transform (with the sign of j reversed) of $f_{XY}(x, y)$. Thus, from the inverse Fourier transform, we have

$$f_{XY}(x, y) = \frac{1}{(2\pi)^2} \int_{-\infty}^{\infty} \int_{-\infty}^{\infty} \Psi_{XY}(\omega_1, \omega_2)\, e^{-j(\omega_1 x + \omega_2 y)}\, d\omega_1\, d\omega_2 \tag{4.69}$$

From Eqs. (4.66) and (4.68), we see that

$$\Psi_X(\omega) = \Psi_{XY}(\omega, 0) \qquad \Psi_Y(\omega) = \Psi_{XY}(0, \omega) \tag{4.70}$$

which are called *marginal characteristic functions*.

Similarly, we can define the joint characteristic function of n r.v.'s $X_1, \ldots, X_n$ by

$$\Psi_{X_1 \cdots X_n}(\omega_1, \ldots, \omega_n) = E\left[e^{j(\omega_1 X_1 + \cdots + \omega_n X_n)} \right] \tag{4.71}$$

As in the case of the moment generating function, if $X_1, \ldots, X_n$ are independent, then

$$\Psi_{X_1 \cdots X_n}(\omega_1, \ldots, \omega_n) = \Psi_{X_1}(\omega_1) \cdots \Psi_{X_n}(\omega_n) \tag{4.72}$$

C. Lemmas for Characteristic Functions:

As with the moment generating function, we have the following two lemmas:

Lemma 4.4: A distribution function is uniquely determined by its characteristic function.

Lemma 4.5: Given cdf's $F(x), F_1(x), F_2(x), \ldots$ with corresponding characteristic functions $\Psi(\omega), \Psi_1(\omega),$ $\Psi_2(\omega), \ldots,$ then $F_n(x) \to F(x)$ at points of continuity of $F(x)$ if and only if $\Psi_n(\omega) \to \Psi(\omega)$ for every ω.

4.9 The Laws of Large Numbers and the Central Limit Theorem

A. The Weak Law of Large Numbers:

Let $X_1, \ldots, X_n$ be a sequence of independent, identically distributed r.v.'s each with a finite mean $E(X_i) = \mu$. Let

$$\bar{X}_n = \frac{1}{n} \sum_{i=1}^{n} X_i = \frac{1}{n}(X_1 + \cdots + X_n) \tag{4.73}$$

Then, for any $\varepsilon > 0$,

$$\lim_{n \to \infty} P\left(\left|\bar{X}_n - \mu\right| > \varepsilon\right) = 0 \tag{4.74}$$

Equation (4.74) is known as the *weak law of large numbers,* and $\bar{X}_n$ is known as the *sample mean.*

B. The Strong Law of Large Numbers:

Let $X_1, \ldots, X_n$ be a sequence of independent, identically distributed r.v.'s each with a finite mean $E(X_i) = \mu$. Then, for any $\varepsilon > 0$,

$$P\left(\lim_{n \to \infty} \left|\bar{X}_n - \mu\right| > \varepsilon\right) = 0 \tag{4.75}$$

where $\bar{X}_n$ is the sample mean defined by Eq. (4.73). Equation (4.75) is known as the *strong law* of large numbers.

Notice the important difference between Eqs. (4.74) and (4.75). Equation (4.74) tells us how a sequence of probabilities converges, and Eq. (4.75) tells us how the sequence of r.v.'s behaves in the limit. The strong law of large numbers tells us that the sequence $(\bar{X}_n)$ is converging to the constant μ.

C. The Central Limit Theorem:

The *central limit theorem* is one of the most remarkable results in probability theory. There are many versions of this theorem. In its simplest form, the central limit theorem is stated as follows:

Let $X_1, \ldots, X_n$ be a sequence of independent, identically distributed r.v.'s each with mean μ and variance σ^2. Let

$$Z_n = \frac{X_1 + \cdots + X_n - n\mu}{\sigma\sqrt{n}} = \frac{\bar{X}_n - \mu}{\sigma/\sqrt{n}} \tag{4.76}$$

where $\bar{X}_n$ is defined by Eq. (4.73). Then the distribution of Z_n tends to the standard normal as $n \to \infty$; that is,

$$\lim_{n \to \infty} Z_n = N(0;1) \tag{4.77}$$

or

$$\lim_{n \to \infty} F_{Z_n}(z) = \lim_{n \to \infty} P(Z_n \leq z) = \Phi(z) \tag{4.78}$$

where $\Phi(z)$ is the cdf of a standard normal r.v. [Eq. (2.73)]. Thus, the central limit theorem tells us that for large n, the distribution of the sum $S_n = X_1 + \cdots + X_n$ is approximately normal regardless of the form of the distribution of the individual X_i's. Notice how much stronger this theorem is than the laws of large numbers. In practice, whenever an observed r.v. is known to be a sum of a large number of r.v.'s, then the central limit theorem gives us some justification for assuming that this sum is normally distributed.

SOLVED PROBLEMS

Functions of One Random Variable

4.1. If X is $N(\mu; \sigma^2)$, then show that $Z = (X - \mu)/\sigma$ is a standard normal r.v.; that is, $N(0; 1)$.

The cdf of Z is

$$F_Z(z) = P(Z \le z) = P\left(\frac{X - \mu}{\sigma} \le z\right) = P(X \le z\sigma + \mu)$$

$$= \int_{-\infty}^{z\sigma + \mu} \frac{1}{\sqrt{2\pi}\sigma} e^{-(x-\mu)^2/(2\sigma^2)} \, dx$$

By the change of variable $y = (x - \mu)/\sigma$ (that is, $x = \sigma y + \mu$), we obtain

$$F_Z(z) = P(Z \le z) = \int_{-\infty}^{z} \frac{1}{\sqrt{2\pi}} e^{-y^2/2} \, dy$$

and

$$f_Z(z) = \frac{dF_z(z)}{dz} = \frac{1}{\sqrt{2\pi}} e^{-z^2/2}$$

which indicates that $Z = N(0; 1)$.

4.2. Verify Eq. (4.6).

Assume that $y = g(x)$ is a continuous monotonically increasing function [Fig. 4-1(a)]. Since $y = g(x)$ is monotonically increasing, it has an inverse that we denote by $x = g^{-1}(y) = h(y)$. Then

$$F_Y(y) = P(Y \le y) = P[X \le h(y)] = F_X[h(y)] \tag{4.79}$$

and

$$f_Y(y) = \frac{d}{dy} F_Y(y) = \frac{d}{dy} \{F_X[h(y)]\}$$

Applying the chain rule of differentiation to this expression yields

$$f_Y(y) = f_X[h(y)] \frac{d}{dy} h(y)$$

which can be written as

$$f_Y(y) = f_X(x) \frac{dx}{dy} \qquad x = h(y) \tag{4.80}$$

If $y = g(x)$ is monotonically decreasing [Fig. 4-1(b)], then

$$F_Y(y) = P(Y \leq y) = P[X > h(y)] = 1 - F_X[h(y)] \tag{4.81}$$

Thus,

$$f_Y(y) = \frac{d}{dy} F_Y(y) = -f_X(x)\frac{dx}{dy} \qquad x = h(y) \tag{4.82}$$

In Eq. (4.82), since $y = g(x)$ is monotonically decreasing, dy/dx (and dx/dy) is negative. Combining Eqs. (4.80) and (4.82), we obtain

$$f_Y(y) = f_X(x)\left|\frac{dx}{dy}\right| = f_X[h(y)]\left|\frac{dh(y)}{dy}\right|$$

which is valid for any continuous monotonic (increasing or decreasing) function $y = g(x)$.

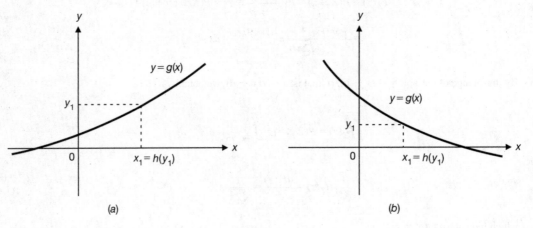

(a)   (b)

Fig. 4-1

4.3. Let X be a r.v. with cdf $F_X(x)$ and pdf $f_X(x)$. Let $Y = aX + b$, where a and b are real constants and $a \neq 0$.

(a) Find the cdf of Y in terms of $F_X(x)$.

(b) Find the pdf of Y in terms of $f_X(x)$.

(a) If $a > 0$, then [Fig. 4-2(a)]

$$F_Y(y) = P(Y \leq y) = P(aX + b \leq y) = P\left(X \leq \frac{y-b}{a}\right) = F_X\left(\frac{y-b}{a}\right) \tag{4.83}$$

If $a < 0$, then [Fig. 4-2(b)]

$$F_Y(y) = P(Y \leq y) = P(aX + b \leq y) = P(aX \leq y - b)$$

$$= P\left(X \geq \frac{y-b}{a}\right) \qquad \text{(since } a < 0, \text{ note the change in the inequality sign)}$$

$$= 1 - P\left(X < \frac{y-b}{a}\right)$$

$$= 1 - P\left(X \leq \frac{y-b}{a}\right) + P\left(X = \frac{y-b}{a}\right)$$

$$= 1 - F_X\left(\frac{y-b}{a}\right) + P\left(X = \frac{y-b}{a}\right) \tag{4.84}$$

Note that if X is continuous, then $P[X = (y - b)/a] = 0$, and

$$F_Y(y) = 1 - F_X\left(\frac{y-b}{a}\right) \qquad a < 0 \tag{4.85}$$

(b)   From Fig. 4-2, we see that $y = g(x) = ax + b$ is a continuous monotonically increasing ($a > 0$) or decreasing ($a < 0$) function. Its inverse is $x = g^{-1}(y) = h(y) = (y - b)/a$, and $dx/dy = 1/a$. Thus, by Eq. (4.6),

$$f_Y(y) = \frac{1}{|a|} f_X\left(\frac{y-b}{a}\right) \tag{4.86}$$

Note that Eq. (4.86) can also be obtained by differentiating Eqs. (4.83) and (4.85) with respect to y.

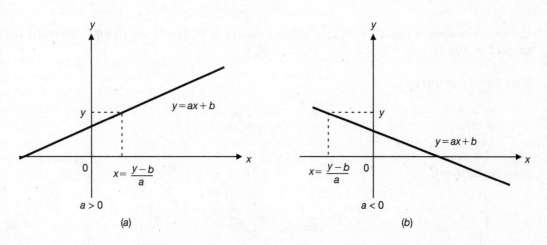

Fig. 4-2

4.4.   Let $Y = aX + b$. Determine the pdf of Y, if X is a uniform r.v. over $(0, 1)$.

The pdf of X is [Eq. (2.56)]

$$f_X(x) = \begin{cases} 1 & 0 < x < 1 \\ 0 & \text{otherwise} \end{cases}$$

Then by Eq. (4.86), we get

$$f_Y(y) = \frac{1}{|a|} f_X\left(\frac{y-b}{a}\right) = \begin{cases} \dfrac{1}{|a|} & y \in R_Y \\ 0 & \text{otherwise} \end{cases} \tag{4.87}$$

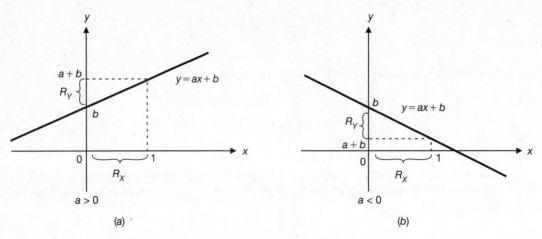

Fig. 4-3

The range R_Y is found as follows: From Fig. 4-3, we see that

For $a > 0$: $R_Y = \{y : b < y < a + b\}$
For $a < 0$: $R_Y = \{y : a + b < y < b\}$

4.5. Let $Y = aX + b$. Show that if $X = N(\mu; \sigma^2)$, then $Y = N(a\mu + b; a^2 \sigma^2)$, and find the values of a and b so that $Y = N(0; 1)$.

Since $X = N(\mu; \sigma^2)$, by Eq. (2.71),

$$f_X(x) = \frac{1}{\sqrt{2\pi}\,\sigma} \exp\left[-\frac{1}{2\sigma^2}(x - \mu)^2\right]$$

Hence, by Eq. (4.86),

$$f_Y(y) = \frac{1}{\sqrt{2\pi}\,|a|\sigma} \exp\left\{-\frac{1}{2\sigma^2}\left[\left(\frac{y - b}{a}\right) - \mu\right]^2\right\}$$

$$= \frac{1}{\sqrt{2\pi}\,|a|\sigma} \exp\left\{-\frac{1}{2a^2\sigma^2}[y - (a\mu + b)]^2\right\}$$

(4.88)

which is the pdf of $N(a\mu + b; a^2\sigma^2)$. Hence, $Y = N(a\mu + b; a^2\sigma^2)$. Next, let $a\mu + b = 0$ and $a^2\sigma^2 = 1$, from which we get $a = 1/\sigma$ and $b = -\mu/\sigma$. Thus, $Y = (X - \mu)/\sigma$ is $N(0; 1)$ (see Prob. 4.1).

4.6. Let X be a r.v. with pdf $f_X(x)$. Let $Y = X^2$. Find the pdf of Y.

The event $A = (Y \le y)$ in R_Y is equivalent to the event $B = (-\sqrt{y} \le X \le \sqrt{y})$ in R_X (Fig. 4-4). If $y \le 0$, then

$$F_Y(y) = P(Y \le y) = 0$$

and $f_Y(y) = 0$. If $y > 0$, then

$$F_Y(y) = P(Y \le y) = P\left(-\sqrt{y} \le X \le \sqrt{y}\right) = F_X\left(\sqrt{y}\right) - F_X\left(-\sqrt{y}\right)$$

(4.89)

and
$$f_Y(y) = \frac{d}{dy}F_Y(y) = \frac{d}{dy}F_X\left(\sqrt{y}\right) - \frac{d}{dy}F_X\left(-\sqrt{y}\right) = \frac{1}{2\sqrt{y}}\left[f_X\left(\sqrt{y}\right) + f_X\left(-\sqrt{y}\right)\right]$$

Thus,
$$f_Y(y) = \begin{cases} \frac{1}{2\sqrt{y}}\left[f_X\left(\sqrt{y}\right) + f_X\left(-\sqrt{y}\right)\right] & y > 0 \\ 0 & y \le 0 \end{cases} \tag{4.90}$$

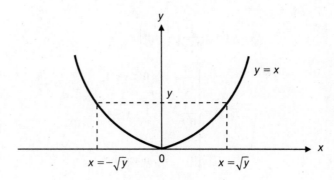

Fig. 4-4

Alternative Solution:

If $y < 0$, then the equation $y = x^2$ has no real solutions; hence, $f_Y(y) = 0$. If $y > 0$, then $y = x^2$ has two solutions, $x_1 = \sqrt{y}$ and $x_2 = -\sqrt{y}$. Now, $y = g(x) = x^2$ and $g'(x) = 2x$. Hence, by Eq. (4.8),

$$f_Y(y) = \begin{cases} \frac{1}{2\sqrt{y}}\left[f_X\left(\sqrt{y}\right) + f_X\left(-\sqrt{y}\right)\right] & y > 0 \\ 0 & y < 0 \end{cases}$$

4.7. Let $Y = X^2$. Find the pdf of Y if $X = N(0; 1)$.

Since $X = N(0; 1)$

$$f_X(x) = \frac{1}{\sqrt{2\pi}}e^{-x^2/2}$$

Since $f_X(x)$ is an even function, by Eq. (4.90), we obtain

$$f_Y(y) = \begin{cases} \frac{1}{\sqrt{y}}f_X\left(\sqrt{y}\right) = \frac{1}{\sqrt{2\pi y}}e^{-y/2} & y > 0 \\ 0 & y < 0 \end{cases} \tag{4.91}$$

4.8. Let $Y = X^2$. Find and sketch the pdf of Y if X is a uniform r.v. over $(-1, 2)$.

The pdf of X is [Eq. (2.56)] [Fig. 4-5(*a*)]

$$f_X(x) = \begin{cases} \frac{1}{3} & -1 < x < 2 \\ 0 & \text{otherwise} \end{cases}$$

In this case, the range of Y is $(0, 4)$, and we must be careful in applying Eq. (4.90). When $0 < y < 1$, both $\sqrt{y}$ and $-\sqrt{y}$ are in $R_X = (-1, 2)$, and by Eq. (4.90),

$$f_Y(y) = \frac{1}{2\sqrt{y}}\left(\frac{1}{3} + \frac{1}{3}\right) = \frac{1}{3\sqrt{y}}$$

When $1 < y < 4$, $\sqrt{y}$ is in $R_X = (-1, 2)$ but $-\sqrt{y} < -1$, and by Eq. (4.90),

$$f_Y(y) = \frac{1}{2\sqrt{y}}\left(\frac{1}{3} + 0\right) = \frac{1}{6\sqrt{y}}$$

Thus,

$$f_Y(y) = \begin{cases} \dfrac{1}{3\sqrt{y}} & 0 < y < 1 \\[2mm] \dfrac{1}{6\sqrt{y}} & 1 < y < 4 \\[2mm] 0 & \text{otherwise} \end{cases}$$

which is sketched in Fig. 4-5(b).

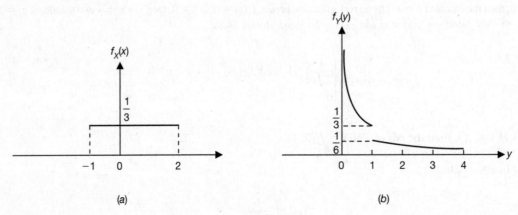

Fig. 4-5

4.9. Let $Y = e^X$. Find the pdf of Y if X is a uniform r.v. over $(0, 1)$.

The pdf of X is

$$f_X(x) = \begin{cases} 1 & 0 < x < 1 \\ 0 & \text{otherwise} \end{cases}$$

The cdf of Y is

$$F_Y(y) = P(Y \le y) = P(e^X \le y) = P(X \le \ln y)$$
$$= \int_{-\infty}^{\ln y} f_X(x)\,dx = \int_0^{\ln y} dx = \ln y \qquad 1 < y < e$$

Thus,
$$f_Y(y) = \frac{d}{dy}F_Y(y) = \frac{d}{dy}\ln y = \frac{1}{y} \qquad 1 < y < e \tag{4.92}$$

Alternative Solution:

The function $y = g(x) = e^x$ is a continuous monotonically increasing function. Its inverse is $x = g^{-1}(y) = h(y) = \ln y$. Thus, by Eq. (4.6), we obtain

$$f_Y(y) = f_X(\ln y)\left|\frac{d}{dy}\ln y\right| = \frac{1}{y}f_X(\ln y) = \begin{cases} \dfrac{1}{y} & 0 < \ln y < 1 \\ 0 & \text{otherwise} \end{cases}$$

or
$$f_Y(y) = \begin{cases} \dfrac{1}{y} & 1 < y < e \\ 0 & \text{otherwise} \end{cases}$$

4.10. Let $Y = e^X$. Find the pdf of Y if $X = N(\mu; \sigma^2)$.

The pdf of X is [Eq. (2.71)]

$$f_X(x) = \frac{1}{\sqrt{2\pi}\,\sigma}\exp\left[-\frac{1}{2\sigma^2}(x-\mu)^2\right] \qquad -\infty < x < \infty$$

Thus, using the technique shown in the alternative solution of Prob. 4.9, we obtain

$$f_Y(y) = \frac{1}{y}f_X(\ln y) = \frac{1}{y\sqrt{2\pi}\sigma}\exp\left[-\frac{1}{2\sigma^2}(\ln y - \mu)^2\right] \qquad 0 < y < \infty \tag{4.93}$$

Note that $X = \ln Y$ is the normal r.v.; hence, the r.v. Y is called the *log-normal* r.v.

4.11. Let X be a r.v. with pdf $f_X(x)$. Let $Y = 1/X$. Find the pdf of Y in terms of $f_X(x)$.

We see that the inverse of $y = 1/x$ is $x = 1/y$ and $dx/dy = -1/y^2$. Thus, by Eq. (4.6)

$$f_Y(y) = \frac{1}{y^2}f_X\left(\frac{1}{y}\right) \tag{4.94}$$

4.12. Let $Y = 1/X$ and X be a Cauchy r.v. with parameter a. Show that Y is also a Cauchy r.v. with parameter $1/a$.

From Prob. 2.83, we have

$$f_X(x) = \frac{a/\pi}{a^2 + x^2} \qquad -\infty < x < \infty$$

By Eq. (4.94)
$$f_Y(y) = \frac{1}{y^2}\frac{a/\pi}{a^2 + (1/y)^2} = \frac{(1/a)\pi}{(1/a)^2 + y^2} \qquad -\infty < y < \infty$$

which indicates that Y is also a Cauchy r.v. with parameter $1/a$.

4.13. Let $Y = \tan X$. Find the pdf of Y if X is a uniform r.v. over $(-\pi/2, \pi/2)$.

The cdf of X is [Eq. (2.57)]

$$F_X(x) = \begin{cases} 0 & x \leq -\pi/2 \\ \dfrac{1}{\pi}(x + \pi/2) & -\pi/2 < x < \pi/2 \\ 1 & x \geq \pi/2 \end{cases}$$

Now
$$F_Y(y) = P(Y \leq y) = P(\tan X \leq y) = P(X \leq \tan^{-1} y)$$
$$= F_X(\tan^{-1} y) = \frac{1}{\pi}\left(\tan^{-1} y + \frac{\pi}{2}\right) = \frac{1}{2} + \frac{1}{\pi}\tan^{-1} y \qquad -\infty < y < \infty$$

Then the pdf of Y is given by

$$f_Y(y) = \frac{d}{dy}F_Y(y) = \frac{1}{\pi(1 + y^2)} \qquad -\infty < y < \infty$$

Note that the r.v. Y is a *Cauchy* r.v. with parameter 1.

4.14. Let X be a continuous r.v. with the cdf $F_X(x)$. Let $Y = F_X(X)$. Show that Y is a uniform r.v. over $(0, 1)$.

Notice from the properties of a cdf that $y = F_X(x)$ is a monotonically nondecreasing function. Since $0 \leq F_X(x) \leq 1$ for all real x, y takes on values only on the interval $(0, 1)$. Using Eq. (4.80) (Prob. 4.2), we have

$$f_Y(y) = f_X(x)\frac{1}{dy/dx} = f_X(x)\frac{1}{dF_X(x)/dx} = \frac{f_X(x)}{f_X(x)} = 1 \qquad 0 < y < 1$$

Hence, Y is a uniform r.v. over $(0, 1)$.

4.15. Let Y be a uniform r.v. over $(0, 1)$. Let $F(x)$ be a function which has the properties of the cdf of a continuous r.v. with $F(a) = 0$, $F(b) = 1$, and $F(x)$ strictly increasing for $a < x < b$, where a and b could be $-\infty$ and ∞, respectively. Let $X = F^{-1}(Y)$. Show that the cdf of X is $F(x)$.

$$F_X(x) = P(X \leq x) = P[F^{-1}(Y) \leq x]$$

Since $F(x)$ is strictly increasing, $F^{-1}(Y) \leq x$ is equivalent to $Y \leq F(x)$, and hence

$$F_X(x) = P(X \leq x) = P[Y \leq F(x)]$$

Now Y is a uniform r.v. over $(0, 1)$, and by Eq. (2.57),

$$F_Y(y) = P(Y \leq y) = y \qquad 0 < y < 1$$

and accordingly,

$$F_X(x) = P(X \leq x) = P[Y \leq F(x)] = F(x) \qquad 0 < F(x) < 1$$

Note that this problem is the converse of Prob. 4.14.

4.16. Let X be a continuous r.v. with the pdf

$$f_X(x) = \begin{cases} e^{-x} & x > 0 \\ 0 & x < 0 \end{cases}$$

Find the transformation $Y = g(X)$ such that the pdf of Y is

$$f_Y(y) = \begin{cases} \dfrac{1}{2\sqrt{y}} & 0 < y < 1 \\ 0 & \text{otherwise} \end{cases}$$

The cdf of X is

$$F_X(x) = \int_{-\infty}^{x} f_X(\xi)\, d\xi = \begin{cases} \int_0^x e^{-\xi}\, d\xi \\ 0 \end{cases} = \begin{cases} 1 - e^{-x} & x > 0 \\ 0 & x < 0 \end{cases}$$

Then from the result of Prob. 4.14, the r.v. $Z = 1 - e^{-X}$ is uniformly distributed over $(0, 1)$. Similarly, the cdf of Y is

$$F_Y(y) = \begin{cases} \int_0^y \dfrac{1}{2\sqrt{\eta}}\, d\eta \\ 0 \end{cases} = \begin{cases} \sqrt{y} & 0 < y < 1 \\ 0 & \text{otherwise} \end{cases}$$

and the r.v. $W = \sqrt{Y}$ is uniformly distributed over $(0, 1)$. Thus, by setting $Z = W$, the required transformation is $Y = (1 - e^{-X})^2$.

Functions of Two Random Variables

4.17. Consider $Z = X + Y$. Show that if X and Y are independent Poisson r.v.'s with parameters λ_1 and λ_2, respectively, then Z is also a Poisson r.v. with parameter $\lambda_1 + \lambda_2$.

We can write the event

$$(X + Y = n) = \bigcup_{i=0}^{n} (X = i, Y = n - i)$$

where events $(X = i, Y = n - i)$, $i = 0, 1, \ldots, n$, are disjoint. Since X and Y are independent, by Eqs. (1.62) and (2.48), we have

$$P(Z = n) = P(X + Y = n) = \sum_{i=0}^{n} P(X = i, Y = n - i) = \sum_{i=0}^{n} P(X = i)\, P(Y = n - i)$$

$$= \sum_{i=0}^{n} e^{-\lambda_1} \frac{\lambda_1^i}{i!} e^{-\lambda_2} \frac{\lambda_2^{n-i}}{(n-i)!} = e^{-(\lambda_1 + \lambda_2)} \sum_{i=0}^{n} \frac{\lambda_1^i \lambda_2^{n-i}}{i!\,(n-i)!}$$

$$= \frac{e^{-(\lambda_1 + \lambda_2)}}{n!} \sum_{i=0}^{n} \frac{n!}{i!\,(n-i)!} \lambda_1^i \lambda_2^{n-i}$$

$$= \frac{e^{-(\lambda_1 + \lambda_2)}}{n!} (\lambda_1 + \lambda_2)^n$$

which indicates that $Z = X + Y$ is a Poisson r.v. with $\lambda_1 + \lambda_2$.

4.18. Consider two r.v.'s X and Y with joint pdf $f_{XY}(x, y)$. Let $Z = X + Y$.

    (*a*)   Determine the pdf of Z.

    (*b*)   Determine the pdf of Z if X and Y are independent.

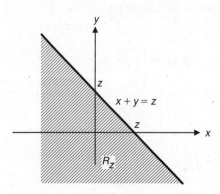

Fig. 4-6

    (*a*)   The range R_Z of Z corresponding to the event $(Z \leq z) = (X + Y \leq z)$ is the set of points (x, y) which lie on and to the left of the line $z = x + y$ (Fig. 4-6). Thus, we have

$$F_Z(z) = P(X + Y \leq z) = \int_{-\infty}^{\infty} \left[\int_{-\infty}^{z-x} f_{XY}(x, y)\, dy \right] dx \tag{4.95}$$

Then

$$f_Z(z) = \frac{d}{dz} F_Z(z) = \int_{-\infty}^{\infty} \left[\frac{d}{dz} \int_{-\infty}^{z-x} f_{XY}(x, y)\, dy \right] dx \tag{4.96}$$

$$= \int_{-\infty}^{\infty} f_{XY}(x, z - x)\, dx$$

    (*b*)   If X and Y are independent, then Eq. (4.96) reduces to

$$f_Z(z) = \int_{-\infty}^{\infty} f_X(x) f_Y(z - x)\, dx \tag{4.97}$$

The integral on the right-hand side of Eq. (4.97) is known as a *convolution* of $f_X(z)$ and $f_Y(z)$. Since the convolution is commutative, Eq. (4.97) can also be written as

$$f_Z(z) = \int_{-\infty}^{\infty} f_Y(y) f_X(z - y)\, dy \tag{4.98}$$

4.19. Using Eqs. (4.19) and (3.30), redo Prob. 4.18(*a*); that is, find the pdf of $Z = X + Y$.

Let $Z = X + Y$ and $W = X$. The transformation $z = x + y$, $w = x$ has the inverse transformation $x = w$, $y = z - w$, and

$$J(x, y) = \begin{vmatrix} \dfrac{\partial z}{\partial x} & \dfrac{\partial z}{\partial y} \\[2mm] \dfrac{\partial w}{\partial x} & \dfrac{\partial w}{\partial y} \end{vmatrix} = \begin{vmatrix} 1 & 1 \\ 1 & 0 \end{vmatrix} = -1$$

By Eq. (4.19), we obtain

$$f_{ZW}(z, w) = f_{XY}(w, z - w)$$

Hence, by Eq. (3.30), we get

$$f_Z(z) = \int_{-\infty}^{\infty} f_{ZW}(z, w)\, dw = \int_{-\infty}^{\infty} f_{XY}(w, z - w)\, dw = \int_{-\infty}^{\infty} f_{XY}(x, z - x)\, dx$$

4.20. Suppose that X and Y are independent standard normal r.v.'s. Find the pdf of $Z = X + Y$.

The pdf's of X and Y are

$$f_X(x) = \frac{1}{\sqrt{2\pi}}\, e^{-x^2/2} \qquad f_Y(y) = \frac{1}{\sqrt{2\pi}}\, e^{-y^2/2}$$

Then, by Eq. (4.97), we have

$$f_Z(z) \int_{-\infty}^{\infty} f_X(x) f_Y(z - x)\, dx = \int_{-\infty}^{\infty} \frac{1}{\sqrt{2\pi}}\, e^{-x^2/2} \frac{1}{\sqrt{2\pi}}\, e^{-(z-x)^2/2} dx$$

$$= \frac{1}{2\pi} \int_{-\infty}^{\infty} e^{-(z^2 - 2zx + 2x^2)/2}\, dx$$

Now, $z^2 - 2zx + 2x^2 = (\sqrt{2}\, x - z/\sqrt{2})^2 + z^2/2$, and we have

$$f_Z(z) = \frac{1}{\sqrt{2\pi}}\, e^{-z^2/4} \frac{1}{\sqrt{2\pi}} \int_{-\infty}^{\infty} e^{-(\sqrt{2}x - z/\sqrt{2})^2/2}\, dx$$

$$= \frac{1}{\sqrt{2\pi}}\, e^{-z^2/4} \frac{1}{\sqrt{2}} \int_{-\infty}^{\infty} \frac{1}{\sqrt{2\pi}}\, e^{-u^2/2}\, du$$

with the change of variables $u = \sqrt{2}\, x - z/\sqrt{2}$. Since the integrand is the pdf of $N(0; 1)$, the integral is equal to unity, and we get

$$f_Z(z) = \frac{1}{\sqrt{2\pi}\sqrt{2}}\, e^{-z^2/4} = \frac{1}{\sqrt{2\pi}\sqrt{2}}\, e^{-z^2/2(\sqrt{2})^2}$$

which is the pdf of $N(0; 2)$. Thus, Z is a normal r.v. with zero mean and variance 2.

4.21. Let X and Y be independent uniform r.v.'s over $(0, 1)$. Find and sketch the pdf of $Z = X + Y$.

Since X and Y are independent, we have

$$f_{XY}(x, y) = f_X(x) f_Y(y) = \begin{cases} 1 & 0 < x < 1, 0 < y < 1 \\ 0 & \text{otherwise} \end{cases}$$

The range of Z is $(0, 2)$, and

$$F_Z(z) = P(X + Y \le z) = \iint_{x+y\le z} f_{XY}(x, y)\, dx\, dy = \iint_{x+y\le z} dx\, dy$$

If $0 < z < 1$ [Fig. 4-7(a)],

$$F_Z(z) = \iint\limits_{x+y<z} dx\,dy = \text{shaded area} = \frac{z^2}{2}$$

and

$$f_Z(z) = \frac{d}{dz} F_Z(z) = z$$

If $1 < z < 2$ [Fig. 4-7(b)],

$$F_Z(z) = \iint\limits_{x+y<z} dx\,dy = \text{shaded area} = 1 - \frac{(2-z)^2}{2}$$

and

$$f_Z(z) = \frac{d}{dz} F_Z(z) = 2 - z$$

Hence,

$$f_Z(z) = \begin{cases} z & 0 < z < 1 \\ 2-z & 1 < z < 2 \\ 0 & \text{otherwise} \end{cases}$$

which is sketched in Fig. 4-7(c). Note that the same result can be obtained by the convolution of $f_X(z)$ and $f_Y(z)$.

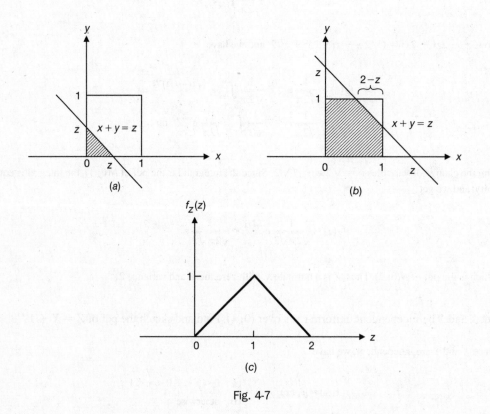

Fig. 4-7

4.22. Let X and Y be independent exponential r.v.'s with common parameter λ and let $Z = X + Y$. Find $f_Z(z)$.

From Eq. (2.60) we have

$$f_X(x) = \lambda e^{-\lambda x} \qquad x > 0, \qquad f_Y(y) = \lambda e^{-\lambda y} \qquad y > 0$$

In order to apply Eq. (4.97) we need to rewrite $f_X(x)$ and $f_Y(y)$ as

$$f_X(x) = \lambda e^{-\lambda x} u(x) \qquad -\infty < x < \infty, \qquad f_Y(y) = \lambda e^{-\lambda y} u(y) \qquad -\infty < y < \infty$$

where $u(\xi)$ is a unit step function defined as

$$u(\xi) = \begin{cases} 1 & \xi > 0 \\ 0 & \xi < 0 \end{cases} \tag{4.99}$$

Now by Eq. (4.97) we have

$$f_Z(z) = \int_{-\infty}^{\infty} \lambda e^{-\lambda x} u(x) \lambda e^{-\lambda(z-x)} u(z-x)\, dx$$

Using Eq. (4.99), we have

$$u(x)u(z-x) = \begin{cases} 1 & 0 < x < z \\ 0 & \text{otherwise} \end{cases}$$

Thus,

$$f_Z(z) = \lambda^2 e^{-\lambda z} \int_0^z dx = \lambda^2 z e^{-\lambda z} u(z)$$

Note that X and Y are gamma r.v.'s with parameter $(1, \lambda)$ and Z is a gamma r.v. with parameter $(2, \lambda)$ (see Prob. 4.23).

4.23. Let X and Y be independent gamma r.v.'s with respective parameters (α, λ) and (β, λ). Show that $Z = X + Y$ is also a gamma r.v. with parameters $(\alpha + \beta, \lambda)$.

From Eq. (2.65),

$$f_X(x) = \begin{cases} \dfrac{\lambda e^{-\lambda x} (\lambda x)^{\alpha-1}}{\Gamma(\alpha)} & x > 0 \\ 0 & x < 0 \end{cases}$$

$$f_Y(y) = \begin{cases} \dfrac{\lambda e^{-\lambda y} (\lambda x)^{\beta-1}}{\Gamma(\beta)} & y > 0 \\ 0 & y < 0 \end{cases}$$

The range of Z is $(0, \infty)$, and using Eq. (4.97), we have

$$f_Z(z) = \frac{1}{\Gamma(\alpha)\Gamma(\beta)} \int_0^z \lambda e^{-\lambda x} (\lambda x)^{\alpha-1} \lambda e^{-\lambda(z-x)} [\lambda(z-x)]^{\beta-1}\, dx$$

$$= \frac{\lambda^{\alpha+\beta}}{\Gamma(\alpha)\Gamma(\beta)} e^{-\lambda z} \int_0^z x^{\alpha-1} (z-x)^{\beta-1}\, dx$$

By the change of variable $w = x/z$, we have

$$f_Z(z) = \frac{\lambda^{\alpha+\beta}}{\Gamma(\alpha)\Gamma(\beta)} e^{-\lambda z} z^{\alpha+\beta-1} \int_0^1 w^{\alpha-1} (1-w)^{\beta-1}\, dw$$

$$= k e^{-\lambda z} z^{\alpha+\beta-1}$$

where k is a constant which does not depend on z. The value of k is determined as follows: Using Eq. (2.22) and definition (2.66) of the gamma function, we have

$$\int_{-\infty}^{\infty} f_Z(z)\, dz = k \int_0^z e^{-\lambda z} z^{\alpha+\beta-1}\, dz$$

$$= \frac{k}{\lambda^{\alpha+\beta}} \int_0^\infty e^{-v} v^{\alpha+\beta-1}\, dv \quad (\lambda z = v)$$

$$= \frac{k}{\lambda^{\alpha+\beta}} \Gamma(\alpha+\beta) = 1$$

Hence, $k = \lambda^{\alpha+\beta}/\Gamma(\alpha+\beta)$ and

$$f_Z(z) = \frac{\lambda^{\alpha+\beta}}{\Gamma(\alpha+\beta)} e^{-\lambda z} z^{\alpha+\beta-1} = \frac{\lambda e^{-\lambda x}(\lambda z)^{\alpha+\beta-1}}{\Gamma(\alpha+\beta)} \quad z > 0$$

which indicates that Z is a gamma r.v. with parameters $(\alpha+\beta, \lambda)$.

4.24. Let X and Y be two r.v.'s with joint pdf $f_{XY}(x, y)$. and let $Z = X - Y$.

    (*a*)   Find $f_Z(z)$.

    (*b*)   Find $f_Z(z)$ if X and Y are independent.

    (*a*)   From Eq. (4.12) and Fig. 4-8 we have

$$F_Z(z) = P(X - Y \le z) = \int_{y=-\infty}^{\infty} \int_{x=-\infty}^{y+z} f_{XY}(x, y)\, dx\, dy$$

Then

$$f_Z(z) = \frac{d\, F_Z(z)}{dz} = \int_{y=-\infty}^{\infty} \left[\frac{\partial}{\partial z} \int_{x=-\infty}^{y+z} f_{X,Y}(x, y)\, dx \right] dy$$

$$= \int_{-\infty}^{\infty} f_{XY}(y+z, y)\, dy$$

(4.100)

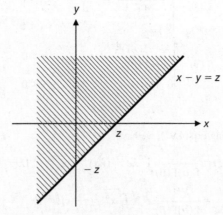

Fig. 4-8

    (*b*)   If X and Y are independent, by Eq. (3.32), Eq. (4.100) reduces to

$$f_Z(z) = \int_{-\infty}^{\infty} f_X(y+z) f_Y(y)\, dy$$

(4.101)

which is the convolution of $f_X(-z)$ with $f_Y(z)$.

4.25. Consider two r.v.'s X and Y with joint pdf $f_{XY}(x, y)$. Determine the pdf of $Z = XY$.

Let $Z = XY$ and $W = X$. The transformation $z = xy$, $w = x$ has the inverse transformation $x = w$, $y = z/w$, and

$$\bar{J}(z, w) = \begin{vmatrix} \dfrac{\partial x}{\partial z} & \dfrac{\partial x}{\partial w} \\ \dfrac{\partial y}{\partial z} & \dfrac{\partial y}{\partial w} \end{vmatrix} = \begin{vmatrix} 0 & 1 \\ \dfrac{1}{w} & -\dfrac{z}{w^2} \end{vmatrix} = -\frac{1}{w}$$

Thus, by Eq. (4.23), we obtain

$$f_{ZW}(z, w) = \left| \frac{1}{w} \right| f_{XY}\left(w, \frac{z}{w} \right) \tag{4.102}$$

and the marginal pdf of Z is

$$f_Z(z) = \int_{-\infty}^{\infty} \left| \frac{1}{w} \right| f_{XY}\left(w, \frac{z}{w} \right) dw \tag{4.103}$$

4.26. Let X and Y be independent uniform r.v.'s over $(0, 1)$. Find the pdf of $Z = XY$.

We have

$$f_{XY}(x, y) = \begin{cases} 1 & 0 < x < 1, 0 < y < 1 \\ 0 & \text{otherwise} \end{cases}$$

The range of Z is $(0, 1)$. Then

$$f_{XY}\left(w, \frac{z}{w} \right) = \begin{cases} 1 & 0 < w < 1, 0 < z/w < 1 \\ 0 & \text{otherwise} \end{cases}$$

or

$$f_{XY}\left(w, \frac{z}{w} \right) = \begin{cases} 1 & 0 < z < w < 1 \\ 0 & \text{otherwise} \end{cases}$$

By Eq. (4.103),

$$f_Z(z) = \int_z^1 \frac{1}{w} dw = -\ln z \qquad 0 < z < 1$$

Thus,

$$f_Z(z) = \begin{cases} -\ln z & 0 < z < 1 \\ 0 & \text{otherwise} \end{cases}$$

4.27. Consider two r.v.'s X and Y with joint pdf $f_{XY}(x, y)$. Determine the pdf of $Z = X/Y$.

Let $Z = X/Y$ and $W = Y$. The transformation $z = x/y$, $w = y$ has the inverse transformation $x = zw$, $y = w$, and

$$\bar{J}(z, w) = \begin{vmatrix} \dfrac{\partial x}{\partial z} & \dfrac{\partial x}{\partial w} \\ \dfrac{\partial y}{\partial z} & \dfrac{\partial y}{\partial w} \end{vmatrix} = \begin{vmatrix} w & z \\ 0 & 1 \end{vmatrix} = w$$

Thus, by Eq. (4.23), we obtain

$$f_{ZW}(z, w) = |w| \, f_{XY}(zw, w) \tag{4.104}$$

and the marginal pdf of Z is

$$f_Z(z) = \int_{-\infty}^{\infty} |w| \, f_{XY}(zw, w) \, dw \tag{4.105}$$

4.28. Let X and Y be independent standard normal r.v.'s. Find the pdf of $Z = X/Y$.

Since X and Y are independent, using Eq. (4.105), we have

$$f_Z(z) = \int_{-\infty}^{\infty} |w| \, f_X(zw) f_Y(w) \, dw = \int_{-\infty}^{\infty} |w| \frac{1}{2\pi} e^{-w^2(1+z^2)/2} \, dw$$

$$= \frac{1}{2\pi} \int_0^{\infty} w e^{-w^2(1+z^2)/2} \, dw - \frac{1}{2\pi} \int_{-\infty}^0 w e^{-w^2(1+z^2)/2} \, dw$$

$$= \frac{1}{\pi(1+z^2)} \qquad -\infty < z < \infty$$

which is the pdf of a Cauchy r.v. with parameter 1.

4.29. Let X and Y be two r.v.'s with joint pdf $f_{XY}(x, y)$ and joint cdf $F_{XY}(x, y)$. Let $Z = \max(X, Y)$.

    (*a*)   Find the cdf of Z.

    (*b*)   Find the pdf of Z if X and Y are independent.

    (*a*)   The region in the xy plane corresponding to the event $\{\max(X, Y) \le z\}$ is shown as the shaded area in Fig. 4-9. Then

$$F_Z(z) = P(Z \le z) = P(X \le z, Y \le z) = F_{XY}(z, z) \tag{4.106}$$

    (*b*)   If X and Y are independent, then

$$F_Z(z) = F_X(z) F_Y(z)$$

and differentiating with respect to z gives

$$f_Z(z) = f_X(z) F_Y(z) + F_X(z) \, f_Y(z) \tag{4.107}$$

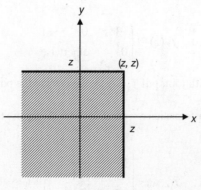

Fig. 4-9

4.30. Let X and Y be two r.v.'s with joint pdf $f_{XY}(x, y)$ and joint cdf $F_{XY}(x, y)$. Let $W = \min(X, Y)$.

   (*a*)   Find the cdf of W.

   (*b*)   Find the pdf of W if X and Y are independent.

   (*a*)   The region in the xy plane corresponding to the event $\{\min(X, Y) \le w\}$ is shown as the shaded area in Fig. 4-10. Then

$$P(W \le w) = P\{(X \le w) \cup (Y \le w)\}$$
$$= P(X \le w) + P(Y \le w) - P\{(X \le w) \cap (Y \le w)\}$$

Thus, $\qquad\qquad F_W(w) = F_X(w) + F_Y(w) - F_{XY}(w, w)$ \hfill (4.108)

   (*b*)   If X and Y are independent, then

$$F_W(w) = F_X(w) + F_Y(w) - F_X(w)F_Y(w)$$

and differentiating with respect to w gives

$$f_W(w) = f_X(w) + f_Y(w) - f_X(w)F_Y(w) - F_X(w)f_Y(w)$$
$$= f_X(w)[1 - F_Y(w)] + f_Y(w)[1 - F_X(w)]$$ \hfill (4.109)

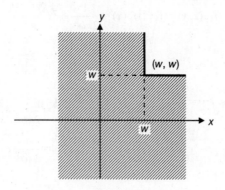

Fig. 4-10

4.31. Let X and Y be two r.v.'s with joint pdf $f_{XY}(x, y)$. Let $Z = X^2 + Y^2$. Find $f_z(z)$.

As shown in Fig. 4-11, $D_z(X^2 + Y^2 \le z)$ represents the area of a circle with radius $\sqrt{z}$.

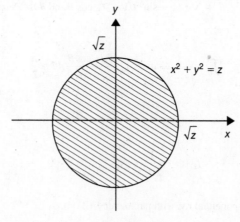

Fig. 4-11

Hence, by Eq. (4.12)

$$F_Z(z) = \int_{y=-\sqrt{z}}^{\sqrt{z}} \int_{x=-\sqrt{z-y^2}}^{\sqrt{z-y^2}} f_{XY}(x, y)\, dx\, dy$$

and

$$
\begin{aligned}
f_Z(z) = \frac{d\,F_Z(z)}{dz} &= \int_{y=-\sqrt{z}}^{\sqrt{z}} \left[\frac{\partial}{\partial z} \int_{x=-\sqrt{z-y^2}}^{\sqrt{z-y^2}} f_{XY}(x, y)\, dx \right] dy \\
&= \int_{y=-\sqrt{z}}^{\sqrt{z}} \frac{\partial}{\partial z} \left[\int_0^{\sqrt{z-y^2}} f_{XY}(x, y)\, dx - \int_0^{-\sqrt{z-y^2}} f_{XY}(x, y)\, dx \right] dy \\
&= \int_{y=-\sqrt{z}}^{\sqrt{z}} \frac{1}{2\sqrt{z-y^2}} \left[f_{XY}\left(\sqrt{z-y^2}, y\right) + f_{XY}\left(-\sqrt{z-y^2}, y\right) \right] dy
\end{aligned}
\tag{4.110}
$$

4.32. Let X and Y be independent normal r.v.'s with $\mu_X = \mu_Y = 0$ and $\sigma_X{}^2 = \sigma_Y{}^2 = \sigma^2$. Let $Z = X^2 + Y^2$. Find $f_Z(z)$.

Since X and Y are independent, from Eqs. (3.32) and (2.71) we have

$$f_{XY}(x, y) = f_X(x) f_Y(y) = \frac{1}{2\pi\sigma^2} e^{-\frac{1}{2\sigma^2}(x^2 + y^2)} \tag{4.111}$$

and

$$f_{XY}\left(\sqrt{z-y^2}, y\right) = \frac{1}{2\pi\sigma^2} e^{-\frac{1}{2\sigma^2}(z-y^2+y^2)} = \frac{1}{2\pi\sigma^2} e^{-\frac{1}{2\sigma^2}z}$$

$$f_{XY}\left(-\sqrt{z-y^2}, y\right) = \frac{1}{2\pi\sigma^2} e^{-\frac{1}{2\sigma^2}(z-y^2+y^2)} = \frac{1}{2\pi\sigma^2} e^{-\frac{1}{2\sigma^2}z}$$

Thus, using Eq. (4.110), we obtain

$$f_Z(z) = \int_{y=-\sqrt{z}}^{\sqrt{z}} \frac{1}{2\sqrt{z-y^2}} \left(2 \frac{1}{2\pi\sigma^2} e^{-\frac{1}{2\pi\sigma^2}z} \right) dy = \frac{1}{\pi\sigma^2} e^{-\frac{1}{2\sigma^2}z} \int_0^{\sqrt{z}} \frac{1}{\sqrt{z-y^2}}\, dy$$

Let $y = \sqrt{z}\,\sin\theta$. Then $\sqrt{z-y^2} = \sqrt{z(1-\sin^2\theta)} = \sqrt{z}\cos\theta$ and $dy = \sqrt{z}\cos\theta\, d\theta$
and

$$\int_0^{\sqrt{z}} \frac{1}{\sqrt{z-y^2}}\, dy = \int_0^{\pi/2} \frac{\sqrt{z}\cos\theta}{\sqrt{z}\cos\theta}\, d\theta = \frac{\pi}{2}$$

Hence,

$$f_Z(z) = \frac{1}{2\sigma^2} e^{-\frac{1}{2\sigma^2}z} \qquad z > 0 \tag{4.112}$$

which indicates that Z is an exponential r.v. with parameter $1/(2\sigma^2)$.

4.33. Let X and Y be two r.v.'s with joint pdf $f_{XY}(x, y)$. Let

$$R = \sqrt{X^2 + Y^2} \qquad \Theta = \tan^{-1}\frac{Y}{X} \tag{4.113}$$

Find $f_{R\Theta}(r, \theta)$ in terms of $f_{XY}(x, y)$.

We assume that $r \geq 0$ and $0 \leq \theta < 2\pi$. With this assumption, the transformation

$$\sqrt{x^2 + y^2} = r \qquad \tan^{-1}\frac{y}{x} = \theta$$

has the inverse transformation

$$x = r \cos \theta \qquad y = r \sin \theta$$

Since

$$\bar{J}(x, y) = \begin{vmatrix} \dfrac{\partial x}{\partial r} & \dfrac{\partial x}{\partial \theta} \\ \dfrac{\partial y}{\partial r} & \dfrac{\partial y}{\partial \theta} \end{vmatrix} = \begin{vmatrix} \cos \theta & -r \sin \theta \\ \sin \theta & r \cos \theta \end{vmatrix} = r$$

by Eq. (4.23) we obtain

$$f_{R\Theta}(r, \theta) = r f_{XY}(r \cos \theta, r \sin \theta) \tag{4.114}$$

4.34. A voltage V is a function of time t and is given by

$$V(t) = X \cos \omega t + Y \sin \omega t \tag{4.115}$$

in which ω is a constant angular frequency and $X = Y = N(0; \sigma^2)$ and they are independent.

(a) Show that $V(t)$ may be written as

$$V(t) = R \cos (\omega t - \Theta) \tag{4.116}$$

(b) Find the pdf's of r.v.'s R and Θ and show that R and Θ are independent.

(a) We have

$$V(t) = X \cos \omega t + Y \sin \omega t$$

$$= \sqrt{X^2 + Y^2} \left(\frac{X}{\sqrt{X^2 + Y^2}} \cos \omega t + \frac{Y}{\sqrt{X^2 + Y^2}} \sin \omega t \right)$$

$$= \sqrt{X^2 + Y^2} \,(\cos \Theta \cos \omega t + \sin \Theta \sin \omega t)$$

$$= R \cos(\omega t - \Theta)$$

Where $\qquad R = \sqrt{X^2 + Y^2} \qquad$ and $\qquad \Theta = \tan^{-1}\frac{Y}{X}$

which is the transformation (4.113).

(b) Since $X = Y = N(0; \sigma^2)$ and they are independent, we have

$$f_{XY}(x, y) = \frac{1}{2\pi\sigma^2} e^{-(x^2+y^2)/(2\sigma^2)}$$

Thus, using Eq. (4.114), we get

$$f_{R\Theta}(r, \theta) = r f_{XY}(r \cos\theta, r \sin\theta) = \frac{r}{2\pi\sigma^2} e^{-r^2/(2\sigma^2)} \qquad (4.117)$$

Now

$$f_R(r) = \int_0^{2\pi} f_{R\Theta}(r, \theta)\, d\theta = \frac{r}{2\pi\sigma^2} e^{-r^2/(2\sigma^2)} \int_0^{2\pi} d\theta = \frac{r}{\sigma^2} e^{-r^2/(2\sigma^2)} \qquad (4.118)$$

$$f_\Theta(\theta) = \int_0^\infty f_{R\Theta}(r, \theta)\, dr = \frac{1}{2\pi\sigma^2} \int_0^\infty r e^{-r^2/(2\sigma^2)}\, dr = \frac{1}{2\pi} \qquad (4.119)$$

and $f_{R\Theta}(r, \theta) = f_R(r) f_\Theta(\theta)$; hence, R and Θ are independent.

Note that R is a *Rayleigh* r.v. (Prob. 2.26), and Θ is a uniform r.v. over $(0, 2\pi)$.

Functions of N Random Variables

4.35. Let X, Y, and Z be independent standard normal r.v.'s. Let $W = (X^2 + Y^2 + Z^2)^{1/2}$. Find the pdf of W.

We have

$$f_{XYZ}(x, y, z) = f_X(x) f_Y(y) f_Z(z) = \frac{1}{(2\pi)^{3/2}} e^{-(x^2 + y^2 + z^2)/2}$$

and

$$F_W(w) = P(W \le w) = P(X^2 + Y^2 + Z^2 \le w^2)$$

$$= \iiint_{R_w} \frac{1}{(2\pi)^{3/2}} e^{-(x^2 + y^2 + z^2)/2}\, dx\, dy\, dz$$

where $R_w = \{(x, y, z): x^2 + y^2 + z^2 \le w^2\}$. Using spherical coordinates (Fig. 4-12), we have

$$x^2 + y^2 + z^2 = r^2$$
$$dx\, dy\, dz = r^2 \sin\theta\, dr\, d\theta\, d\varphi$$

and

$$F_W(w) = \frac{1}{(2\pi)^{3/2}} \int_0^{2\pi} \int_0^\pi \int_0^w e^{-r^2/2} r^2 \sin\theta\, dr\, d\theta\, d\varphi$$

$$= \frac{1}{(2\pi)^{3/2}} \int_0^{2\pi} d\varphi \int_0^\pi \sin\theta\, d\theta \int_0^w e^{-r^2/2} r^2\, dr \qquad (4.120)$$

$$= \frac{1}{(2\pi)^{3/2}} (2\pi)(2) \int_0^w r^2 e^{-r^2/2}\, dr$$

Thus, the pdf of W is

$$f_W(w) = \frac{d}{dw} F_W(w) = \begin{cases} \sqrt{\dfrac{2}{\pi}} w^2 e^{-w^2/2} & w > 0 \\ 0 & w < 0 \end{cases} \qquad (4.121)$$

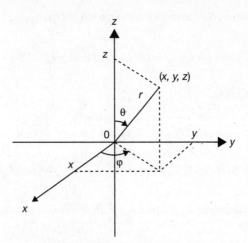

Fig. 4-12 Spherical coordinates.

4.36. Let $X_1, \ldots, X_n$ be n independent r.v.'s each with the identical pdf $f(x)$. Let $Z = \max(X_1, \ldots, X_n)$. Find the pdf of Z.

The probability $P(z < Z < z + dz)$ is equal to the probability that one of the r.v.'s falls in $(z, z + dz)$ and all others are less than z. The probability that one of X_i $(i = 1, \ldots, n)$ falls in $(z, z + dz)$ and all others are all less than z is

$$f(z) \, dz \left(\int_{-\infty}^{z} f(x) dx \right)^{n-1}$$

Since there are n ways of choosing the variables to be maximum, we have

$$f_Z(z) = n f(z) \left(\int_{-\infty}^{z} f(x) dx \right)^{n-1} = n f(z) [F(z)]^{n-1} \qquad (4.122)$$

When $n = 2$, Eq. (4.122) reduces to

$$f_Z(z) = 2 f(z) \int_{-\infty}^{z} f(x) dx = 2 f(z) F(z) \qquad (4.123)$$

which is the same as Eq. (4.107) (Prob. 4.29) with $f_X(z) = f_Y(z) = f(z)$ and $F_X(z) = F_Y(z) = F(z)$.

4.37. Let $X_1, \ldots, X_n$ be n independent r.v.'s each with the identical pdf $f(x)$. Let $W = \min(X_1, \ldots, X_n)$. Find the pdf of W.

The probability $P(w < W < w + dw)$ is equal to the probability that one of the r.v.'s falls in $(w, w + dw)$ and all others are greater than w. The probability that one of X_i $(i = 1, \ldots, n)$ falls in $(w, w + dw)$ and all others are greater than w is

$$f(w) \, dw \left(\int_{w}^{\infty} f(x) \, dx \right)^{n-1}$$

Since there are n ways of choosing the variables to be minimum, we have

$$f_W(w) = nf(w)\left(\int_w^\infty f(x)\,dx\right)^{n-1} = nf(z)[1 - F(z)]^{n-1} \tag{4.124}$$

When $n = 2$, Eq. (4.124) reduces to

$$f_W(w) = 2f(w)\int_w^\infty f(x)\,dx = 2f(w)[1 - F(w)] \tag{4.125}$$

which is the same as Eq. (4.109) (Prob. 4.30) with $f_X(w) = f_Y(w) = f(w)$ and $F_X(w) = F_Y(w) = F(w)$.

4.38. Let $X_i, i = 1, \ldots, n$, be n independent gamma r.v.'s with respective parameters $(\alpha_i, \lambda), i = 1, \ldots, n$. Let

$$Y = X_1 + \cdots + X_n = \sum_{i=1}^n X_i$$

Show that Y is also a gamma r.v. with parameters $(\Sigma_{i=1}^n \alpha_i, \lambda)$.

We prove this proposition by induction. Let us assume that the proposition is true for $n = k$; that is,

$$Z = X_1 + \cdots + X_k = \sum_{i=1}^k X_i$$

is a gamma r.v. with parameters $(\beta, \lambda) = \left(\sum_{i=1}^k \alpha_i, \lambda\right).$

Let

$$W = Z + X_{k+1} = \sum_{i=1}^{k+1} X_i$$

Then, by the result of Prob. 4.23, we see that W is a gamma r.v. with parameters $(\beta + \alpha_{k+1}, \lambda) = (\Sigma_{i=1}^{k+1} \alpha_i, \lambda)$. Hence, the proposition is true for $n = k + 1$. Next, by the result of Prob. 4.23, the proposition is true for $n = 2$. Thus, we conclude that the proposition is true for any $n \geq 2$.

4.39. Let $X_1, \ldots, X_n$ be n independent exponential r.v.'s each with parameter λ. Let

$$Y = X_1 + \cdots + X_n = \sum_{i=1}^n X_i$$

Show that Y is a gamma r.v. with parameters (n, λ).

We note that an exponential r.v. with parameter λ is a gamma r.v. with parameters $(1, \lambda)$. Thus, from the result of Prob. 4.38 and setting $\alpha_i = 1$, we conclude that Y is a gamma r.v. with parameters (n, λ).

4.40. Let $Z_1, \ldots, Z_n$ be n independent standard normal r.v.'s. Let

$$Y = Z_1^2 + \cdots + Z_n^2 = \sum_{i=1}^n Z_i^2$$

Find the pdf of Y.

Let $Y_i = Z_i^2$. Then by Eq. (4.91) (Prob. 4.7), the pdf of Y_i is

$$f_{Y_1}(y) = \begin{cases} \dfrac{1}{\sqrt{2\pi y}}\, e^{-y/2} & y > 0 \\ 0 & y < 0 \end{cases}$$

Now, using Eq. (2.99), we can rewrite

$$\frac{1}{\sqrt{2xy}}\, e^{-y/2} = \frac{\frac{1}{2} e^{-y/2} (y/2)^{1/2-1}}{\sqrt{\pi}} = \frac{\frac{1}{2} e^{-y/2} (y/2)^{1/2-1}}{\Gamma\left(\dfrac{1}{2}\right)}$$

and we recognize the above as the pdf of a gamma r.v. with parameters $(\frac{1}{2}, \frac{1}{2})$ [Eq. (2.65)]. Thus, by the result of Prob. 4.38, we conclude that Y is the gamma r.v. with parameters $(n/2, \frac{1}{2})$ and

$$f_Y(y) = \begin{cases} \dfrac{\frac{1}{2} e^{-y/2} (y/2)^{n/2-1}}{\Gamma(n/2)} = \dfrac{e^{-y/2}\, y^{n/2-1}}{2^{n/2}\Gamma(n/2)} & y > 0 \\ 0 & y < 0 \end{cases} \tag{4.126}$$

When n is an even integer, $\Gamma(n/2) = [(n/2) - 1]!$, whereas when n is odd, $\Gamma(n/2)$ can be obtained from $\Gamma(\alpha) = (\alpha - 1)\Gamma(\alpha - 1)$ [Eq. (2.97)] and $\Gamma(\frac{1}{2}) = \sqrt{\pi}$ [Eq. (2.99)].

Note that Equation (4.126) is referred to as the *chi-square* (χ^2) density function with n degrees of freedom, and Y is known as the *chi-square* (χ^2) r.v. with n degrees of freedom. It is important to recognize that the sum of the squares of n independent standard normal r.v.'s is a chi-square r.v. with n degrees of freedom. The chi-square distribution plays an important role in statistical analysis.

4.41. Let X_1, X_2, and X_3 be independent standard normal r.v.'s. Let

$$Y_1 = X_1 + X_2 + X_3$$
$$Y_2 = X_1 - X_2$$
$$Y_3 = X_2 - X_3$$

Determine the joint pdf of Y_1, Y_2, and Y_3.

Let

$$y_1 = x_1 + x_2 + x_3$$
$$y_2 = x_1 - x_2 \tag{4.127}$$
$$y_3 = x_2 - x_3$$

By Eq. (4.32), the Jacobian of transformation (4.127) is

$$J(x_1, x_2, x_3) = \begin{vmatrix} 1 & 1 & 1 \\ 1 & -1 & 0 \\ 0 & 1 & -1 \end{vmatrix} = 3$$

Thus, solving the system (4.127), we get

$$x_1 = \frac{1}{3}(y_1 + 2y_2 + y_3)$$

$$x_2 = \frac{1}{3}(y_1 - y_2 + y_3)$$

$$x_3 = \frac{1}{3}(y_1 - y_2 - 2y_3)$$

Then by Eq. (4.31), we obtain

$$f_{Y_1 Y_2 Y_3}(y_1, y_2, y_3) = \frac{1}{3} f_{X_1 X_2 X_3}\left(\frac{y_1 + 2y_2 + y_3}{3}, \frac{y_1 - y_2 + y_3}{3}, \frac{y_1 - y_2 - 2y_3}{3}\right) \tag{4.128}$$

Since X_1, X_2, and X_3 are independent,

$$f_{X_1 X_2 X_3}(x_1, x_2, x_3) = \prod_{i=1}^{3} f_{X_i}(x_i) = \frac{1}{(2\pi)^{3/2}} e^{-(x_1^2 + x_2^2 + x_3^2)/2}$$

Hence,

$$f_{Y_1 Y_2 Y_3}(y_1, y_2, y_3) = \frac{1}{3(2\pi)^{3/2}} e^{-q(y_1, y_2, y_3)/2}$$

where

$$q(y_1, y_2, y_3) = \left(\frac{y_1 + 2y_2 + y_3}{3}\right)^2 + \left(\frac{y_1 - 2y_2 + y_3}{3}\right)^2 + \left(\frac{y_1 - 2y_2 - 2y_3}{3}\right)^2$$

$$= \frac{1}{3}y_1^2 + \frac{2}{3}y_2^2 + \frac{2}{3}y_3^2 + \frac{2}{3}y_2 y_3$$

Expectation

4.42. Let X be a uniform r.v. over $(0, 1)$ and $Y = e^X$.

(a) Find $E(Y)$ by using $f_Y(y)$.

(b) Find $E(Y)$ by using $f_X(x)$.

(a) From Eq. (4.92) (Prob. 4.9),

$$f_Y(y) = \begin{cases} \dfrac{1}{y} & 1 < y < e \\ 0 & \text{otherwise} \end{cases}$$

Hence,

$$E(Y) = \int_{-\infty}^{\infty} y f_Y(y)\, dy = \int_{1}^{e} dy = e - 1$$

(b) The pdf of X is

$$f_X(x) = \begin{cases} 1 & 0 < x < 1 \\ 0 & \text{otherwise} \end{cases}$$

Then, by Eq. (4.33),

$$E(Y) = \int_{-\infty}^{\infty} e^x f_X(x)\, dx = \int_{0}^{1} e^x dx = e - 1$$

4.43. Let $Y = aX + b$, where a and b are constants. Show that

(a)
$$E(Y) = E(aX + b) = aE(X) + b \qquad\qquad (4.129)$$

(b)
$$\text{Var}(Y) = \text{Var}(aX + b) = a^2\,\text{Var}(X) \qquad\qquad (4.130)$$

We verify for the continuous case. The proof for the discrete case is similar.

(a) By Eq. (4.33),

$$E(Y) = E(aX + b) = \int_{-\infty}^{\infty} (ax + b)\, f_X(x)\, dx$$

$$= a\int_{-\infty}^{\infty} x f_X(x)\, dx + b\int_{-\infty}^{\infty} f_X(x)\, dx = aE(X) + b$$

(b) Using Eq. (4.129), we have

$$\text{Var}(Y) = \text{Var}(aX + b) = E\{(aX + b - [aE(X) + b])^2\}$$

$$= E\{a^2[X - E(X)]^2\} = a^2 E\{[X - E(X)]^2\} = a^2\,\text{Var}(X)$$

4.44. Verify Eq. (4.39).

Using Eqs. (3.58) and (3.38), we have

$$E[E(Y\,|\,X)] = \int_{-\infty}^{\infty} E(Y\,|\,x) f_X(x)\, dx = \int_{-\infty}^{\infty} \left[\int_{-\infty}^{\infty} y f_{Y|X}(Y\,|\,x)\, dy \right] f_X(x)\, dx$$

$$= \int_{-\infty}^{\infty} \int_{-\infty}^{\infty} y\, \frac{f_{XY}(x, y)}{f_X(x)}\, f_X(x)\, dx\, dy = \int_{-\infty}^{\infty} y \left[\int_{-\infty}^{\infty} f_{XY}(x, y)\, dx \right] dy$$

$$= \int_{-\infty}^{\infty} y f_Y(y)\, dy = E[Y]$$

4.45. Let $Z = aX + bY$, where a and b are constants. Show that

$$E(Z) = E(aX + bY) = aE(X) + bE(Y) \qquad\qquad (4.131)$$

We verify for the continuous case. The proof for the discrete case is similar.

$$E(Z) = E(aX + bY) = \int_{-\infty}^{\infty} \int_{-\infty}^{\infty} (ax + by) f_{XY}(x, y)\, dx\, dy$$

$$= a\int_{-\infty}^{\infty} \int_{-\infty}^{\infty} x f_{XY}(x, y)\, dx\, dy + b\int_{-\infty}^{\infty} \int_{-\infty}^{\infty} y f_{XY}(x, y)\, dx\, dy$$

$$= a\int_{-\infty}^{\infty} x \left[\int_{-\infty}^{\infty} f_{XY}(x, y)\, dy \right] dx + b\int_{-\infty}^{\infty} y \left[\int_{-\infty}^{\infty} f_{XY}(x, y)\, dx \right] dy$$

$$= a\int_{-\infty}^{\infty} x f_X(x)\, dx + b\int_{-\infty}^{\infty} y f_Y(y)\, dy = aE(X) + bE(Y)$$

Note that Eq. (4.131) (the linearity of E) can be easily extended to n r.v.'s:

$$E\left(\sum_{i=1}^{n} a_i X_i \right) = \sum_{i=1}^{n} a_i E(X_i) \qquad\qquad (4.132)$$

4.46. Let $Y = aX + b$.

(a) Find the covariance of X and Y.

(b) Find the correlation coefficient of X and Y.

(a) By Eq. (4.131), we have

$$E(XY) = E[X(aX + b)] = aE(X^2) + bE(X)$$
$$E(Y) = E(aX + b) = aE(X) + b$$

Thus, the covariance of X and Y is [Eq. (3.51)]

$$\begin{aligned}
\text{Cov}(X, Y) = \sigma_{XY} &= E(XY) - E(X)E(Y)\\
&= aE(X^2) + bE(X) - E(X)[aE(X) + b]\\
&= a\{E(X^2) - [E(X)]^2\} = a\sigma_X^2
\end{aligned} \tag{4.133}$$

(b) By Eq. (4.130), we have $\sigma_Y = |a|\,\sigma_X$. Thus, the correlation coefficient of X and Y is [Eq. (3.53)]

$$\rho_{XY} = \frac{\sigma_{XY}}{\sigma_X\sigma_Y} = \frac{a\sigma_X^2}{\sigma_X|a|\sigma_X} = \frac{a}{|a|} = \begin{cases} 1 & a > 0\\ -1 & a < 0 \end{cases} \tag{4.134}$$

4.47. Verify Eq. (4.36).

Since X and Y are independent, we have

$$\begin{aligned}
E[g(X)h(y)] &= \int_{-\infty}^{\infty}\int_{-\infty}^{\infty} g(x)h(y)f_{XY}(x, y)\,dx\,dy\\
&= \int_{-\infty}^{\infty}\int_{-\infty}^{\infty} g(x)h(y)f_X(x)f_Y(y)\,dx\,dy\\
&= \int_{-\infty}^{\infty} g(x)f_X(x)\,dx \int_{-\infty}^{\infty} h(y)f_Y(y)\,dy\\
&= E[g(X)]E[h(Y)]
\end{aligned}$$

The proof for the discrete case is similar.

4.48. Let X and Y be defined by

$$X = \cos \Theta \qquad Y = \sin \Theta$$

where Θ is a random variable uniformly distributed over $(0, 2\pi)$.

(a) Show that X and Y are uncorrelated.

(b) Show that X and Y are not independent.

(a) We have

$$f_\Theta(\theta) = \begin{cases} \dfrac{1}{2\pi} & 0 < \theta < 2\pi\\ 0 & \text{otherwise} \end{cases}$$

Then, $$E(X) = \int_{-\infty}^{\infty} xf_X(x)\,dx = \int_0^{2\pi} \cos\theta\, f_\Theta(\theta)\,d\theta = \frac{1}{2\pi}\int_0^{2\pi} \cos\theta\,d\theta = 0$$

Similarly, $$E(Y) = \frac{1}{2\pi} \int_0^{2\pi} \sin\theta \, d\theta = 0$$

$$E(XY) = \frac{1}{2\pi} \int_0^{2\pi} \cos\theta \sin\theta \, d\theta = \frac{1}{4\pi} \int_0^{2\pi} \sin 2\theta \, d\theta = 0 = E(X)E(Y)$$

Thus, by Eq. (3.52), X and Y are uncorrelated.

(b) $$E(X^2) = \frac{1}{2\pi} \int_0^{2\pi} \cos^2\theta \, d\theta = \frac{1}{4\pi} \int_0^{2\pi} (1 + \cos 2\theta) \, d\theta = \frac{1}{2}$$

$$E(Y^2) = \frac{1}{2\pi} \int_0^{2\pi} \sin^2\theta \, d\theta = \frac{1}{4\pi} \int_0^{2\pi} (1 - \cos 2\theta) \, d\theta = \frac{1}{2}$$

$$E(X^2Y^2) = \frac{1}{2\pi} \int_0^{2\pi} \cos^2\theta \sin^2\theta \, d\theta = \frac{1}{16\pi} \int_0^{2\pi} (1 - \cos 4\theta) \, d\theta = \frac{1}{8}$$

Hence,

$$E(X^2 \, Y^2) = \frac{1}{8} \neq \frac{1}{4} = E(X^2)E(Y^2)$$

If X and Y were independent, then by Eq. (4.36), we would have $E(X^2Y^2) = E(X^2)E(Y^2)$. Therefore, X and Y are not independent.

4.49. Let $X_1, \ldots, X_n$ be n r.v.'s. Show that

$$\mathrm{Var}\left(\sum_{i=1}^{n} a_i X_i\right) = \sum_{i=1}^{n} \sum_{j=1}^{n} a_i a_j \, \mathrm{Cov}(X_i, X_j) \qquad (4.135)$$

If $X_1, \ldots, X_n$ are pairwise independent, then

$$\mathrm{Var}\left(\sum_{i=1}^{n} a_i X_i\right) = \sum_{i=1}^{n} a_i^2 \, \mathrm{Var}(X_i) \qquad (4.136)$$

Let $$Y = \sum_{i=1}^{n} a_i X_i$$

Then by Eq. (4.132), we have

$$\mathrm{Var}(Y) = E\{[Y - E(Y)]^2\} = E\left\{\left(\sum_{i=1}^{n} a_i[X_i - E(X_i)]\right)^2\right\}$$

$$= E\left\{\sum_{i=1}^{n} \sum_{j=1}^{n} a_i a_j [X_i - E(X_i)][X_j - E(X_j)]\right\}$$

$$= \sum_{i=1}^{n} \sum_{j=1}^{n} a_i a_j E\{[X_i - E(X_i)][X_j - E(X_j)]\}$$

$$= \sum_{i=1}^{n} \sum_{j=1}^{n} a_i a_j \, \mathrm{Cov}(X_i, X_j)$$

If $X_1, \ldots, X_n$ are pairwise independent, then (Prob. 3.32)

$$\text{Cov}(X_i, X_j) = \begin{cases} \text{Var}(X_i) & i = j \\ 0 & i \neq j \end{cases}$$

and Eq. (4.135) reduces to

$$\text{Var}\left(\sum_{i=1}^{n} a_i X_i\right) = \sum_{i=1}^{n} a_i^2 \, \text{Var}(X_i)$$

4.50 Verify Jensen's inequality (4.40),

$$E[g(x)] \geq g(E[X]) \qquad g(x) \text{ is a convex function}$$

Expanding $g(x)$ in a Taylor's series expansion around $\mu = E(x)$, we have

$$g(x) = g(\mu) + g'(\mu)(x - \mu) + \frac{1}{2} g''(\xi)(x - \xi)^2$$

where ξ is some value between x and μ. If $g(x)$ is convex, then $g''(\zeta) \geq 0$ and we obtain

$$g(x) \geq g(\mu) + g'(\mu)(x - \mu)$$

Hence,

$$g(X) \geq g(\mu) + g'(\mu)(X - \mu) \tag{4.137}$$

Taking expectation, we get

$$E[g(x)] \geq g(\mu) + g'(\mu)E(X - \mu) = g(\mu) = g(E[X])$$

4.51. Verify Cauchy-Schwarz inequality (4.41),

$$E\left(\left|XY\right|\right) \leq \sqrt{E(X^2)E(Y^2)}$$

We have

$$E([\alpha|X| - |Y|]^2) = E(X^2)\alpha^2 - 2E(|XY|)\alpha + E(Y^2) \geq 0 \tag{4.138}$$

The discriminant of the quadratic in α appearing in Eq. (4.138) must be nonpositive because the quadratic cannot have two distinct real roots. Therefore,

$$(2E[|XY|])^2 - 4E(X^2)E(Y^2) \leq 0$$

and we obtain

$$(E[|XY|])^2 \leq E(X^2)E(Y^2)$$

or $$(E[|XY|] \leq \sqrt{E(X^2)E(Y^2)}$$

Probability Generating Functions

4.52. Let X be a Bernoulli r.v. with parameter p.

    (*a*)   Find the probability generating function $G_X(z)$ of X.

    (*b*)   Find the mean and variance of X.

    (*a*)   From Eq. (2.32)

$$p_X(x) = p^x (1-p)^{1-x} = p^x q^{1-x} \qquad q = 1-p \qquad x = 0,1$$

By Eq. (4.42)

$$G_X(Z) = \sum_{x=0}^{1} p_X(x)z^x = p_X(0) + p_X(1)z = q + pz \qquad q = 1-p \tag{4.139}$$

    (*b*)   Differentiating Eq. (4.139), we have

$$G_X'(z) = p \qquad G_X''(z) = 0$$

Using Eqs. (4.49) and (4.55), we obtain

$$\mu = E(X) = G_X'(1) = p$$
$$\sigma^2 = \text{Var}(X) = G_X'(1) + G_X''(1) - [G_X'(1)]^2 = p - p^2 = p(1-p)$$

4.53. Let X be a binomial r.v. with parameters (n, p).

    (*a*)   Find the probability generating function $G_X(z)$ of X.

    (*b*)   Find $P(X = 0)$ and $P(X = 1)$.

    (*c*)   Find the mean and variance of X.

    (*a*)   From Eq. (2.36)

$$p_X(x) = \binom{n}{x} p^x q^{n-x} \qquad q = 1-p \qquad x = 0,1,\ldots$$

By Eq. (4.42)

$$G_X(z) = \sum_{x=0}^{\infty} \binom{n}{x} p^x q^{n-x} z^x = \sum_{x=0}^{\infty} \binom{n}{x} (pz)^x q^{n-x} = (pz+q)^n \qquad q = 1-p \tag{4.140}$$

    (*b*)   From Eqs. (4.47) and (4.140)

$$P(X = 0) = G_X(0) = q^n = (1-p)^n$$

Differentiating Eq. (4.140), we have

$$G_X'(z) = n\,p(pz + q)^{n-1} \tag{4.141}$$

Then from Eq. (4.48)

$$P(X = 1) = G_X'(0) = np\,q^{n-1} = np(1-p)^{n-1}$$

(c) Differentiating Eq. (4.141) again, we have

$$G_X''(z) = n(n-1) p^2 (pz + q)^{n-2} \tag{4.142}$$

Thus, using Eqs. (4.49) and (4.55), we obtain

$$\mu = E(X) = G_X'(1) = np(p+q)^{n-1} = np \qquad \text{since } (p+q) = 1.$$
$$\sigma^2 = \text{Var}(X) = G_X'(1) + G_x''(1) - [G_X'(1)]^2 = np + n(n-1) p^2 - n^2 p^2 = np (1-p)$$

4.54. Let $X_1, X_2, \ldots, X_n$ be independent Bernoulli r.v.'s with the same parameter p, and let $Y = X_1 + X_2 + \cdots + X_n$. Show that Y is a binomial r.v. with parameters (n, p).

By Eq. (4.139)

$$G_X(z) = q + pz \qquad q = 1 - p$$

Now applying property 5 Eq. (4.52), we have

$$G_Y(z) = \prod_{i=1}^{n} G_{X_i}(z) = (q + pz)^n \qquad q = 1 - p$$

Comparing with Eq. (4.140), we conclude that Y is a binomial r.v. with parameters (n, p).

4.55. Let X be a geometric r.v. with parameter p.
(a) Find the probability generating function $G_X(z)$ of X.
(b) Find the mean and variance of X.

(a) From Eq. (2.40) we have

$$p_X(X) = (1-p)^{x-1} p = q^{x-1} p \qquad q = 1 - p \qquad x = 1, 2, \ldots$$

Then by Eq. (4.42)

$$G_X(z) = \sum_{x=1}^{\infty} q^{x-1} p z^x = \frac{p}{q} \sum_{x=1}^{\infty} (zq)^x$$

$$= \frac{p}{q} \left[\sum_{x=0}^{\infty} (zq)^x - 1 \right] = \frac{p}{q} \left(\frac{1}{1-zq} - 1 \right) = \frac{zp}{1-zq} \qquad |zq| < 1$$

Thus,

$$G_X(z) = \frac{zp}{1-zq} \qquad |z| < \frac{1}{q} \qquad q = 1 - p \tag{4.143}$$

(b) Differentiating Eq. (4.143), we have

$$G_X'(z) = \frac{p}{1-zq} + \frac{zpq}{(1-zq)^2} = \frac{p}{(1-zq)^2} \tag{4.144}$$

$$G_X''(z) = \frac{2pq}{(1-zq)^3} \tag{4.145}$$

Thus, using Eqs. (4.49) and (4.55), we obtain

$$\mu = E(X) = G_X'(1) = \frac{p}{(1-q)^2} = \frac{p}{p^2} = \frac{1}{p}$$

$$\sigma^2 = \text{Var}(X)\, G_X'(1) + G_X''(1) - [G_X'(1)]^2 = \frac{1}{p} + \frac{2p(1-p)}{p^3} - \frac{1}{p^2} = \frac{1-p}{p^2}$$

4.56. Let X be a negative binomial r.v. with parameters p and k.

(a) Find the probability generating function $G_X(z)$ of X.

(b) Find the mean and variance of X.

(a) From Eq. (2.45)

$$p_X(x) = P(X=x) = \binom{x-1}{k-1} p^k (1-p)^{x-k} = \binom{x-1}{k-1} p^k q^{x-k} \qquad q = 1-p \qquad x = k, k+1, \ldots$$

By Eq. (4.42)

$$\begin{aligned}
G_X(z) &= \sum_{x=k}^{\infty} \binom{x-1}{k-1} p^k q^{x-k}\, z^x \\
&= p^k z^k \left[1 + kqz + \frac{k(k+1)}{2!}(qz)^2 + \frac{k(k+1)(k+2)}{3!}(qz)^3 + \cdots \right] \\
&= p^k z^k (1-qz)^{-k} = \left(\frac{zp}{1-zq} \right)^k \qquad |z| < \frac{1}{q}
\end{aligned}$$

(4.146)

(b) Differentiating Eq. (4.146), we have

$$G_X'(z) = k\, p^k z^{k-1} \left(\frac{1}{1-zq} \right)^{k+1} = k\, p^k \frac{z^{k-1}}{(1-zq)^{k+1}}$$

$$G_X''(z) = k\, p^k \left[\frac{(k-1)z^{k-2} + 2qz^{k-1}}{(1-zq)^{k+2}} \right]$$

Then,

$$G_X'(1) = k\, p^k \frac{1}{(1-q)^{k+1}} = k\, p^k \frac{1}{p^{k+1}} = \frac{k}{p}$$

$$G_X''(1) = k\, p^k \left[\frac{(k-1) + 2(1-p)}{p^{k+2}} \right] = \frac{k(k+1-2p)}{p^2} = \frac{k(k+1)}{p^2} - \frac{2k}{p}$$

Thus, by Eqs. (4.49) and (4.55), we obtain

$$\mu = E(X) = G_X'(1) = \frac{k}{p}$$

$$\sigma^2 = \text{Var}(X) = G_X'(1) + G_X''(1) - [G_X'(1)]^2 = \frac{k}{p} + \frac{k(k+1)}{p^2} - \frac{2k}{p} - \frac{k^2}{p^2} = \frac{k(1-p)}{p^2}$$

4.57. Let $X_1, X_2, \ldots, X_n$ be independent geometric r.v.'s with the same parameter p, and let $Y = X_1 + X_2 + \cdots + X_k$. Show that Y is a negative binomial r.v. with parameters p and k.

By Eq. (4.143) (Prob.4.55)

$$G_X(z) = \frac{zp}{1 - zp} \qquad |z| < \frac{1}{q} \qquad q = 1 - p$$

Now applying Eq. (4.52), we have

$$G_Y(z) = \prod_{i=1}^n G_{X_i}(z) = \left(\frac{zp}{1 - zp}\right)^k \qquad |z| < \frac{1}{q} \qquad q = 1 - p$$

Comparing with Eq. (4.146) (Prob. 4.56), we conclude that Y is a negative binomial r.v. with parameters p and k.

4.58. Let X be a Poisson r.v. with parameter λ.

   (a)   Find the probability generating function of $G_X(z)$ of X.

   (b)   Find the mean and variance of X.

   (a)   From Eq. (2.48)

$$p_X(x) = e^{-\lambda} \frac{\lambda^x}{x!} \qquad x = 0, 1, \ldots$$

By Eq. (4.42)

$$G_X(z) = \sum_{x=0}^{\infty} p_X(x) z^x = \sum_{x=0}^{\infty} e^{-\lambda} \frac{(\lambda z)^x}{x!} = e^{-\lambda} \sum_{x=0}^{\infty} \frac{(\lambda z)^x}{x!} = e^{-\lambda} e^{\lambda z} = e^{(z-1)\lambda} \tag{4.147}$$

   (b)   Differentiating Eq. (4.147), we have

$$G_X'(z) = \lambda e^{(z-1)\lambda}, \qquad G_X''(z) = \lambda^2 e^{(z-1)\lambda}$$

Then,

$$G_X'(1) = \lambda, \, G_X''(1) = \lambda^2$$

Thus, by Eq. (4.49) and Eq. (4.55), we obtain

$$\mu = E(X) = G_X'(1) = \lambda$$
$$\sigma^2 = \text{Var}(X) = G_X'(1) + G_X''(1) - [G_X'(1)]^2 = \lambda + \lambda^2 - \lambda^2 = \lambda$$

4.59. Let $X_1, X_2, \ldots, X_n$ be independent Poisson r.v.'s with the same parameter λ and let $Y = X_1 + X_2 + \cdots + X_n$. Show that Y is a Poisson r.v. with parameter $n\lambda$.

Using Eq. (4.147) and Eq. (4.52), we have

$$G_Y(z) = \prod_{i=1}^n G_{X_i}(z) = \prod_{i=1}^n e^{(z-1)\lambda} = e^{(z-1)n\lambda} \tag{4.148}$$

which indicates that Y is a Poisson r.v. with parameter $n\lambda$.

Moment Generating Functions

4.60. Let the moment of a discrete r.v. X be given by

$$E(X^k) = 0.8 \qquad k = 1, 2, \ldots$$

(*a*) Find the moment generating function of X.

(*b*) Find $P(X = 0)$ and $P(X = 1)$.

(*a*) By Eq. (4.57), the moment generating function of X is

$$M_X(t) = 1 + tE(X) + \frac{t^2}{2!}E(X^2) + \cdots + \frac{t^k}{k!}E(X^k) + \cdots$$

$$= 1 + 0.8\left(t + \frac{t^2}{2!} + \cdots + \frac{t^k}{k!} + \cdots\right) = 1 + 0.8\sum_{k=1}^{\infty}\frac{t^k}{k!}$$

$$= 0.2 + 0.8\sum_{k=0}^{\infty}\frac{t^k}{k!} = 0.2 + 0.8e^t \tag{4.149}$$

(*b*) By definition (4.56),

$$M_X(t) = E(e^{tX}) = \sum_i e^{tx_i}p_X(x_i) \tag{4.150}$$

Thus, equating Eqs. (4.149) and (4.150), we obtain

$$p_X(0) = P(X = 0) = 0.2 \qquad p_X(1) = P(X = 1) = 0.8$$

4.61. Let X be a Bernoulli r.v.

(*a*) Find the moment generating function of X.

(*b*) Find the mean and variance of X.

(*a*) By definition (4.56) and Eq. (2.32),

$$M_X(t) = E(e^{tX}) = \sum_i e^{tx_i}p_X(x_i)$$

$$= e^{t(0)}p_X(0) + e^{t(1)}p_X(1) = (1 - p) + pe^t \tag{4.151}$$

which can also be obtained by substituting z by e^t in Eq. (4.139).

(*b*) By Eq. (4.58),

$$E(X) = M_X'(0) = pe^t\Big|_{t=0} = p$$

$$E(X^2) = M_X''(0) = pe^t\Big|_{t=0} = p$$

Hence, $\qquad \mathrm{Var}(X) = E(X^2) - [E(X)]^2 = p - p^2 = p(1 - p)$

4.62. Let X be a binomial r.v. with parameters (n, p).

    (*a*)   Find the moment generating function of X.

    (*b*)   Find the mean and variance of X.

    (*a*)   By definition (4.56) and Eq. (2.36), and letting $q = 1 - p$, we get

$$
\begin{aligned}
M_X(t) = E(e^{tX}) &= \sum_{k=0}^{n} e^{tk} \binom{n}{k} p^k q^{n-k} \\
&= \sum_{k=0}^{n} \binom{n}{k} (e^t p)^k q^{n-k} = (q + pe^t)^n
\end{aligned}
\tag{4.152}
$$

which can also be obtained by substituting z by e^t in Eq. (4.140).

    (*b*)   The first two derivatives of $M_X(t)$ are

$$
\begin{aligned}
M_X'(t) &= n(q + pe^t)^{n-1} pe^t \\
M_X''(t) &= n(q + pe^t)^{n-1} pe^t + n(n-1)(q + pe^t)^{n-2}(pe^t)^2
\end{aligned}
$$

Thus, by Eq. (4.58),

$$
\begin{aligned}
\mu_X = E(X) &= M_X'(0) = np \\
E(X^2) = M_X''(0) &= np + n(n-1)p^2
\end{aligned}
$$

Hence, $\qquad\qquad\qquad \sigma_X^2 = E(X^2) - [E(X)]^2 = np(1 - p)$

4.63. Let X be a Poisson r.v. with parameter λ.

    (*a*)   Find the moment generating function of X.

    (*b*)   Find the mean and variance of X.

    (*a*)   By definition (4.56) and Eq. (2.48),

$$
\begin{aligned}
M_X(t) = E(e^{tX}) &= \sum_{i=0}^{\infty} e^{ti} e^{-\lambda} \frac{\lambda^i}{i!} \\
&= e^{-\lambda} \sum_{i=0}^{\infty} \frac{(\lambda e^t)^i}{i!} = e^{-\lambda} e^{\lambda e^t} = e^{\lambda(e^t - 1)}
\end{aligned}
\tag{4.153}
$$

which can also be obtained by substituting z by e^t in Eq. (4.147).

    (*b*)   The first two derivatives of $M_X(t)$ are

$$
\begin{aligned}
M_X'(t) &= \lambda e^t e^{\lambda(e^t - 1)} \\
M_X''(t) &= (\lambda e^t)^2 e^{\lambda(e^t - 1)} + \lambda e^t e^{\lambda(e^t - 1)}
\end{aligned}
$$

Thus, by Eq. (4.58),

$$
E(X) = M_X'(0) = \lambda \qquad\qquad E(X^2) = M_X''(0) = \lambda^2 + \lambda
$$

Hence, $\qquad\qquad$ $\text{Var}(X) = E(X^2) - [E(X)]^2 = \lambda^2 + \lambda - \lambda^2 = \lambda$

4.64. Let X be an exponential r.v. with parameter λ.

    (*a*)   Find the moment generating function of X.

    (*b*)   Find the mean and variance of X.

    (*a*)   By definition (4.56) and Eq. (2.60),

$$M_X(t) = E(e^{tX}) = \int_0^\infty \lambda e^{-\lambda x} e^{tx}\, dx$$
$$= \frac{\lambda}{t-\lambda} e^{(t-\lambda)x}\Big|_0^\infty = \frac{\lambda}{\lambda - t} \qquad \lambda > t \tag{4.154}$$

    (*b*)   The first two derivatives of $M_X(t)$ are

$$M_X'(t) = \frac{\lambda}{(\lambda - t)^2} \qquad M_X''(t) = \frac{2\lambda}{(\lambda - t)^3}$$

    Thus, by Eq. (4.58),

$$E(X) = M_X'(0) = \frac{1}{\lambda} \qquad E(X^2) = M_X''(0) = \frac{2}{\lambda^2}$$

Hence, $$\text{Var}(X) = E(X^2) - [E(X)]^2 = \frac{2}{\lambda^2} - \left(\frac{1}{\lambda}\right)^2 = \frac{1}{\lambda^2}$$

4.65. Let X be a gamma r.v. with parameters (α, λ).

    (*a*)   Find the moment generating function of X.

    (*b*)   Find the mean and variance of X.

    (*a*)   By definition (4.56) and Eq. (2.65)

$$M_X(t) = E(e^{tX}) = \int_0^\infty \frac{e^{tx} x^{\alpha-1} \lambda^\alpha e^{-\lambda x}}{\Gamma(\alpha)}\, dx = \frac{\lambda^\alpha}{\Gamma(\alpha)} \int_0^\infty x^{\alpha-1} e^{-(\lambda - t)x}\, dx$$

    Let $y = (\lambda - t)x, dy = (\lambda - t)dx$. Then

$$M_X(t) = \frac{\lambda^\alpha}{\Gamma(\alpha)} \int_0^\infty \left(\frac{y}{\lambda - t}\right)^{\alpha-1} e^{-y} \frac{dy}{(\lambda - t)} = \frac{\lambda^\alpha}{(\lambda - t)^\alpha \Gamma(\alpha)} \int_0^\infty y^{\alpha-1} e^{-y}\, dy$$

    Since $\int_0^\infty y^{\alpha-1} e^{-y}\, dy = \Gamma(\alpha)$ (Eq. (2.66)) we obtain

$$M_X(t) = \left(\frac{\lambda}{\lambda - t}\right)^\alpha \tag{4.155}$$

    (*b*)   The first two derivatives of $M_X(t)$ are

$$M_X'(t) = \alpha \lambda^\alpha (\lambda - t)^{-(\alpha+1)}, M_X''(t) = \alpha(\alpha + 1) \lambda^\alpha (\lambda - t)^{-(\alpha+2)}$$

Thus, by Eq. (4.58)

$$\mu = E(X) = M_X'(0) = \frac{\alpha}{\lambda}, \quad E(X^2) = M_X''(0) = \frac{\alpha(\alpha+1)}{\lambda^2}$$

Hence,

$$\sigma^2 = \text{Var}(X) = E(X^2) - [E(X)]^2 = \frac{\alpha(\alpha+1)}{\lambda^2} - \frac{\alpha^2}{\lambda^2} = \frac{\alpha}{\lambda^2}$$

4.66. Find the moment generating function of the standard normal r.v. $X = N(0; 1)$, and calculate the first three moments of X.

By definition (4.56) and Eq. (2.71),

$$M_X(t) = E(e^{tX}) = \int_{-\infty}^{\infty} \frac{1}{\sqrt{2\pi}} e^{-x^2/2} e^{tx} \, dx$$

Combining the exponents and completing the square, that is,

$$-\frac{x^2}{2} + tx = -\frac{(x-t)^2}{2} + \frac{t^2}{2}$$

we obtain

$$M_X(t) = e^{t^2/2} \int_{-\infty}^{\infty} \frac{1}{\sqrt{2\pi}} e^{-(x-t)^2/2} \, dx = e^{t^2/2} \qquad (4.156)$$

since the integrand is the pdf of $N(t; 1)$.

Differentiating $M_X(t)$ with respect to t three times, we have

$$M_X'(t) = te^{t^2/2} \qquad M_X''(t) = (t^2+1)e^{t^2/2} \qquad M_X^{(3)}(t) = (t^3+3t)e^{t^2/2}$$

Thus, by Eq. (4.58),

$$E(X) = M_X'(0) = 0 \qquad E(X^2) = M_X''(0) = 1 \qquad E(X^3) = M_X^{(3)}(0) = 0$$

4.67. Let $Y = aX + b$. Let $M_X(t)$ be the moment generating function of X. Show that the moment generating function of Y is given by

$$M_Y(t) = e^{tb} M_X(at) \qquad (4.157)$$

By Eqs. (4.56) and (4.129),

$$M_Y(t) = E(e^{tY}) = E[e^{t(aX+b)}]$$
$$= e^{tb} E(e^{atX}) = e^{tb} M_X(at)$$

4.68. Find the moment generating function of a normal r.v. $N(\mu; \sigma^2)$.

If X is $N(0; 1)$, then from Prob. 4.1 (or Prob. 4.43), we see that $Y = \sigma X + \mu$ is $N(\mu; \sigma^2)$. Then by setting $a = \sigma$ and $b = \mu$ in Eq. (4.157) (Prob. 4.67) and using Eq. (4.156), we get

$$M_Y(t) = e^{\mu t} M_X(\sigma t) = e^{\mu t} e^{(\sigma t)^2/2} = e^{\mu t + \sigma^2 t^2/2} \tag{4.158}$$

4.69. Suppose that r.v. $X = N(0;1)$. Find the moment generating function of $Y = X^2$.

By definition (4.56) and Eq. (2.71)

$$M_Y(t) = E\big(e^{tY}\big) = E\big(e^{tX^2}\big)$$

$$= \int_{-\infty}^{\infty} e^{tx^2} \frac{1}{\sqrt{2\pi}} e^{-x^2/2} dx = \frac{1}{\sqrt{2\pi}} \int_{-\infty}^{\infty} e^{-\left(\frac{1}{2}-t\right)x^2} dx$$

Using $\displaystyle \int_{-\infty}^{\infty} e^{-ax^2} dx = \sqrt{\frac{\pi}{a}}$ we obtain

$$M_Y(t) = \frac{1}{\sqrt{2\pi}} \sqrt{\frac{\pi}{\frac{1}{2}-t}} = \frac{1}{\sqrt{1-2t}} \tag{4.159}$$

4.70. Let $X_1, \ldots, X_n$ be n independent r.v.'s and let the moment generating function of X_i be $M_{X_i}(t)$. Let $Y = X_1 + \cdots + X_n$. Find the moment generating function of Y.

By definition (4.56),

$$\begin{aligned} M_Y(t) = E(e^{tY}) &= E[e^{t(X_1+\cdots+X_n)}] = E(e^{tX_1}\cdots e^{tX_n}) \\ &= E(e^{tX_1})\cdots E(e^{tX_n}) \quad \text{(independence)} \\ &= M_{X_1}(t)\cdots M_{X_n}(t) \end{aligned} \tag{4.160}$$

4.71. Show that if $X_1, \ldots, X_n$ are independent Poisson r.v.'s X_i having parameter λ_i, then $Y = X_1 + \cdots + X_n$ is also a Poisson r.v. with parameter $\lambda = \lambda_1 + \cdots + \lambda_n$.

Using Eqs. (4.160) and (4.153), the moment generating function of Y is

$$M_Y(t) = \prod_{i=1}^{n} e^{\lambda_i(e^t - 1)} = e^{(\Sigma \lambda_i)(e^t - t)} = e^{\lambda(e^t - 1)}$$

which is the moment generating function of a Poisson r.v. with parameter λ. Hence, Y is a Poisson r.v. with parameter $\lambda = \Sigma \lambda_i = \lambda_1 + \cdots + \lambda_n$.

Note that Prob. 4.17 is a special case for $n = 2$.

4.72. Show that if $X_1, \ldots, X_n$ are independent normal r.v.'s and $X_i = N(\mu_i; \sigma_i^2)$, then $Y = X_1 + \cdots + X_n$ is also a normal r.v. with mean $\mu = \mu_1 + \cdots + \mu_n$ and variance $\sigma^2 = \sigma_1^2 + \cdots + \sigma_n^2$.

Using Eqs. (4.160) and (4.158), the moment generating function of Y is

$$M_Y(t) = \prod_{i=1}^{n} e^{\left(\mu_i t + \sigma_i^2 t^2/2\right)} = e^{(\Sigma \mu_i)t + (\Sigma \sigma_i^2)t^2/2} = e^{\mu t + \sigma^2 t^2/2}$$

which is the moment generating function of a normal r.v. with mean μ and variance σ^2. Hence, Y is a normal r.v. with mean $\mu = \mu_1 + \cdots + \mu_n$ and variance $\sigma^2 = \sigma_1^2 + \cdots + \sigma_n^2$.

Note that Prob. 4.20 is a special case for $n = 2$ with $\mu_i = 0$ and $\sigma_i^2 = 1$.

4.73. Find the moment generating function of a gamma r.v. Y with parameters (n, λ).

From Prob. 4.39, we see that if $X_1, \ldots, X_n$ are independent exponential r.v.'s, each with parameter λ, then $Y = X_1 + \ldots + X_n$ is a gamma r.v. with parameters (n, λ). Thus, by Eqs. (4.160) and (4.154), the moment generating function of Y is

$$M_Y(t) = \prod_{i=1}^{n} \left(\frac{\lambda}{\lambda - t}\right) = \left(\frac{\lambda}{\lambda - t}\right)^n \tag{4.161}$$

4.74. Suppose that $X_1, X_2, \ldots, X_n$ be independent standard normal r.v.'s and $X_i = N(0; 1)$. Let $Y = X_1^2 + X_2^2 + \ldots + X_n^2$

(a) Find the moment generating function of Y.

(b) Find the mean and variance of Y.

(a) By Eqs. (4.159) and (4.160)

$$M_Y(t) = E(e^{tY}) = \prod_{i=1}^{n} (1 - 2t)^{-1/2} = (1 - 2t)^{-n/2} \tag{4.162}$$

(b) Differentiating Eq. (4.162), we obtain

$$M_Y{}'(t) = n(1 - 2t)^{-\frac{n}{2}-1}, \qquad M_Y{}''(t) = n(n + 2)(1 - 2t)^{-\frac{n}{2}-2}$$

Thus, by Eq. (4.58)

$$E(Y) = M_Y{}'(0) = n \tag{4.163}$$

$$E(Y^2) = M_Y{}''(0) = n(n + 2) \tag{4.164}$$

Hence,

$$\text{Var}(Y) = E(Y^2) - [E(Y)]^2 = n(n + 2) - n^2 = 2n \tag{4.165}$$

Characteristic Functions

4.75. The r.v. X can take on the values $x_1 = -1$ and $x_2 = +1$ with pmf's $p_X(x_1) = p_X(x_2) = 0.5$. Determine the characteristic function of X.

By definition (4.66), the characteristic function of X is

$$\Psi_X(\omega) = 0.5e^{-j\omega} + 0.5e^{j\omega} = \frac{1}{2}(e^{j\omega} + e^{-j\omega}) = \cos\omega$$

4.76. Find the characteristic function of a Cauchy r.v. X with parameter α and pdf given by

$$f_X(x) = \frac{a}{\pi(x^2 + a^2)} \qquad -\infty < x < \infty$$

By direct integration (or from the Table of Fourier transforms in Appendix B), we have the following Fourier transform pair:

$$e^{-a|x|} \leftrightarrow \frac{2a}{\omega^2 + a^2}$$

Now, by the duality property of the Fourier transform, we have the following Fourier transform pair:

$$\frac{2a}{x^2 + a^2} \leftrightarrow 2\pi e^{-a|-\omega|} = 2\pi e^{-a|\omega|}$$

or (by the linearity property of the Fourier transform)

$$\frac{a}{\pi(x^2 + a^2)} \leftrightarrow e^{-a|\omega|}$$

Thus, the characteristic function of X is

$$\Psi_X(\omega) = e^{-a|\omega|} \tag{4.166}$$

Note that the moment generating function of the Cauchy r.v. X does not exist, since $E(X^n) \to \infty$ for $n \geq 2$.

4.77. The characteristic function of a r.v. X is given by

$$\Psi_X(\omega) = \begin{cases} 1 - |\omega| & |\omega| < 1 \\ 0 & |\omega| > 1 \end{cases}$$

Find the pdf of X.

From formula (4.67), we obtain the pdf of X as

$$\begin{aligned}
f_X(x) &= \frac{1}{2\pi} \int_{-\infty}^{\infty} \Psi_X(\omega) e^{-j\omega x} d\omega \\
&= \frac{1}{2\pi}\left[\int_{-1}^{0}(1+\omega)e^{-j\omega x}d\omega + \int_{0}^{1}(1-\omega)e^{-j\omega x}d\omega \right] \\
&= \frac{1}{2\pi x^2}(2 - e^{jx} - e^{-jx}) = \frac{1}{\pi x^2}(1 - \cos x) \\
&= \frac{1}{2\pi}\left[\frac{\sin(x/2)}{x/2} \right]^2 \qquad -\infty < x < \infty
\end{aligned}$$

4.78. Find the characteristic function of a normal r.v. $X = N(\mu; \sigma^2)$.

The moment generating function of $N(\mu; \sigma^2)$ is [Eq. (4.158)]

$$M_X(t) = e^{\mu t + \sigma^2 t^2/2}$$

Thus, the characteristic function of $N(\mu; \sigma^2)$ is obtained by setting $t = j\omega$ in $M_X(t)$; that is,

$$\Psi_X(\omega) = e^{\mu t + \sigma^2 t^2/2}\bigg|_{t = j\omega} = e^{j\omega\mu - \sigma^2\omega^2/2} \tag{4.167}$$

4.79. Let $Y = aX + b$. Show that if $\Psi_X(\omega)$ is the characteristic function of X, then the characteristic function of Y is given by

$$\Psi_Y(\omega) = e^{j\omega b}\Psi_X(a\omega) \tag{4.168}$$

By definition (4.66),

$$\Psi_Y(\omega) = E\left(e^{j\omega Y}\right) = E\left[e^{j\omega(aX+b)}\right]$$
$$= e^{j\omega b}E\left(e^{ja\omega X}\right) = e^{j\omega b}\Psi_X(a\omega)$$

4.80. Using the characteristic equation technique, redo part (b) of Prob. 4.18.

Let $Z = X + Y$, where X and Y are independent. Then

$$\Psi_Z(\omega) = E\left(e^{j\omega Z}\right) = E\left(e^{j\omega(X+Y)}\right) = E\left(e^{j\omega X}\right)E\left(e^{j\omega Y}\right)$$
$$= \Psi_X(\omega)\Psi_Y(\omega) \tag{4.169}$$

Applying the convolution theorem of the Fourier transform (Appendix B), we obtain

$$f_Z(z) = \mathcal{F}^{-1}[\Psi_Z(\omega)] = \mathcal{F}^{-1}[\Psi_X(\omega)\Psi_Y(\omega)]$$
$$= f_X(z) * f_Y(z) = \int_{-\infty}^{\infty} f_X(x)f_Y(z-x)\,dx$$

The Laws of Large Numbers and the Central Limit Theorem

4.81. Verify the weak law of large numbers (4.74); that is,

$$\lim_{n \to \infty} P(|\bar{X}_n - \mu| > \varepsilon) = 0 \qquad \text{for any } \varepsilon$$

where $\bar{X}_n = \dfrac{1}{n}(X_1 + \cdots + X_n)$ and $E(X_i) = \mu$, $\text{Var}(X_i) = \sigma^2$.

Using Eqs. (4.132) and (4.136), we have

$$E(\bar{X}_n) = \mu \quad \text{and} \quad \text{Var}(\bar{X}_n) = \frac{\sigma^2}{n} \tag{4.170}$$

Then it follows from Chebyshev's inequality [Eq. (2.116)] (Prob. 2.39) that

$$P(|\bar{X}_n - \mu| > \varepsilon) \le \frac{\sigma^2}{n\varepsilon^2} \tag{4.171}$$

Since $\lim_{n \to \infty} \sigma^2/(n\varepsilon^2) = 0$, we get

$$\lim_{n \to \infty} P(|\bar{X}_n - \mu| > \varepsilon) = 0$$

4.82. Let X be a r.v. with pdf $f_X(x)$ and let $X_1, \ldots, X_n$ be a set of independent r.v.'s each with pdf $f_X(x)$. Then the set of r.v.'s $X_1, \ldots, X_n$ is called a *random sample of size n* of X. The *sample mean* is defined by

$$\bar{X}_n = \frac{1}{n}(X_1 + \cdots + X_n) = \frac{1}{n}\sum_{i=1}^{n} X_i \tag{4.172}$$

Let $X_1, \ldots, X_n$ be a random sample of X with mean μ and variance σ^2. How many samples of X should be taken if the probability that the sample mean will not deviate from the true mean μ by more than $\sigma/10$ is at least 0.95?

Setting $\varepsilon = \sigma/10$ in Eq. (4.171), we have

$$P\left(|\bar{X}_n - \mu| > \frac{\sigma}{10}\right) = 1 - P\left(|\bar{X}_n - \mu| \le \frac{\sigma}{10}\right) \le \frac{\sigma^2}{n\sigma^2/100} = \frac{100}{n}$$

or

$$P\left(|\bar{X}_n - \mu| \le \frac{\sigma}{10}\right) \ge 1 - \frac{100}{n}$$

Thus, if we want this probability to be at least 0.95, we must have $100/n \le 0.05$ or $n \ge 100/0.05 = 2000$.

4.83. Verify the central limit theorem (4.77).

Let $X_1, \ldots, X_n$ be a sequence of independent, identically distributed r.v.'s with $E(X_i) = \mu$ and $Var(X_i) = \sigma^2$. Consider the sum $S_n = X_1 + \cdots + X_n$. Then by Eqs. (4.132) and (4.136), we have $E(S_n) = n\mu$ and $Var(S_n) = n\sigma^2$. Let

$$Z_n = \frac{S_n - n\mu}{\sqrt{n}\sigma} = \frac{1}{\sqrt{n}}\sum_{i=1}^{n}\left(\frac{X_i - \mu}{\sigma}\right) \tag{4.173}$$

Then by Eqs. (4.129) and (4.130), we have $E(Z_n) = 0$ and $Var(Z_n) = 1$. Let $M(t)$ be the moment generating function of the standardized r.v. $Y_i = (X_i - \mu)/\sigma$. Since $E(Y_i) = 0$ and $E(Y_i^2) = Var(Y_i) = 1$, by Eq. (4.58), we have

$$M(0) = 1 \qquad M'(0) = E(Y_i) = 0 \qquad M''(0) = E(Y_i^2) = 1$$

Given that $M'(t)$ and $M''(t)$ are continuous functions of t, a Taylor (or Maclaurin) expansion of $M(t)$ about $t = 0$ can be expressed as

$$M(t) = M(0) + M'(0)t + M''(t_1)\frac{t^2}{2!} = 1 + M''(t_1)\frac{t^2}{2} \qquad 0 \le t_1 \le t$$

By adding and subtracting $t^2/2$, we have

$$M(t) = 1 + \frac{1}{2}t^2 + \frac{1}{2}[M''(t_1) - 1]t^2 \tag{4.174}$$

Now, by Eqs. (4.157) and (4.160), the moment generating function of Z_n is

$$M_{Z_n}(t) = \left[M\left(\frac{t}{\sqrt{n}} \right) \right]^n \tag{4.175}$$

Using Eqs. (4.174), Eq. (4.175) can be written as

$$M_{Z_n}(t) = \left[1 + \frac{1}{2}\left(\frac{t}{\sqrt{n}} \right)^2 + \frac{1}{2}[M''(t_1) - 1]\left(\frac{t}{\sqrt{n}} \right)^2 \right]^n$$

where now t_1 is between 0 and $t/\sqrt{n}$. Since $M''(t)$ is continuous at $t = 0$ and $t_1 \to 0$ as $n \to \infty$, we have

$$\lim_{n \to \infty} [M''(t_1) - 1] = M''(0) - 1 = 1 - 1 = 0$$

Thus, from elementary calculus, $\lim_{n \to \infty} (1 + x/n)^n = e^x$, and we obtain

$$\lim_{n \to \infty} M_{Z_n}(t) = \lim_{n \to \infty} \left\{ 1 + \frac{t^2}{2n} + \frac{1}{2n}[M''(t_1) - 1]t^2 \right\}^n$$

$$= \lim_{n \to \infty} \left(1 + \frac{t^2/2}{n} \right)^n = e^{t^2/2}$$

The right-hand side is the moment generating function of the standard normal r.v. $Z = N(0; 1)$ [Eq. (4.156)]. Hence, by Lemma 4.3 of the moment generating function,

$$\lim_{n \to \infty} Z_n = N(0;1)$$

4.84. Let $X_1, \ldots, X_n$ be n independent Cauchy r.v.'s with identical pdf shown in Prob. 4.76. Let

$$Y_n = \frac{1}{n}(X_1 + \cdots + X_n) = \frac{1}{n}\sum_{n=1}^{n} X_i$$

(a)　Find the characteristic function of Y_n.

(b)　Find the pdf of Y_n.

(c)　Does the central limit theorem hold?

(a)　From Eq. (4.166), the characteristic function of X_i is

$$\Psi_{X_i}(\omega) = e^{-\alpha |\omega|}$$

Let $Y = X_1 + \cdots + X_n$. Then the characteristic function of Y is

$$\Psi_Y(\omega) = E(e^{j\omega Y}) = E[e^{j\omega(X_1 + \cdots + X_n)}] = \prod_{i=1}^{n} \Psi_{X_i}(\omega) = e^{-na|\omega|} \tag{4.176}$$

Now $Y_n = (1/n)Y$. Thus, by Eq. (4.168), the characteristic function of Y_n is

$$\Psi_{Y_n}(\omega) = \Psi_Y\left(\frac{\omega}{n}\right) = e^{-na|\omega/n|} = e^{-a|\omega|} \tag{4.177}$$

(b) Equation (4.177) indicates that Y_n is also a Cauchy r.v. with parameter a, and its pdf is the same as that of X_i.

(c) Since the characteristic function of Y_n is independent of n and so is its pdf, Y_n does not tend to a normal r.v. as $n \to \infty$, and so the central limit theorem does not hold in this case.

4.85. Let Y be a binomial r.v. with parameters (n, p). Using the central limit theorem, derive the approximation formula

$$P(Y \le y) \approx \Phi\left(\frac{y - np}{\sqrt{np(1 - p)}}\right) \tag{4.178}$$

where $\Phi(z)$ is the cdf of a standard normal r.v. [Eq. (2.73)].

We saw in Prob. 4.54 that if $X_1, \ldots, X_n$ are independent Bernoulli r.v.'s, each with parameter p, then $Y = X_1 + \ldots + X_n$ is a binomial r.v. with parameters (n, p). Since X_i's are independent, we can apply the central limit theorem to the r.v. Z_n defined by

$$Z_n = \frac{1}{\sqrt{n}} \sum_{i=1}^{n} \left(\frac{X_i - E(X_i)}{\text{Var}(X_i)}\right) = \frac{1}{\sqrt{n}} \sum_{i=1}^{n} \left(\frac{X_i - p}{\sqrt{p(1-p)}}\right) \tag{4.179}$$

Thus, for large n, Z_n is normally distributed and

$$P(Z_n \le x) \approx \Phi(x) \tag{4.180}$$

Substituting Eq. (4.179) into Eq. (4.180) gives

$$P\left[\frac{1}{\sqrt{np(1-p)}}\left(\sum_{i=1}^{n}(X_i - p)\right) \le x\right] = P\left[Y \le x\sqrt{np(1-p)} + np\right] \approx \Phi(x)$$

or

$$P(Y \le y) \approx \Phi\left(\frac{y - np}{\sqrt{np(1-p)}}\right)$$

Because we are approximating a discrete distribution by a continuous one, a slightly better approximation is given by

$$P(Y \le y) \approx \Phi\left(\frac{y + \dfrac{1}{2} - np}{\sqrt{np(1-p)}}\right) \tag{4.181}$$

Formula (4.181) is referred to as a *continuity correction* of Eq. (4.178).

4.86. Let Y be a Poisson r.v. with parameter λ. Using the central limit theorem, derive approximation formula:

$$P(Y \leq y) \approx \Phi\left(\frac{y - \lambda}{\sqrt{\lambda}}\right) \tag{4.182}$$

We saw in Prob. 4.71 that if $X_1, \ldots, X_n$ are independent Poisson r.v.'s X_i having parameter λ_i, then $Y = X_1 + \cdots + X_n$ is also a Poisson r.v. with parameter $\lambda = \lambda_1 + \cdots + \lambda_n$. Using this fact, we can view a Poisson r.v. Y with parameter λ as a sum of independent Poisson r.v.'s $X_i, i = 1, \cdots, n$, each with parameter λ/n; that is,

$$Y = X_1 + \cdots + X_n$$

$$E(X_i) = \frac{\lambda}{n} = \text{Var}(X_i)$$

The central limit theorem then implies that the r.v. Z is defined by

$$Z = \frac{Y - E(Y)}{\sqrt{\text{Var}(Y)}} = \frac{Y - \lambda}{\sqrt{\lambda}} \tag{4.183}$$

is approximately normal and

$$P(Z \leq z) \approx \Phi(z) \tag{4.184}$$

Substituting Eq. (4.183) into Eq. (4.184) gives

$$P\left(\frac{Y - \lambda}{\sqrt{\lambda}} \leq z\right) = P(Y \leq \sqrt{\lambda}z + \lambda) \approx \Phi(z)$$

or

$$P(Y \leq y) \approx \Phi\left(\frac{y - \lambda}{\sqrt{\lambda}}\right)$$

Again, using a continuity correction, a slightly better approximation is given by

$$P(Y \leq y) \approx \Phi\left(\frac{y + \dfrac{1}{2} - \lambda}{\sqrt{\lambda}}\right) \tag{4.185}$$

SUPPLEMENTARY PROBLEMS

4.87. Let $Y = 2X + 3$. Find the pdf of Y if X is a uniform r.v. over $(-1, 2)$.

4.88. Let X be a r.v. with pdf $f_X(x)$. Let $Y = |X|$. Find the pdf of Y in terms of $f_X(x)$.

4.89. Let $Y = \sin X$, where X is uniformly distributed over $(0, 2\pi)$. Find the pdf of Y.

4.90. Let X and Y be independent r.v.'s, each uniformly distributed over $(0, 1)$. Let $Z = X + Y$, $W = X - Y$. Find the marginal pdf's of Z and W.

4.91. Let X and Y be independent exponential r.v.'s with parameters α and β, respectively. Find the pdf of (a) $Z = X - Y$; (b) $Z = X/Y$; (c) $Z = \max(X, Y)$; (d) $Z = \min(X, Y)$.

4.92. Let X denote the number of heads obtained when three independent tossings of a fair coin are made. Let $Y = X^2$. Find $E(Y)$.

4.93. Let X be a uniform r.v. over $(-1, 1)$. Let $Y = X^n$.

    (a)   Calculate the covariance of X and Y.

    (b)   Calculate the correlation coefficient of X and Y.

4.94. What is the pmf of r.v. X whose probability generating function is $G_X(z) = \dfrac{1}{2-z}$?

4.95. Let $Y = aX + b$. Express the probability generating function of Y, $G_Y(z)$, in terms of the probability generating function of X, $G_X(z)$.

4.96. Let the moment generating function of a discrete r.v. X be given by

$$M_X(t) = 0.25e^t + 0.35e^{3t} + 0.40e^{5t}$$

Find $P(X = 3)$.

4.97. Let X be a geometric r.v. with parameter p.

    (a)   Determine the moment generating function of X.

    (b)   Find the mean of X for $p = \dfrac{2}{3}$.

4.98. Let X be a uniform r.v. over (a, b).

    (a)   Determine the moment generating function of X.

    (b)   Using the result of (a), find $E(X)$, $E(X^2)$, and $E(X^3)$.

4.99. Consider a r.v. X with pdf

$$f_X(x) = \frac{1}{\sqrt{32\pi}} e^{-(x+7)^2/32} \qquad -\infty < x < \infty$$

Find the moment generating function of X.

4.100. Let $X = N(0; 1)$. Using the moment generating function of X, determine $E(X^n)$.

4.101. Let X and Y be independent binomial r.v.'s with parameters (n, p) and (m, p), respectively. Let $Z = X + Y$. What is the distribution of Z?

4.102. Let (X, Y) be a continuous bivariate r.v. with joint pdf

$$f_{XY}(x, y) = \begin{cases} e^{-(x+y)} & x > 0, y > 0 \\ 0 & \text{otherwise} \end{cases}$$

    (*a*)   Find the joint moment generating function of X and Y.

    (*b*)   Find the joint moments m_{10}, m_{01}, and m_{11}.

4.103. Let (X, Y) be a bivariate normal r.v. defined by Eq. (3.88). Find the joint moment generating function of X and Y.

4.104. Let $X_1, \ldots, X_n$ be n independent r.v.'s and $X_i > 0$. Let

$$Y = X_1 \cdots X_n = \prod_{i=1}^{n} X_i$$

Show that for large n, the pdf of Y is approximately log-normal.

4.105. Let $Y = (X - \lambda)/\sqrt{\lambda}$, where X is a Poisson r.v. with parameter λ. Show that $Y \approx N(0; 1)$ when λ is sufficiently large.

4.106. Consider an experiment of tossing a fair coin 1000 times. Find the probability of obtaining more that 520 heads (*a*) by using formula (4.178), and (*b*) by formula (4.181).

4.107. The number of cars entering a parking lot is Poisson distributed with a rate of 100 cars per hour. Find the time required for more than 200 cars to have entered the parking lot with probability 0.90 (*a*) by using formula (4.182), and (*b*) by formula (4.185).

ANSWERS TO SUPPLEMENTARY PROBLEMS

4.87. $f_Y(y) = \begin{cases} \dfrac{1}{6} & 1 < y < 7 \\ 0 & \text{otherwise} \end{cases}$

4.88. $f_Y(y) = \begin{cases} f_X(y) + f_X(-y) & y > 0 \\ 0 & y < 0 \end{cases}$

4.89. $f_Y(y) = \begin{cases} \dfrac{1}{\pi\sqrt{1 - y^2}} & -1 < y < 1 \\ 0 & \text{otherwise} \end{cases}$

4.90. $f_Z(z) = \begin{cases} z & 0 < z < 1 \\ -z + 2 & 1 < z < 2 \\ 0 & \text{otherwise} \end{cases}$ $f_W(w) = \begin{cases} w + 1 & -1 < w < 0 \\ -w + 1 & 0 < w < 1 \\ 0 & \text{otherwise} \end{cases}$

4.91. (a) $f_Z(z) = \begin{cases} \dfrac{\alpha\beta}{\alpha+\beta}e^{-\alpha z} & z > 0 \\[3mm] \dfrac{\alpha\beta}{\alpha+\beta}e^{\beta z} & z < 0 \end{cases}$ (b) $f_Z(z) = \begin{cases} \dfrac{\alpha\beta}{(\alpha z+\beta)^2} & z > 0 \\[3mm] 0 & z < 0 \end{cases}$

(c) $f_Z(z) = \begin{cases} \alpha e^{-\alpha z}(1-e^{-\beta z}) + \beta e^{-\beta z}(1-e^{-\alpha z}) & z > 0 \\ 0 & z < 0 \end{cases}$

(d) $f_Z(z) = \begin{cases} (\alpha+\beta)e^{-(\alpha+\beta)z} & z > 0 \\ 0 & z < 0 \end{cases}$

4.92. 3

4.93. (a) $\text{Cov}(X,Y) = \begin{cases} \dfrac{1}{n+2} & n = \text{odd} \\[3mm] 0 & n = \text{even} \end{cases}$ (b) $\rho_{XY} = \begin{cases} \dfrac{\sqrt{3(2n+1)}}{n+2} & n = \text{odd} \\[3mm] 0 & n = \text{even} \end{cases}$

4.94. $p_X(x) = \left(\dfrac{1}{2}\right)^{x+1}$

4.95. $G_Y(z) = z^a G_X(z^b)$.

4.96. 0.35

4.97. (a) $M_X(t) = \dfrac{pe^t}{1-qe^t} \quad t < -\ln q, \; q = 1-p$ (b) $E(X) = \dfrac{3}{2}$

4.98. (a) $M_X(t) = \dfrac{e^{tb} - e^{ta}}{t(b-a)}$

(b) $E(X) = \dfrac{1}{2}(b+a), \quad E(X^2) = \dfrac{1}{3}(b^2+ab+a^2), \quad E(X^3) = \dfrac{1}{4}(b^3+b^2a+ba^2+a^3)$

4.99. $M_X(t) = e^{-7t + 8t^2}$

4.100. $E(X^n) = \begin{cases} 0 & n = 1, 3, 5, \ldots \\ 1 \cdot 3 \cdots (n-1) & n = 2, 4, 6, \ldots \end{cases}$

4.101. *Hint:* Use the moment generating functions.

Z is a binomial r.v. with parameters $(n + m, p)$.

4.102. (a) $M_{XY}(t_1, t_2) = \dfrac{1}{(1-t_1)(1-t_2)}$ (b) $m_{10} = 1, m_{01} = 1, m_{11} = 1$

4.103. $M_{XY}(t_1, t_2) = e^{t_1 \mu_X + t_2 \mu_Y + (t_1^2 \sigma_X^2 + 2 t_1 t_2 \sigma_X \sigma_Y \rho + t_2^2 \sigma_Y^2)/2}$

4.104. *Hint:* Take the natural logarithm of Y and use the central limit theorem and the result of Prob. 4.10.

4.105. *Hint:* Find the moment generating function of Y and let $\lambda \to \infty$.

4.106. (*a*) 0.1038 (*b*) 0.0974

4.107. (*a*) 2.189 h (*b*) 2.1946 h

CHAPTER 5

Random Processes

5.1 Introduction

In this chapter, we introduce the concept of a random (or stochastic) process. The theory of random processes was first developed in connection with the study of fluctuations and noise in physical systems. A random process is the mathematical model of an empirical process whose development is governed by probability laws. Random processes provides useful models for the studies of such diverse fields as statistical physics, communication and control, time series analysis, population growth, and management sciences.

5.2 Random Processes

A. Definition:

A *random process* is a family of r.v.'s $\{X(t), t \in T\}$ defined on a given probability space, indexed by the parameter t, where t varies over an index set T.

Recall that a random variable is a function defined on the sample space S (Sec. 2.2). Thus, a random process $\{X(t), t \in T\}$ is really a function of two arguments $\{X(t, \zeta), t \in T, \zeta \in S\}$. For a fixed $t(= t_k)$, $X(t_k, \zeta) = X_k(\zeta)$ is a r.v. denoted by $X(t_k)$, as ζ varies over the sample space S. On the other hand, for a fixed sample point $\zeta_i \in S$, $X(t, \zeta_i) = X_i(t)$ is a single function of time t, called a *sample function* or a *realization* of the process. (See Fig. 5-1.) The totality of all sample functions is called an *ensemble*.

Of course if both ζ and t are fixed, $X(t_k, \zeta_i)$ is simply a real number. In the following we use the notation $X(t)$ to represent $X(t, \zeta)$.

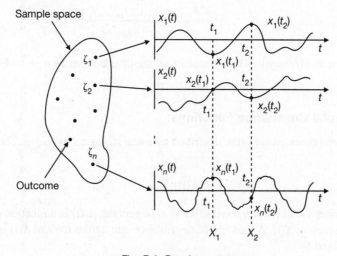

Fig. 5-1 Random process.

B. Description of Random Process:

In a random process $\{X(t), t \in T\}$, the index set T is called the *parameter set* of the random process. The values assumed by $X(t)$ are called *states*, and the set of all possible values forms the *state space E* of the random process. If the index set T of a random process is discrete, then the process is called a *discrete-parameter* (or *discrete-time*) process. A discrete-parameter process is also called a *random sequence* and is denoted by $\{X_n, n = 1, 2, ...\}$. If T is continuous, then we have a *continuous-parameter* (or *continuous-time*) process. If the state space E of a random process is discrete, then the process is called a *discrete-state* process, often referred to as a *chain*. In this case, the state space E is often assumed to be $\{0, 1, 2, ...\}$. If the state space E is continuous, then we have a *continuous-state* process.

A complex random process $X(t)$ is defined by

$$X(t) = X_1(t) + jX_2(t)$$

where $X_1(t)$ and $X_2(t)$ are (real) random processes and $j = \sqrt{-1}$. Throughout this book, all random processes are real random processes unless specified otherwise.

5.3 Characterization of Random Processes

A. Probabilistic Descriptions:

Consider a random process $X(t)$. For a fixed time t_1, $X(t_1) = X_1$ is a r.v., and its cdf $F_X(x_1; t_1)$ is defined as

$$F_X(x_1; t_1) = P\{X(t_1) \leq x_1\} \tag{5.1}$$

$F_X(x_1; t_1)$ is known as the *first-order distribution* of $X(t)$. Similarly, given t_1 and t_2, $X(t_1) = X_1$ and $X(t_2) = X_2$ represent two r.v.'s. Their joint distribution is known as the *second-order distribution* of $X(t)$ and is given by

$$F_X(x_1, x_2; t_1, t_2) = P\{X(t_1) \leq x_1, X(t_2) \leq x_2\} \tag{5.2}$$

In general, we define the *nth-order distribution* of $X(t)$ by

$$F_X(x_1, ..., x_n; t_1, ..., t_n) = P\{X(t_1) \leq x_1, ..., X(t_n) \leq x_n\} \tag{5.3}$$

If $X(t)$ is a discrete-time process, then $X(t)$ is specified by a collection of pmf's:

$$p_X(x_1, ..., x_n; t_1, ..., t_n) = P\{X(t_1) = x_1, ..., X(t_n) = x_n\} \tag{5.4}$$

If $X(t)$ is a continuous-time process, then $X(t)$ is specified by a collection of pdf's:

$$f_X(x_1, ..., x_n; t_1, ..., t_n) = \frac{\partial^n F_X(x_1, ..., x_n; t_1, ..., t_n)}{\partial x_1 \cdots \partial x_n} \tag{5.5}$$

The complete characterization of $X(t)$ requires knowledge of all the distributions as $n \to \infty$. Fortunately, often much less is sufficient.

B. Mean, Correlation, and Covariance Functions:

As in the case of r.v.'s, random processes are often described by using statistical averages. The *mean* of $X(t)$ is defined by

$$\mu_X(t) = E[X(t)] \tag{5.6}$$

where $X(t)$ is treated as a random variable for a fixed value of t. In general, $\mu_X(t)$ is a function of time, and it is often called the *ensemble average* of $X(t)$. A measure of dependence among the r.v.'s of $X(t)$ is provided by its *autocorrelation function*, defined by

$$R_X(t, s) = E[X(t)X(s)] \tag{5.7}$$

Note that

$$R_X(t, s) = R_X(s, t) \tag{5.8}$$

and

$$R_X(t, t) = E[X^2(t)] \tag{5.9}$$

The *autocovariance function* of $X(t)$ is defined by

$$\begin{aligned} K_X(t, s) = \mathrm{Cov}[X(t), X(s)] &= E\{[X(t) - \mu_X(t)][X(s) - \mu_X(s)]\} \\ &= R_X(t, s) - \mu_X(t)\mu_X(s) \end{aligned} \tag{5.10}$$

It is clear that if the mean of $X(t)$ is zero, then $K_X(t, s) = R_X(t, s)$. Note that the *variance* of $X(t)$ is given by

$$\sigma_X^2(t) = \mathrm{Var}[X(t)] = E\{[X(t) - \mu_X(t)]^2\} = K_X(t, t) \tag{5.11}$$

If $X(t)$ is a complex random process, then its autocorrelation function $R_X(t, s)$ and autocovariance function $K_X(t, s)$ are defined, respectively, by

$$R_X(t, s) = E[X(t)X^*(s)] \tag{5.12}$$

and

$$K_X(t, s) = E\{[X(t) - \mu_X(t)][X(s) - \mu_X(s)]^*\} \tag{5.13}$$

where * denotes the complex conjugate.

5.4 Classification of Random Processes

If a random process $X(t)$ possesses some special probabilistic structure, we can specify less to characterize $X(t)$ completely. Some simple random processes are characterized completely by only the first-and second-order distributions.

A. Stationary Processes:

A random process $\{X(t), t \in T\}$ is said to be *stationary* or *strict-sense stationary* if, for all n and for every set of time instants $(t_i \in T, i = 1, 2, ..., n\}$,

$$F_X(x_1, ..., x_n; t_1, ..., t_n) = F_X(x_1, ..., x_n; t_1 + \tau, ..., t_n + \tau) \tag{5.14}$$

for any τ. Hence, the distribution of a stationary process will be unaffected by a shift in the time origin, and $X(t)$ and $X(t + \tau)$ will have the same distributions for any τ. Thus, for the first-order distribution,

$$F_X(x; t) = F_X(x; t + \tau) = F_X(x) \tag{5.15}$$

and

$$f_X(x; t) = f_X(x) \tag{5.16}$$

Then

$$\mu_X(t) = E[X(t)] = \mu \tag{5.17}$$

$$\mathrm{Var}[X(t)] = \sigma^2 \tag{5.18}$$

where μ and σ^2 are constants. Similarly, for the second-order distribution,

$$F_X(x_1, x_2; t_1, t_2) = F_X(x_1, x_2; t_2 - t_1) \tag{5.19}$$

and

$$f_X(x_1, x_2; t_1, t_2) = f_X(x_1, x_2; t_2 - t_1) \tag{5.20}$$

Nonstationary processes are characterized by distributions depending on the points $t_1, t_2, ..., t_n$.

B. Wide-Sense Stationary Processes:

If stationary condition (5.14) of a random process $X(t)$ does not hold for all n but holds for $n \leq k$, then we say that the process $X(t)$ is *stationary to order k*. If $X(t)$ is stationary to order 2, then $X(t)$ is said to be *wide-sense stationary* (WSS) or *weak stationary*. If $X(t)$ is a WSS random process, then we have

1. $E[X(t)] = \mu$ (constant)   (5.21)
2. $R_X(t, s) = E[X(t)X(s)] = R_X(|s - t|)$ (5.22)

Note that a strict-sense stationary process is also a WSS process, but, in general, the converse is not true.

C. Independent Processes:

In a random process $X(t)$, if $X(t_i)$ for $i = 1, 2, \ldots, n$ are independent r.v.'s, so that for $n = 2, 3, \ldots$,

$$F_X(x_1, \ldots, x_n; t_1, \ldots, t_n) = \prod_{i=1}^{n} F_X(x_i; t_i)$$
         (5.23)

then we call $X(t)$ an *independent random process*. Thus, a first-order distribution is sufficient to characterize an independent random process $X(t)$.

D. Processes with Stationary Independent Increments:

A random process $\{X(t), t \geq 0\}$ is said to have *independent increments* if whenever $0 < t_1 < t_2 < \cdots < t_n$,

$$X(0), X(t_1) - X(0), X(t_2) - X(t_1), \ldots, X(t_n) - X(t_{n-1})$$

are independent. If $\{X(t), t \geq 0\}$ has independent increments and $X(t) - X(s)$ has the same distribution as $X(t + h) - X(s + h)$ for all $s, t, h \geq 0, s < t$, then the process $X(t)$ is said to have *stationary independent increments*.

Let $\{X(t), t \geq 0\}$ be a random process with stationary independent increments and assume that $X(0) = 0$. Then (Probs. 5.21 and 5.22)

$$E[X(t)] = \mu_1 t$$
         (5.24)

where $\mu_1 = E[X(1)]$ and

$$\text{Var}[X(t)] = \sigma_1^2 t$$
         (5.25)

where $\sigma_1^2 = \text{Var}[X(1)]$.

From Eq. (5.24), we see that processes with stationary independent increments are nonstationary. Examples of processes with stationary independent increments are Poisson processes and Wiener processes, which are discussed in later sections.

E. Markov Processes:

A random process $\{X(t), t \in T\}$ is said to be a *Markov process* if

$$P\{X(t_{n+1}) \leq x_{n+1} | X(t_1) = x_1, X(t_2) = x_2, \ldots, X(t_n) = x_n\} = P\{X(t_{n+1}) \leq x_{n+1} | X(t_n) = x_n\}$$
     (5.26)

whenever $t_1 < t_2 < \cdots < t_n < t_{n+1}$.

A discrete-state Markov process is called a *Markov chain*. For a discrete-parameter Markov chain $\{X_n, n \geq 0\}$ (see Sec. 5.5), we have for every n

$$P(X_{n+1} = j | X_0 = i_0, X_1 = i_1, \ldots, X_n = i) = P(X_{n+1} = j | X_n = i)$$
         (5.27)

Equation (5.26) or Eq. (5.27) is referred to as the *Markov property* (which is also known as the *memoryless property*). This property of a Markov process states that the future state of the process depends only on the present state and not on the past history. Clearly, any process with independent increments is a Markov process.

Using the Markov property, the *n*th-order distribution of a Markov process $X(t)$ can be expressed as (Prob. 5.25)

$$F_X(x_1, \ldots, x_n; t_1, \ldots, t_n) = F_X(x_1; t_1) \prod_{k=2}^{n} P\{X(t_k) \le x_k \,|\, X(t_{k-1}) = x_{k-1}\} \tag{5.28}$$

Thus, all finite-order distributions of a Markov process can be expressed in terms of the second-order distributions.

F. Normal Processes:

A random process $\{X(t), t \in T\}$ is said to be a *normal* (or *Gaussian*) process if for any integer n and any subset $\{t_1, \ldots, t_n\}$ of T, the n r.v.'s $X(t_1), \ldots, X(t_n)$ are jointly normally distributed in the sense that their joint characteristic function is given by

$$\Psi_{X(t_1)\cdots X(t_n)}(\omega_1, \ldots, \omega_n) = E\{\exp j[\omega_1 X(t_1) + \cdots + \omega_n X(t_n)]\}$$

$$= \exp\left\{ j \sum_{i=1}^{n} \omega_i \, E[X(t_i) - \frac{1}{2} \sum_{i=1}^{n} \sum_{k=1}^{n} \omega_i \omega_k \, \mathrm{Cov}[X(t_i), X(t_k)] \right\} \tag{5.29}$$

where $\omega_1, \ldots, \omega_n$ are any real numbers (see Probs. 5.59 and 5.60). Equation (5.29) shows that a normal process is completely characterized by the second-order distributions. Thus, if a normal process is wide-sense stationary, then it is also strictly stationary.

G. Ergodic Processes:

Consider a random process $\{X(t), -\infty < t < \infty\}$ with a typical sample function $x(t)$. The time average of $x(t)$ is defined as

$$\langle \mathrm{x(t)} \rangle = \lim_{T \to \infty} \frac{1}{T} \int_{-T/2}^{T/2} x(t)\, dt \tag{5.30}$$

Similarly, the time autocorrelation function $\bar{R}_X(\tau)$ of $x(t)$ is defined as

$$\bar{R}_X(\tau) = \langle x(t) x(t+\tau) \rangle = \lim_{T \to \infty} \frac{1}{T} \int_{-T/2}^{T/2} x(t) x(t+\tau)\, dt \tag{5.31}$$

A random process is said to be *ergodic* if it has the property that the time averages of sample functions of the process are equal to the corresponding statistical or ensemble averages. The subject of *ergodicity* is extremely complicated. However, in most physical applications, it is assumed that stationary processes are ergodic.

5.5   Discrete-Parameter Markov Chains

In this section we treat a discrete-parameter Markov chain $\{X_n, n \ge 0\}$ with a discrete state space $E = \{0, 1, 2, \ldots\}$, where this set may be finite or infinite. If $X_n = i$, then the Markov chain is said to be in state i at time n (or the *n*th step). A discrete-parameter Markov chain $\{X_n, n \ge 0\}$ is characterized by [Eq. (5.27)]

$$P(X_{n+1} = j \,|\, X_0 = i_0, X_1 = i_1, \ldots, X_n = i) = P(X_{n+1} = j \,|\, X_n = i) \tag{5.32}$$

where $P\{x_{n+1} = j \mid X_n = i\}$ are known as one-step transition probabilities. If $P\{x_{n+1} = j \mid X_n = i\}$ is independent of n, then the Markov chain is said to possess *stationary transition probabilities* and the process is referred to as a *homogeneous* Markov chain. Otherwise the process is known as a *nonhomogeneous* Markov chain. Note that the concepts of a Markov chain's having stationary transition probabilities and being a stationary random process should not be confused. The Markov process, in general, is not stationary. We shall consider only homogeneous Markov chains in this section.

A. Transition Probability Matrix:

Let $\{X_n, n \geq 0\}$ be a homogeneous Markov chain with a discrete infinite state space $E = \{0, 1, 2, \dots\}$. Then

$$p_{ij} = P\{X_{n+1} = j \mid X_n = i\} \qquad i \geq 0, j \geq 0 \qquad (5.33)$$

regardless of the value of n. A *transition probability matrix* of $\{X_n, n \geq 0\}$ is defined by

$$P = [p_{ij}] = \begin{bmatrix} p_{00} & p_{01} & p_{02} & \cdots \\ p_{10} & p_{11} & p_{12} & \cdots \\ p_{20} & p_{21} & p_{22} & \cdots \\ \vdots & \vdots & \vdots & \end{bmatrix}$$

where the elements satisfy

$$p_{ij} \geq 0 \qquad \sum_{j=0}^{\infty} p_{ij} = 1 \qquad i = 0, 1, 2, \dots \qquad (5.34)$$

In the case where the state space E is finite and equal to $\{1, 2, \dots, m\}$, P is $m \times m$ dimensional; that is,

$$P = [p_{ij}] = \begin{bmatrix} p_{11} & p_{12} & \cdots & p_{1m} \\ p_{21} & p_{22} & \cdots & p_{2m} \\ \vdots & \vdots & \ddots & \vdots \\ p_{m1} & p_{m2} & \cdots & p_{mm} \end{bmatrix}$$

where

$$p_{ij} \geq 0 \qquad \sum_{j=1}^{m} p_{ij} = 1 \qquad i = 1, 2, \dots, m \qquad (5.35)$$

A square matrix whose elements satisfy Eq. (5.34) or (5.35) is called a *Markov* matrix or *stochastic* matrix.

B. Higher-Order Transition Probabilities—Chapman-Kolmogorov Equation:

Tractability of Markov chain models is based on the fact that the probability distribution of $\{X_n, n \geq 0\}$ can be computed by matrix manipulations.

   Let $P = [p_{ij}]$ be the transition probability matrix of a Markov chain $\{X_n, n \geq 0\}$. Matrix powers of P are defined by

$$P^2 = PP$$

with the (i, j)th element given by

$$p_{ij}^{(2)} = \sum_{k} p_{ik} p_{kj}$$

Note that when the state space E is infinite, the series above converges, since by Eq. (5.34),

$$\sum_k p_{ik} p_{kj} \le \sum_k p_{ik} = 1$$

Similarly, $P^3 = PP^2$ has the (i, j)th element

$$p_{ij}^{(3)} = \sum_k p_{ik} p_{kj}^{(2)}$$

and in general, $P^{n+1} = PP^n$ has the (i, j)th element

$$p_{ij}^{(n+1)} = \sum_k p_{ik} p_{kj}^{(n)} \tag{5.36}$$

Finally, we define $P^0 = I$, where I is the identity matrix.

The n-step transition probabilities for the homogeneous Markov chain $\{X_n, n \ge 0\}$ are defined by

$$P(X_n = j \mid X_0 = i)$$

Then we can show that (Prob. 5.90)

$$p_{ij}^{(n)} = P(X_n = j \mid X_0 = i) \tag{5.37}$$

We compute $p_{ij}^{(n)}$ by taking matrix powers.

The matrix identity

$$P^{n+m} = P^n P^m \qquad n, m \ge 0$$

when written in terms of elements

$$p_{ij}^{(n+m)} = \sum_k p_{ik}^{(n)} p_{kj}^{(m)} \tag{5.38}$$

is known as the *Chapman-Kolmogorov equation*. It expresses the fact that a transition from i to j in $n + m$ steps can be achieved by moving from i to an intermediate k in n steps (with probability $p_{ik}^{(n)}$, and then proceeding to j from k in m steps (with probability $p_{kj}^{(m)}$). Furthermore, the events "go from i to k in n steps" and "go from k to j in m steps" are independent. Hence, the probability of the transition from i to j in $n + m$ steps via i, k, j is $p_{ik}^{(n)} p_{kj}^{(m)}$. Finally, the probability of the transition from i to j is obtained by summing over the intermediate state k.

C. The Probability Distribution of $\{X_n, n \ge 0\}$:

Let $p_i(n) = P(X_n = i)$ and

$$\mathbf{p}(n) = [p_0(n) \quad p_1(n) \quad p_2(n) \quad \cdots]$$

where

$$\sum_k p_k(n) = 1$$

Then $p_i(0) = P(X_0 = i)$ are the *initial-state* probabilities,

$$\mathbf{p}(0) = [p_0(0) \quad p_1(0) \quad p_2(0) \quad \cdots]$$

is called the *initial-state probability vector*, and $\mathbf{p}(n)$ is called the *state probability vector after n transitions* or the *probability distribution* of X_n. Now it can be shown that (Prob. 5.29)

$$\mathbf{p}(n) = \mathbf{p}(0) P^n \tag{5.39}$$

which indicates that the probability distribution of a homogeneous Markov chain is completely determined by the one-step transition probability matrix P and the initial-state probability vector $\mathbf{p}(0)$.

D.   Classification of States:

1.   Accessible States:

State j is said to be *accessible* from state i if for some $n \geq 0, p_{ij}^{(n)} > 0$, and we write $i \rightarrow j$. Two states i and j accessible to each other are said to *communicate*, and we write $i \leftrightarrow j$. If all states communicate with each other, then we say that the Markov chain is *irreducible*.

2.   Recurrent States:

Let T_j be the time (or the number of steps) of the first visit to state j after time zero, unless state j is never visited, in which case we set $T_j = \infty$. Then T_j is a discrete r.v. taking values in $\{1, 2, ..., \infty\}$.

Let

$$f_{ij}^{(m)} = P(T_j = m \mid X_0 = i) = P(X_m = j, X_k \neq j, k = 1, 2, ..., m - 1 \mid X_0 = i) \tag{5.40}$$

and $f_{ij}^{(0)} = 0$ since $T_j \geq 1$. Then

$$f_{ij}^{(1)} = P(T_j = 1 \mid X_0 = i) = P(X_1 = j \mid X_0 = i) = p_{ij} \tag{5.41}$$

and

$$f_{ij}^{(m)} = \sum_{k \neq j} p_{ik} f_{kj}^{(m-1)} \qquad m = 2, 3, ... \tag{5.42}$$

The probability of visiting j in finite time, starting from i, is given by

$$f_{ij} = \sum_{n=0}^{\infty} f_{ij}^{(n)} = P(T_j < \infty \mid X_0 = i) \tag{5.43}$$

Now state j is said to be *recurrent* if

$$f_{jj} = P(T_j < \infty \mid X_0 = j) = 1 \tag{5.44}$$

That is, starting from j, the probability of eventual return to j is one. A recurrent state j is said to be *positive recurrent* if

$$E(T_j \mid X_0 = j) < \infty \tag{5.45}$$

and state j is said to be *null recurrent* if

$$E(T_j \mid X_0 = j) = \infty \tag{5.46}$$

Note that

$$E(T_j \mid X_0 = j) = \sum_{n=0}^{\infty} n f_{jj}^{(n)} \tag{5.47}$$

3.   Transient States:

State j is said to be *transient* (or *nonrecurrent*) if

$$f_{jj} = P(T_j < \infty \mid X_0 = j) < 1 \tag{5.48}$$

In this case there is positive probability of never returning to state j.

4. Periodic and Aperiodic States:

We define the period of state j to be

$$d(j) = \gcd\{n \geq 1 : p_{jj}^{(n)} > 0\}$$

where gcd stands for greatest common divisor.

If $d(j) > 1$, then state j is called *periodic* with period $d(j)$. If $d(j) = 1$, then state j is called *aperiodic*. Note that whenever $p_{jj} > 0$, j is aperiodic.

5. Absorbing States:

State j is said to be an *absorbing state* if $p_{jj} = 1$; that is, once state j is reached, it is never left.

E. Absorption Probabilities:

Consider a Markov chain $X(n) = \{X_n, n \geq 0\}$ with finite state space $E = \{1, 2, ..., N\}$ and transition probability matrix P. Let $A = \{1, ..., m\}$ be the set of absorbing states and $B = \{m + 1, ..., N\}$ be a set of nonabsorbing states. Then the transition probability matrix P can be expressed as

$$P = \begin{bmatrix} 1 & 0 & \cdots & 0 & 0 & \cdots & 0 \\ 0 & 1 & \cdots & 0 & 0 & \cdots & 0 \\ \vdots & \vdots & & \vdots & \vdots & & \vdots \\ 0 & & \cdots & 1 & 0 & \cdots & 0 \\ p_{m+1,1} & & \cdots & p_{m+1,m} & p_{m+1,m+1} & \cdots & p_{m+1,N} \\ \vdots & \vdots & & \vdots & \vdots & & \vdots \\ p_{N,1} & & \cdots & p_{N,m} & p_{N,m+1} & \cdots & p_{N,N} \end{bmatrix} = \begin{bmatrix} I & O \\ R & Q \end{bmatrix} \qquad (5.49a)$$

where I is an $m \times m$ identity matrix, O is an $m \times (N - m)$ zero matrix, and

$$R = \begin{bmatrix} p_{m+1,1} & \cdots & p_{m+1,m} \\ \vdots & & \vdots \\ p_{N,1} & \cdots & p_{N,m} \end{bmatrix} \qquad Q = \begin{bmatrix} p_{m+1,m+1} & \cdots & p_{m+1,N} \\ \vdots & & \vdots \\ p_{N,m+1} & \cdots & p_{N,N} \end{bmatrix} \qquad (5.49b)$$

Note that the elements of R are the one-step transition probabilities from nonabsorbing to absorbing states, and the elements of Q are the one-step transition probabilities among the nonabsorbing states.

Let $U = [u_{kj}]$, where

$$u_{kj} = P\{X_n = j(\in A) \mid X_0 = k(\in B)\}$$

It is seen that U is an $(N - m) \times m$ matrix and its elements are the absorption probabilities for the various absorbing states. Then it can be shown that (Prob. 5.40)

$$U = (I - Q)^{-1} R = \Phi R \qquad (5.50)$$

The matrix $\Phi = (I - Q)^{-1}$ is known as the *fundamental matrix* of the Markov chain $X(n)$. Let T_k denote the total time units (or steps) to absorption from state k. Let

$$\mathbf{T} = [T_{m+1} \quad T_{m+2} \quad \cdots \quad T_N]$$

Then it can be shown that (Prob. 5.74)

$$E(T_k) = \sum_{i=m+1}^{N} \phi_{ki} \qquad k = m + 1, ..., N \qquad (5.51)$$

where ϕ_{ki} is the (k, i)th element of the fundamental matrix Φ.

F.　Stationary Distributions:

Let P be the transition probability matrix of a homogeneous Markov chain $\{X_n, n \geq 0\}$. If there exists a probability vector $\hat{\mathbf{p}}$ such that

$$\hat{\mathbf{p}}P = \hat{\mathbf{p}} \tag{5.52}$$

then $\hat{\mathbf{p}}$ is called a *stationary distribution* for the Markov chain. Equation (5.52) indicates that a stationary distribution $\hat{\mathbf{p}}$ is a (left) *eigenvector* of P with *eigenvalue* 1. Note that any nonzero multiple of $\hat{\mathbf{p}}$ is also an eigenvector of P. But the stationary distribution $\hat{\mathbf{p}}$ is fixed by being a probability vector; that is, its components sum to unity.

G.　Limiting Distributions:

A Markov chain is called *regular* if there is a finite positive integer m such that after m time-steps, every state has a nonzero chance of being occupied, no matter what the initial state. Let $A > O$ denote that every element a_{ij} of A satisfies the condition $a_{ij} > 0$. Then, for a regular Markov chain with transition probability matrix P, there exists an $m > 0$ such that $P^m > O$. For a regular homogeneous Markov chain we have the following theorem:

THEOREM 5.5.1

Let $\{X_n, n \geq 0\}$ be a regular homogeneous finite-state Markov chain with transition matrix P. Then

$$\lim_{n \to \infty} P^n = \hat{P} \tag{5.53}$$

where $\hat{P}$ is a matrix whose rows are identical and equal to the stationary distribution $\hat{\mathbf{p}}$ for the Markov chain defined by Eq. (5.52).

5.6　Poisson Processes

A.　Definitions:

Let t represent a time variable. Suppose an experiment begins at $t = 0$. Events of a particular kind occur randomly, the first at T_1, the second at T_2, and so on. The r.v. T_i denotes the time at which the ith event occurs, and the values t_i of T_i ($i = 1, 2, \ldots$) are called *points of occurrence* (Fig. 5-2).

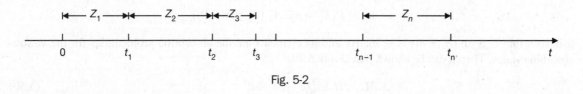

Fig. 5-2

Let

$$Z_n = T_n - T_{n-1} \tag{5.54}$$

and $T_0 = 0$. Then Z_n denotes the time between the $(n-1)$st and the nth events (Fig. 5-2). The sequence of ordered r.v.'s $\{Z_n, n \geq 1\}$ is sometimes called an *interarrival process*. If all r.v.'s Z_n are independent and identically distributed, then $\{Z_n, n \geq 1\}$ is called a *renewal process* or a *recurrent process*. From Eq. (5.54), we see that

$$T_n = Z_1 + Z_2 + \cdots + Z_n$$

where T_n denotes the time from the beginning until the occurrence of the nth event. Thus, $\{T_n, n \geq 0\}$ is sometimes called an *arrival process*.

B. Counting Processes:

A random process $\{X(t), t \geq 0\}$ is said to be a *counting process* if $X(t)$ represents the total number of "events" that have occurred in the interval $(0, t)$. From its definition, we see that for a counting process, $X(t)$ must satisfy the following conditions:

1. $X(t) \geq 0$ and $X(0) = 0$.
2. $X(t)$ is integer valued.
3. $X(s) \leq X(t)$ if $s < t$.
4. $X(t) - X(s)$ equals the number of events that have occurred on the interval (s, t).

A typical sample function (or realization) of $X(t)$ is shown in Fig. 5-3.

A counting process $X(t)$ is said to possess independent increments if the numbers of events which occur in disjoint time intervals are independent. A counting process $X(t)$ is said to possess stationary increments if the number of events in the interval $(s + h, t + h)$—that is, $X(t + h) - X(s + h)$—has the same distribution as the number of events in the interval (s, t)—that is, $X(t) - X(s)$—for all $s < t$ and $h > 0$.

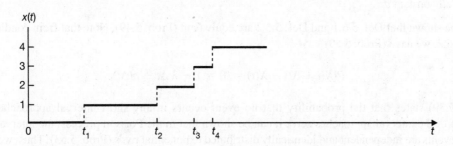

Fig. 5-3 A sample function of a counting process.

C. Poisson Processes:

One of the most important types of counting processes is the *Poisson process* (or *Poisson counting process*), which is defined as follows:

DEFINITION 5.6.1

A counting process $X(t)$ is said to be a Poisson process with *rate* (or *intensity*) $\lambda (> 0)$ if

1. $X(0) = 0$.
2. $X(t)$ has independent increments.
3. The number of events in any interval of length t is Poisson distributed with mean λt; that is, for all $s, t > 0$,

$$P[X(t + s) - X(s) = n] = e^{-\lambda t} \frac{(\lambda t)^n}{n!} \qquad n = 0, 1, 2, \ldots \tag{5.55}$$

It follows from condition 3 of Def. 5.6.1 that a Poisson process has stationary increments and that

$$E[X(t)] = \lambda t \tag{5.56}$$

Then by Eq. (2.51) (Sec. 2.7E), we have

$$\text{Var}[X(t)] = \lambda t \tag{5.57}$$

Thus, the expected number of events in the unit interval $(0, 1)$, or any other interval of unit length, is just λ (hence the name of the rate or intensity).

An alternative definition of a Poisson process is given as follows:

DEFINITION 5.6.2

A counting process $X(t)$ is said to be a Poisson process with rate (or intensity) $\lambda (> 0)$ if

1. $X(0) = 0$.
2. $X(t)$ has independent and stationary increments.
3. $P[X(t + \Delta t) - X(t) = 1] = \lambda \Delta t + o(\Delta t)$
4. $P[X(t + \Delta t) - X(t) \geq 2] = o(\Delta t)$

where $o(\Delta t)$ is a function of Δt which goes to zero faster than does Δt; that is,

$$\lim_{\Delta t \to 0} \frac{o(\Delta t)}{\Delta t} = 0 \tag{5.58}$$

Note: Since addition or multiplication by a scalar does not change the property of approaching zero, even when divided by Δt, $o(\Delta t)$ satisfies useful identities such as $o(\Delta t) + o(\Delta t) = o(\Delta t)$ and $ao(\Delta t) = o(\Delta t)$ for all constant a.

It can be shown that Def. 5.6.1 and Def. 5.6.2 are equivalent (Prob. 5.49). Note that from conditions 3 and 4 of Def. 5.6.2, we have (Prob. 5.50)

$$P[X(t + \Delta t) - X(t) = 0] = 1 - \lambda \Delta t + o(\Delta t) \tag{5.59}$$

Equation (5.59) states that the probability that no event occurs in any short interval approaches unity as the duration of the interval approaches zero. It can be shown that in the Poisson process, the intervals between successive events are independent and identically distributed exponential r.v.'s (Prob. 5.53). Thus, we also identify the Poisson process as a renewal process with exponentially distributed intervals.

The autocorrelation function $R_X(t, s)$ and the autocovariance function $K_X(t, s)$ of a Poisson process $X(t)$ with rate λ are given by (Prob. 5.52)

$$R_X(t, s) = \lambda \min(t, s) + \lambda^2 ts \tag{5.60}$$

$$K_X(t, s) = \lambda \min(t, s) \tag{5.61}$$

5.7 Wiener Processes

Another example of random processes with independent stationary increments is a *Wiener process*.

DEFINITION 5.7.1

A random process $\{X(t), t \geq 0\}$ is called a Wiener process if

1. $X(t)$ has stationary independent increments.
2. The increment $X(t) - X(s)(t > s)$ is normally distributed.
3. $E[X(t)] = 0$.
4. $X(0) = 0$.

The Wiener process is also known as the *Brownian motion process*, since it originates as a model for Brownian motion, the motion of particles suspended in a fluid. From Def. 5.7.1, we can verify that a Wiener process is a normal process (Prob. 5.61) and

$$E[X(t)] = 0 \tag{5.62}$$

$$\mathrm{Var}[X(t)] = \sigma^2 t \tag{5.63}$$

where σ^2 is a parameter of the Wiener process which must be determined from observations. When $\sigma^2 = 1$, $X(t)$ is called a *standard* Wiener (or standard Brownian motion) process.

The autocorrelation function $R_X(t, s)$ and the autocovariance function $K_X(t, s)$ of a Wiener process $X(t)$ are given by (see Prob. 5.23)

$$R_X(t, s) = K_X(t, s) = \sigma^2 \min(t, s) \qquad s, t \geq 0 \tag{5.64}$$

DEFINITION 5.7.2

A random process $\{X(t), t \geq 0\}$ is called a *Wiener process with drift coefficient* μ if

1. $X(t)$ has stationary independent increments.
2. $X(t)$ is normally distributed with mean μt.
3. $X(0) = 0$.

From condition 2, the pdf of a standard Wiener process with drift coefficient μ is given by

$$f_{X(t)}(x) = \frac{1}{\sqrt{2\pi t}} e^{-(x-\mu t)^2/(2t)} \tag{5.65}$$

5.8 Martingales

Martingales have their roots in gaming theory. A martingale is a random process that models a fair game. It is a powerful tool with many applications, especially in the field of mathematical finance.

A. Conditional Expectation and Filtrations:

The conditional expectation $E(Y \mid X_1, \ldots, X_n)$ is a r.v. (see Sec. 4.5 D) characterized by two properties:

1. The value of $E(Y \mid X_1, \ldots, X_n)$ depends only on the values of $X_1, \ldots, X_n$, that is,

$$E(Y \mid X_1, \ldots, X_n = g(X_1, \ldots, X_n) \tag{5.66}$$

2. $$E[E(Y \mid X_1, \ldots, X_n)] = E(Y) \tag{5.67}$$

If $X_1, \ldots, X_n$ is a sequence of r.v.'s , we will use F_n to denote the information contained in $X_1, \ldots, X_n$ and we write $E(Y \mid F_n)$ for $E(Y \mid X_1, \ldots, X_n)$, that is,

$$E(Y \mid X_1, \ldots, X_n) = E(Y \mid F_n) \tag{5.68}$$

We also define information carried by r.v.'s $X_1, \ldots, X_n$ in terms of the associated event space (σ-field), $\sigma(X_1, \ldots, X_n)$. Thus,

$$F_n = \sigma(X_1, \ldots, X_n) \tag{5.69}$$

and we say that F_n is an event space generated by $X_1, \ldots, X_n$. We have

$$F_n \subset F_m \qquad \text{if} \qquad 1 \leq n \leq m \tag{5.70}$$

A collection $\{F_n, n = 1, 2, \ldots\}$ satisfying Eq. (5.70) is called a *filtration*.

Note that if a r.v. Z can be written as a function of $X_1, \ldots, X_n$, it is called *measurable* with respect to $X_1, \ldots, X_n$, or F_n-*measurable*.

Properties of Conditional Expectations:

1. Linearity:

$$E(aY_1 + b\,Y_2 | F_n) = a\,E(Y_1 | F_n) + b\,E(Y_2 | F_n) \tag{5.71}$$

where a and b are constants.

2. Positivity:

If $Y \geq 0$, then
$$E(Y | F_n) \geq 0 \tag{5.72}$$

3. Measurablity:

If Y is F_n-measurable, then
$$E(Y | F_n) = Y \tag{5.73}$$

4. Stability:

If Z is F_n-measurable, then
$$E(YZ | F_n) = Z\,E(Y | F_n) \tag{5.74}$$

5. Independence Law:

If Y is independent of F_n, then
$$E(Y | F_n) = E(Y) \tag{5.75}$$

6. Tower Property:

$$E[E(Y | F_n) | F_m] = E(Y | F_m) \quad \text{if } m \leq n \tag{5.76}$$

7. Projection Law:

$$E[E(Y | F_n)] = E(Y) \tag{5.77}$$

8. Jensen's Inequality:

If g is a convex function and $E(|Y|) < \infty$, then

$$E(g(Y) | F_n) \geq g(E\,(Y | F_n)) \tag{5.78}$$

B. Martingale:

Definition:

A discrete random process $\{M_n, n \geq 0\}$ is a *martingale* with respect to F_n if

(1) $E(|M_n|) < \infty$ for all $n \geq 0$

(2) $E(M_{n+1} | F_n) = M_n$ for all n (5.79)

It immediately follows from Eq. (5.79) that for a martingale

(2') $E(M_m | F_n) = M_n$ for $m \geq n$ (5.80)

A discrete random process $\{M_n, n \geq 0\}$ is a *submartingale* (*supermartingale*) with respect to F_n if

(1) $E(|M_n|) < \infty$ for all $n \geq 0$

(2) $E(M_{n+1} | F_n) \geq (\leq) M_n$ for all n (5.81)

While a martingale models a fair game, the submartingale and supermartingale model favorable and unfavorable games, respectively.

Theorem 5.8.1

Let $\{M_n, n \geq 0\}$ be a martingale. Then for any given n

$$E(M_n) = E(M_{n-1}) = \ldots = E(M_0) \tag{5.82}$$

Equation (5.82) indicates that in a martingale all the r.v.'s have the same expectation (Prob. 5.67).

Theorem 5.8.2 (Doob decomposition)

Let $X = \{X_n, n \geq 0\}$ be a submartingale with respect to F_n. Then there exists a martingale $M = \{M_n, n \geq 0\}$ and a process $A = \{A_n, n \geq 0\}$ such that

(1) M is a martingale with respect to F_n;
(2) A is an increasing process $A_{n+1} \geq A_n$;
(3) A_n is F_{n-1}-measurable for all n;
(4) $X_n = M_n + A_n$.

(For the proof of this theorem see Prob. 5.78.)

C. Stopping Time and the Optional Stopping Theorem:

Definition:

A r.v. T is called a *stopping time* with respect to F_n if

1. T takes values from the set $\{0, 1, 2, \ldots, \infty\}$
2. The event $\{T = n\}$ is F_n-measurable.

EXAMPLE 5.1: A gambler has $100 and plays the slot machine at $1 per play.

1. The gambler stops playing when his capital is depleted. The number $T = n_1$ of plays that it takes the gambler to stop play is a stopping time.
2. The gambler stops playing when his capital reaches $200. The number $T = n_2$ of plays that it takes the gambler to stop play is a stopping time.
3. The gambler stops playing when his capital reaches $200, or is depleted, whichever comes first. The number $T = \min(n_1, n_2)$ of plays that it takes the gambler to stop play is a stopping time.

EXAMPLE 5.2 A typical example of the event T is not a stopping time; it is the moment the stock price attains its maximum over a certain period. To determine whether T is a point of maximum, we have to know the future values of the stock price and event $\{T = n\} \notin F_n$.

Lemma 5.8.1

1. If T_1 and T_2 are stopping times, then so is $T_1 + T_2$.
2. If T_1 and T_2 are stopping times, then $T = \min(n_1, n_2)$ and $T = \max(n_1, n_2)$ are also stopping times.
3. $\min(T, n)$ is a stopping time for any fixed n.

Let I_A denote the *indicator function* of A, that is, the r.v. which equals 1 if A occurs and 0 otherwise. Note that $I_{\{T > n\}}$, the indicator function of the event $\{T > n\}$, is F_n-measurable (since we need only the information up through time n to determine if we have stopped by time n).

Optimal Stopping Theorem:

Suppose $\{M_n, n \geq 0\}$ is a martingale and T is a stopping time. If

(1) $E(T) < \infty$ (5.83)

(2) $E(|M_T|) < \infty$ (5.84)

(3) $\lim_{n \to \infty} E\left(|M_n| I_{\{T > n\}}\right) = 0$ (5.85)

Then

$$E(M_T) = E(M_0) \quad\quad\quad (5.86)$$

Note that Eqs. (5.84) and (5.85) are always satisfied if the martingale is bounded and $P(T < \infty) = 1$.

D. Martingale in Continuous Time

A continuous-time filtration is a family $\{F_t, t \geq 0\}$ contained in the event space F such that $F_s \subset F_t$ for $s < t$. The continuous random process $X(t)$ is a martingale with respect to F_t if

(1) $E(|X(t)|) < \infty$ (5.87)

(2) $E(X(t)|F_s) = X(s)$ for $t \geq s$ (5.88)

Similarly, continuous-time submartingales and supermartingales can be defined by replacing equal (=) sign by $\geq$ and $\leq$, respectively, in Eq. (5.88).

SOLVED PROBLEMS

Random Processes

5.1. Let $X_1, X_2, \ldots$ be independent Bernoulli r.v.'s (Sec. 2.7A) with $P(X_n = 1) = p$ and $P(X_n = 0) = q = 1 - p$ for all n. The collection of r.v.'s $\{X_n, n \geq 1\}$ is a random process, and it is called a *Bernoulli process*.

  (*a*) Describe the Bernoulli process.

  (*b*) Construct a typical sample sequence of the Bernoulli process.

  (*a*) The Bernoulli process $\{X_n, n \geq 1\}$ is a discrete-parameter, discrete-state process. The state space is $E = \{0, 1\}$, and the index set is $T = \{1, 2, \ldots\}$.

  (*b*) A sample sequence of the Bernoulli process can be obtained by tossing a coin consecutively. If a head appears, we assign 1, and if a tail appears, we assign 0. Thus, for instance,

n	1	2	3	4	5	6	7	8	9	10	...
Coin tossing	H	T	T	H	H	H	T	H	H	T	...
x_n	1	0	0	1	1	1	0	1	1	0	...

The sample sequence $\{x_n\}$ obtained above is plotted in Fig. 5-4.

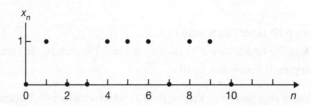

Fig. 5-4 A sample function of a Bernoulli process.

5.2. Let $Z_1, Z_2, \ldots$ be independent identically distributed r.v.'s with $P(Z_n = 1) = p$ and $P(Z_n = -1) = q = 1 - p$ for all n. Let

$$X_n = \sum_{i=1}^{n} Z_i \qquad n = 1, 2, \ldots \tag{5.89}$$

and $X_0 = 0$. The collection of r.v.'s $\{X_n, n \geq 0\}$ is a random process, and it is called the *simple random walk* $X(n)$ in one dimension.

  (*a*) Describe the simple random walk $X(n)$.

  (*b*) Construct a typical sample sequence (or realization) of $X(n)$.

(a) The simple random walk $X(n)$ is a discrete-parameter (or time), discrete-state random process. The state space is $E = \{\ldots, -2, -1, 0, 1, 2, \ldots\}$, and the index parameter set is $T = \{0, 1, 2, \ldots\}$.

(b) A sample sequence $x(n)$ of a simple random walk $X(n)$ can be produced by tossing a coin every second and letting $x(n)$ increase by unity if a head appears and decrease by unity if a tail appears. Thus, for instance,

n	0	1	2	3	4	5	6	7	8	9	10	...
Coin tossing		H	T	T	H	H	H	T	H	H	T	...
$x(n)$	0	1	0	−1	0	1	2	1	2	3	2	...

The sample sequence $x(n)$ obtained above is plotted in Fig. 5-5. The simple random walk $X(n)$ specified in this problem is said to be *unrestricted* because there are no bounds on the possible values of X_n.

The simple random walk process is often used in the following primitive gambling model: Toss a coin. If a head appears, you win one dollar; if a tail appears, you lose one dollar (see Prob. 5.38).

5.3. Let $\{X_n, n \geq 0\}$ be a simple random walk of Prob. 5.2. Now let the random process $X(t)$ be defined by

$$X(t) = X_n \qquad n \leq t < n + 1$$

(a) Describe $X(t)$.

(b) Construct a typical sample function of $X(t)$.

(a) The random process $X(t)$ is a continuous-parameter (or time), discrete-state random process. The state space is $E = \{\ldots, -2, -1, 0, 1, 2, \ldots\}$, and the index parameter set is $T = \{t, t \geq 0\}$.

(b) A sample function $x(t)$ of $X(t)$ corresponding to Fig. 5-5 is shown in Fig. 5-6.

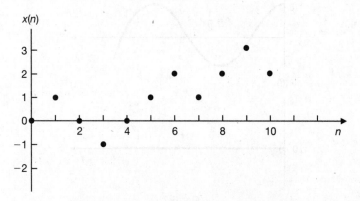

Fig. 5-5 A sample function of a random walk.

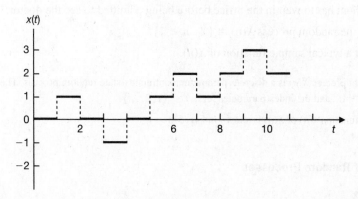

Fig. 5-6

5.4. Consider a random process $X(t)$ defined by

$$X(t) = Y \cos \omega t \qquad t \geq 0$$

where ω is a constant and Y is a uniform r.v. over $(0, 1)$.

   (a)   Describe $X(t)$.

   (b)   Sketch a few typical sample functions of $X(t)$.

   (a)   The random process $X(t)$ is a continuous-parameter (or time), continuous-state random process. The state space is $E = \{x: -1 < x < 1\}$ and the index parameter set is $T = \{t: t \geq 0\}$.

   (b)   Three sample functions of $X(t)$ are sketched in Fig. 5-7.

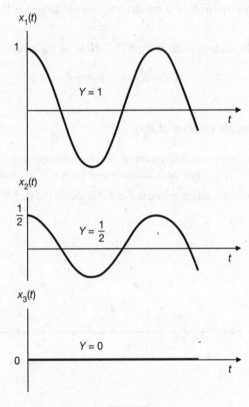

Fig. 5-7

5.5. Consider patients coming to a doctor's office at random points in time. Let X_n denote the time (in hours) that the nth patient has to wait in the office before being admitted to see the doctor.

   (a)   Describe the random process $X(n) = \{X_n, n \geq 1\}$.

   (b)   Construct a typical sample function of $X(n)$.

   (a)   The random process $X(n)$ is a discrete-parameter, continuous-state random process. The state space is $E = \{x: x \geq 0\}$, and the index parameter set is $T = \{1, 2, \ldots\}$.

   (b)   A sample function $x(n)$ of $X(n)$ is shown in Fig. 5-8.

Characterization of Random Processes

5.6. Consider the Bernoulli process of Prob. 5.1. Determine the probability of occurrence of the sample sequence obtained in part (b) of Prob. 5.1.

Since X_n's are independent, we have

$$P(X_1 = x_1, X_2 = x_2, \ldots, X_n = x_n) = P(X_1 = x_1) P(X_2 = x_2) \ldots P(X_n = x_n) \qquad (5.90)$$

Thus, for the sample sequence of Fig. 5-4,

$$P(X_1 = 1, X_2 = 0, X_3 = 0, X_4 = 1, X_5 = 1, X_6 = 1, X_7 = 0, X_8 = 1, X_9 = 1, X_{10} = 0) = p^6 q^4$$

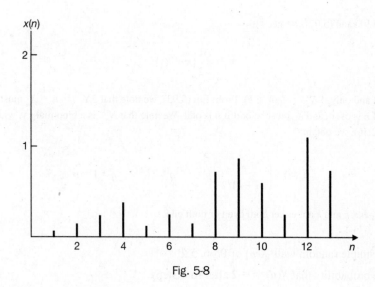

Fig. 5-8

5.7. Consider the random process $X(t)$ of Prob. 5.4. Determine the pdf's of $X(t)$ at $t = 0, \pi/4\omega, \pi/2\omega, \pi/\omega$.

For $t = 0, X(0) = Y \cos 0 = Y$. Thus,

$$f_{X(0)}(x) = \begin{cases} 1 & 0 < x < 1 \\ 0 & \text{otherwise} \end{cases}$$

For $t = \pi/4\omega, X(\pi/4\omega) = Y \cos \pi/4 = 1/\sqrt{2}\, Y$. Thus,

$$f_{X(\pi/4\omega)}(x) = \begin{cases} \sqrt{2} & 0 < x < 1/\sqrt{2} \\ 0 & \text{otherwise} \end{cases}$$

For $t = \pi/2\omega, X(\pi/2\omega) = Y \cos \pi/2 = 0$; that is, $X(\pi/2\omega) = 0$ irrespective of the value of Y. Thus, the pmf of $X(\omega/2\omega)$ is

$$P_{X(\pi/2\omega)}(x) = P(X = 0) = 1$$

For $t = \pi/\omega, X(\pi/\omega) = Y \cos \pi = -Y$. Thus,

$$f_{X(\pi/\omega)}(x) = \begin{cases} 1 & -1 < x < 0 \\ 0 & \text{otherwise} \end{cases}$$

5.8. Derive the first-order probability distribution of the simple random walk $X(n)$ of Prob. 5.2.

The first-order probability distribution of the simple random walk $X(n)$ is given by

$$p_n(k) = P(X_n = k)$$

where k is an integer. Note that $P(X_0 = 0) = 1$. We note that $p_n(k) = 0$ if $n < |k|$ because the simple random walk cannot get to level k in less than $|k|$ steps. Thus, $n \geq |k|$.

Let N_n^+ and N_n^- be the r.v.'s denoting the numbers of $+1$s and -1s, respectively, in the first n steps. Then

$$n = N_n^+ + N_n^- \tag{5.91}$$
$$X_n = N_n^+ - N_n^- \tag{5.92}$$

Adding Eqs. (5.91) and (5.92), we get

$$N_n^+ = \frac{1}{2}(n + X_n) \tag{5.93}$$

Thus, $X_n = k$ if and only if $N_n^+ = \frac{1}{2}(n + k)$. From Eq. (5.93), we note that $2N_n^+ = n + X_n$ must be even. Thus, X_n must be even if n is even, and X_n must be odd if n is odd. We note that N_n^+ is a binomial r.v. with parameters (n, p). Thus, by Eq. (2.36), we obtain

$$p_n(k) = \binom{n}{(n+k)/2} p^{(n+k)/2} q^{(n-k)/2} \qquad q = 1 - p \tag{5.94}$$

where $n \geq |k|$, and n and k are either both even or both odd.

5.9. Consider the simple random walk $X(n)$ of Prob. 5.2.

(a) Find the probability that $X(n) = -2$ after four steps.

(b) Verify the result of part (a) by enumerating all possible sample sequences that lead to the value $X(n) = -2$ after four steps.

(a) Setting $k = -2$ and $n = 4$ in Eq. (5.94), we obtain

$$P(X_4 = -2) = p_4(-2) = \binom{4}{1} pq^3 = 4pq^3 \qquad q = 1 - p$$

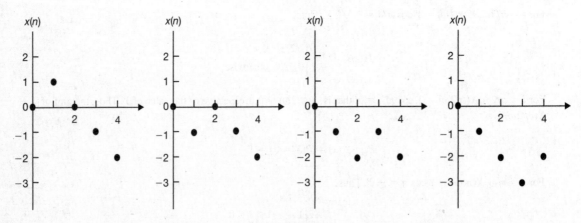

Fig. 5-9

(b) All possible sample functions that lead to the value $X_4 = -2$ after four steps are shown in Fig. 5-9. For each sample sequence, $P(X_4 = -2) = pq^3$. There are only four sample functions that lead to the value $X_4 = -2$ after four steps. Thus, $P(X_4 = -2) = 4pq^3$.

5.10 Find the mean and variance of the simple random walk $X(n)$ of Prob. 5.2.

From Eq. (5.89), we have

$$X_n = X_{n-1} + Z_n \qquad n = 1, 2, \ldots \tag{5.95}$$

and $X_0 = 0$ and Z_n $(n = 1, 2, \ldots)$ are independent and identically distributed (iid) r.v.'s with

$$P(Z_n = +1) = p \qquad P(Z_n = -1) = q = 1 - p$$

From Eq. (5.95), we observe that

$$\begin{aligned}
X_1 &= X_0 + Z_1 = Z_1 \\
X_2 &= X_1 + Z_2 = Z_1 + Z_2 \\
&\;\;\vdots \\
X_n &= Z_1 + Z_2 + \cdots + Z_n
\end{aligned} \tag{5.96}$$

Then, because the Z_n are iid r.v.'s and $X_0 = 0$, by Eqs. (4.132) and (4.136), we have

$$E(X_n) = E\left(\sum_{k=1}^{n} Z_k \right) = n E(Z_k)$$

$$\mathrm{Var}(X_n) = \mathrm{Var}\left(\sum_{k=1}^{n} Z_k \right) = n\, \mathrm{Var}(Z_k)$$

Now
$$E(Z_k) = (1)p + (-1)q = p - q \tag{5.97}$$

$$E(Z_k^2) = (1)^2 p + (-1)^2 q = p + q = 1 \tag{5.98}$$

Thus,
$$\mathrm{Var}(Z_k) = E(Z_k^2) - [E(Z_k)]^2 = 1 - (p - q)^2 = 4pq \tag{5.99}$$

Hence,
$$E(X_n) = n(p - q) \qquad q = 1 - p \tag{5.100}$$

$$\mathrm{Var}(X_n) = 4npq \qquad q = 1 - p \tag{5.101}$$

Note that if $p = q = \frac{1}{2}$, then

$$E(X_n) = 0 \tag{5.102}$$
$$\mathrm{Var}(X_n) = n \tag{5.103}$$

5.11. Find the autocorrelation function $R_x(n, m)$ of the simple random walk $X(n)$ of Prob. 5.2.

From Eq. (5.96), we can express X_n as

$$X_n = \sum_{i=0}^{n} Z_i \qquad n = 1, 2, \ldots \tag{5.104}$$

where $Z_0 = X_0 = 0$ and Z_i $(i \geq 1)$ are iid r.v.'s with

$$P(Z_i = +1) = p \qquad P(Z_i = -1) = q = 1 - p$$

By Eq. (5.7),

$$R_x(n, m) = E[X(n)X(m)] = E(X_n X_m)$$

Then by Eq. (5.104),

$$R_X(n, m) = \sum_{i=0}^{n} \sum_{k=0}^{m} E(Z_i Z_k) = \sum_{i=0}^{\min(n,m)} E(Z_i^2) + \sum_{\substack{i=0 \ k=0 \\ i \neq k}}^{n} \sum^{m} E(Z_i)E(Z_k) \tag{5.105}$$

Using Eqs. (5.97) and (5.98), we obtain

$$R_x(n, m) = \min(n, m) + [nm - \min(n, m)](p - q)^2 \tag{5.106}$$

or

$$R_X(n, m) = \begin{cases} m + (nm - m)(p - q)^2 & m < n \\ n + (nm - n)(p - q)^2 & n < m \end{cases} \tag{5.107}$$

Note that if $p = q = \frac{1}{2}$, then

$$R_x(n, m) = \min(n, m) \qquad n, m > 0 \tag{5.108}$$

5.12. Consider the random process $X(t)$ of Prob. 5.4; that is,

$$X(t) = Y \cos \omega t \qquad t \geq 0$$

where ω is a constant and Y is a uniform r.v. over $(0, 1)$.

(a) Find $E[X(t)]$.

(b) Find the autocorrelation function $R_x(t, s)$ of $X(t)$.

(c) Find the autocovariance function $K_x(t, s)$ of $X(t)$.

(a) From Eqs. (2.58) and (2.110), we have $E(Y) = \frac{1}{2}$ and $E(Y^2) = \frac{1}{3}$. Thus,

$$E[X(t)] = E(Y \cos \omega t) = E(Y) \cos \omega t = \frac{1}{2} \cos \omega t \tag{5.109}$$

(b) By Eq. (5.7), we have

$$R_X(t, s) = E[X(t)X(s)] = E(Y^2 \cos \omega t \cos \omega s)$$
$$= E(Y^2) \cos \omega t \cos \omega s = \frac{1}{3} \cos \omega t \cos \omega s \tag{5.110}$$

(c) By Eq. (5.10), we have

$$K_X(t, s) = R_X(t, s) - E[X(t)]E[X(s)]$$
$$= \frac{1}{3} \cos \omega t \cos \omega s - \frac{1}{4} \cos \omega t \cos \omega s \tag{5.111}$$
$$= \frac{1}{12} \cos \omega t \cos \omega s$$

5.13. Consider a discrete-parameter random process $X(n) = \{X_n, n \geq 1\}$ where the X_n's are iid r.v.'s with common cdf $F_x(x)$, mean μ, and variance σ^2.

(a) Find the joint cdf of $X(n)$.

(b) Find the mean of $X(n)$.

(c) Find the autocorrelation function $R_x(n, m)$ of $X(n)$.

(d) Find the autocovariance function $K_x(n, m)$ of $X(n)$.

(a)   Since the X_n's are iid r.v.'s with common cdf $F_x(x)$, the joint cdf of $X(n)$ is given by

$$F_X(x_1, ..., x_n) = \prod_{i=1}^{n} F_X(x_i) = [F_X(x)]^n \qquad (5.112)$$

(b)   The mean of $X(n)$ is

$$\mu_x(n) = E(X_n) = \mu \qquad \text{for all } n \qquad (5.113)$$

(c)   If $n \neq m$, by Eqs. (5.7) and (5.113),

$$R_x(n, m) = E(X_n X_m) = E(X_n)E(X_m) = \mu^2$$

If $n = m$, then by Eq. (2.31),

$$E(X_n^2) = \text{Var}(X_n) + [E(X_n)]^2 = \sigma^2 + \mu^2$$

Hence,

$$R_X(n, m) = \begin{cases} \mu^2 & n \neq m \\ \sigma^2 + \mu^2 & n = m \end{cases} \qquad (5.114)$$

(d)   By Eq. (5.10),

$$K_X(n, m) = R_X(n, m) - \mu_X(n)\mu_X(m) = \begin{cases} 0 & n \neq m \\ \sigma^2 & n = m \end{cases} \qquad (5.115)$$

Classification of Random Processes

5.14. Show that a random process which is stationary to order n is also stationary to all orders lower than n.

Assume that Eq. (5.14) holds for some particular n; that is,

$$P\{X(t_1) \leq x_1, ..., X(t_n) \leq x_n\} = P\{X(t_1 + \tau) \leq x_1, ..., X(t_n + \tau) \leq x_n\}$$

for any τ. Letting $x_n \to \infty$, we have [see Eq. (3.63)]

$$P\{X(t_1) \leq x_1, ..., X(t_{n-1}) \leq x_{n-1}\} = P\{X(t_1 + \tau) \leq x_1, ..., X(t_{n-1} + \tau) \leq x_{n-1}\}$$

and the process is stationary to order $n - 1$. Continuing the same procedure, we see that the process is stationary to all orders lower than n.

5.15. Show that if $\{X(t), t \in T\}$ is a strict-sense stationary random process, then it is also WSS.

Since $X(t)$ is strict-sense stationary, the first- and second-order distributions are invariant through time translation for all $\tau \in T$. Then we have

$$\mu_x(t) = E[X(t)] = E[X(t + \tau)] = \mu_x(t + \tau)$$

And, hence, the mean function $\mu_x(t)$ must be constant; that is,

$$E[X(t)] = \mu \text{ (constant)}$$

Similarly, we have

$$E[X(s)X(t)] = E[X(s + \tau)X(t + \tau)]$$

so that the autocorrelation function would depend on the time points s and t only through the difference $|t - s|$. Thus, $X(t)$ is WSS.

5.16. Let $\{X_n, n \geq 0\}$ be a sequence of iid r.v.'s with mean 0 and variance 1. Show that $\{X_n, n \geq 0\}$ is a WSS process.

By Eq. (5.113),

$$E(X_n) = 0 \text{ (constant)} \qquad \text{for all } n$$

and by Eq. (5.114),

$$R_X(n, n+k) = E(X_n X_{n+k}) = \begin{cases} E(X_n)E(X_{n+k}) = 0 & k \neq 0 \\ E(X_n^2) = \text{Var}(X_n) = 1 & k = 0 \end{cases}$$

which depends only on k. Thus, $\{X_n\}$ is a WSS process.

5.17. Show that if a random process $X(t)$ is WSS, then it must also be covariance stationary.

If $X(t)$ is WSS, then

$$E[X(t)] = \mu \text{ (constant)} \qquad \text{for all } t$$
$$R_x(t, t + \tau)] = R_x(\tau) \qquad \text{for all } t$$

Now $\qquad K_X(t, t + \tau) = \text{Cov}[X(t)X(t + \tau)] = R_X(t, t + \tau) - E[X(t)]E[X(t + \tau)]$
$$= R_X(\tau) - \mu^2$$

which indicates that $K_X(t, t + \tau)$ depends only on τ; thus, $X(t)$ is covariance stationary.

5.18. Consider a random process $X(t)$ defined by

$$X(t) = U \cos \omega t + V \sin \omega t \qquad -\infty < t < \infty \qquad (5.116)$$

where ω is constant and U and V are r.v.'s.

(a) Show that the condition

$$E(U) = E(V) = 0 \qquad (5.117)$$

is necessary for $X(t)$ to be stationary.

(b) Show that $X(t)$ is WSS if and only if U and V are uncorrelated with equal variance; that is,

$$E(UV) = 0 \qquad E(U^2) = E(V^2) = \sigma^2 \qquad (5.118)$$

(a) Now

$$\mu_X(t) = E[X(t)] = E(U) \cos \omega t + E(V) \sin \omega t$$

must be independent of t for $X(t)$ to be stationary. This is possible only if $\mu_x(t) = 0$, that is, $E(U) = E(V) = 0$.

(b) If $X(t)$ is WSS, then

$$E[X^2(0)] = E\left[X^2\left(\frac{\pi}{2\omega}\right)\right] = R_{XX}(0) = \sigma_X^2$$

But $X(0) = U$ and $X(\pi/2\omega) = V$; thus,

$$E(U^2) = E(V^2) = \sigma_x^2 = \sigma^2$$

Using the above result, we obtain

$$
\begin{aligned}
R_x(t, t + \tau) &= E[X(t)X(t + \tau)] \\
&= E\{(U \cos \omega t + V \sin \omega t)[U \cos \omega(t + \tau) + V \sin \omega(t + \tau)]\} \\
&= \sigma^2 \cos \omega\tau + E(UV) \sin(2\omega t + \omega\tau) \quad\quad\quad (5.119)
\end{aligned}
$$

which will be a function of τ only if $E(UV) = 0$. Conversely, if $E(UV) = 0$ and $E(U^2) = E(V^2) = \sigma^2$, then from the result of part (a) and Eq. (5.119), we have

$$
\begin{aligned}
\mu_x(t) &= 0 \\
R_x(t, t + \tau) &= \sigma^2 \cos \omega\tau = R_x(\tau)
\end{aligned}
$$

Hence, $X(t)$ is WSS.

5.19. Consider a random process $X(t)$ defined by

$$X(t) = U \cos t + V \sin t \quad\quad -\infty < t < \infty$$

where U and V are independent r.v.'s, each of which assumes the values -2 and 1 with the probabilities $\frac{1}{3}$ and $\frac{2}{3}$, respectively. Show that $X(t)$ is WSS but not strict-sense stationary.

We have

$$
E(U) = E(V) = \frac{1}{3}(-2) + \frac{2}{3}(1) = 0
$$

$$
E(U^2) = E(V^2) = \frac{1}{3}(-2)^2 + \frac{2}{3}(1)^2 = 2
$$

Since U and V are independent,

$$E(UV) = E(U)E(V) = 0$$

Thus, by the results of Prob. 5.18, $X(t)$ is WSS. To see if $X(t)$ is strict-sense stationary, we consider $E[X^3(t)]$.

$$
\begin{aligned}
E[X^3(t)] &= E[(U \cos t + V \sin t)^3] \\
&= E(U^3) \cos^3 t + 3E(U^2 V) \cos^2 t \sin t + 3E(UV^2) \cos t \sin^2 t + E(V^3) \sin^3 t
\end{aligned}
$$

Now

$$E(U^3) = E(V^3) = \frac{1}{3}(-2)^3 + \frac{2}{3}(1)^3 = -2$$

$$E(U^2 V) = E(U^2)E(V) = 0 \quad\quad E(UV^2) = E(U)E(V^2) = 0$$

Thus,

$$E[X^3(t)] = -2(\cos^3 t + \sin^3 t)$$

which is a function of t. From Eq. (5.16), we see that all the moments of a strict-sense stationary process must be independent of time. Thus, $X(t)$ is not strict-sense stationary.

5.20. Consider a random process $X(t)$ defined by

$$X(t) = A \cos(\omega t + \Theta) \quad\quad -\infty < t < \infty$$

where A and ω are constants and Θ is a uniform r.v. over $(-\pi, \pi)$. Show that $X(t)$ is WSS.

From Eq. (2.56), we have

$$f_\Theta(\theta) = \begin{cases} \dfrac{1}{2\pi} & -\pi < \theta < \pi \\ 0 & \text{otherwise} \end{cases}$$

Then

$$\mu_X(t) = \frac{A}{2\pi} \int_{-\pi}^{\pi} \cos(\omega t + \theta)\, d\theta = 0 \tag{5.120}$$

Setting $s = t + \tau$ in Eq. (5.7), we have

$$\begin{aligned} R_{XX}(t, t + \tau) &= \frac{A^2}{2\pi} \int_{-\pi}^{\pi} \cos(\omega t + \theta) \cos[\omega(t + \tau) + \theta)]\, d\theta \\ &= \frac{A^2}{2\pi} \int_{-\pi}^{\pi} \frac{1}{2}[\cos \omega t + \cos(2\omega t + 2\theta + \omega\tau)]\, d\theta \\ &= \frac{A^2}{2} \cos \omega\tau \end{aligned} \tag{5.121}$$

Since the mean of $X(t)$ is a constant and the autocorrelation of $X(t)$ is a function of time difference only, we conclude that $X(t)$ is WSS.

5.21. Let $\{X(t), t \ge 0\}$ be a random process with stationary independent increments, and assume that $X(0) = 0$. Show that

$$E[X(t)] = \mu_1 t \tag{5.122}$$

where $\mu_1 = E[X(1)]$.

Let

$$f(t) = E[X(t)] = E[X(t) - X(0)]$$

Then, for any t and s and using Eq. (4.132) and the property of the stationary independent increments, we have

$$\begin{aligned} f(t + s) &= E[X(t + s) - X(0)] \\ &= E[X(t + s) - X(s) + X(s) - X(0)] \\ &= E[X(t + s) - X(s)] + E[X(s) - X(0)] \\ &= E[X(t) - X(0)] + E[X(s) - X(0)] \\ &= f(t) + f(s) \end{aligned} \tag{5.123}$$

The only solution to the above functional equation is $f(t) = ct$, where c is a constant. Since $c = f(1) = E[X(1)]$, we obtain

$$E[X(t)] = \mu_1 t \qquad \mu_1 = E[X(1)]$$

5.22. Let $\{X(t), t \ge 0\}$ be a random process with stationary independent increments, and assume that $X(0) = 0$. Show that

(a)

$$\text{Var}[X(t)] = \sigma_1^2 t \tag{5.124}$$

(b)

$$\text{Var}[X(t) - X(s)] = \sigma_1^2(t - s) \qquad t > s \tag{5.125}$$

where $\sigma_1^2 = \text{Var}[X(1)]$.

(a)   Let

$$g(t) = \text{Var}[X(t)] = \text{Var}[X(t) - X(0)]$$

Then, for any t and s and using Eq. (4.136) and the property of the stationary independent increments, we get

$$\begin{aligned} g(t + s) &= \text{Var}[X(t + s) - X(0)] \\ &= \text{Var}[X(t + s) - X(s) + X(s) - X(0)] \\ &= \text{Var}[X(t + s) - X(s)] + \text{Var}[X(s) - X(0)] \\ &= \text{Var}[X(t) - X(0)] + \text{Var}[X(s) - X(0)] \\ &= g(t) + g(s) \end{aligned}$$

which is the same functional equation as Eq. (5.123). Thus, $g(t) = kt$, where k is a constant. Since $k = g(1) = \text{Var}[X(1)]$, we obtain

$$\text{Var}[X(t)] = \sigma_1^2 t \qquad \sigma_1^2 = \text{Var}[X(1)]$$

(*b*)   Let $t > s$. Then

$$\begin{aligned}
\text{Var}[X(t)] &= \text{Var}[X(t) - X(s) + X(s) - X(0)] \\
&= \text{Var}[X(t) - X(s)] + \text{Var}[X(s) - X(0)] \\
&= \text{Var}[X(t) - X(s)] + \text{Var}[X(s)]
\end{aligned}$$

Thus, using Eq. (5.124), we obtain

$$\text{Var}[X(t) - X(s)] = \text{Var}[X(t)] - \text{Var}[X(s)] = \sigma_1^2(t - s)$$

5.23. Let $\{X(t), t \geq 0\}$ be a random process with stationary independent increments, and assume that $X(0) = 0$. Show that

$$\text{Cov}[X(t), X(s)] = K_X(t, s) = \sigma_1^2 \min(t, s) \tag{5.126}$$

where $\sigma_1^2 = \text{Var}[X(1)]$.

By definition (2.28),

$$\begin{aligned}
\text{Var}[X(t) - X(s)] &= E(\{X(t) - X(s) - E[X(t) - X(s)]\}^2) \\
&= E[(\{X(t) - E[X(t)]\} - \{X(s) - E[X(s)]\})^2] \\
&= E(\{X(t) - E[X(t)]\}^2 - 2\{X(t) - E[X(t)]\}\{X(s) - E[X(s)]\} + \{X(s) - E[X(s)]\}^2) \\
&= \text{Var}[X(t)] - 2\,\text{Cov}[X(t), X(s)] + \text{Var}[X(s)]
\end{aligned}$$

Thus, $\text{Cov}[X(t), X(s)] = \frac{1}{2}\{\text{Var}[X(t)] + \text{Var}[X(s)] - \text{Var}[X(t) - X(s)]\}$

Using Eqs. (5.124) and (5.125), we obtain

$$K_X(t, s) = \begin{cases} \dfrac{1}{2}\sigma_1^2[t + s - (t - s)] = \sigma_1^2 s & t > s \\[2mm] \dfrac{1}{2}\sigma_1^2[t + s - (s - t)] = \sigma_1^2 t & s > t \end{cases}$$

or $K_X(t, s) = \sigma_1^2 \min(t, s)$

where $\sigma_1^2 = \text{Var}[X(1)]$.

5.24. (*a*)   Show that a simple random walk $X(n)$ of Prob. 5.2 is a Markov chain.

(*b*)   Find its one-step transition probabilities.

(*a*)   From Eq. (5.96) (Prob. 5.10), $X(n) = \{X_n, n \geq 0\}$ can be expressed as

$$X_0 = 0 \qquad X_n = \sum_{i=1}^{n} Z_i \qquad n \geq 1$$

where Z_n ($n = 1, 2, \dots$) are iid r.v.'s with

$$P(Z_n = k) = a_k \qquad (k = 1, -1) \qquad \text{and} \qquad a_1 = p \qquad a_{-1} = q = 1 - p$$

Then $X(n) = \{X_n, n \geq 0\}$ is a Markov chain, since

$$
\begin{aligned}
P(X_{n+1} = i_{n+1} | X_0 = 0, X_1 = i_1, \ldots, X_n = i_n) \\
= P(Z_{n+1} + i_n = i_{n+1} | X_0 = 0, X_1 = i_1, \ldots, X_n = i_n) \\
= P(Z_{n+1} = i_{n+1} - i_n) = a_{i_{n+1} - i_n} = P(X_{n+1} = i_{n+1} | X_n = i_n)
\end{aligned}
$$

since Z_{n+1} is independent of $X_0, X_1, \ldots, X_n$.

(b)   The one-step transition probabilities are given by

$$
p_{jk} = P(X_n = k | X_{n-1} = j) = \begin{cases} p & k = j+1 \\ q = 1-p & k = j-1 \\ 0 & \text{otherwise} \end{cases}
$$

which do not depend on n. Thus, a simple random walk $X(n)$ is a homogeneous Markov chain.

5.25. Show that for a Markov process $X(t)$, the second-order distribution is sufficient to characterize $X(t)$.

Let $X(t)$ be a Markov process with the nth-order distribution

$$
F_X(x_1, x_2, \ldots, x_n; t_1, t_2, \ldots, t_n) = P\{X(t_1) \leq x_1, X(t_2) \leq x_2, \ldots, X(t_n) \leq x_n\}
$$

Then, using the Markov property (5.26), we have

$$
\begin{aligned}
F_X(x_1, x_2, \ldots, x_n; t_1, t_2, \ldots, t_n) &= P\{X(t_n) \leq x_n | X(t_1) \leq x_1, X(t_2) \leq x_2, \ldots, X(t_{n-1}) \leq x_{n-1}\} \\
&\quad \times P\{X(t_1) \leq x_1, X(t_2) \leq x_2, \ldots, X(t_{n-1}) \leq x_{n-1}\} \\
&= P\{X(t_n) \leq x_n | X(t_{n-1}) \leq x_{n-1}\} F_X(x_1, \ldots, x_{n-1}; t_1, \ldots, t_{n-1})
\end{aligned}
$$

Applying the above relation repeatedly for lower-order distribution, we can write

$$
F_X(x_1, x_2, \ldots, x_n; t_1, t_2, \ldots, t_n) = F_X(x_1, t_1) \prod_{k=2}^{n} P\{X(t_k) \leq x_k | X(t_{k-1}) \leq x_{k-1}\} \tag{5.127}
$$

Hence, all finite-order distributions of a Markov process can be completely determined by the second-order distribution.

5.26. Show that if a normal process is WSS, then it is also strict-sense stationary.

By Eq. (5.29), a normal random process $X(t)$ is completely characterized by the specification of the mean $E[X(t)]$ and the covariance function $K_X(t, s)$ of the process. Suppose that $X(t)$ is WSS. Then, by Eqs. (5.21) and (5.22), Eq. (5.29) becomes

$$
\Psi_{X(t_1) \cdots X(t_n)}(\omega_1, \ldots, \omega_n) = \exp\left\{ j \sum_{i=1}^{n} \mu \omega_i - \frac{1}{2} \sum_{i=1}^{n} \sum_{k=1}^{n} K_X(t_i - t_k) \omega_i \omega_k \right\} \tag{5.128}
$$

Now we translate all of the time instants $t_1, t_2, \ldots, t_n$ by the same amount τ. The joint characteristic function of the new r.v.'s $X(t_i + \tau)$, $i = 1, 2, \ldots, n$, is then

$$
\begin{aligned}
\Psi_{X(t_1+\tau) \cdots X(t_n+\tau)}(\omega_1, \ldots, \omega_n) &= \exp\left\{ j \sum_{i=1}^{n} \mu \omega_i - \frac{1}{2} \sum_{i=1}^{n} \sum_{k=1}^{n} K_X[t_i + \tau - (t_k + \tau)] \omega_i \omega_k \right\} \\
&= \exp\left\{ j \sum_{i=1}^{n} \mu \omega_i - \frac{1}{2} \sum_{i=1}^{n} \sum_{k=1}^{n} K_X(t_i - t_k) \omega_i \omega_k \right\} \\
&= \Psi_{X(t_1) \cdots X(t_n)}(\omega_1, \ldots, \omega_n) \tag{5.129}
\end{aligned}
$$

which indicates that the joint characteristic function (and hence the corresponding joint pdf) is unaffected by a shift in the time origin. Since this result holds for any n and any set of time instants ($t_i \in T, i = 1, 2, ..., n$), it follows that if a normal process is WSS, then it is also strict-sense stationary.

5.27. Let $\{X(t), -\infty < t < \infty\}$ be a zero-mean, stationary, normal process with the autocorrelation function

$$R_X(\tau) = \begin{cases} 1 - \dfrac{|\tau|}{T} & -T \le \tau \le T \\ 0 & \text{otherwise} \end{cases} \tag{5.130}$$

Let $\{X(t_i), i = 1, 2, ..., n\}$ be a sequence of n samples of the process taken at the time instants

$$t_i = i\frac{T}{2} \qquad i = 1, 2, ..., n$$

Find the mean and the variance of the sample mean

$$\hat{\mu}_n = \frac{1}{n} \sum_{i=1}^{n} X(t_i) \tag{5.131}$$

Since $X(t)$ is zero-mean and stationary, we have

$$E[X(t_i)] = 0$$

and $\quad R_X(t_i, t_k) = E[X(t_i)X(t_k)] = R_X(t_k - t_i) = R_X\left[(k - i)\dfrac{T}{2}\right]$

Thus, $\qquad E(\hat{\mu}_n) = E\left[\dfrac{1}{n}\sum_{i=1}^{n} X(t_i)\right] = \dfrac{1}{n}\sum_{i=1}^{n} E[X(t_i)] = 0 \tag{5.132}$

and $\qquad \mathrm{Var}(\hat{\mu}_n) = E\{[\hat{\mu}_n - E(\hat{\mu}_n)]^2\} = E(\hat{\mu}_n^2)$

$$= E\left\{\left[\frac{1}{n}\sum_{i=1}^{n} X(t_i)\right]\left[\frac{1}{n}\sum_{k=1}^{n} X(t_k)\right]\right\}$$

$$= \frac{1}{n^2}\sum_{i=1}^{n}\sum_{k=1}^{n} E[X(t_i)X(t_k)] = \frac{1}{n^2}\sum_{i=1}^{n}\sum_{k=1}^{n} R_X\left[(k - i)\frac{T}{2}\right]$$

By Eq. (5.130),

$$R_X[(k - i)T/2] = \begin{cases} 1 & k = i \\ \dfrac{1}{2} & |k - i| = 1 \\ 0 & |k - i| > 2 \end{cases}$$

Thus, $\qquad \mathrm{Var}(\hat{\mu}_n) = \dfrac{1}{n^2}\left[n(1) + 2(n - 1)\left(\dfrac{1}{2}\right) + 0\right] = \dfrac{1}{n^2}(2n - 1) \tag{5.133}$

Discrete-Parameter Markov Chains

5.28. Show that if P is a Markov matrix, then P^n is also a Markov matrix for any positive integer n.

Let $\qquad P = [p_{ij}] = \begin{bmatrix} p_{11} & p_{12} & \cdots & p_{1m} \\ p_{12} & p_{22} & \cdots & p_{2m} \\ \vdots & \vdots & \ddots & \vdots \\ p_{m1} & p_{m2} & \cdots & p_{mm} \end{bmatrix}$

Then by the property of a Markov matrix [Eq. (5.35)], we can write

$$
\begin{bmatrix}
p_{11} & p_{12} & \cdots & p_{1m} \\
p_{12} & p_{22} & \cdots & p_{2m} \\
\vdots & \vdots & \ddots & \vdots \\
p_{m1} & p_{m2} & \cdots & p_{mm}
\end{bmatrix}
\begin{bmatrix} 1 \\ 1 \\ \vdots \\ 1 \end{bmatrix}
=
\begin{bmatrix} 1 \\ 1 \\ \vdots \\ 1 \end{bmatrix}
$$

or
$$P\mathbf{a} = \mathbf{a} \tag{5.134}$$

where
$$\mathbf{a}^T = [1 \quad 1 \ldots 1]$$

Premultiplying both sides of Eq. (5.134) by P, we obtain

$$P^2\mathbf{a} = P\mathbf{a} = \mathbf{a}$$

which indicates that P^2 is also a Markov matrix. Repeated premultiplication by P yields

$$P^n\mathbf{a} = \mathbf{a}$$

which shows that P^n is also a Markov matrix.

5.29. Verify Eq. (5.39); that is,

$$\mathbf{p}(n) = \mathbf{p}(0)P^n$$

We verify Eq. (5.39) by induction. If the state of X_0 is i, state X_1 will be j only if a transition is made from i to j. The events $\{X_0 = i, i = 1, 2, \ldots\}$ are mutually exclusive, and one of them must occur. Hence, by the law of total probability [Eq. (1.44)],

$$P(X_1 = j) = \sum_i P(X_0 = i)P(X_1 = j | X_0 = i)$$

or
$$p_j(1) = \sum_i p_i(0)p_{ij} \qquad j = 1, 2, \ldots \tag{5.135}$$

In terms of vectors and matrices, Eq. (5.135) can be expressed as

$$\mathbf{p}(1) = \mathbf{p}(0)P \tag{5.136}$$

Thus, Eq. (5.39) is true for $n = 1$. Assume now that Eq. (5.39) is true for $n = k$; that is,

$$\mathbf{p}(k) = \mathbf{p}(0)P^k$$

Again, by the law of total probability,

$$P(X_{k+1} = j) = \sum_i P(X_k = i)P(X_{k+1} = j | X_k = i)$$

or
$$p_j(k + 1) = \sum_i p_i(k)p_{ij} \qquad j = 1, 2, \ldots \tag{5.137}$$

In terms of vectors and matrices, Eq. (5.137) can be expressed as

$$\mathbf{p}(k + 1) = \mathbf{p}(k)P = \mathbf{p}(0)P^kP = \mathbf{p}(0)P^{k+1} \tag{5.138}$$

which indicates that Eq. (5.39) is true for $k + 1$. Hence, we conclude that Eq. (5.39) is true for all $n \geq 1$.

5.30. Consider a two-state Markov chain with the transition probability matrix

$$P = \begin{bmatrix} 1 - a & a \\ b & 1 - b \end{bmatrix} \qquad 0 < a < 1, 0 < b < 1 \tag{5.139}$$

(a) Show that the n-step transition probability matrix P^n is given by

$$P^n = \frac{1}{a+b}\left\{\begin{bmatrix} b & a \\ b & a \end{bmatrix} + (1-a-b)^n \begin{bmatrix} a & -a \\ -b & b \end{bmatrix}\right\} \tag{5.140}$$

(b) Find P^n when $n \to \infty$.

(a) From matrix analysis, the characteristic equation of P is

$$c(\lambda) = |\lambda I - P| = \begin{vmatrix} \lambda - (1-a) & -a \\ -b & \lambda - (1-b) \end{vmatrix}$$

$$= (\lambda - 1)(\lambda - 1 + a + b) = 0$$

Thus, the eigenvalues of P are $\lambda_1 = 1$ and $\lambda_2 = 1 - a - b$. Then, using the spectral decomposition method, P^n can be expressed as

$$P^n = \lambda_1^n E_1 + \lambda_2^n E_2 \tag{5.141}$$

where E_1 and E_2 are constituent matrices of P, given by

$$E_1 = \frac{1}{\lambda_1 - \lambda_2}[P - \lambda_2 I] \qquad E_2 = \frac{1}{\lambda_2 - \lambda_1}[P - \lambda_1 I] \tag{5.142}$$

Substituting $\lambda_1 = 1$ and $\lambda_2 = 1 - a - b$ in the above expressions, we obtain

$$E_1 = \frac{1}{a+b}\begin{bmatrix} b & a \\ b & a \end{bmatrix} \qquad E_2 = \frac{1}{a+b}\begin{bmatrix} a & -a \\ -b & b \end{bmatrix}$$

Thus, by Eq. (5.141), we obtain

$$P^n = E_1 + (1-a-b)^n E_2$$

$$= \frac{1}{a+b}\left\{\begin{bmatrix} b & a \\ b & a \end{bmatrix} + (1-a-b)^n \begin{bmatrix} a & -a \\ -b & b \end{bmatrix}\right\} \tag{5.143}$$

(b) If $0 < a < 1, 0 < b < 1$, then $0 < 1 - a < 1$ and $|1 - a - b| < 1$. So $\lim_{n\to\infty} (1 - a - b)^n = 0$ and

$$\lim_{n\to\infty} P^n = \frac{1}{a+b}\begin{bmatrix} b & a \\ b & a \end{bmatrix} \tag{5.144}$$

Note that a limiting matrix exists and has the same rows (see Prob. 5.47).

5.31. An example of a two-state Markov chain is provided by a communication network consisting of the sequence (or cascade) of stages of binary communication channels shown in Fig. 5-10. Here X_n

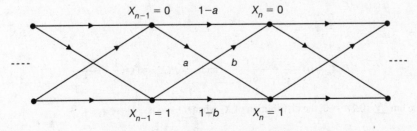

Fig. 5-10 Binary communication network.

denotes the digit leaving the *n*th stage of the channel and X_0 denotes the digit entering the first stage. The transition probability matrix of this communication network is often called the *channel matrix* and is given by Eq. (5.139); that is,

$$P = \begin{bmatrix} 1-a & a \\ b & 1-b \end{bmatrix} \qquad 0 < a < 1, 0 < b < 1$$

Assume that $a = 0.1$ and $b = 0.2$, and the initial distribution is $P(X_0 = 0) = P(X_0 = 1) = 0.5$.

(a) Find the distribution of X_n.

(b) Find the distribution of X_n when $n \to \infty$.

(a) The channel matrix of the communication network is

$$P = \begin{bmatrix} 0.9 & 0.1 \\ 0.2 & 0.8 \end{bmatrix}$$

and the initial distribution is

$$\mathbf{p}(0) = [0.5 \quad 0.5]$$

By Eq. (5.39), the distribution of X_n is given by

$$\mathbf{p}(n) = \mathbf{p}(0)P^n = \begin{bmatrix} 0.5 & 0.5 \end{bmatrix} \begin{bmatrix} 0.9 & 0.1 \\ 0.2 & 0.8 \end{bmatrix}^n$$

Letting $a = 0.1$ and $b = 0.2$ in Eq. (5.140), we get

$$\begin{bmatrix} 0.9 & 0.1 \\ 0.2 & 0.8 \end{bmatrix}^n = \frac{1}{0.3} \begin{bmatrix} 0.2 & 0.1 \\ 0.2 & 0.1 \end{bmatrix} + \frac{(0.7)^n}{0.3} \begin{bmatrix} 0.1 & -0.1 \\ -0.2 & 0.2 \end{bmatrix}$$

$$= \begin{bmatrix} \dfrac{2 + (0.7)^n}{3} & \dfrac{1 - (0.7)^n}{3} \\ \dfrac{2 - 2(0.7)^n}{3} & \dfrac{1 + 2(0.7)^n}{3} \end{bmatrix}$$

Thus, the distribution of X_n is

$$\mathbf{p}(n) = \begin{bmatrix} 0.5 & 0.5 \end{bmatrix} \begin{bmatrix} \dfrac{2 + (0.7)^n}{3} & \dfrac{1 - (0.7)^n}{3} \\ \dfrac{2 - 2(0.7)^n}{3} & \dfrac{1 + 2(0.7)^n}{3} \end{bmatrix}$$

$$= \begin{bmatrix} \dfrac{2}{3} - \dfrac{(0.7)^n}{6} & \dfrac{1}{3} + \dfrac{(0.7)^n}{6} \end{bmatrix}$$

that is,

$$P(X_n = 0) = \frac{2}{3} - \frac{(0.7)^n}{6} \qquad \text{and} \qquad P(X_n = 1) = \frac{1}{3} + \frac{(0.7)^n}{6}$$

(b) Since $\lim_{n \to \infty} (0.7)^n = 0$, the distribution of X_n when $n \to \infty$ is

$$P(X_\infty = 0) = \frac{2}{3} \qquad \text{and} \qquad P(X_\infty = 1) = \frac{1}{3}$$

5.32. Verify the transitivity property of the Markov chain; that is, if $i \rightarrow j$ and $j \rightarrow k$, then $i \rightarrow k$.

By definition, the relations $i \rightarrow j$ and $j \rightarrow k$ imply that there exist integers n and m such that $p_{ij}^{(n)} > 0$ and $p_{jk}^{(m)} > 0$. Then, by the Chapman-Kolmogorov equation (5.38), we have

$$p_{ik}^{(n+m)} = \sum_r p_{ir}^{(n)} p_{rk}^{(m)} \geq p_{ij}^{(n)} p_{jk}^{(m)} > 0 \tag{5.145}$$

Therefore, $i \rightarrow k$.

5.33. Verify Eq. (5.42).

If the Markov chain $\{X_n\}$ goes from state i to state j in m steps, the first step must take the chain from i to some state k, where $k \neq j$. Now after that first step to k, we have $m - 1$ steps left, and the chain must get to state j, from state k, on the last of those steps. That is, the first visit to state j must occur on the $(m - 1)$st step, starting now in state k. Thus, we must have

$$f_{ij}^{(m)} = \sum_{k \neq j} p_{ik} f_{kj}^{(m-1)} \quad m = 2, 3, \ldots$$

5.34. Show that in a finite-state Markov chain, not all states can be transient.

Suppose that the states are $0, 1, \ldots, m$, and suppose that they are all transient. Then by definition, after a finite amount of time (say T_0), state 0 will never be visited; after a finite amount of time (say T_1), state 1 will never be visited; and so on. Thus, after a finite time $T = \max\{T_0, T_1, \ldots, T_m\}$, no state will be visited. But as the process must be in some state after time T, we have a contradiction. Thus, we conclude that not all states can be transient and at least one of the states must be recurrent.

5.35. A state transition diagram of a finite-state Markov chain is a line diagram with a vertex corresponding to each state and a directed line between two vertices i and j if $p_{ij} > 0$. In such a diagram, if one can move from i and j by a path following the arrows, then $i \rightarrow j$. The diagram is useful to determine whether a finite-state Markov chain is irreducible or not, or to check for periodicities. Draw the state transition diagrams and classify the states of the Markov chains with the following transition probability matrices:

$$(a) \quad P = \begin{bmatrix} 0 & 0.5 & 0.5 \\ 0.5 & 0 & 0.5 \\ 0.5 & 0.5 & 0 \end{bmatrix} \qquad (b) \quad P = \begin{bmatrix} 0 & 0 & 0.5 & 0.5 \\ 1 & 0 & 0 & 0 \\ 0 & 1 & 0 & 0 \\ 0 & 1 & 0 & 0 \end{bmatrix}$$

$$(c) \quad P = \begin{bmatrix} 0.3 & 0.4 & 0 & 0 & 0.3 \\ 0 & 1 & 0 & 0 & 0 \\ 0 & 0 & 0 & 0.6 & 0.4 \\ 0 & 0 & 0 & 0 & 1 \\ 0 & 0 & 1 & 0 & 0 \end{bmatrix}$$

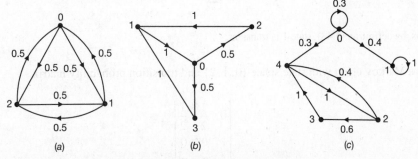

Fig. 5-11 State transition diagram.

(a) The state transition diagram of the Markov chain with P of part (a) is shown in Fig. 5-11(a). From Fig. 5-11(a), it is seen that the Markov chain is irreducible and aperiodic. For instance, one can get back to state 0 in two steps by going from 0 to 1 to 0. However, one can also get back to state 0 in three steps by going from 0 to 1 to 2 to 0. Hence 0 is aperiodic. Similarly, we can see that states 1 and 2 are also aperiodic.

(b) The state transition diagram of the Markov chain with P of part (b) is shown in Fig. 5-11(b). From Fig. 5-11(b), it is seen that the Markov chain is irreducible and periodic with period 3.

(c) The state transition diagram of the Markov chain with P of part (c) is shown in Fig. 5-11(c). From Fig. 5-11(c), it is seen that the Markov chain is not irreducible, since states 0 and 4 do not communicate, and state 1 is absorbing.

5.36. Consider a Markov chain with state space $\{0, 1\}$ and transition probability matrix

$$P = \begin{bmatrix} 1 & 0 \\ \dfrac{1}{2} & \dfrac{1}{2} \end{bmatrix}$$

(a) Show that state 0 is recurrent.

(b) Show that state 1 is transient.

(a) By Eqs. (5.41) and (5.42), we have

$$f_{00}^{(1)} = p_{00} = 1 \qquad f_{10}^{(1)} = p_{10} = \frac{1}{2}$$

$$f_{00}^{(2)} = p_{01}\, f_{10}^{(1)} = (0)\frac{1}{2} = 0$$

$$f_{00}^{(n)} = 0 \qquad n \geq 2$$

Then, by Eqs. (5.43),

$$f_{00} = P(T_0 < \infty \,|\, X_0 = 0) = \sum_{n=0}^{\infty} f_{00}^{(n)} = 1 + 0 + 0 + \cdots = 1$$

Thus, by definition (5.44), state 0 is recurrent.

(b) Similarly, we have

$$f_{11}^{(1)} = p_{11} = \frac{1}{2} \qquad f_{01}^{(1)} = p_{01} = 0$$

$$f_{11}^{(2)} = p_{10}\, f_{01}^{(1)} = \left(\frac{1}{2}\right)0 = 0$$

$$f_{11}^{(n)} = 0 \qquad n \geq 2$$

and

$$f_{11} = P(T_1 < \infty \,|\, X_0 = 1) = \sum_{n=0}^{\infty} f_{11}^{(n)} = \frac{1}{2} + 0 + 0 + \cdots = \frac{1}{2} < 1$$

Thus, by definition (5.48), state 1 is transient.

5.37. Consider a Markov chain with state space $\{0, 1, 2\}$ and transition probability matrix

$$P = \begin{bmatrix} 0 & \dfrac{1}{2} & \dfrac{1}{2} \\ 1 & 0 & 0 \\ 1 & 0 & 0 \end{bmatrix}$$

Show that state 0 is periodic with period 2.

The characteristic equation of P is given by

$$c(\lambda) = |\lambda I - P| = \begin{vmatrix} \lambda & -\dfrac{1}{2} & -\dfrac{1}{2} \\ -1 & \lambda & 0 \\ -1 & 0 & \lambda \end{vmatrix} = \lambda^3 - \lambda = 0$$

Thus, by the Cayley-Hamilton theorem (in matrix analysis), we have $P^3 = P$. Thus, for $n \geq 1$,

$$P^{(2n)} = P^2 = \begin{bmatrix} 0 & \dfrac{1}{2} & \dfrac{1}{2} \\ 1 & 0 & 0 \\ 1 & 0 & 0 \end{bmatrix} \begin{bmatrix} 0 & \dfrac{1}{2} & \dfrac{1}{2} \\ 1 & 0 & 0 \\ 1 & 0 & 0 \end{bmatrix} = \begin{bmatrix} 1 & 0 & 0 \\ 0 & \dfrac{1}{2} & \dfrac{1}{2} \\ 0 & \dfrac{1}{2} & \dfrac{1}{2} \end{bmatrix}$$

$$P^{(2n+1)} = P = \begin{bmatrix} 0 & \dfrac{1}{2} & \dfrac{1}{2} \\ 1 & 0 & 0 \\ 1 & 0 & 0 \end{bmatrix}$$

Therefore, $\qquad\qquad d(0) = \gcd \{n \geq 1 : p_{00}^{(n)} > 0\} = \gcd\{2, 5, 6, \dots\} = 2$

Thus, state 0 is periodic with period 2.

Note that the state transition diagram corresponding to the given P is shown in Fig. 5-12. From Fig. 5-12, it is clear that state 0 is periodic with period 2.

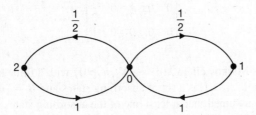

Fig. 5-12

5.38. Let two gamblers, A and B, initially have k dollars and m dollars, respectively. Suppose that at each round of their game, A wins one dollar from B with probability p and loses one dollar to B with probability $q = 1 - p$. Assume that A and B play until one of them has no money left. (This is known as the *Gambler's Ruin* problem.) Let X_n be A's capital after round n, where $n = 0, 1, 2, \dots$ and $X_0 = k$.

(a) Show that $X(n) = \{X_n, n \geq 0\}$ is a Markov chain with absorbing states.

(b) Find its transition probability matrix P.

(a) The total capital of the two players at all times is

$$k + m = N$$

Let Z_n $(n \geq 1)$ be independent r.v.'s with $P(Z_n = 1) = p$ and $P(Z_n = -1) = q = 1 - p$ for all n.

Then

$$X_n = X_{n-1} + Z_n \qquad n = 1, 2, \dots$$

and $X_0 = k$. The game ends when $X_n = 0$ or $X_n = N$. Thus, by Probs. 5.2 and 5.24, $X(n) = \{X_n, n \geq 0\}$ is a Markov chain with state space $E = \{0, 1, 2, \dots, N\}$, where states 0 and N are absorbing states. The Markov chain $X(n)$ is also known as a simple random walk with *absorbing barriers*.

(b)   Since

$$p_{i,i+1} = P(X_{n+1} = i + 1 \,|\, X_n = i) = p$$
$$p_{i,i-1} = P(X_{n+1} = i - 1 \,|\, X_n = i) = q$$
$$p_{i,i} = P(X_{n+1} = i \,|\, X_n = i) = 0 \qquad i \neq 0, N$$
$$p_{0,0} = P(X_{n+1} = 0 \,|\, X_n = 0) = 1$$
$$p_{N,N} = P(X_{n+1} = N \,|\, X_n = N) = 1$$

the transition probability matrix P is

$$P = \begin{bmatrix}
1 & 0 & 0 & 0 & \cdots & \cdots & \cdots & 0 \\
q & 0 & p & 0 & \cdots & \cdots & \cdots & 0 \\
0 & q & 0 & p & \cdots & \cdots & \cdots & 0 \\
\vdots & & & & \ddots & & & \vdots \\
\vdots & & & & & \ddots & & \vdots \\
0 & 0 & 0 & 0 & \cdots & q & 0 & p \\
0 & 0 & 0 & 0 & \cdots & 0 & 0 & 1
\end{bmatrix} \tag{5.146}$$

For example, when $p = q = \frac{1}{2}$ and $N = 4$,

$$P = \begin{bmatrix}
1 & 0 & 0 & 0 & 0 \\
\frac{1}{2} & 0 & \frac{1}{2} & 0 & 0 \\
0 & \frac{1}{2} & 0 & \frac{1}{2} & 0 \\
0 & 0 & \frac{1}{2} & 0 & \frac{1}{2} \\
0 & 0 & 0 & 0 & 1
\end{bmatrix}$$

5.39. Consider a homogeneous Markov chain $X(n) = \{X_n, n \geq 0\}$ with a finite state space $E = \{0, 1, \ldots, N\}$, of which $A = \{0, 1, \ldots, m\}$, $m \geq 1$, is a set of absorbing states and $B = \{m + 1, \ldots, N\}$ is a set of nonabsorbing states. It is assumed that at least one of the absorbing states in A is accessible from any nonabsorbing states in B. Show that absorption of $X(n)$ in one or another of the absorbing states is certain.

If $X_0 \in A$, then there is nothing to prove, since $X(n)$ is already absorbed. Let $X_0 \in B$. By assumption, there is at least one state in A which is accessible from any state in B. Now assume that state $k \in A$ is accessible from $j \in B$. Let $n_{jk} (< \infty)$ be the smallest number n such that $p_{jk}^{(n)} > 0$. For a given state j, let n_j be the largest of n_{jk} as k varies and n' be the largest of n_j as j varies. After n' steps, no matter what the initial state of $X(n)$, there is a probability $p > 0$ that $X(n)$ is in an absorbing state. Therefore,

$$P\{X_{n'} \in B\} = 1 - p$$

and $0 < 1 - p < 1$. It follows by homogeneity and the Markov property that

$$P\{X_{k(n')} \in B\} = (1 - p)^k \qquad k = 1, 2, \ldots$$

Now since $\lim_{k \to \infty} (1 - p)^k = 0$, we have

$$\lim_{n \to \infty} P\{X_n \in B\} = 0 \qquad \text{or} \qquad \lim_{n \to \infty} P\{X_n \in \bar{B} = A\} = 1$$

which shows that absorption of $X(n)$ in one or another of the absorption states is certain.

5.40. Verify Eq. (5.50).

Let $X(n) = \{X_n, n \geq 0\}$ be a homogeneous Markov chain with a finite state space $E = \{0, 1, ..., N\}$, of which $A = \{0, 1, ..., m\}, m \geq 1$, is a set of absorbing states and $B = \{m + 1, ..., N\}$ is a set of nonabsorbing states. Let state $k \in B$ at the first step go to $i \in E$ with probability p_{ki}. Then

$$u_{kj} = P\{X_n = j(\in A) | X_0 = k(\in B)\}$$

$$= \sum_{i=1}^{N} p_{ki} P\{X_n = j(\in A) | X_0 = i\} \tag{5.147}$$

Now
$$P\{X_n = j(\in A), X_0 = i\} = \begin{cases} 1 & i = j \\ 0 & i \in A, i \neq j \\ u_{ij} & i \in B, i = m+1, ..., N \end{cases}$$

Then Eq. (5.147) becomes

$$u_{kj} = p_{kj} + \sum_{i=m+1}^{N} p_{ki} u_{ij} \qquad k = m+1, ..., N; j = 1, ..., m \tag{5.148}$$

But $p_{kj}, k = m + 1, ..., N; j = 1, ..., m$, are the elements of R, whereas $p_{ki}, k = m + 1, ..., N; i = m + 1, ..., N$ are the elements of Q [see Eq. (5.49a)]. Hence, in matrix notation, Eq. (5.148) can be expressed as

$$U = R + QU \qquad \text{or} \qquad (I - Q)U = R \tag{5.149}$$

Premultiplying both sides of the second equation of Eq. (5.149) with $(I - Q)^{-1}$, we obtain

$$U = (I - Q)^{-1} R = \Phi R$$

5.41. Consider a simple random walk $X(n)$ with absorbing barriers at state 0 and state $N = 3$ (see Prob. 5.38).

(a) Find the transition probability matrix P.

(b) Find the probabilities of absorption into states 0 and 3.

(a) The transition probability matrix P is [Eq. (5.146)]

$$P = \begin{array}{c} \\ 0 \\ 1 \\ 2 \\ 3 \end{array} \begin{array}{cccc} 0 & 1 & 2 & 3 \end{array} \\ \begin{bmatrix} 1 & 0 & 0 & 0 \\ q & 0 & p & 0 \\ 0 & q & 0 & p \\ 0 & 0 & 0 & 1 \end{bmatrix}$$

(b) Rearranging the transition probability matrix P as [Eq. (5.49a)],

$$P = \begin{array}{c} \\ 0 \\ 3 \\ 1 \\ 2 \end{array} \begin{array}{cccc} 0 & 3 & 1 & 2 \end{array} \\ \begin{bmatrix} 1 & 0 & 0 & 0 \\ 0 & 1 & 0 & 0 \\ q & 0 & 0 & p \\ 0 & p & q & 0 \end{bmatrix}$$

and by Eq. (5.49b), the matrices Q and R are given by

$$R = \begin{bmatrix} p_{10} & p_{13} \\ p_{20} & p_{23} \end{bmatrix} = \begin{bmatrix} q & 0 \\ 0 & p \end{bmatrix} \qquad Q = \begin{bmatrix} p_{11} & p_{12} \\ p_{21} & p_{22} \end{bmatrix} = \begin{bmatrix} 0 & p \\ q & 0 \end{bmatrix}$$

Then
$$I - Q = \begin{bmatrix} 1 & -p \\ -q & 1 \end{bmatrix}$$

and
$$\Phi = (I - Q)^{-1} = \frac{1}{1 - pq}\begin{bmatrix} 1 & p \\ q & 1 \end{bmatrix} \tag{5.150}$$

By Eq. (5.50),

$$U = \begin{bmatrix} u_{10} & u_{13} \\ u_{20} & u_{23} \end{bmatrix} = \Phi R = \frac{1}{1 - pq}\begin{bmatrix} 1 & p \\ q & 1 \end{bmatrix}\begin{bmatrix} q & 0 \\ 0 & p \end{bmatrix} = \frac{1}{1 - pq}\begin{bmatrix} q & p^2 \\ q^2 & p \end{bmatrix} \tag{5.151}$$

Thus, the probabilities of absorption into state 0 from states 1 and 2 are given, respectively, by

$$u_{10} = \frac{q}{1 - pq} \quad \text{and} \quad u_{20} = \frac{q^2}{1 - pq}$$

and the probabilities of absorption into state 3 from states 1 and 2 are given, respectively, by

$$u_{13} = \frac{p^2}{1 - pq} \quad \text{and} \quad u_{23} = \frac{p}{1 - pq}$$

Note that

$$u_{10} + u_{13} = \frac{q + p^2}{1 - pq} = \frac{1 - p + p^2}{1 - p(1 - p)} = 1$$

$$u_{20} + u_{23} = \frac{q^2 + p}{1 - pq} = \frac{q^2 + (1 - q)}{1 - (1 - q)q} = 1$$

which confirm the proposition of Prob. 5.39.

5.42. Consider the simple random walk $X(n)$ with absorbing barriers at 0 and 3 (Prob. 5.41). Find the expected time (or steps) to absorption when $X_0 = 1$ and when $X_0 = 2$.

The fundamental matrix Φ of $X(n)$ is [Eq. (5.150)]

$$\Phi = \begin{bmatrix} \phi_{11} & \phi_{12} \\ \phi_{21} & \phi_{22} \end{bmatrix} = \frac{1}{1 - pq}\begin{bmatrix} 1 & p \\ q & 1 \end{bmatrix}$$

Let T_i be the time to absorption when $X_0 = i$. Then by Eq. (5.51), we get

$$E(T_1) = \frac{1}{1 - pq}(1 + p) \quad E(T_2) = \frac{1}{1 - pq}(q + 1) \tag{5.152}$$

5.43. Consider the gambler's game described in Prob. 5.38. What is the probability of A's losing all his money?

Let $P(k), k = 0, 1, 2, ..., N$, denote the probability that A loses all his money when his initial capital is k dollars. Equivalently, $P(k)$ is the probability of absorption at state 0 when $X_0 = k$ in the simple random walk $X(n)$ with absorbing barriers at states 0 and N. Now if $0 < k < N$, then

$$P(k) = pP(k + 1) + qP(k - 1) \quad k = 1, 2, ..., N - 1 \tag{5.153}$$

where $pP(k+1)$ is the probability that A wins the first round and subsequently loses all his money and $qP(k-1)$ is the probability that A loses the first round and subsequently loses all his money. Rewriting Eq. (5.153), we have

$$P(k+1) - \frac{1}{p}P(k) + \frac{q}{p}P(k-1) = 0 \qquad k = 1, 2, ..., N-1 \tag{5.154}$$

which is a second-order homogeneous linear constant-coefficient difference equation. Next, we have

$$P(0) = 1 \qquad \text{and} \qquad P(N) = 0 \tag{5.155}$$

since if $k = 0$, absorption at 0 is a sure event, and if $k = N$, absorption at N has occurred and absorption at 0 is impossible. Thus, finding $P(k)$ reduces to solving Eq. (5.154) subject to the boundary conditions given by Eq. (5.155). Let $P(k) = r^k$. Then Eq. (5.154) becomes

$$r^{k+1} - \frac{1}{p}r^k + \frac{q}{p}r^{k-1} = 0 \qquad p+q = 1$$

Setting $k = 1$ (and noting that $p + q = 1$), we get

$$r^2 - \frac{1}{p}r + \frac{q}{p} = (r-1)\left(r - \frac{q}{p}\right) = 0$$

from which we get $r = 1$ and $r = q/p$. Thus,

$$P(k) = c_1 + c_2\left(\frac{q}{p}\right)^k \qquad q \neq p \tag{5.156}$$

where c_1 and c_2 are arbitrary constants. Now, by Eq. (5.155),

$$P(0) = 1 \rightarrow c_1 + c_2 = 1$$

$$P(N) = 0 \rightarrow c_1 + c_2\left(\frac{q}{p}\right)^N = 0$$

Solving for c_1 and c_2, we obtain

$$c_1 = \frac{-(q/p)^N}{1-(q/p)^N} \qquad c_2 = \frac{1}{1-(q/p)^N}$$

Hence,

$$P(k) = \frac{(q/p)^k - (q/p)^N}{1-(q/p)^N} \qquad q \neq p \tag{5.157}$$

Note that if $N \gg k$,

$$P(k) = \begin{cases} 1 & q > p \\ \left(\dfrac{q}{p}\right)^k & p > q \end{cases} \tag{5.158}$$

Setting $r = q/p$ in Eq. (5.157), we have

$$P(k) = \frac{r^k - r^N}{1-r^N} \xrightarrow[r \to 1]{} 1 - \frac{k}{N}$$

Thus, when $p = q = \frac{1}{2}$,

$$P(k) = 1 - \frac{k}{N} \tag{5.159}$$

5.44. Show that Eq. (5.157) is consistent with Eq. (5.151).

Substituting $k = 1$ and $N = 3$ in Eq. (5.134), and noting that $p + q = 1$, we have

$$P(1) = \frac{(q/p) - (q/p)^3}{1 - (q/p)^3} = \frac{q(p^2 - q^2)}{(p^3 - q^3)}$$

$$= \frac{q(p + q)}{p^2 + pq + q^2} = \frac{q}{(p + q)^2 - pq} = \frac{q}{1 - pq}$$

Now from Eq. (5.151), we have

$$u_{10} = \frac{q}{1 - pq} = P(1)$$

5.45. Consider the simple random walk $X(n)$ with state space $E = \{0, 1, 2, \ldots, N\}$, where 0 and N are absorbing states (Prob. 5.38). Let r.v. T_k denote the time (or number of steps) to absorption of $X(n)$ when $X_0 = k$, $k = 0, 1, \ldots, N$. Find $E(T_k)$.

Let $Y(k) = E(T_k)$. Clearly, if $k = 0$ or $k = N$, then absorption is immediate, and we have

$$Y(0) = Y(N) = 0 \tag{5.160}$$

Let the probability that absorption takes m steps when $X_0 = k$ be defined by

$$P(k, m) = P(T_k = m) \qquad m = 1, 2, \ldots \tag{5.161}$$

Then, we have (Fig. 5-13)

$$P(k, m) = pP(k + 1, m - 1) + qP(k - 1, m - 1) \tag{5.162}$$

and

$$Y(k) = E(T_k) = \sum_{m=1}^{\infty} mP(k, m) = p \sum_{m=1}^{\infty} mP(k + 1, m - 1) + q \sum_{m=1}^{\infty} mP(k - 1, m - 1)$$

Setting $m - 1 = i$, we get

$$Y(k) = p \sum_{i=0}^{\infty} (i + 1)P(k + 1, i) + q \sum_{i=0}^{\infty} (i + 1)P(k - 1, i)$$

$$= p \sum_{i=0}^{\infty} iP(k + 1, i) + q \sum_{i=0}^{\infty} iP(k - 1, i) + p \sum_{i=0}^{\infty} P(k + 1, i) + q \sum_{i=0}^{\infty} P(k - 1, i)$$

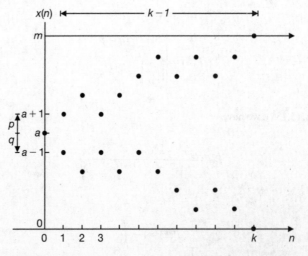

Fig. 5-13 Simple random walk with absorbing barriers.

Now by the result of Prob. 5.39, we see that absorption is certain; therefore,

$$\sum_{i=0}^{\infty} P(k+1, i) = \sum_{i=0}^{\infty} P(k-1, i) = 1$$

Thus,

$$Y(k) = pY(k+1) + qY(k-1) + p + q$$

or

$$Y(k) = pY(k+1) + qY(k-1) + 1 \qquad k = 1, 2, \ldots, N-1 \tag{5.163}$$

Rewriting Eq. (5.163), we have

$$Y(k+1) - \frac{1}{p} Y(k) + \frac{q}{p} Y(k-1) = -\frac{1}{p} \tag{5.164}$$

Thus, finding $P(k)$ reduces to solving Eq. (5.164) subject to the boundary conditions given by Eq. (5.160). Let the general solution of Eq. (5.164) be

$$Y(k) = Y_h(k) + Y_p(k)$$

where $Y_h(k)$ is the homogeneous solution satisfying

$$Y_h(k+1) - \frac{1}{p} Y_h(k) + \frac{q}{p} Y_h(k-1) = 0 \tag{5.165}$$

and $Y_p(k)$ is the particular solution satisfying

$$Y_p(k+1) - \frac{1}{p} Y_p(k) + \frac{q}{p} Y_p(k-1) = -\frac{1}{p} \tag{5.166}$$

Let $Y_p(k) = \alpha k$, where α is a constant. Then Eq. (5.166) becomes

$$(k+1)\alpha - \frac{1}{p} k\alpha + \frac{q}{p}(k-1)\alpha = -\frac{1}{p}$$

from which we get $\alpha = 1/(q-p)$ and

$$Y_p(k) = \frac{k}{q-p} \qquad q \neq p \tag{5.167}$$

Since Eq. (5.165) is the same as Eq. (5.154), by Eq. (5.156), we obtain

$$Y_h(k) = c_1 + c_2 \left(\frac{q}{p}\right)^k \qquad q \neq p \tag{5.168}$$

where c_1 and c_2 are arbitrary constants. Hence, the general solution of Eq. (5.164) is

$$Y(k) = c_1 + c_2 \left(\frac{q}{p}\right)^k + \frac{k}{q-p} \qquad q \neq p \tag{5.169}$$

Now, by Eq. (5.160),

$$Y(0) = 0 \rightarrow c_1 + c_2 = 0$$

$$Y(N) = 0 \rightarrow c_1 + c_2 \left(\frac{q}{p}\right)^N + \frac{N}{q-p} = 0$$

Solving for c_1 and c_2, we obtain

$$c_1 = \frac{-N/(q-p)}{1-(q/p)^N} \qquad c_2 = \frac{N/(q-p)}{1-(q/p)^N}$$

Substituting these values in Eq. (5.169), we obtain (for $p \neq q$)

$$Y(k) = E(T_k) = \frac{1}{q-p}\left(k - N\left[\frac{1-(q/p)^k}{1-(q/p)^N}\right]\right)$$

(5.170)

When $p = q = \frac{1}{2}$, we have

$$Y(k) = E(T_k) = k(N-k) \qquad p = q = \frac{1}{2}$$

(5.171)

5.46. Consider a Markov chain with two states and transition probability matrix

$$p = \begin{bmatrix} 0 & 1 \\ 1 & 0 \end{bmatrix}$$

(a)   Find the stationary distribution $\hat{\mathbf{p}}$ of the chain.

(b)   Find $\lim_{n \to \infty} P^n$.

(a)   By definition (5.52),

$$\hat{\mathbf{p}}P = \hat{\mathbf{p}}$$

or

$$[p_1 \quad p_2]\begin{bmatrix} 0 & 1 \\ 1 & 0 \end{bmatrix} = [p_1 \quad p_2]$$

which yields $p_1 = p_2$. Since $p_1 + p_2 = 1$, we obtain

$$\hat{\mathbf{p}} = \begin{bmatrix} \dfrac{1}{2} & \dfrac{1}{2} \end{bmatrix}$$

(b)   Now

$$P^n = \begin{cases} \begin{bmatrix} 0 & 1 \\ 1 & 0 \end{bmatrix} & n = 1, 3, 5, \dots \\[2mm] \begin{bmatrix} 1 & 0 \\ 0 & 1 \end{bmatrix} & n = 2, 4, 6, \dots \end{cases}$$

and $\lim_{n \to \infty} P^n$ does not exist.

5.47. Consider a Markov chain with two states and transition probability matrix

$$P = \begin{bmatrix} \dfrac{3}{4} & \dfrac{1}{4} \\[2mm] \dfrac{1}{2} & \dfrac{1}{2} \end{bmatrix}$$

(a)   Find the stationary distribution $\hat{\mathbf{p}}$ of the chain.

(b)   Find $\lim_{n \to \infty} P^n$.

(c)   Find $\lim_{n \to \infty} P^n$ by first evaluating P^n.

(a)   By definition (5.52), we have

$$\hat{\mathbf{p}}P = \hat{\mathbf{p}}$$

or

$$[p_1 \quad p_2]\begin{bmatrix} \dfrac{3}{4} & \dfrac{1}{4} \\[2mm] \dfrac{1}{2} & \dfrac{1}{2} \end{bmatrix} = [p_1 \quad p_2]$$

which yields

$$\frac{3}{4}p_1 + \frac{1}{2}p_2 = p_1$$

$$\frac{1}{4}p_1 + \frac{1}{2}p_2 = p_2$$

Each of these equations is equivalent to $p_1 = 2p_2$. Since $p_1 + p_2 = 1$, we obtain

$$\hat{\mathbf{p}} = \begin{bmatrix} \dfrac{2}{3} & \dfrac{1}{3} \end{bmatrix}$$

(b)   Since the Markov chain is regular, by Eq. (5.53), we obtain

$$\lim_{n \to \infty} P^n = \lim_{n \to \infty} \begin{bmatrix} \dfrac{3}{4} & \dfrac{1}{4} \\ \dfrac{1}{2} & \dfrac{1}{2} \end{bmatrix}^n = \begin{bmatrix} \hat{\mathbf{p}} \\ \hat{\mathbf{p}} \end{bmatrix} = \begin{bmatrix} \dfrac{2}{3} & \dfrac{1}{3} \\ \dfrac{2}{3} & \dfrac{1}{3} \end{bmatrix}$$

(c)   Setting $a = \frac{1}{4}$ and $b = \frac{1}{2}$ in Eq. (5.143) (Prob. 5.30), we get

$$P^n = \begin{bmatrix} \dfrac{2}{3} & \dfrac{1}{3} \\ \dfrac{2}{3} & \dfrac{1}{3} \end{bmatrix} - \left(\dfrac{1}{4}\right)^n \begin{bmatrix} \dfrac{1}{3} & -\dfrac{1}{3} \\ -\dfrac{2}{3} & \dfrac{2}{3} \end{bmatrix}$$

Since $\lim_{n \to \infty} (\frac{1}{4})^n = 0$, we obtain

$$\lim_{n \to \infty} P^n = \lim_{n \to \infty} \begin{bmatrix} \dfrac{3}{4} & \dfrac{1}{4} \\ \dfrac{1}{2} & \dfrac{1}{2} \end{bmatrix}^n = \begin{bmatrix} \dfrac{2}{3} & \dfrac{1}{3} \\ \dfrac{2}{3} & \dfrac{1}{3} \end{bmatrix}$$

Poisson Processes

5.48. Let T_n denote the arrival time of the nth customer at a service station. Let Z_n denote the time interval between the arrival of the nth customer and the $(n-1)$st customer; that is,

$$Z_n = T_n - T_{n-1} \qquad n \ge 1 \tag{5.172}$$

and $T_0 = 0$. Let $\{X(t), t \ge 0\}$ be the counting process associated with $\{T_n, n \ge 0\}$. Show that if $X(t)$ has stationary increments, then $Z_n, n = 1, 2, \ldots$, are identically distributed r.v.'s.

We have

$$P(Z_n > z) = 1 - P(Z_n \le z) = 1 - F_{Z_n}(z)$$

By Eq. (5.172), $\qquad P(Z_n > z) = P(T_n - T_{n-1} > z) = P(T_n > T_{n-1} + z)$

Suppose that the observed value of T_{n-1} is t_{n-1}. The event $(T_n > T_{n-1} + z \,|\, T_{n-1} = t_{n-1})$ occurs if and only if $X(t)$ does not change count during the time interval $(t_{n-1}, t_{n-1} + z)$ (Fig. 5-14). Thus,

$$P(Z_n > z \,|\, T_{n-1} = t_{n-1}) = P(T_n > T_{n-1} + z \,|\, T_{n-1} = t_{n-1})$$
$$= P[X(t_{n-1} + z) - X(t_{n-1}) = 0]$$

or $\qquad P(Z_n \le z \,|\, T_{n-1} = t_{n-1}) = 1 - P[X(t_{n-1} + z) - X(t_{n-1}) = 0] \tag{5.173}$

Since $X(t)$ has stationary increments, the probability on the right-hand side of Eq. (5.173) is a function only of the time difference z. Thus,

$$P(Z_n \leq z \,|\, T_{n-1} = t_{n-1}) = 1 - P[X(z) = 0] \tag{5.174}$$

which shows that the conditional distribution function on the left-hand side of Eq. (5.174) is independent of the particular value of n in this case, and hence we have

$$F_{Z_n}(z) = P(Z_n \leq z) = 1 - P[X(z) = 0] \tag{5.175}$$

which shows that the cdf of Z_n is independent of n. Thus, we conclude that the Z_n's are identically distributed r.v.'s.

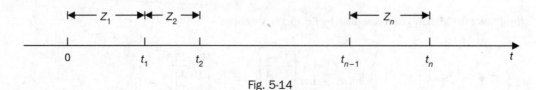

Fig. 5-14

5.49. Show that Definition 5.6.2 implies Definition 5.6.1.

Let $p_n(t) = P[X(t) = n]$. Then, by condition 2 of Definition 5.6.2, we have

$$p_0(t + \Delta t) = P[X(t + \Delta t) = 0] = P[X(t) = 0, X(t + \Delta t) - X(0) = 0]$$
$$= P[X(t) = 0] \, P[X(t + \Delta t) - X(t) = 0]$$

Now, by Eq. (5.59), we have

$$P[X(t + \Delta t) - X(t) = 0] = 1 - \lambda \, \Delta t + o(\Delta t)$$

Thus,
$$p_0(t + \Delta t) = p_0(t)[1 - \lambda \, \Delta t + o(\Delta t)]$$

or
$$\frac{p_0(t + \Delta t) - p_0(t)}{\Delta t} = -\lambda p_0(t) + \frac{o(\Delta t)}{\Delta t}$$

Letting $\Delta t \to 0$, and by Eq. (5.58), we obtain

$$p'_0(t) = -\lambda p_0(t) \tag{5.176}$$

Solving the above differential equation, we get

$$p_0(t) = k e^{-\lambda t}$$

where k is an integration constant. Since $p_0(0) = P[X(0) = 0] = 1$, we obtain

$$p_0(t) = e^{-\lambda t} \tag{5.177}$$

Similarly, for $n > 0$,

$$P_n(t + \Delta t) = P[X(t + \Delta t) = n]$$
$$= P[X(t) = n, X(t + \Delta t) - X(0) = 0]$$
$$+ P[X(t) = n - 1, X(t + \Delta t) - X(0) = 1] + \sum_{k=2}^{n} P[X(t) = n - k, X(t + \Delta t) - X(0) = k]$$

Now, by condition 4 of Definition 5.6.2, the last term in the above expression is $o(\Delta t)$. Thus, by conditions 2 and 3 of Definition 5.6.2, we have

$$p_n(t + \Delta t) = p_n(t)[1 - \lambda \, \Delta t + o(\Delta t)] + p_{n-1}(t)[\lambda \, \Delta t + o(\Delta t)] + o(\Delta t)$$

Thus
$$\frac{p_n(t + \Delta t) - p_n(t)}{\Delta t} = -\lambda p_n(t) + \lambda p_{n-1}(t) + \frac{o(\Delta t)}{\Delta t}$$

and letting $\Delta t \rightarrow 0$ yields

$$p'_n(t) + \lambda p_n(t) = \lambda p_{n-1}(t) \tag{5.178}$$

Multiplying both sides by $e^{\lambda t}$, we get

$$e^{\lambda t}[p'_n(t) + \lambda p_n(t)] = \lambda e^{\lambda t} p_{n-1}(t)$$

Hence,

$$\frac{d}{dt}[e^{\lambda t} p_n(t)] = \lambda e^{\lambda t} p_{n-1}(t) \tag{5.179}$$

Then by Eq. (5.177), we have

$$\frac{d}{dt}[e^{\lambda t} p_1(t)] = \lambda$$

or

$$p_1(t) = (\lambda t + c)e^{-\lambda t}$$

where c is an integration constant. Since $p_1(0) = P[X(0) = 1] = 0$, we obtain

$$p_1(t) = \lambda t e^{-\lambda t} \tag{5.180}$$

To show that

$$p_n(t) = e^{-\lambda t} \frac{(\lambda t)^n}{n!}$$

we use mathematical induction. Assume that it is true for $n - 1$; that is,

$$p_{n-1}(t) = e^{-\lambda t} \frac{(\lambda t)^{n-1}}{(n-1)!}$$

Substituting the above expression into Eq. (5.179), we have

$$\frac{d}{dt}[e^{\lambda t} p_n(t)] = \frac{\lambda^n t^{n-1}}{(n-1)!}$$

Integrating, we get

$$e^{\lambda t} p_n(t) = \frac{(\lambda t)^n}{n!} + c_1$$

Since $p_n(0) = 0, c_1 = 0$, and we obtain

$$p_n(t) = e^{-\lambda t} \frac{(\lambda t)^n}{n!} \tag{5.181}$$

which is Eq. (5.55) of Definition 5.6.1. Thus we conclude that Definition 5.6.2 implies Definition 5.6.1.

5.50. Verify Eq. (5.59).

We note first that $X(t)$ can assume only nonnegative integer values; therefore, the same is true for the counting increment $X(t + \Delta t) - X(t)$. Thus, summing over all possible values of the increment, we get

$$\sum_{k=0}^{\infty} P[X(t + \Delta t) - X(t) = k] = P[X(t + \Delta t) - X(t) = 0]$$

$$+ P[X(t + \Delta t) - X(t) = 1] + P[X(t + \Delta t) - X(t) \geq 2]$$

$$= 1$$

Substituting conditions 3 and 4 of Definition 5.6.2 into the above equation, we obtain

$$P[X(t + \Delta t) - X(t) = 0] = 1 - \lambda \Delta t + o(\Delta t)$$

5.51. (a) Using the Poisson probability distribution in Eq. (5.181), obtain an analytical expression for the correction term $o(\Delta t)$ in the expression (condition 3 of Definition 5.6.2)

$$P[X(t + \Delta t) - X(t) = 1] = \lambda \Delta t + o(\Delta t) \tag{5.182}$$

(b) Show that this correction term does have the property of Eq. (5.58); that is,

$$\lim_{\Delta t \to 0} \frac{o(\Delta t)}{\Delta t} = 0$$

(a) Since the Poisson process $X(t)$ has stationary increments, Eq. (5.182) can be rewritten as

$$P[X(\Delta t) = 1] = p_1(\Delta t) = \lambda \Delta t + o(\Delta t) \tag{5.183}$$

Using Eq. (5.181) [or Eq. (5.180)], we have

$$p_1(\Delta t) = \lambda \Delta t \, e^{-\lambda \Delta t} = \lambda \Delta t (1 + e^{-\lambda \Delta t} - 1)$$
$$= \lambda \Delta t + \lambda \Delta t (e^{-\lambda \Delta t} - 1)$$

Equating the above expression with Eq. (5.183), we get

$$\lambda \Delta t + o(\Delta t) = \lambda \Delta t + \lambda \Delta t (e^{-\lambda \Delta t} - 1)$$

from which we obtain

$$o(\Delta t) = \lambda \Delta t (e^{-\lambda \Delta t} - 1) \tag{5.184}$$

(b) From Eq. (5.184), we have

$$\lim_{\Delta t \to 0} \frac{o(\Delta t)}{\Delta t} = \lim_{\Delta t \to 0} \frac{\lambda \Delta t (e^{-\lambda \Delta t} - 1)}{\Delta t} = \lim_{\Delta t \to 0} \lambda (e^{-\lambda \Delta t} - 1) = 0$$

5.52. Find the autocorrelation function $R_X(t, s)$ and the autocovariance function $K_X(t, s)$ of a Poisson process $X(t)$ with rate λ.

From Eqs. (5.56) and (5.57),

$$E[X(t)] = \lambda t \qquad \text{Var}[X(t)] = \lambda t$$

Now, the Poisson process $X(t)$ is a random process with stationary independent increments and $X(0) = 0$. Thus, by Eq. (5.126) (Prob. 5.23), we obtain

$$K_X(t, s) = \sigma_1^2 \min(t, s) = \lambda \min(t, s) \tag{5.185}$$

since $\sigma_1^2 = \text{Var}[X(1)] = \lambda$. Next, since $E[X(t)] \, E[X(s)] = \lambda^2 ts$, by Eq. (5.10), we obtain

$$R_X(t, s) = \lambda \min(t, s) + \lambda^2 ts \tag{5.186}$$

5.53. Show that the time intervals between successive events (or interarrival times) in a Poisson process $X(t)$ with rate λ are independent and identically distributed exponential r.v.'s with parameter λ.

Let $Z_1, Z_2, \ldots$ be the r.v.'s representing the lengths of interarrival times in the Poisson process $X(t)$. First, notice that $\{Z_1 > t\}$ takes place if and only if no event of the Poisson process occurs in the interval $(0, t)$, and thus by Eq. (5.177),

$$P(Z_1 > t) = P\{X(t) = 0\} = e^{-\lambda t}$$

or

$$F_{Z_1}(t) = P(Z_1 \leq t) = 1 - e^{-\lambda t}$$

Hence, Z_1 is an exponential r.v. with parameter λ [Eq. (2.61)]. Let $f_1(t)$ be the pdf of Z_1. Then we have

$$
\begin{aligned}
P(Z_2 > t) &= \int P(Z_2 > t) \big| Z_1 = \tau) f_1(\tau) \, d\tau \\
&= \int P[X(t + \tau) - X(\tau) = 0] f_1(\tau) \, d\tau \\
&= e^{-\lambda t} \int f_1(\tau) \, d\tau = e^{-\lambda t}
\end{aligned}
\tag{5.187}
$$

which indicates that Z_2 is also an exponential r.v. with parameter λ and is independent of Z_1. Repeating the same argument, we conclude that $Z_1, Z_2, \ldots$ are iid exponential r.v.'s with parameter λ.

5.54. Let T_n denote the time of the nth event of a Poisson process $X(t)$ with rate λ. Show that T_n is a gamma r.v. with parameters (n, λ).

Clearly,

$$
T_n = Z_1 + Z_2 + \cdots + Z_n
$$

where $Z_n, n = 1, 2, \ldots$, are the interarrival times defined by Eq. (5.172). From Prob. 5.53, we know that Z_n are iid exponential r.v.'s with parameter λ. Now, using the result of Prob. 4.39, we see that T_n is a gamma r.v. with parameters (n, λ), and its pdf is given by [Eq. (2.65)]:

$$
f_{T_n}(t) =
\begin{cases}
\lambda e^{-\lambda t} \dfrac{(\lambda t)^{n-1}}{(n-1)!} & t > 0 \\
0 & t < 0
\end{cases}
\tag{5.188}
$$

The random process $\{T_n, n \geq 1\}$ is often called an *arrival process*.

5.55. Suppose t is not a point at which an event occurs in a Poisson process $X(t)$ with rate λ. Let $W(t)$ be the r.v. representing the time until the next occurrence of an event. Show that the distribution of $W(t)$ is independent of t and $W(t)$ is an exponential r.v. with parameter λ.

Let s ($0 \leq s < t$) be the point at which the last event [say the $(n - 1)$st event] occurred (Fig. 5-15). The event $\{W(t) > \tau\}$ is equivalent to the event

$$
\{Z_n > t - s + \tau \,|\, Z_n > t - s\}
$$

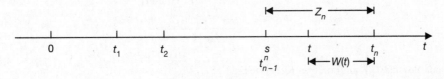

Fig. 5-15

Thus, using Eq. (5.187), we have

$$
\begin{aligned}
P[W(t) > \tau] &= P(Z_n > t - s + \tau \,|\, Z_n > t - s) \\
&= \frac{P(Z_n > t - s + \tau)}{P(Z_n > t - s)} = \frac{e^{-\lambda(t - s + \tau)}}{e^{-\lambda(t - s)}} = e^{-\lambda \tau}
\end{aligned}
$$

and
$$
P[W(t) \leq \tau] = 1 - e^{\lambda \tau}
\tag{5.189}
$$

which indicates that $W(t)$ is an exponential r.v. with parameter λ and is independent of t. Note that $W(t)$ is often called a *waiting time*.

5.56. Patients arrive at the doctor's office according to a Poisson process with rate $\lambda = \frac{1}{10}$ minute. The doctor will not see a patient until at least three patients are in the waiting room.

(a)   Find the expected waiting time until the first patient is admitted to see the doctor.

(b)   What is the probability that nobody is admitted to see the doctor in the first hour?

(a)   Let T_n denote the arrival time of the nth patient at the doctor's office. Then

$$T_n = Z_1 + Z_2 + \cdots + Z_n$$

where $Z_n, n = 1, 2, \ldots$, are iid exponential r.v.'s with parameter $\lambda = \frac{1}{10}$. By Eqs. (4.132) and (2.62),

$$E(T_n) = E\left(\sum_{i=i}^{n} Z_i \right) = \sum_{i=i}^{n} E(Z_i) = n\frac{1}{\lambda} \tag{5.190}$$

The expected waiting time until the first patient is admitted to see the doctor is

$$E(T_3) = 3(10) = 30 \text{ minutes}$$

(b)   Let $X(t)$ be the Poisson process with parameter $\lambda = \frac{1}{10}$. The probability that nobody is admitted to see the doctor in the first hour is the same as the probability that at most two patients arrive in the first 60 minutes. Thus, by Eq. (5.55),

$$P[X(60) - X(0) \leq 2] = P[X(60) - X(0) = 0] + P[X(60) - X(0) = 1] + P[X(60) - X(0) = 2]$$

$$= e^{-60/10} + e^{-60/10}\left(\frac{60}{10} \right) + e^{-60/10}\frac{1}{2}\left(\frac{60}{10} \right)^2$$

$$= e^{-6}(1 + 6 + 18) \approx 0.062$$

5.57. Let T_n denote the time of the nth event of a Poisson process $X(t)$ with rate λ. Suppose that one event has occurred in the interval $(0, t)$. Show that the conditional distribution of arrival time T_1 is uniform over $(0, t)$.

For $\tau \leq t$,

$$P[T_1 \leq \tau \,|\, X(t) = 1] = \frac{P[T_1 \leq \tau, X(t) = 1]}{P[X(t) = 1]}$$

$$= \frac{P[X(\tau) = 1, X(t) - X(\tau) = 0]}{P[X(t) = 1]}$$

$$= \frac{P[X(\tau) = 1]P[X(t) - X(\tau) = 0]}{P[X(t) = 1]}$$

$$= \frac{\lambda \tau e^{-\lambda \tau} e^{-\lambda(t-\tau)}}{\lambda t e^{-\lambda t}} = \frac{\tau}{t} \tag{5.191}$$

which indicates that T_1 is uniform over $(0, t)$ [see Eq. (2.57)].

5.58. Consider a Poisson process $X(t)$ with rate λ, and suppose that each time an event occurs, it is classified as either a type 1 or a type 2 event. Suppose further that the event is classified as a type 1 event with probability p and a type 2 event with probability $1 - p$. Let $X_1(t)$ and $X_2(t)$ denote the number of type 1 and type 2 events, respectively, occurring in $(0, t)$. Show that $\{X_1(t), t \geq 0\}$ and $\{X_2(t), t \geq 0\}$ are both Poisson processes with rates λp and $\lambda(1 - p)$, respectively. Furthermore, the two processes are independent.

We have

$$X(t) = X_1(t) + X_2(t)$$

First we calculate the joint probability $P[X_1(t) = k, X_2(t) = m]$.

$$P[X_1(t) = k, X_2(t) = m] = \sum_{n=0}^{\infty} P[X_1(t) = k, X_2(t) = m \mid X(t) = n]P[X(t) = n]$$

Note that

$$P[X_1(t) = k, X_2(t) = m \mid X(t) = n] = 0 \qquad \text{when } n \neq k + m$$

Thus, using Eq. (5.181), we obtain

$$P[X_1(t) = k, X_2(t) = m] = P[X_1(t) = k, X_2(t) = m \mid X(t) = k + m]P[X(t) = k + m]$$

$$= P[X_1(t) = k, X_2(t) = m \mid X(t) = k + m]e^{-\lambda t}\frac{(\lambda t)^{k+m}}{(k+m)!}$$

Now, given that $k + m$ events occurred, since each event has probability p of being a type 1 event and probability $1 - p$ of being a type 2 event, it follows that

$$P[X_1(t) = k, X_2(t) = m \mid X(t) = k + m] = \binom{k+m}{k}p^k(1-p)^m$$

Thus,

$$P[X_1(t) = k, X_2(t) = m] = \binom{k+m}{k}p^k(1-p)^m e^{-\lambda t}\frac{(\lambda t)^{k+m}}{(k+m)!}$$

$$= \frac{(k+m)!}{k!m!}p^k(1-p)^m e^{-\lambda t}\frac{(\lambda t)^{k+m}}{(k+m)!}$$

$$= e^{-\lambda pt}\frac{(\lambda pt)^k}{k!}e^{-\lambda(1-p)t}\frac{[\lambda(1-p)t]^m}{m!} \tag{5.192}$$

Then

$$P[X_1(t) = k] = \sum_{m=1}^{\infty} P[X_1(t) = k, X_2(t) = m]$$

$$= e^{-\lambda pt}\frac{(\lambda pt)^k}{k!}e^{-\lambda(1-p)t}\sum_{m=1}^{\infty}\frac{[\lambda(1-p)t]^m}{m!}$$

$$= e^{-\lambda pt}\frac{(\lambda pt)^k}{k!}e^{-\lambda(1-p)t}e^{\lambda(1-p)t}$$

$$= e^{-\lambda pt}\frac{(\lambda pt)^k}{k!} \tag{5.193}$$

which indicates that $X_1(t)$ is a Poisson process with rate λp. Similarly, we can obtain

$$P[X_2(t) = m] = \sum_{k=1}^{\infty} P[X_1(t) = k, X_2(t) = m]$$

$$= e^{-\lambda(1-p)t}\frac{[\lambda(1-p)t]^m}{m!} \tag{5.194}$$

and so $X_2(t)$ is a Poisson process with rate $\lambda(1 - p)$. Finally, from Eqs. (5.193), (5.194), and (5.192), we see that

$$P[X_1(t) = k, X_2(t) = m] = P[X_1(t) = k]P[X_2(t) = m]$$

Hence, $X_1(t)$ and $X_2(t)$ are independent.

Wiener Processes

5.59. Let $X_1, \ldots, X_n$ be jointly normal r.v.'s. Show that the joint characteristic function of $X_1, \ldots, X_n$ is given by

$$\Psi_{X_1 \cdots X_n}(\omega_1, \ldots, \omega_n) = \exp\left(j \sum_{i=1}^{n} \omega_i \mu_i - \frac{1}{2} \sum_{i=1}^{n} \sum_{k=1}^{n} \omega_i \omega_k \sigma_{ik} \right) \tag{5.195}$$

where $\mu_i = E(X_i)$ and $\sigma_{ik} = \text{Cov}(X_i, X_k)$.

Let

$$Y = a_1 X_1 + a_2 X_2 + \cdots + a_n X_n$$

By definition (4.66), the characteristic function of Y is

$$\Psi_Y(\omega) = E[e^{j\omega(a_1 X_1 + \cdots + a_n X_n)}] = \Psi_{X_1 \cdots X_n}(\omega a_1, \ldots, \omega a_n) \tag{5.196}$$

Now, by the results of Prob. 4.72, we see that Y is a normal r.v. with mean and variance given by [Eqs. (4.132) and (4.135)]

$$\mu_Y = E(Y) = \sum_{i=1}^{n} a_i E(X_i) = \sum_{i=1}^{n} a_i \mu_i \tag{5.197}$$

$$\sigma_Y^2 = \text{Var}(Y) = \sum_{i=1}^{n} \sum_{k=1}^{n} a_i a_k \text{Cov}(X_i, X_k) = \sum_{i=1}^{n} \sum_{k=1}^{n} a_i a_k \sigma_{ik} \tag{5.198}$$

Thus, by Eq. (4.167),

$$\Psi_Y(\omega) = \exp\left[j\omega\mu_Y - \frac{1}{2}\sigma_Y^2\omega^2 \right]$$

$$= \exp\left(j\omega \sum_{i=1}^{n} a_i\mu_i - \frac{1}{2}\omega^2 \sum_{i=1}^{n} \sum_{k=1}^{n} a_i a_k \sigma_{ik} \right) \tag{5.199}$$

Equating Eqs. (5.199) and (5.196) and setting $\omega = 1$, we get

$$\Psi_{X_1 \cdots X_n}(a_1, \ldots a_n) = \exp\left(j \sum_{i=1}^{n} a_i\mu_i - \frac{1}{2} \sum_{i=1}^{n} \sum_{k=1}^{n} a_i a_k \sigma_{ik} \right)$$

By replacing a_i's with ω_i's, we obtain Eq. (5.195); that is,

$$\Psi_{X_1 \cdots X_n}(\omega_1, \ldots, \omega_n) = \exp\left(j \sum_{i=1}^{n} \omega_i\mu_i - \frac{1}{2} \sum_{i=1}^{n} \sum_{k=1}^{n} \omega_i \omega_k \sigma_{ik} \right)$$

Let

$$\boldsymbol{\mu} = \begin{bmatrix} \mu_1 \\ \vdots \\ \mu_n \end{bmatrix} \qquad \boldsymbol{\omega} = \begin{bmatrix} \omega_1 \\ \vdots \\ \omega_n \end{bmatrix} \qquad K = [\sigma_{ik}] = \begin{bmatrix} \sigma_{11} & \cdots & \sigma_{1n} \\ \vdots & \ddots & \vdots \\ \sigma_{n1} & \cdots & \sigma_{nn} \end{bmatrix}$$

Then we can write

$$\sum_{i=1}^{n} \omega_i\mu_i = \boldsymbol{\omega}^T \boldsymbol{\mu} \qquad \sum_{i=1}^{n} \sum_{k=1}^{n} \omega_i \omega_k \sigma_{ik} = \boldsymbol{\omega}^T K \boldsymbol{\omega}$$

and Eq. (5.195) can be expressed more compactly as

$$\Psi_{X_1\dots X_n}(\omega_1,\dots,\omega_n) = \exp\left(j\omega^T\mu - \frac{1}{2}\omega^T K\omega \right) \tag{5.200}$$

5.60. Let $X_1, \dots, X_n$ be jointly normal r.v.'s Let

$$Y_1 = a_{11}X_1 + \dots + a_{1n}X_n$$
$$\vdots \tag{5.201}$$
$$Y_m = a_{m1}X_1 + \dots + a_{mn}X_n$$

where a_{ik} $(i = 1, \dots, m; j = 1, \dots, n)$ are constants. Show that $Y_1, \dots, Y_m$ are also jointly normal r.v.'s.

Let
$$\mathbf{X} = \begin{bmatrix} X_1 \\ \vdots \\ X_n \end{bmatrix} \qquad \mathbf{Y} = \begin{bmatrix} Y_1 \\ \vdots \\ Y_m \end{bmatrix} \qquad A = [a_{ik}] = \begin{bmatrix} a_{11} & \cdots & a_{1n} \\ \vdots & \ddots & \vdots \\ a_{m1} & \cdots & a_{mn} \end{bmatrix}$$

Then Eq. (5.201) can be expressed as

$$\mathbf{Y} = A\mathbf{X}$$

Let
$$\mu_{\mathbf{X}} = E(\mathbf{X}) = \begin{bmatrix} \mu_1 \\ \vdots \\ \mu_n \end{bmatrix} \qquad \omega = \begin{bmatrix} \omega_1 \\ \vdots \\ \omega_m \end{bmatrix} \qquad K_{\mathbf{X}} = [\sigma_{ik}] = \begin{bmatrix} a_{11} & \cdots & a_{1n} \\ \vdots & \ddots & \vdots \\ a_{n1} & \cdots & a_{nn} \end{bmatrix} \tag{5.202}$$

Then the characteristic function for $\mathbf{Y}$ can be written as

$$\Psi_{\mathbf{Y}}(\omega_1, \dots, \omega_m) = E(e^{j\omega^T\mathbf{Y}}) = E(e^{j\omega^T A\mathbf{X}})$$
$$= E[e^{j(A^T\omega)^T\mathbf{X}}] = \Psi_{\mathbf{X}}(A^T\omega)$$

Since $\mathbf{X}$ is a normal random vector, by Eq. (5.177) we can write

$$\Psi_{\mathbf{X}}(A^T\omega) = \exp\left[j(A^T\omega)^T\mu_{\mathbf{X}} - \frac{1}{2}(A^T\omega)^T K_{\mathbf{X}}(A^T\omega) \right]$$
$$= \exp\left[j\omega^T A\mu_{\mathbf{X}} - \frac{1}{2}\omega^T A K_{\mathbf{X}} A^T\omega \right]$$

Thus,
$$\Psi_{\mathbf{Y}}(\omega_1,\cdots,\omega_m) = \exp\left(j\omega^T\mu_{\mathbf{Y}} - \frac{1}{2}\omega^T K_{\mathbf{Y}}\omega \right) \tag{5.203}$$

where
$$\mu_{\mathbf{Y}} = A\mu_{\mathbf{X}} \qquad K_{\mathbf{Y}} = A K_{\mathbf{X}} A^T \tag{5.204}$$

Comparing Eqs. (5.200) and (5.203), we see that Eq. (5.203) is the characteristic function of a random vector $\mathbf{Y}$. Hence, we conclude that $Y_1, \dots, Y_m$ are also jointly normal r.v.'s

Note that on the basis of the above result, we can say that a random process $\{X(t), t \in T\}$ is a normal process if every finite linear combination of the r.v.'s $X(t_i), t_i \in T$ is normally distributed.

5.61. Show that a Wiener process $X(t)$ is a normal process.

Consider an arbitrary linear combination

$$\sum_{i=1}^{n} a_i X(t_i) = a_1 X(t_1) + a_2 X(t_2) + \dots + a_n X(t_n) \tag{5.205}$$

where $0 \leq t_1 < \ldots < t_n$ and a_i are real constants. Now we write

$$\sum_{i=1}^{n} a_i X(t_i) = (a_1 + \cdots + a_n)[X(t_1) - X(0)] + (a_2 + \cdots + a_n)[X(t_2) - X(t_1)]$$

$$+ \cdots + (a_{n-1} + a_n)[X(t_{n-1}) - X(t_{n-2})] + a_n[X(t_n) - X(t_{n-1})] \qquad (5.206)$$

Now from conditions 1 and 2 of Definition 5.7.1, the right-hand side of Eq. (5.206) is a linear combination of independent normal r.v.'s. Thus, based on the result of Prob. 5.60, the left-hand side of Eq. (5.206) is also a normal r.v.; that is, every finite linear combination of the r.v.'s $X(t_i)$ is a normal r.v. Thus, we conclude that the Wiener process $X(t)$ is a normal process.

5.62. A random process $\{X(t), t \in T\}$ is said to be *continuous in probability* if for every $\varepsilon > 0$ and $t \in T$,

$$\lim_{h \to 0} P\{|X(t+h) - X(t)| > \varepsilon\} = 0 \qquad (5.207)$$

Show that a Wiener process $X(t)$ is continuous in probability.

From Chebyshev inequality (2.116), we have

$$P\{|X(t+h) - X(t)| > \varepsilon\} \leq \frac{\text{Var}[X(t+h) - X(t)]}{\varepsilon^2} \qquad \varepsilon > 0$$

Since $X(t)$ has stationary increments, we have

$$\text{Var}[X(t+h) - X(t)] = \text{Var}[X(h)] = \sigma^2 h$$

in view of Eq. (5.63). Hence,

$$\lim_{h \to 0} P\{|X(t+h) - X(t)| > \varepsilon\} = \lim_{h \to 0} \frac{\sigma^2 h}{\varepsilon^2} = 0$$

Thus, the Wiener process $X(t)$ is continuous in probability.

Martingales

5.63. Let $Y = X_1 + X_2 + X_3$ where X_i is the outcome of the ith toss of a fair coin. Verify the tower property Eq. (5.76).

Let $X_i = 1$ when it is a head and $X_i = 0$ when it is a tail. Since the coin is fair, we have

$$P(X_i = 1) = P(X_i = 0) = \frac{1}{2} \quad \text{and} \quad E(X_i) = \frac{1}{2}$$

and X_i's are independent. Now

$$E[E(Y|F_2)|F_1] = E[E(Y|X_2, X_1)|X_1]$$
$$= E(X_1 + X_2 + E(X_3)|X_1)$$
$$= X_1 + E(X_2) + E(X_3) = X_1 + 1$$

and

$$E(Y|F_1) = E(Y|X_1) = E(X_1 + X_2 + X_3|X_1)$$
$$= X_1 + E(X_2 + X_3)$$
$$= X_1 + E(X_2) + E(X_3) = X_1 + 1$$

Thus,

$$E[E(Y|F_2)|F_1] = E(Y|F_1)$$

5.64. Let $X_1, X_2, \ldots$ be i.i.d. r.v.'s with mean μ. Let

$$S = \sum_{i=1}^{n} X_i = X_1 + X_2 + \cdots + X_n$$

Let F_n denote the information contained in $X_1, \ldots, X_n$. Show that

$$E(S_n|F_m) = S_m + (n - m)\mu \qquad m < n \qquad (5.208)$$

Let $m < n$, then by Eq. (5.71)

$$E(S_n|F_m) = E(X_1 + \cdots + X_m|F_m) + E(X_{m+1} + \cdots + X_n|F_n)$$

Since $X_1 + X_2 + \cdots + X_m$ is measurable with respect to F_m, by Eq. (5.73)

$$E(X_1 + \cdots + X_m|F_m) = X_1 + \cdots + X_m = S_m$$

Since $X_{m+1} + \cdots + X_n$ is independent of $X_1, \ldots, X_m$, by Eq. (5.75)

$$E(X_{m+1} + \cdots + X_n|F_n) = E(X_{m+1} + \cdots + X_n) = (n - m)\mu$$

Thus, we obtain

$$E(S_n|F_m) = S_m + (n - m)\,\mu \qquad m < n$$

5.65. Let $X_1, X_2, \ldots$ be i.i.d. r.v.'s with $E(X_i) = 0$ and $E(X_i^2) = \sigma^2$ for all i. Let $S = \sum_{i=1}^{n} X_i = X_1 + X_2 + \cdots + X_n$. Let F_n denote the information contained in $X_1, \ldots, X_n$. Show that

$$E(S_n^2|F_m) = S_m^2 + (n - m)\,\sigma^2 \qquad m < n \qquad (5.209)$$

Let $m < n$, then by Eq. (5.71)

$$\begin{aligned}
E(S_n^2|F_m) &= E([S_m + (S_n - S_m)]^2|F_m) \\
&= E(S_m^2|F_m) + 2\,E[S_m(S_n - S_m)|F_m] + E([(S_n - S_m)]^2|F_m)
\end{aligned}$$

Since S_m is dependent only on $X_1, \ldots, X_m$, by Eqs. (5.73) and (5.75)

$$E(S_m^2\,|F_m) = S_m^2, E([(S_n - S_m)^2\,|F_m]) = E(S_n - S_m)^2 = \text{Var}\,(S_n - S_m) = (n - m)\,\sigma^2$$

since $E(X_i) = \mu = 0$, $\text{Var}\,(X_i) = E(X_i^2) = \sigma^2$ and $\text{Var}\,(S_n - S_m) = \text{Var}\,(X_{m+1} + \ldots + X_n) = (n - m)\sigma^2$. Next, by Eq. (5.74)

$$E[S_m(S_n - S_m)|F_m] = S_m E[(S_n - S_m)|F_m] = S_m E(S_n - S_m) = 0$$

Thus, we obtain

$$E(S_n^2|F_m) = S_m^2 + (n - m)\,\sigma^2 \qquad m < n$$

5.66. Verify Eq. (5.80), that is

$$E(M_m | F_n) = M_n \qquad \text{for } m \geq n$$

By condition (2) of martingale, Eq. (5.79), we have

$$E(M_{n+1} | F_n) = M_n \qquad \text{for all } n$$

Then by tower property Eq. (5.76)

$$E(M_{n+2} | F_n) = E[E(M_{n+2} | F_{n+1}) | F_n] = E(M_{n+1} | F_n) = M_n$$

and so on, and we obtain Eq. (5.80), that is

$$E(M_m | F_n) = M_n \qquad \text{for } m \geq n$$

5.67. Verify Eq. (5.82), that is

$$E(M_n) = E(M_{n-1}) = \cdots = E(M_0)$$

Since $\{M_n, n \geq 0\}$ is a martingale, we have

$$E(M_{n+1} | F_n) = M_n \qquad \text{for all } n$$

Applying Eq. (5.77), we have

$$E[E(M_{n+1} | F_n)] = E(M_{n+1}) = E(M_n)$$

Thus, by induction we obtain

$$E(M_n) = E(M_{n-1}) = \cdots = E(M_0)$$

5.68. Let $X_1, X_2, \ldots$ be a sequence of independent r.v.'s with $E[|X_n|] = < \infty$ and $E(X_n) = 0$ for all n. Set $S_0 = 0, S_n = \sum_{i=1}^{n} X_i = X_1 + X_2 + \cdots + X_n$. Show that $\{S_n, n \geq 0\}$ is a martingale.

$$E[|S_n|] \leq E(|X_1| + \cdots + |X_n|) = E(|X_1|) + \cdots + E(|X_n|) < \infty$$
$$E(S_{n+1} | F_n) = E(S_n + X_{n+1} | F_n)$$
$$= S_n + E(X_{n+1} | F_n) = S_n + E(X_{n+1}) = S_n$$

since $E(X_n) = 0$ for all n.

Thus, $\{S_n, n \geq 0\}$ is a martingale.

5.69. Consider the same problem as Prob. 5.68 except $E(X_n) \geq 0$ for all n. Show that $\{S_n, n \geq 0\}$ is a submartingale.

Assume max $E(|X_n|) = k < \infty$, then

$$E[|S_n|] \leq E(|X_1| + \ldots + |X_n|) = E(|X_1|) + \ldots + E(|X_n|) \leq nk < \infty$$
$$E(S_{n+1} | F_n) = E(S_n + X_{n+1} | F_n)$$
$$= S_n + E(X_{n+1} | F_n) = S_n + E(X_{n+1}) \geq S_n$$

since $E(X_n) \geq 0$ for all n.

Thus, $\{S_n, n \geq 0\}$ is a submartingale.

5.70. Let $X_1, X_2, \ldots$ be a sequence of Bernoulli r.v.'s with

$$X_i = \begin{cases} 1 & \text{with probability } p \\ 0 & \text{with probability } q = 1 - p \end{cases}$$

Let $S_n = \sum_{i=1}^{n} X_i = X_1 + X_2 + \ldots + X_n$. Show that (1) if $p = \frac{1}{2}$ then $\{S_n\}$ is a martingale. (2) if $p > \frac{1}{2}$ then $\{S_n\}$ is a submartingale, and (3) if $p < \frac{1}{2}$ then $\{S_n\}$ is a supermartingale.

$$E(X_i) = p(1) + (1 - p)(-1) = 2p - 1$$

(1) If $p = \frac{1}{2}$, $E(X_i) = 0$, and

$$E[|S_n|] \leq E(|X_i| + \cdots + |X_n|) = E(|X_1|) + \cdots + E(|X_n|) = 0 < \infty$$
$$E(S_{n+1}|F_n) = E(S_n + X_{n+1}|F_n)$$
$$= S_n + E(X_{n+1}|F_n) = S_n + E(X_{n+1}) = S_n$$

Thus, $\{S_n\}$ is a martingale.

(2) If $p > \frac{1}{2}$, $0 < E(X_i) \leq 1$, and

$$E[|S_n|] \leq E(|X_1| + \cdots + |X_n|) = E(|X_1|) + \cdots + E(|X_n|) \leq n < \infty$$
$$E(S_{n+1}|F_n) = E(S_n + X_{n+1}|F_n)$$
$$= S_n + E(X_{n+1}|F_n) = S_n + E(X_{n+1}) > S_n$$

Thus, $\{S_n\}$ is a submartingale.

(3) If $p < \frac{1}{2}$, $E(X_i) < 0$, and

$$E(S_{n+1}|F_n) = E(S_n + X_{n+1}|F_n)$$
$$= S_n + E(X_{n+1}|F_n) = S_n + E(X_{n+1}) < S_n$$

Thus, $\{S_n\}$ is a supermartingale.

Note that this problem represents a tossing a coin game, "heads" you win \$1 and "tails" you lose \$1. Thus, if $p = \frac{1}{2}$, it is a fair coin and if $p > \frac{1}{2}$, the game is favorable, and if $p < \frac{1}{2}$, the game is unfavorable.

5.71. Let $X_1, X_2, \ldots$ be a sequence of i.i.d. r.v.'s with $E(X_i) = \mu > 0$. Set

$$S_0 = 0, S_n = \sum_{i=1}^{n} X_i = X_1 + X_2 + \cdots + X_n \text{ and}$$
$$M_n = S_n - n\mu \tag{5.210}$$

Show that $\{M_n, n \geq 0\}$ is a martingale.

$$E(|M_n|) = E(|S_n - n\mu|) \leq E(|S_n|) + n\mu \leq E\left(\sum_{i=1}^{n} |X_i|\right) + n\mu = 2n\mu < \infty$$

Next, using Eq. (5.208) of Prob. 5.64, we have

$$E(M_{n+1}|F_n) = E(S_{n+1} - (n+1)\mu|F_n)$$
$$= E(S_{n+1}|F_n) - (n+1)\mu$$
$$= S_n + \mu - (n+1)\mu = S_n - n\mu = M_n$$

Thus, $\{M_n, n \geq 0\}$ is a martingale.

5.72. Let $X_1, X_2, \ldots$ be i.i.d. r.v.'s with $E(X_i) = 0$ and $E(X_i^2) = \sigma^2$ for all i. Let $S_0 = 0$, $S_n = \sum_{i=1}^{n} X_i = X_1 + X_2 + \cdots + X_n$, and

$$M_n = S_n^2 - n\sigma^2 \tag{5.211}$$

Show that $\{M_n, n \geq 0\}$ is a martingale,

$$M_n = S_n^2 - n\sigma^2 = \left(\sum_{i=1}^{n} X_i\right)^2 - n\sigma^2 = \sum_{i=1}^{n} X_i^2 + 2\sum_{i<j} X_i X_j - n\sigma^2$$

Using the triangle inequality, we have

$$E(|M_n|) \leq \sum_{i=1}^{n} E(X_i^2) + 2\sum_{i<j} E(|X_i X_j|) + n\sigma^2$$

Using Cauchy-Schwarz inequality (Eq. (4.41)), we have

$$E(|X_i X_j|) \leq \sqrt{E(X_i^2)E(X_i^2)} = \sigma^2$$

Thus,

$$E(|M_n|) \leq n\sigma^2 + \frac{n(n-1)}{2}\sigma^2 + n\sigma^2 = \frac{n(n+3)}{2}\sigma^2 < \infty$$

Next,

$$\begin{aligned}
E(M_{n+1}|F_n) &= E[(X_{n+1} + S_n)^2 - (n+1)\sigma^2|F_n] \\
&= E[X_{n+1}^2 + 2X_{n+1}S_n + S_n^2 - (n+1)\sigma^2|F_n] \\
&= M_n + E(X_{n+1}^2) + 2E(X_{n+1})S_n - \sigma^2 \\
&= M_n + \sigma^2 - \sigma^2 = M_n
\end{aligned}$$

Thus, $\{M_n, n \geq 0\}$ is a martingale.

5.73. Let $X_1, X_2, \ldots$ be a sequence of i.i.d. r.v.'s with $E(X_i) = \mu$ and $E(|X_i|) < \infty$ for all i. Show that

$$M_n = \frac{1}{\mu^n} \prod_{i=1}^{n} X_i \tag{5.212}$$

is a martingale.

$$E(|M_n|) = E\left(\left|\frac{1}{\mu^n}\prod_{i=1}^{n} X_i\right|\right) = \frac{1}{\mu^n}\prod_{i=1}^{n} E(X_i) = \left|\frac{\mu^n}{\mu^n}\right| = 1 < \infty$$

$$E(M_{n+1}|F_n) = E\left(M_n \frac{1}{\mu} X_{n+1}\Big|F_n\right)$$

$$= M_n \frac{1}{\mu} E(X_{n+1}) = M_n \frac{\mu}{\mu} = M_n$$

Thus, $\{M_n\}$ is a martingale.

5.74. An urn contains initially a red and black ball. At each time $n \geq 1$, a ball is taken randomly, its color noted, and both this ball and another ball of the same color are put back into the urn. Continue similarly after n draws, the urn contains $n + 2$ balls. Let X_n denote the number of black balls after n draws. Let $M_n = X_n / (n + 2)$ be the fraction of black balls after n draws. Show that $\{M_n, n \geq 0\}$ is a martingale. (This is known as Polya's Urn.)

$X_0 = 1$ and X_n is a (time-homogeneous) Markov chain with transition

$$P\left(X_{n+1} = k + 1 \mid X_n = k\right) = \frac{k}{n+2} \quad \text{and} \quad P\left(X_{n+1} = k \mid X_n = k\right) = \frac{n+2+k}{n+2}$$

and X_n takes values in $\{1, 2, \ldots, n + 1\}$ and $E\left(X_{n+1} \mid X_n\right) = X_n + \dfrac{X_n}{n+2}$.

Now,

$$E\left[\left|M_n\right|\right] = E\left[\frac{\left|X_n\right|}{n+2}\right] \leq \frac{n+1}{n+2} < \infty$$

and

$$E\left(M_{n+1} \mid F_n\right) = E\left(\frac{1}{n+3} X_{n+1} \mid X_n\right)$$

$$= \frac{1}{n+3} E\left(X_{n+1} \mid X_n\right) = \frac{1}{n+3}\left(X_n + \frac{X_n}{n+2}\right) = \frac{X_n}{n+2} = M_n$$

Thus, $\{M_n, n \geq 0\}$ is a martingale.

5.75. Let $X_1, X_2, \ldots$ be a sequence of independent r.v.'s with

$$P\{X = 1\} = P\{X = -1\} = \tfrac{1}{2}$$

We can think of X_i as the result of a tossing a fair coin game where one wins \$1 if heads come up and loses \$1 if tails come up. The one way of betting strategy is to keep doubling the bet until one eventually wins. At this point one stops. (This strategy is the original martingale game.) Let S_n denote the winnings (or losses) up through n tosses. $S_0 = 0$. Whenever one wins, one stops playing, so $P(S_{n+1} = 1 \mid S_n = 1) = 1$. Show that $\{S_n, n \geq 0\}$ is a martingale—that is, the game is fair.

Suppose the first n tosses of the coin have turned up tails. So the loss S_n is given by

$$S_n = -(1 + 2 + 4 + \cdots + 2^{n-1}) = -(2^n - 1)$$

At this time, one double the bet again and bet 2^n on the next toss. This gives

$$P(S_{n+1} = 1 \mid S_n = -(2^n - 1)) = \tfrac{1}{2}, \quad P(S_{n+1} = -(2^n - 1) \mid S_n = -(2^n - 1)) = \tfrac{1}{2}$$

and

$$E\left(S_{n+1} \mid F_n\right) = E\left(S_{n+1}\right) = \frac{1}{2}(1) + \frac{1}{2}\left[-\left(2^{n+1} - 1\right)\right]$$

$$= \frac{1}{2} - 2^n + \frac{1}{2} = -\left(2^n - 1\right) = S_n$$

Thus, $\{S_n, n \geq 0\}$ is a martingale.

5.76. Let $\{X_n, n \geq 0\}$ be a martingale with respect to the filtration F_n and let g be a convex function such that $E[g(X_n)] < \infty$ for all $n \geq 0$. Then show that the sequence $\{Z_n, n \geq 0\}$ defined by

$$Z_n = g(X_n) \tag{5.213}$$

is a submartingale with respect to F_n.

$$E(|Z_n|) = E(|g(Xn)|) < \infty$$

By Jensen's inequality Eq. (4.40) and the martingale property of X_n, we have

$$E(Z_{n+1}|F_n) = E[g(X_{n+1})|F_n] \geq g[E(X_{n+1}|F_n)] = g(X_n) = Z_n$$

Thus, $\{Z_n, n \geq 0\}$ is a submartingale.

5.77. Let F_n be a filtration and $E(X) < \infty$. Define

$$X_n = E(X|F_n) \tag{5.214}$$

Show that $\{X_n, n \geq 0\}$ is a martingale with respect to F_n.

$$E(|X_n|) = E(|E(X|F_n)|) \leq |E[E(X|F_n)]| = |E(X)| < \infty$$
$$E(X_{n+1}|F_n) = E[E(X|F_{n+1})|F_n]$$
$$= E(X|F_n) \qquad \text{by Eq. (5.76)}$$
$$= X_n$$

Thus, $\{X_n, n \geq 0\}$ is a martingale with respect to F_n.

5.78. Prove Theorem 5.8.2 (Doob decomposition).

Since X is a submartingale, we have

$$E(X_{n+1}|F_n) \geq X_n \tag{5.215}$$

Let

$$d_n = E(X_{n+1} - X_n|F_n) = E(X_{n+1}|F_n) - X_n \geq 0 \tag{5.216}$$

and d_n is F_n-measurable.

Set $A_0 = 0, A_n = \sum_{i=1}^{n-1} d_i = d_1 + d_2 + \cdots + d_{n-1}$, and $M_n = X_n - A_n$. Then it is easily seen that (2), (3), and (4) of Theorem 5.82 are satisfied. Next,

$$E(M_{n+1}|F_n) = E(X_{n+1} - A_{n+1}|F_n) = E(X_{n+1}|F_n) - A_{n+1}$$
$$= X_n + d_n - \sum_{i=1}^{n} d_i = X_n - \sum_{i=1}^{n-1} d_i = X_n - A_n = M_n$$

Thus, (1) of Theorem 5.82 is also verified.

5.79. Let $\{M_n, n \geq 0\}$ be a martingale. Suppose that the stopping time T is bounded, that is $T \leq k$. Then show that

$$E(M_T) = E(M_0) \tag{5.217}$$

Note that $I_{\{T=j\}}$, the indicator function of the event $\{T = j\}$, is F_n-measurable (since we need only the information up to time n to determine if we have stopped by time n). Then we can write

$$M_T = \sum_{j=0}^{k} M_j \, I_{\{T=j\}}$$

and

$$E\left(M_T \big| F_{k-1}\right) = E\left(M_k \, I_{\{T=k\}} \big| F_{k-1}\right) + \sum_{j=0}^{k-1} E\left(M_j \, I_{\{T=j\}} \big| F_{k-1}\right)$$

For $j \le k - 1, M_j I_{\{T=j\}}$ is F_{k-1}-measurable, thus,

$$E(M_j \, I_{\{T=j\}} \big| F_{k-1}) = M_j \, I_{\{T=j\}}$$

Since T is known to be no more than k, the event $\{T = k\}$ is the same as the event $\{T > k - 1$ which is F_{k-1}-measurable. Thus,

$$E\left(M_k \, I_{\{T=k\}} \big| F_{k-1}\right) = E\left(M_k \, I_{\{T>k-1\}} \big| F_{k-1}\right)$$
$$= I_{\{T>k-1\}} E\left(M_k \big| F_{k-1}\right) = I_{\{T>k-1\}} M_{k-1}$$

since $\{M_n\}$ be a martingale. Hence,

$$E\left(M_T \big| F_{k-1}\right) = I_{\{T>k-1\}} M_{k-1} + \sum_{j=0}^{k-1} E\left(M_j \, I_{\{T=j\}}\right)$$

In a similar way, we can derive

$$E\left(M_T \big| F_{k-2}\right) = I_{\{T>k-2\}} M_{k-2} + \sum_{j=0}^{k-2} E\left(M_j \, I_{\{T=j\}}\right)$$

And continue this process until we get

$$E(M_T \big| F_0) = M_0$$

and finally

$$E[E(M_T \big| F_0)] = E(M_T) = E(M_0)$$

5.80. Verify the Optional Stopping Theorem.

Consider the stopping times $T_n = \min\{T, n\}$. Note that

$$M_T = M_{T_n} + M_T I_{\{T > n\}} - M_n I_{\{T>n\}} \tag{5.218}$$

Hence,

$$E(M_T) = E(M_{T_n}) + E(M_T I_{\{T>n\}}) - E(M_n I_{\{T>n\}}) \tag{5.219}$$

Since T_n is a bounded stopping time, by Eq. (5.217), we have

$$E(M_{T_n}) = E(M_0) \tag{5.220}$$

and $\lim_{n \to \infty} P(T > n) = 0$, then if $E(|M_T|) < \infty$, (condition (1), Eq. (5.83)) we have. $\lim_{n \to \infty} (|M_T| I_{\{T > n\}}) = 0$. Thus, by condition (3), Eq. (5.85), we get. $\lim_{n \to \infty} (|M_T| I_{\{T > n\}}) = 0$. Hence, by Eqs. (5.219) and (5.220), we obtain

$$E(M_T) = E(M_0)$$

5.81. Let two gamblers, A and B, initially have a dollars and b dollars, respectively. Suppose that at each round of tossing a fair coin A wins one dollar from B if "heads" comes up, and gives one dollar to B if "tails" comes up. The game continues until either A or B runs out of money.

(a) What is the probability that when the game ends, A has all the cash?

(b) What is the expected duration of the game?

(a) Let $X_1, X_2, \ldots$ be the sequence of play-by-play increments in A's fortune; thus, $X_i = \pm 1$ according to whether ith toss is "heads" or "tails." The total change in A's fortune after n plays is $S_n = \sum_{i=1}^{n} X_i$. The game continues until time T where $T = \min\{n : s_n = -a \text{ or } +b\}$. It is easily seen that T is a stopping time with respect to $F_n = \sigma(X_1, X_2, \ldots, X_n)$ and $\{S_n\}$ is a martingale with respect to F_n. (See Prob. 5.68.) Thus, by the Optional Stopping Theorem, for each $n < \infty$

$$0 = E(S_0) = E(S_{\min(T, n)})$$
$$= -aP(T \le n \text{ and } S_T = -a) + bP(T \le n \text{ and } S_T = b) + E(S_n I_{\{T > n\}})$$

As $n \to \infty$, the probability of the event $\{T > n\}$ converges to zero. Since S_n must be between $-a$ and b on the event $\{T > n\}$, it follows that $E(S_n I_{\{T > n\}})$ converges to zero as $n \to \infty$. Thus, letting $n \to \infty$, we obtain

$$-aP(S_T = -a) + bP(S_T = b) = 0 \tag{5.221}$$

Since S_T must be $-a$ or b, we have

$$P(S_T = -a) + P(S_T = b) = 1 \tag{5.222}$$

Solving Eqs. (5.221) and (5.222) for $P(S_T = -a)$ and $P(S_T = b)$, we obtain (cf. Prob. 5.43)

$$P(S_T = -a) = \frac{b}{a + b}, \quad P(S_T = b) = \frac{a}{a + b} \tag{5.223}$$

Thus, the probability that when the game ends, A has all the cash is $a/(a + b)$.

(b) It is seen that $\{S_n^2 - n\}$ is a martingale (see Prob. 5.72, $\sigma^2 = 1$). Then the Optional Stopping Theorem implies that, for each $n = 1, 2, \ldots$,

$$E\left(S_{\min(T, n)}^2 - \min(T, n)\right) = 0 \tag{5.224}$$

Thus,

$$E(\min(T, n)) = E\left(S_{\min(T, n)}^2\right) = E\left(S_T^2 I_{\{T \le n\}}\right) + E\left(S_n^2 I_{\{T > n\}}\right) \tag{5.225}$$

Now, as $n \to \infty$, $\min(T, n) \to T$ and $S_T^2 I_{\{T \le n\}} \to S_T^2$, and $\lim_{n \to \infty} E[\min(T, n)] = E(T)$

$$\lim_{n \to \infty} E\left(S_T^2 I_{\{T \le n\}}\right) = E\left(S_T^2\right) = a^2\left(\frac{a}{a + b}\right) + b^2\left(\frac{a}{a + b}\right) = ab$$

Since S_n^s is bounded on the event $\{T > n\}$, and since the probability of this event converges to zero as $n \to \infty$, $E(S_n^2 I_{\{T > n\}}) \to 0$ as $n \to \infty$. Thus, as $n \to \infty$, Eq. (5.225) reduces to

$$E(T) = ab \tag{5.226}$$

5.82. Let $X(t)$ be a Poisson process with rate $\lambda > 0$. Show that $x(t) - \lambda t$ is a martingale.

We have

$$E(|X(t) - \lambda t|) \leq E[X(t)] + \lambda t = 2\lambda t < \infty$$

since $X(t) \geq 0$ and by Eq. (5.56), $E[X(t)] = \lambda t$.

$$\begin{aligned}
E[X(t) - \lambda t \,|\, F_s] &= E[X(s) - \lambda t + X(t) - X(s) \,|\, F_s] \\
&= E[X(s) - \lambda t \,|\, F_s] + E[X(t) - X(s) \,|\, F_s] \\
&= X(s) - \lambda t + E[X(t) - X(s)] \\
&= X(s) - \lambda t + \lambda(t - s) = X(s) - \lambda s
\end{aligned}$$

Thus, $x(t) - \lambda t$ is a martingale.

SUPPLEMENTARY PROBLEMS

5.83. Consider a random process $X(n) = \{X_n, n \geq 1\}$, where

$$X_n = Z_1 + Z_2 + \cdots + Z_n$$

and Z_n are iid r.v.'s with zero mean and variance σ^2. Is $X(n)$ stationary?

5.84. Consider a random process $X(t)$ defined by

$$X(t) = Y \cos(\omega t + \Theta)$$

where Y and Θ are independent r.v.'s and are uniformly distributed over $(-A, A)$ and $(-\pi, \pi)$, respectively.

(a) Find the mean of $X(t)$.

(b) Find the autocorrelation function $R_X(t, s)$ of $X(t)$.

5.85. Suppose that a random process $X(t)$ is wide-sense stationary with autocorrelation

$$R_X(t, t + \tau) = e^{-|\tau|/2}$$

(a) Find the second moment of the r.v. $X(5)$.

(b) Find the second moment of the r.v. $X(5) - X(3)$.

5.86. Consider a random process $X(t)$ defined by

$$X(t) = U \cos t + (V + 1) \sin t \qquad -\infty < t < \infty$$

where U and V are independent r.v.'s for which

$$E(U) = E(V) = 0 \qquad E(U^2) = E(V^2) = 1$$

(a) Find the autocovariance function $K_X(t, s)$ of $X(t)$.

(b) Is $X(t)$ WSS?

5.87. Consider the random processes

$$X(t) = A_0 \cos(\omega_0 t + \Theta) \qquad Y(t) = A_1 \cos(\omega_1 t + \Phi)$$

where A_0, A_1, ω_0, and ω_1 are constants, and r.v.'s Θ and Φ are independent and uniformly distributed over $(-\pi, \pi)$.

(a) Find the cross-correlation function of $R_{XY}(t, t + \tau)$ of $X(t)$ and $Y(t)$.

(b) Repeat (a) if $\Theta = \Phi$.

5.88. Given a Markov chain $\{X_n, n \geq 0\}$, find the joint pmf

$$P(X_0 = i_0, X_1 = i_1, \ldots, X_n = i_n)$$

5.89. Let $\{X_n, n \geq 0\}$ be a homogeneous Markov chain. Show that

$$P(X_{n+1} = k_1, \ldots, X_{n+m} = k_m | X_0 = i_0, \ldots, X_n = i) = P(X_1 = k_1, \ldots, X_m = k_m | X_0 = i)$$

5.90. Verify Eq. (5.37).

5.91. Find P^n for the following transition probability matrices:

(a) $P = \begin{bmatrix} 1 & 0 \\ 0.5 & 0.5 \end{bmatrix}$ (b) $P = \begin{bmatrix} 1 & 0 & 0 \\ 0 & 1 & 0 \\ 0 & 0 & 1 \end{bmatrix}$ (c) $P = \begin{bmatrix} 1 & 0 & 0 \\ 0 & 1 & 0 \\ 0.3 & 0.2 & 0.5 \end{bmatrix}$

5.92. A certain product is made by two companies, A and B, that control the entire market. Currently, A and B have 60 percent and 40 percent, respectively, of the total market. Each year, A loses $\frac{2}{3}$ of its market share to B, while B loses $\frac{1}{2}$ of its share to A. Find the relative proportion of the market that each hold after 2 years.

5.93. Consider a Markov chain with state $\{0, 1, 2\}$ and transition probability matrix

$$P = \begin{bmatrix} 0 & \frac{1}{2} & \frac{1}{2} \\ \frac{1}{2} & 0 & \frac{1}{2} \\ 1 & 0 & 0 \end{bmatrix}$$

Is state 0 periodic?

5.94. Verify Eq. (5.51).

5.95. Consider a Markov chain with transition probability matrix

$$P = \begin{bmatrix} 0.6 & 0.2 & 0.2 \\ 0.4 & 0.5 & 0.1 \\ 0.6 & 0 & 0.4 \end{bmatrix}$$

Find the steady-state probabilities.

5.96. Let $X(t)$ be a Poisson process with rate λ. Find $E[X^2(t)]$.

5.97. Let $X(t)$ be a Poisson process with rate λ. Find $E\{[X(t) - X(s)]^2\}$ for $t > s$.

5.98. Let $X(t)$ be a Poisson process with rate λ. Find

$$P[X(t - d) = k \,|\, X(t) = j] \qquad d > 0$$

5.99. Let T_n denote the time of the nth event of a Poisson process with rate λ. Find the variance of T_n.

5.100. Assume that customers arrive at a bank in accordance with a Poisson process with rate $\lambda = 6$ per hour, and suppose that each customer is a man with probability $\frac{2}{3}$ and a woman with probability $\frac{1}{3}$. Now suppose that 10 men arrived in the first 2 hours. How many woman would you expect to have arrived in the first 2 hours?

5.101. Let $X_1, \ldots, X_n$ be jointly normal r.v.'s. Let

$$Y_i = X_i + c_i \qquad i = 1, \ldots, n$$

where c_i are constants. Show that $Y_1, \ldots, Y_n$ are also jointly normal r.v.'s.

5.102. Derive Eq. (5.63).

5.103. Let $X_1, X_2, \ldots$ be a sequence of Bernoulli r.v.'s in Prob. 5.70. Let $M_n = S_n - n(2p - 1)$. Show that $\{M_n\}$ is a martingale.

5.104. Let $X_1, X_2, \ldots$ be i.i.d. r.v.'s where X_i can take only two values $\frac{3}{2}$ and $\frac{1}{2}$ with equal probability. Let $M_0 = 1$ and $M_n = \prod_{i=1}^{n} X_i$. Show that $\{M_n, n \geq 0\}$ is a martingale.

5.105. Consider $\{X_n\}$ of Prob. 5.70 and $S_n = \sum_{i=1}^{n} X_i$. Let $Y_n = \left(\dfrac{q}{p}\right)^{S_n}$. Show that $\{Y_n\}$ is a martingale.

5.106. Let $X(t)$ be a Wiener's process (or Brownian motion). Show that $\{X(t)\}$ is a martingale.

ANSWERS TO SUPPLEMENTARY PROBLEMS

5.83. No.

5.84. *(a)* $E[X(t)] = 0$;   *(b)* $R_X(t, s) = \frac{1}{6}A^2 \cos \omega(t - s)$

5.85. *(a)* $E[X^2(5)] = 1$;   *(b)* $E\{[X(5) - X(3)]^2\} = 2(1 - e^{-1})$

5.86. *(a)* $K_X(t, s) = \cos(s - t)$;   *(b)* No.

5.87. *(a)* $R_{XY}(t, t + \tau)] = 0$

     (b) $R_{XY}(t, t + \tau) = \dfrac{A_0 A_1}{2} \cos[(\omega_1 - \omega_0)t + \omega_1 \tau]$

5.88. *Hint:* Use Eq. (5.32).

$$p_{i_0}(0)\, p_{i_0 i_1} p_{i_1 i_2} \cdots p_{i_{n-1} i_n}$$

5.89. Use the Markov property (5.27) and the homogeneity property.

5.90. *Hint:* Write Eq. (5.39) in terms of components.

5.91. (a) $P^n = \begin{bmatrix} 1 & 0 \\ 1 & 0 \end{bmatrix} + (0.5)^n \begin{bmatrix} 0 & 0 \\ -1 & 1 \end{bmatrix}$ (b) $P^n = \begin{bmatrix} 1 & 0 & 0 \\ 0 & 1 & 0 \\ 0 & 0 & 1 \end{bmatrix}$

(c) $P^n = \begin{bmatrix} 1 & 0 & 0 \\ 0 & 1 & 0 \\ 0.6 & 0.4 & 0 \end{bmatrix} + (0.5)^n \begin{bmatrix} 0 & 0 & 0 \\ 0 & 0 & 0 \\ -0.6 & -0.4 & 1 \end{bmatrix}$

5.92. A has 43.3 percent and B has 56.7 percent.

5.93. *Hint:* Draw the state transition diagram.

No.

5.94. *Hint:* Let $\tilde{N} = [N_{jk}]$, where N_{jk} is the number of times the state $k (\in B)$ is occupied until absorption takes place when $X(n)$ starts in state $j (\in B)$. Then $T_j = \sum_{k=m+1}^{N} N_{jk}$; calculate $E(N_{jk})$.

5.95. $\hat{\mathbf{p}} = \begin{bmatrix} \dfrac{5}{9} & \dfrac{2}{9} & \dfrac{2}{9} \end{bmatrix}$

5.96. $\lambda t + \lambda^2 t^2$

5.97. *Hint:* Use the independent stationary increments condition and the result of Prob. 5.76.

$$\lambda(t - s) + \lambda^2(t - s)^2$$

5.98. $\dfrac{j!}{k!(j-k)!} \left(\dfrac{t-d}{t} \right)^k \left(\dfrac{d}{t} \right)^{j-k}$

5.99. n/λ^2

5.100. 4

5.101. *Hint:* See Prob. 5.60.

5.102. *Hint:* Use condition (1) of a Wiener process and Eq. (5.102) of Prob. 5.22.

5.103. *Hint:* Note that M_n is the random number S_n minus its expected value.

5.106. *Hint:* Use definition 5.7.1.

CHAPTER 6

Analysis and Processing of Random Processes

6.1 Introduction

In this chapter, we introduce the methods for analysis and processing of random processes. First, we introduce the definitions of stochastic continuity, stochastic derivatives, and stochastic integrals of random processes. Next, the notion of power spectral density is introduced. This concept enables us to study wide-sense stationary processes in the frequency domain and define a white noise process. The response of linear systems to random processes is then studied. Finally, orthogonal and spectral representations of random processes are presented.

6.2 Continuity, Differentiation, Integration

In this section, we shall consider only the continuous-time random processes.

A. Stochastic Continuity:

A random process $X(t)$ is said to be *continuous in mean square* or *mean square (m.s.) continuous* if

$$\lim_{\varepsilon \to 0} E\{[X(t+\varepsilon) - X(t)]^2\} = 0 \tag{6.1}$$

The random process $X(t)$ is m.s. continuous if and only if its autocorrelation function is continuous (Prob. 6.1). If $X(t)$ is WSS, then it is m.s. continuous if and only if its autocorrelation function $R_X(\tau)$ is continuous at $\tau = 0$. If $X(t)$ is m.s. continuous, then its mean is continuous; that is,

$$\lim_{\varepsilon \to 0} \mu_X(t+\varepsilon) = \mu_X(t) \tag{6.2}$$

which can be written as

$$\lim_{\varepsilon \to 0} E[X(t+\varepsilon)] = E[\lim_{\varepsilon \to 0} X(t+\varepsilon)] \tag{6.3}$$

Hence, if $X(t)$ is m.s. continuous, then we may interchange the ordering of the operations of expectation and limiting. Note that m.s. continuity of $X(t)$ does not imply that the sample functions of $X(t)$ are continuous. For instance, the Poisson process is m.s. continuous (Prob. 6.46), but sample functions of the Poisson process have a countably infinite number of discontinuities (see Fig. 5-3).

B.   Stochastic Derivatives:

A random process $X(t)$ is said to have a *m.s. derivative* $X'(t)$ if

$$\underset{\varepsilon \to 0}{\text{l.i.m.}} \frac{X(t + \varepsilon) - X(t)}{\varepsilon} = X'(t) \tag{6.4}$$

where l.i.m. denotes *l*imit *i*n the *m*ean (square); that is,

$$\lim_{\varepsilon \to 0} E\left\{\left[\frac{X(t + \varepsilon) - X(t)}{\varepsilon} - X'(t)\right]^2\right\} = 0 \tag{6.5}$$

The m.s. derivative of $X(t)$ exists if $\partial^2 R_X(t, s)/\partial t\, \partial s$ exists (Prob. 6.6). If $X(t)$ has the m.s. derivative $X'(t)$, then its mean and autocorrelation function are given by

$$E[X'(t)] = \frac{d}{dt} E[X(t)] = \mu'_X(t) \tag{6.6}$$

$$R_{X'}(t, s) = \frac{\partial^2 R_X(t, s)}{\partial t\, \partial s} \tag{6.7}$$

Equation (6.6) indicates that the operations of differentiation and expectation may be interchanged. If $X(t)$ is a normal random process for which the m.s. derivative $X'(t)$ exists, then $X'(t)$ is also a normal random process (Prob. 6.10).

C.   Stochastic Integrals:

A m.s. *integral* of a random process $X(t)$ is defined by

$$Y(t) = \int_{t_0}^{t} X(\alpha)\, d\alpha = \underset{\Delta t_i \to 0}{\text{l.i.m.}} \sum_i X(t_i)\, \Delta t_i \tag{6.8}$$

where $t_0 < t_1 < \cdots < t$ and $\Delta t_i = t_{i+1} - t_i$.

The m.s. integral of $X(t)$ exists if the following integral exists (Prob. 6.11):

$$\int_{t_0}^{t} \int_{t_0}^{t} R_X(\alpha, \beta)\, d\alpha\, d\beta \tag{6.9}$$

This implies that if $X(t)$ is m.s. continuous, then its m.s. integral $Y(t)$ exists (see Prob. 6.1). The mean and the autocorrelation function of $Y(t)$ are given by

$$\mu_Y(t) = E\left[\int_{t_0}^{t} X(\alpha)\, d\alpha\right] = \int_{t_0}^{t} E[X(\alpha)]\, d\alpha = \int_{t_0}^{t} \mu_X(\alpha)\, d\alpha \tag{6.10}$$

$$R_Y(t, s) = E\left[\int_{t_0}^{t} X(\alpha)\, d\alpha \int_{t_0}^{s} X(\beta)\, d\beta\right]$$
$$= \int_{t_0}^{t} \int_{t_0}^{s} E[X(\alpha)X(\beta)]\, d\beta\, d\alpha = \int_{t_0}^{t} \int_{t_0}^{s} R_X(\alpha, \beta)\, d\beta\, d\alpha \tag{6.11}$$

Equation (6.10) indicates that the operations of integration and expectation may be interchanged. If $X(t)$ is a normal random process, then its integral $Y(t)$ is also a normal random process. This follows from the fact that $\Sigma_i X(t_i)\, \Delta t_i$ is a linear combination of the jointly normal r.v.'s. (see Prob. 5.60).

6.3 Power Spectral Densities

In this section we assume that all random processes are WSS.

A. Autocorrelation Functions:

The autocorrelation function of a continuous-time random process $X(t)$ is defined as [Eq. (5.7)]

$$R_X(\tau) = E[X(t)X(t + \tau)] \tag{6.12}$$

Properties of $R_X(\tau)$:

1. $R_X(-\tau) = R_X(\tau)$ (6.13)
2. $|R_X(\tau)| \leq R_X(0)$ (6.14)
3. $R_X(0) = E[X^2(t)] \geq 0$ (6.15)

Property 3 [Eq. (6.15)] is easily obtained by setting $\tau = 0$ in Eq. (6.12). If we assume that $X(t)$ is a voltage waveform across a 1-Ω resistor, then $E[X^2(t)]$ is the average value of power delivered to the 1-Ω resistor by $X(t)$. Thus, $E[X^2(t)]$ is often called the *average power* of $X(t)$. Properties 1 and 2 are verified in Prob. 6.13.

In case of a discrete-time random process $X(n)$, the autocorrelation function of $X(n)$ is defined by

$$R_X(k) = E[X(n)X(n + k)] \tag{6.16}$$

Various properties of $R_X(k)$ similar to those of $R_X(\tau)$ can be obtained by replacing τ by k in Eqs. (6.13) to (6.15).

B. Cross-Correlation Functions:

The cross-correlation function of two continuous-time jointly WSS random processes $X(t)$ and $Y(t)$ is defined by

$$R_{XY}(\tau) = E[X(t)Y(t + \tau)] \tag{6.17}$$

Properties of $R_{XY}(\tau)$:

1. $R_{XY}(-\tau) = R_{YX}(\tau)$ (6.18)
2. $|R_{XY}(\tau)| \leq \sqrt{R_X(0)R_Y(0)}$ (6.19)
3. $|R_{XY}(\tau)| \leq \frac{1}{2}[R_X(0) + R_Y(0)]$ (6.20)

These properties are verified in Prob. 6.14. Two processes $X(t)$ and $Y(t)$ are called (*mutually*) *orthogonal* if

$$R_{XY}(\tau) = 0 \quad \text{for all } \tau \tag{6.21}$$

Similarly, the cross-correlation function of two discrete-time jointly WSS random processes $X(n)$ and $Y(n)$ is defined by

$$R_{XY}(k) = E[X(n)Y(n + k)] \tag{6.22}$$

and various properties of $R_{XY}(k)$ similar to those of $R_{XY}(\tau)$ can be obtained by replacing τ by k in Eqs. (6.18) to (6.20).

C. Power Spectral Density:

The *power spectral density* (or *power spectrum*) $S_X(\omega)$ of a continuous-time random process $X(t)$ is defined as the Fourier transform of $R_X(\tau)$:

$$S_X(\omega) = \int_{-\infty}^{\infty} R_X(\tau)e^{-j\omega\tau}\, d\tau \tag{6.23}$$

Thus, taking the inverse Fourier transform of $S_X(\omega)$, we obtain

$$R_X(\tau) = \frac{1}{2\pi}\int_{-\infty}^{\infty} S_X(\omega)e^{j\omega\tau}\, d\omega \tag{6.24}$$

Equations (6.23) and (6.24) are known as the *Wiener-Khinchin relations*.

Properties of $S_X(\omega)$:

1. $S_X(\omega)$ is real and $S_X(\omega) \geq 0$. $\tag{6.25}$
2. $S_X(-\omega) = S_X(\omega)$ $\tag{6.26}$
3. $E[X^2(t)] = R_X(0) = \dfrac{1}{2\pi}\displaystyle\int_{-\infty}^{\infty} S_X(\omega)\, d\omega$ $\tag{6.27}$

Similarly, the power spectral density $S_X(\Omega)$ of a discrete-time random process $X(n)$ is defined as the Fourier transform of $R_X(k)$:

$$S_X(\Omega) = \sum_{k=-\infty}^{\infty} R_X(k)e^{-j\Omega k} \tag{6.28}$$

Thus, taking the inverse Fourier transform of $S_X(\Omega)$, we obtain

$$R_X(k) = \frac{1}{2\pi}\int_{-\pi}^{\pi} S_X(\Omega)e^{j\Omega k}\, d\Omega \tag{6.29}$$

Properties of $S_X(\Omega)$:

1. $S_X(\Omega + 2\pi) = S_X(\Omega)$ $\tag{6.30}$
2. $S_X(\Omega)$ is real and $S_X(\Omega) \geq 0$. $\tag{6.31}$
3. $S_X(-\Omega) = S_X(\Omega)$ $\tag{6.32}$
4. $E[X^2(n)] = R_X(0) = \dfrac{1}{2\pi}\displaystyle\int_{-\pi}^{\pi} S_X(\Omega)\, d\Omega$ $\tag{6.33}$

Note that property 1 [Eq. (6.30)] follows from the fact that $e^{-j\Omega k}$ is periodic with period 2π. Hence, it is sufficient to define $S_X(\Omega)$ only in the range $(-\pi, \pi)$.

D. Cross Power Spectral Densities:

The *cross power spectral density* (or *cross power spectrum*) $S_{XY}(\omega)$ of two continuous-time random processes $X(t)$ and $Y(t)$ is defined as the Fourier transform of $R_{XY}(\tau)$:

$$S_{XY}(\omega) = \int_{-\infty}^{\infty} R_{XY}(\tau)e^{-j\omega\tau}\, d\tau \tag{6.34}$$

Thus, taking the inverse Fourier transform of $S_{XY}(\omega)$, we get

$$R_{XY}(\tau) = \frac{1}{2\pi}\int_{-\infty}^{\infty} S_{XY}(\omega)e^{j\omega\tau}\, d\omega \tag{6.35}$$

Properties of $S_{XY}(\omega)$:
Unlike $S_X(\omega)$, which is a real-valued function of ω, $S_{XY}(\omega)$, in general, is a complex-valued function.

1. $S_{XY}(\omega) = S_{YX}(-\omega)$ (6.36)
2. $S_{XY}(-\omega) = S_{XY}^*(\omega)$ (6.37)

Similarly, the cross power spectral density $S_{XY}(\Omega)$ of two discrete-time random processes $X(n)$ and $Y(n)$ is defined as the Fourier transform of $R_{XY}(k)$:

$$S_{XY}(\Omega) = \sum_{k=-\infty}^{\infty} R_{XY}(k)e^{-j\Omega k}$$ (6.38)

Thus, taking the inverse Fourier transform of $S_{XY}(\Omega)$, we get

$$R_{XY}(k) = \frac{1}{2\pi} \int_{-\pi}^{\pi} S_{XY}(\Omega)e^{j\Omega k} \, d\Omega$$ (6.39)

Properties of $S_{XY}(\Omega)$:
Unlike $S_X(\Omega)$, which is a real-valued function of ω, $S_{XY}(\Omega)$, in general, is a complex-valued function.

1. $S_{XY}(\Omega + 2\pi) = S_{XY}(\Omega)$ (6.40)
2. $S_{XY}(\Omega) = S_{YX}(-\Omega)$ (6.41)
3. $S_{XY}(-\Omega) = S_{XY}^*(\Omega)$ (6.42)

6.4 White Noise

A continuous-time *white noise* process, $W(t)$, is a WSS zero-mean continuous-time random process whose autocorrelation function is given by

$$R_W(\tau) = \sigma^2 \delta(\tau)$$ (6.43)

where $\delta(\tau)$ is a unit impulse function (or Dirac δ function) defined by

$$\int_{-\infty}^{\infty} \delta(\tau)\phi(\tau) \, d\tau = \phi(0)$$ (6.44)

where $\phi(\tau)$ is any function continuous at $\tau = 0$. Taking the Fourier transform of Eq. (6.43), we obtain

$$S_W(\omega) = \sigma^2 \int_{-\infty}^{\infty} \delta(\tau)e^{-j\omega\tau} \, d\tau = \sigma^2$$ (6.45)

which indicates that $X(t)$ has a constant power spectral density (hence the name white noise). Note that the average power of $W(t)$ is not finite.

Similarly, a WSS zero-mean discrete-time random process $W(n)$ is called a discrete-time white noise if its autocorrelation function is given by

$$R_W(k) = \sigma^2 \delta(k)$$ (6.46)

where $\delta(k)$ is a unit impulse sequence (or unit sample sequence) defined by

$$\delta(k) = \begin{cases} 1 & k = 0 \\ 0 & k \neq 0 \end{cases}$$ (6.47)

Taking the Fourier transform of Eq. (6.46), we obtain

$$S_W(\Omega) = \sigma^2 \sum_{k=-\infty}^{\infty} \delta(k) e^{-j\Omega k} = \sigma^2 \qquad -\pi < \Omega < \pi \tag{6.48}$$

Again the power spectral density of $W(n)$ is a constant. Note that $S_W(\Omega + 2\pi) = S_W(\Omega)$ and the average power of $W(n)$ is $\sigma^2 = \text{Var}[W(n)]$, which is finite.

6.5   Response of Linear Systems to Random Inputs

A.   Linear Systems:

A system is a mathematical model of a physical process that relates the input (or excitation) signal x to the output (or response) signal y. Then the system is viewed as a transformation (or mapping) of x into y. This transformation is represented by the operator $\mathbf{T}$ as (Fig. 6-1)

$$y = \mathbf{T}x \tag{6.49}$$

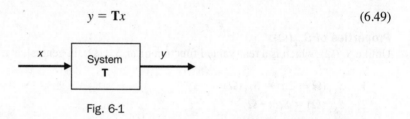

Fig. 6-1

If x and y are continuous-time signals, then the system is called a continuous-time system, and if x and y are discrete-time signals, then the system is called a discrete-time system. If the operator $\mathbf{T}$ is a linear operator satisfying

$$\mathbf{T}\{x_1 + x_2\} = \mathbf{T}x_1 + \mathbf{T}x_2 = y_1 + y_2 \qquad \text{(Additivity)}$$
$$\mathbf{T}\{ax\} = \alpha \mathbf{T}x = \alpha y \qquad \qquad \text{(Homogeneity)}$$

where α is a scalar number, then the system represented by $\mathbf{T}$ is called a linear system. A system is called time-invariant if a time shift in the input signal causes the same time shift in the output signal. Thus, for a continuous-time system,

$$\mathbf{T}\{x(t - t_0)\} = y(t - t_0)$$

for any value of t_0, and for a discrete-time system,

$$\mathbf{T}\{x(n - n_0)\} = y(n - n_0)$$

for any integer n_0. For a continuous-time linear time-invariant (LTI) system, Eq. (6.49) can be expressed as

$$y(t) = \int_{-\infty}^{\infty} h(\lambda) x(t - \lambda)\, d\lambda \tag{6.50}$$

where
$$h(t) = \mathbf{T}\{\delta(t)\} \tag{6.51}$$

is known as the *impulse response* of a continuous-time LTI system. The right-hand side of Eq. (6.50) is commonly called the *convolution integral* of $h(t)$ and $x(t)$, denoted by $h(t) * x(t)$. For a discrete-time LTI system, Eq. (6.49) can be expressed as

$$y(n) = \sum_{i=-\infty}^{\infty} h(i) x(n - i) \tag{6.52}$$

where
$$h(n) = \mathbf{T}\{\delta(n)\} \tag{6.53}$$

is known as the *impulse response* (or *unit sample response*) of a discrete-time LTI system. The right-hand side of Eq. (6.52) is commonly called the *convolution sum* of $h(n)$ and $x(n)$, denoted by $h(n) * x(n)$.

B. Response of a Continuous-Time Linear System to Random Input:

When the input to a continuous-time linear system represented by Eq. (6.49) is a random process $\{X(t), t \in T_x\}$, then the output will also be a random process $\{Y(t), t \in T_y\}$; that is,

$$\mathbf{T}\{X(t), t \in T_x\} = \{Y(t), t \in T_y\} \tag{6.54}$$

For any input sample function $x_i(t)$, the corresponding output sample function is

$$y_i(t) = \mathbf{T}\{x_i(t)\} \tag{6.55}$$

If the system is LTI, then by Eq. (6.50), we can write

$$Y(t) = \int_{-\infty}^{\infty} h(\lambda)X(t - \lambda)\, d\lambda \tag{6.56}$$

Note that Eq. (6.56) is a stochastic integral. Then

$$E[Y(t)] = \int_{-\infty}^{\infty} h(\lambda)E[X(t - \lambda)]\, d\lambda \tag{6.57}$$

The autocorrelation function of $Y(t)$ is given by (Prob. 6.24)

$$R_Y(t, s) = \int_{-\infty}^{\infty}\int_{-\infty}^{\infty} h(\alpha)h(\beta)R_X(t - \alpha, s - \beta)\, d\alpha\, d\beta \tag{6.58}$$

If the input $X(t)$ is WSS, then from Eq. (6.57),

$$E[Y(t)] = \mu_X \int_{-\infty}^{\infty} h(\lambda)\, d\lambda = \mu_X H(0) \tag{6.59}$$

where $H(0) = H(\omega)\big|_{\omega = 0}$ and $H(\omega)$ is the frequency response of the system defined by the Fourier transform of $h(t)$; that is,

$$H(\omega) = \int_{-\infty}^{\infty} h(t)e^{-j\omega t}\, dt \tag{6.60}$$

The autocorrelation function of $Y(t)$ is, from Eq. (6.58),

$$R_Y(t, s) = \int_{-\infty}^{\infty}\int_{-\infty}^{\infty} h(\alpha)h(\beta)R_X(s - t + \alpha - \beta)\, d\alpha\, d\beta \tag{6.61}$$

Setting $s = t + \tau$, we get

$$R_Y(t, t + \tau) = \int_{-\infty}^{\infty}\int_{-\infty}^{\infty} h(\alpha)h(\beta)R_X(\tau + \alpha - \beta)\, d\alpha\, d\beta = R_Y(\tau) \tag{6.62}$$

From Eqs. (6.59) and (6.62), we see that the output $Y(t)$ is also WSS. Taking the Fourier transform of Eq. (6.62), the power spectral density of $Y(t)$ is given by (Prob. 6.25)

$$S_Y(\omega) = \int_{-\infty}^{\infty} R_Y(\tau)e^{-j\omega \tau}\, d\tau = |H(\omega)|^2 S_X(\omega) \tag{6.63}$$

Thus, we obtain the important result that *the power spectral density of the output is the product of the power spectral density of the input and the magnitude squared of the frequency response of the system.*

When the autocorrelation function of the output $R_Y(\tau)$ is desired, it is easier to determine the power spectral density $S_Y(\omega)$ and then evaluate the inverse Fourier transform (Prob. 6.26). Thus,

$$R_Y(\tau) = \frac{1}{2\pi} \int_{-\infty}^{\infty} S_Y(\omega) e^{j\omega\tau}\, d\omega = \frac{1}{2\pi} \int_{-\infty}^{\infty} |H(\omega)|^2 S_X(\omega) e^{j\omega\tau}\, d\omega \tag{6.64}$$

By Eq. (6.15), the average power in the output $Y(t)$ is

$$E[Y^2(t)] = R_Y(0) = \frac{1}{2\pi} \int_{-\infty}^{\infty} |H(\omega)|^2 S_X(\omega)\, d\omega \tag{6.65}$$

C. Response of a Discrete-Time Linear System to Random Input:

When the input to a discrete-time LTI system is a discrete-time random process $X(n)$, then by Eq. (6.52), the output $Y(n)$ is

$$Y(n) = \sum_{i=-\infty}^{\infty} h(i) X(n-i) \tag{6.66}$$

The autocorrelation function of $Y(n)$ is given by

$$R_Y(n, m) = \sum_{i=-\infty}^{\infty} \sum_{l=-\infty}^{\infty} h(i) h(l) R_X(n-i, m-l) \tag{6.67}$$

When $X(n)$ is WSS, then from Eq. (6.66),

$$E[Y(n)] = \mu_X \sum_{i=-\infty}^{\infty} h(i) = \mu_X H(0) \tag{6.68}$$

where $H(0) = H(\Omega)|_{\Omega = 0}$ and $H(\Omega)$ is the frequency response of the system defined by the Fourier transform of $h(n)$:

$$H(\Omega) = \sum_{n=-\infty}^{\infty} h(n) e^{-j\Omega n} \tag{6.69}$$

The autocorrelation function of $Y(n)$ is, from Eq. (6.67),

$$R_Y(n, m) = \sum_{i=-\infty}^{\infty} \sum_{l=-\infty}^{\infty} h(i) h(l) R_X(m-n+i-l) \tag{6.70}$$

Setting $m = n + k$, we get

$$R_Y(n, n+k) = \sum_{i=-\infty}^{\infty} \sum_{l=-\infty}^{\infty} h(i) h(l) R_X(k+i-l) = R_Y(k) \tag{6.71}$$

From Eqs. (6.68) and (6.71), we see that the output $Y(n)$ is also WSS. Taking the Fourier transform of Eq. (6.71), the power spectral density of $Y(n)$ is given by (Prob. 6.28)

$$S_Y(\Omega) = |H(\Omega)|^2 S_X(\Omega) \tag{6.72}$$

which is the same as Eq. (6.63).

6.6 Fourier Series and Karhunen-Loéve Expansions

A. Stochastic Periodicity:

A continuous-time random process $X(t)$ is said to be *m.s. periodic* with period T if

$$E\{[X(t + T) - X(t)]^2\} = 0 \tag{6.73}$$

If $X(t)$ is WSS, then $X(t)$ is m.s. periodic if and only if its autocorrelation function is periodic with period T; that is,

$$R_X(\tau + T) = R_X(\tau) \tag{6.74}$$

B. Fourier Series:

Let $X(t)$ be a WSS random process with periodic $R_X(\tau)$ having period T. Expanding $R_X(\tau)$ into a Fourier series, we obtain

$$R_X(\tau) = \sum_{n=-\infty}^{\infty} c_n e^{jn\omega_0\tau} \qquad \omega_0 = 2\pi/T \tag{6.75}$$

where

$$c_n = \frac{1}{T} \int_0^T R_X(\tau) e^{-jn\omega_0\tau} \, d\tau \tag{6.76}$$

Let $\hat{X}(t)$ be expressed as

$$\hat{X}(t) = \sum_{n=-\infty}^{\infty} X_n e^{jn\omega_0 t} \qquad \omega_0 = 2\pi/T \tag{6.77}$$

where X_n are r.v.'s given by

$$X_n = \frac{1}{T} \int_0^T X(t) e^{-jn\omega_0 t} \, dt \tag{6.78}$$

Note that, in general, X_n are complex-valued r.v.'s. For complex-valued r.v.'s, the correlation between two r.v.'s X and Y is defined by $E(XY^*)$. Then $\hat{X}(t)$ is called the m.s. Fourier series of $X(t)$ such that (Prob. 6.34)

$$E\{|X(t) - \hat{X}(t)|^2\} = 0 \tag{6.79}$$

Furthermore, we have (Prob. 6.33)

$$E(X_n) = \mu_X \delta(n) = \begin{cases} \mu_X & n = 0 \\ 0 & n \neq 0 \end{cases} \tag{6.80}$$

$$E(X_n X_m^*) = c_n \delta(n - m) = \begin{cases} c_n & n = m \\ 0 & n \neq m \end{cases} \tag{6.81}$$

C. Karhunen-Loéve Expansion

Consider a random process $X(t)$ which is not periodic. Let $\hat{X}(t)$ be expressed as

$$\hat{X}(t) = \sum_{n=1}^{\infty} X_n \phi_n(t) \qquad 0 < t < T \tag{6.82}$$

where a set of functions $\{\phi_n(t)\}$ is orthonormal on an interval $(0, T)$ such that

$$\int_0^T \phi_n(t)\phi_m^*(t)\,dt = \delta(n - m) \tag{6.83}$$

and X_n are r.v.'s given by

$$X_n = \int_0^T X(t)\phi_n^*(t)\,dt \tag{6.84}$$

Then $\hat{X}(t)$ is called the *Karhunen-Loéve expansion* of $X(t)$ such that (Prob. 6.38)

$$E\{\,|X(t) - \hat{X}(t)|^2\} = 0 \tag{6.85}$$

Let $R_X(t, s)$ be the autocorrelation function of $X(t)$, and consider the following integral equation:

$$\int_0^T R_X(t, s)\phi_n(s)\,ds = \lambda_n\,\phi_n(t) \qquad 0 < t < T \tag{6.86}$$

where λ_n and $\phi_n(t)$ are called the eigenvalues and the corresponding eigenfunctions of the integral equation (6.86). It is known from the theory of integral equations that if $R_X(t, s)$ is continuous, then $\phi_n(t)$ of Eq. (6.86) are orthonormal as in Eq. (6.83), and they satisfy the following identity:

$$R_X(t, s) = \sum_{n=1}^\infty \lambda_n\phi_n(t)\phi_n^*(s) \tag{6.87}$$

which is known as *Mercer's theorem*.

With the above results, we can show that Eq. (6.85) is satisfied and the coefficient X_n are orthogonal r.v.'s (Prob. 6.37); that is,

$$E(X_n X_m^*) = \lambda_n\,\delta(n - m) = \begin{cases} \lambda_n & n = m \\ 0 & n \neq m \end{cases} \tag{6.88}$$

6.7 Fourier Transform of Random Processes

A. Continuous-Time Random Processes:

The Fourier transform of a continuous-time random process $X(t)$ is a random process $\tilde{X}(\omega)$ given by

$$\tilde{X}(\omega) = \int_{-\infty}^\infty X(t)e^{-j\omega t}\,dt \tag{6.89}$$

which is the stochastic integral, and the integral is interpreted as a m.s. limit; that is,

$$E\left\{\left|\tilde{X}(\omega) - \int_{-\infty}^\infty X(t)e^{-j\omega t}\,dt\right|^2\right\} = 0 \tag{6.90}$$

Note that $\tilde{X}(\omega)$ is a complex random process. Similarly, the inverse Fourier transform

$$X(t) = \frac{1}{2\pi}\int_{-\infty}^\infty \tilde{X}(\omega)e^{j\omega t}\,d\omega \tag{6.91}$$

is also a stochastic integral and should also be interpreted in the m.s. sense. The properties of continuous-time Fourier transforms (Appendix B) also hold for random processes (or random signals). For instance, if $Y(t)$ is the output of a continuous-time LTI system with input $X(t)$, then

$$\tilde{Y}(\omega) = \tilde{X}(\omega)H(\omega) \tag{6.92}$$

where $H(\omega)$ is the frequency response of the system.

Let $\tilde{R}_X(\omega_1, \omega_2)$ be the two-dimensional Fourier transform of $R_X(t, s)$; that is,

$$\tilde{R}_X(\omega_1, \omega_2) = \int_{-\infty}^{\infty}\int_{-\infty}^{\infty} R_X(t, s)e^{-j(\omega_1 t + \omega_2 s)}\, dt\, ds \tag{6.93}$$

Then the autocorrelation function of $\tilde{X}(\omega)$ is given by (Prob. 6.41)

$$R_{\tilde{X}}(\omega_1, \omega_2) = E[\tilde{X}(\omega_1)\,\tilde{X}^*(\omega_2)] = \tilde{R}_X(\omega_1, -\omega_2) \tag{6.94}$$

If $X(t)$ is real, then

$$E[\tilde{X}(\omega_1)\,\tilde{X}(\omega_2)] = \tilde{R}_X(\omega_1, \omega_2) \tag{6.95}$$
$$\tilde{X}(-\omega) = \tilde{X}^*(\omega) \tag{6.96}$$
$$\tilde{R}_X(-\omega_1, -\omega_2) = \tilde{R}_X^*(\omega_1, \omega_2) \tag{6.97}$$

If $X(t)$ is a WSS random process with autocorrelation function $R_X(t, s) = R_X(t - s) = R_X(\tau)$ and power spectral density $S_X(\omega)$, then (Prob. 6.42)

$$\tilde{R}_X(\omega_1, \omega_2) = 2\pi S_X(\omega_1)\,\delta(\omega_1 + \omega_2) \tag{6.98}$$
$$R_{\tilde{X}}(\omega_1, \omega_2) = 2\pi S_X(\omega_1)\delta(\omega_1 - \omega_2) \tag{6.99}$$

Equation (6.99) shows that the Fourier transform of a WSS random process is nonstationary white noise.

B. Discrete-Time Random Processes:

The Fourier transform of a discrete-time random process $X(n)$ is a random process $\tilde{X}(\Omega)$ given by (in m.s. sense)

$$\tilde{X}(\Omega) = \sum_{n=-\infty}^{\infty} X(n)e^{-j\Omega n} \tag{6.100}$$

Similarly, the inverse Fourier transform

$$X(n) = \frac{1}{2\pi}\int_{-\pi}^{\pi} \tilde{X}(\Omega)e^{j\Omega n}\, d\Omega \tag{6.101}$$

should also be interpreted in the m.s. sense. Note that $\tilde{X}(\Omega) + 2\pi) = \tilde{X}(\Omega)$ and the properties of discrete-time Fourier transforms (Appendix B) also hold for discrete-time random signals. For instance, if $Y(n)$ is the output of a discrete-time LTI system with input $X(n)$, then

$$\tilde{Y}(\Omega) = \tilde{X}(\Omega)H(\Omega) \tag{6.102}$$

where $H(\Omega)$ is the frequency response of the system.

Let $\tilde{R}_X(\Omega_1, \Omega_2)$ be the two-dimensional Fourier transform of $R_X(n, m)$:

$$\tilde{R}_X(\Omega_1, \Omega_2) = \sum_{n=-\infty}^{\infty}\sum_{m=-\infty}^{\infty} R_X(n, m)e^{-j(\Omega_1 n + \Omega_2 m)} \tag{6.103}$$

Then the autocorrelation function of $\tilde{X}(\Omega)$ is given by (Prob. 6.44)

$$R_{\tilde{X}}(\Omega_1, \Omega_2) = E[\tilde{X}(\Omega_1)\,\tilde{X}^*(\Omega_2)] = \tilde{R}_X(\Omega_1, -\Omega_2) \tag{6.104}$$

If $X(n)$ is a WSS random process with autocorrelation function $R_X(n, m) = R_X(n - m) = R_X(k)$ and power spectral density $S_X(\Omega)$, then

$$\tilde{R}_X(\Omega_1, \Omega_2) = 2\pi S_X(\Omega_1)\delta(\Omega_1 + \Omega_2) \tag{6.105}$$

$$R_{\tilde{X}}(\Omega_1, \Omega_2) = 2\pi S_X(\Omega_1)\delta(\Omega_1 - \Omega_2) \tag{6.106}$$

Equation (6.106) shows that the Fourier transform of a discrete-time WSS random process is nonstationary white noise.

SOLVED PROBLEMS

Continuity, Differentiation, Integration

6.1. Show that the random process $X(t)$ is m.s. continuous if and only if its autocorrelation function $R_X(t, s)$ is continuous.

We can write

$$
\begin{aligned}
E\{[X(t + \varepsilon) - X(t)]^2\} &= E[X^2(t + \varepsilon) - 2X(t + \varepsilon)X(t) + X^2(t)] \\
&= R_X(t + \varepsilon, t + \varepsilon) - 2R_X(t + \varepsilon, t) + R_X(t, t) \tag{6.107}
\end{aligned}
$$

Thus, if $R_X(t, s)$ is continuous, then

$$\lim_{\varepsilon \to 0} E\{[X(t + \varepsilon) - X(t)]^2\} = \lim_{\varepsilon \to 0}\{R_X(t + \varepsilon, t + \varepsilon) - 2R_X(t + \varepsilon, t) + R_X(t, t)\} = 0$$

and $X(t)$ is m.s. continuous. Next, consider

$$
\begin{aligned}
R_X(t + \varepsilon_1, t + \varepsilon_2) - R_X(t, t) = {} &E\{[X(t + \varepsilon_1) - X(t)][X(t + \varepsilon_2) - X(t)]\} \\
&+ E\{[X(t + \varepsilon_1) - X(t)]X(t)\} + E\{[X(t + \varepsilon_2) - X(t)]X(t)\}
\end{aligned}
$$

Applying Cauchy-Schwarz inequality (3.97) (Prob. 3.35), we obtain

$$
\begin{aligned}
R_X(t + \varepsilon_1, t + \varepsilon_2) - R_X(t, t) \leq {} &(E\{[X(t + \varepsilon_1) - X(t)]^2\}E\{[X(t + \varepsilon_2) - X(t)]^2\})^{1/2} \\
&+ (E\{[X(t + \varepsilon_1) - X(t)]^2\}E[X^2(t)])^{1/2} + (E\{[X(t + \varepsilon_2) - X(t)]^2\}E[X^2(t)])^{1/2}
\end{aligned}
$$

Thus, if $X(t)$ is m.s. continuous, then by Eq. (6.1) we have

$$\lim_{\varepsilon_1, \varepsilon_2 \to 0} R_X(t + \varepsilon_1, t + \varepsilon_2) - R_X(t, t) = 0$$

that is, $R_X(t, s)$ is continuous. This completes the proof.

6.2. Show that a WSS random process $X(t)$ is m.s. continuous if and only if its autocorrelation function $R_X(\tau)$ is continuous at $\tau = 0$.

If $X(t)$ is WSS, then Eq. (6.107) becomes

$$E\{[X(t + \varepsilon) - X(t)]^2\} = 2[R_X(0) - R_X(\varepsilon)] \tag{6.108}$$

Thus if $R_X(\tau)$ is continuous at $\tau = 0$, that is,

$$\lim_{\varepsilon \to 0}[R_X(\varepsilon) - R_X(0)] = 0$$

then

$$\lim_{\varepsilon \to 0} E\{[X(t + \varepsilon) - X(t)]^2\} = 0$$

that is, $X(t)$ is m.s. continuous. Similarly, we can show that if $X(t)$ is m.s. continuous, then by Eq. (6.108), $R_X(\tau)$ is continuous at $\tau = 0$.

6.3. Show that if $X(t)$ is m.s. continuous, then its mean is continuous; that is,

$$\lim_{\varepsilon \to 0} \mu_X(t + \varepsilon) = \mu_X(t)$$

We have

$$\text{Var}[X(t + \varepsilon) - X(t)] = E\{[X(t + \varepsilon) - X(t)]^2\} - \{E[X(t + \varepsilon) - X(t)]\}^2 \geq 0$$

Thus, $\quad E\{[X(t + \varepsilon) - X(t)]^2\} \geq \{E[X(t + \varepsilon) - X(t)]\}^2 = [\mu_X(t + \varepsilon) - \mu_X(t)]^2$

If $X(t)$ is m.s. continuous, then as $\varepsilon \to 0$, the left-hand side of the above expression approaches zero. Thus,

$$\lim_{\varepsilon \to 0} [\mu_X(t + \varepsilon) - \mu_X(t)] = 0 \quad \text{or} \quad \lim_{\varepsilon \to 0} [\mu_X(t + \varepsilon) = \mu_X(t)$$

6.4. Show that the Wiener process $X(t)$ is m.s. continuous.

From Eq. (5.64), the autocorrelation function of the Wiener process $X(t)$ is given by

$$R_X(t, s) = \sigma^2 \min(t, s)$$

Thus, we have

$$|R_X(t + \varepsilon_1, t + \varepsilon_2) - R_X(t, t)| = \sigma^2 |\min(t + \varepsilon_1, t + \varepsilon_2) - t| \leq \sigma^2 \max(\varepsilon_1, \varepsilon_2)$$

Since $\quad \lim_{\varepsilon_1, \varepsilon_2 \to 0} \max(\varepsilon_1, \varepsilon_2) = 0$

$R_X(t, s)$ is continuous. Hence, the Wiener process $X(t)$ is m.s. continuous.

6.5. Show that every m.s. continuous random process is continuous in probability.

A random process $X(t)$ is continuous in probability if, for every t and $a > 0$ (see Prob. 5.62),

$$\lim_{\varepsilon \to 0} P\{|X(t + \varepsilon) - X(t)| > a\} = 0$$

Applying Chebyshev inequality (2.116) (Prob. 2.39), we have

$$P\{|X(t + \varepsilon) - X(t)| > a\} \leq \frac{E[|X(t + \varepsilon) - X(t)|^2]}{a^2}$$

Now, if $X(t)$ is m.s. continuous, then the right-hand side goes to 0 as $\varepsilon \to 0$, which implies that the left-hand side must also go to 0 as $\varepsilon \to 0$. Thus, we have proved that if $X(t)$ is m.s. continuous, then it is also continuous in probability.

6.6. Show that a random process $X(t)$ has a m.s. derivative $X'(t)$ if $\partial^2 R_X(t, s)/\partial t\, \partial s$ exists at $s = t$.

Let $\qquad\qquad\qquad\qquad Y(t; \varepsilon) = \dfrac{X(t + \varepsilon) - X(t)}{\varepsilon} \qquad\qquad\qquad\qquad$ (6.109)

By the Cauchy criterion (see the note at the end of this solution), the m.s. derivative $X'(t)$ exists if

$$\lim_{\varepsilon_1, \varepsilon_2 \to 0} E\{[Y(t; \varepsilon_2) - Y(t; \varepsilon_1)]^2\} = 0 \qquad\qquad (6.110)$$

Now $\qquad E\{[Y(t; \varepsilon_2) - Y(t; \varepsilon_1)]^2\} = E[Y^2(t; \varepsilon_2) - 2Y(t; \varepsilon_2)Y(t; \varepsilon_1) + Y^2(t; \varepsilon_1)]$

$$= E[Y^2(t; \varepsilon_2)] - 2E[Y(t; \varepsilon_2)Y(t; \varepsilon_1)] + E[Y^2(t; \varepsilon_1)] \qquad (6.111)$$

and
$$E[Y(t; \varepsilon_2)Y(t; \varepsilon_1)] = \frac{1}{\varepsilon_1 \varepsilon_2} E\{[X(t + \varepsilon_2) - X(t)][X(t + \varepsilon_1) - X(t)]\}$$

$$= \frac{1}{\varepsilon_1 \varepsilon_2}[R_X(t + \varepsilon_2, t + \varepsilon_1) - R_X(t + \varepsilon_2, t) - R_X(t, t + \varepsilon_1) + R_X(t, t)]$$

$$= \frac{1}{\varepsilon_2}\left\{\frac{R_X(t + \varepsilon_2, t + \varepsilon_1) - R_X(t + \varepsilon_2, t)}{\varepsilon_1} - \frac{R_X(t, t + \varepsilon_1) - R_X(t, t)}{\varepsilon_1}\right\}$$

Thus,
$$\lim_{\varepsilon_1, \varepsilon_2 \to 0} E[Y(t; \varepsilon_2)Y(t; \varepsilon_1)] = \frac{\partial^2 R_X(t, s)}{\partial t \, \partial s}\bigg|_{s=t} = R_2 \tag{6.112}$$

provided $\partial^2 R_X(t, s)/\partial t \, \partial s$ exists at $s = t$. Setting $\varepsilon_1 = \varepsilon_2$ in Eq. (6.112), we get

$$\lim_{\varepsilon_1 \to 0} E[Y^2(t; \varepsilon_1)] = \lim_{\varepsilon_2 \to 0} E[Y^2(t; \varepsilon_2)] = R_2$$

and by Eq. (6.111), we obtain

$$\lim_{\varepsilon_1, \varepsilon_2 \to 0} E\{[Y(t; \varepsilon_2) - Y(t; \varepsilon_1)]^2\} = R_2 - 2R_2 + R_2 = 0 \tag{6.113}$$

Thus, we conclude that $X(t)$ has a m.s. derivative $X'(t)$ if $\partial^2 R_X(t, s)/\partial t \, \partial s$ exists at $s = t$. If $X(t)$ is WSS, then the above conclusion is equivalent to the existence of $\partial^2 R_X(\tau)/\partial^2 \tau$ at $\tau = 0$.

Note: In real analysis, a function $g(\varepsilon)$ of some parameter ε converges to a finite value if

$$\lim_{\varepsilon_1, \varepsilon_2 \to 0} [g(\varepsilon_2) - g(\varepsilon_1)] = 0$$

This is known as the *Cauchy criterion*.

6.7.   Suppose a random process $X(t)$ has a m.s. derivative $X'(t)$.

   (*a*)   Find $E[X'(t)]$.

   (*b*)   Find the cross-correlation function of $X(t)$ and $X'(t)$.

   (*c*)   Find the autocorrelation function of $X'(t)$.

   (*a*)   We have

$$E[X'(t)] = E\left[\text{l.i.m.}_{\varepsilon \to 0} \frac{X(t + \varepsilon) - X(t)}{\varepsilon}\right]$$

$$= \lim_{\varepsilon \to 0} E\left[\frac{X(t + \varepsilon) - X(t)}{\varepsilon}\right]$$

$$= \lim_{\varepsilon \to 0} \frac{\mu_X(t + \varepsilon) - \mu_X(t)}{\varepsilon} = \mu_X'(t) \tag{6.114}$$

   (*b*)   From Eq. (6.17), the cross-correlation function of $X(t)$ and $X'(t)$ is

$$R_{XX'}(t, s) = E[X(t)X'(s)] = E\left[X(t)\text{l.i.m.}_{\varepsilon \to 0} \frac{X(s + \varepsilon) - X(s)}{\varepsilon}\right]$$

$$= \lim_{\varepsilon \to 0} \frac{E[X(t)X(s + \varepsilon)] - E[X(t)X(s)]}{\varepsilon}$$

$$= \lim_{\varepsilon \to 0} \frac{R_X(t, s + \varepsilon) - R_X(t, s)}{\varepsilon} = \frac{\partial R_X(t, s)}{\partial s} \tag{6.115}$$

(c)   Using Eq. (6.115), the autocorrelation function of $X'(t)$ is

$$R_{X'}(t, s) = E[X'(t)X'(s)] = E\left\{\left[\underset{\varepsilon \to 0}{\text{l.i.m.}} \frac{X(t+\varepsilon) - X(t)}{\varepsilon}\right]X'(s)\right\}$$

$$= \lim_{\varepsilon \to 0} \frac{E[X(t+\varepsilon)X'(s)] - E[X(t)X'(s)]}{\varepsilon}$$

$$= \lim_{\varepsilon \to 0} \frac{R_{XX'}(t+\varepsilon, s) - R_{XX'}(t, s)}{\varepsilon}$$

$$= \frac{\partial R_{XX'}(t, s)}{\partial t} = \frac{\partial^2 R_X(t, s)}{\partial t \, \partial s} \tag{6.116}$$

6.8.   If $X(t)$ is a WSS random process and has a m.s. derivative $X'(t)$, then show that

(a) $R_{XX'}(\tau) = \dfrac{d}{d\tau} R_X(\tau)$ $\tag{6.117}$

(b) $R_{X'}(\tau) = -\dfrac{d^2}{d\tau^2} R_X(\tau)$ $\tag{6.118}$

(a)   For a WSS process $X(t)$, $R_X(t, s) = R_X(s - t)$. Thus, setting $s - t = \tau$ in Eq. (6.115) of Prob. 6.7, we obtain $\partial R_X(s - t)/\partial s = dR_X(\tau)/d\tau$ and

$$R_{XX'}(t, t + \tau) = R_{XX'}(\tau) = \frac{dR_X(\tau)}{d\tau}$$

(b)   Now $\partial R_X(s - t)/\partial t = -dR_X(\tau)/d\tau$. Thus, $\partial^2 R_X(s - t)/\partial t \, \partial s = d^2 R_X(\tau)/d\tau^2$, and by Eq. (6.116) of Prob. 6.7, we have

$$R_{X'}(t, t + \tau) = R_{X'}(\tau) = -\frac{d^2}{d\tau^2} R_X(\tau)$$

6.9.   Show that the Wiener process $X(t)$ does not have a m.s. derivative.

From Eq. (5.64), the autocorrelation function of the Wiener process $X(t)$ is given by

$$R_X(t, s) = \sigma^2 \min(t, s) = \begin{cases} \sigma^2 s & t > s \\ \sigma^2 t & t < s \end{cases}$$

Thus, $\dfrac{\partial}{\partial s} R_X(t, s) = \sigma^2 u(t - s) = \begin{cases} \sigma^2 & t > s \\ 0 & t < s \end{cases}$ $\tag{6.119}$

where $u(t - s)$ is a unit step function defined by

$$u(t - s) = \begin{cases} 1 & t > s \\ 0 & t < s \end{cases}$$

and it is not continuous at $s = t$ (Fig. 6-2). Thus, $\partial^2 R_X(t, s)/\partial t \, \partial s$ does not exist at $s = t$, and the Wiener process $X(t)$ does not have a m.s. derivative.

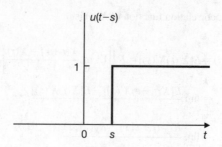

Fig. 6-2 Shifted unit step function.

Note that although a m.s. derivative does not exist for the Wiener process, we can define a generalized derivative of the Wiener process (see Prob. 6.20).

6.10. Show that if $X(t)$ is a normal random process for which the m.s. derivative $X'(t)$ exists, then $X'(t)$ is also a normal random process.

Let $X(t)$ be a normal random process. Now consider

$$Y_\varepsilon(t) = \frac{X(t+\varepsilon) - X(t)}{\varepsilon}$$

Then, n r.v.'s $Y_\varepsilon(t_1), Y_\varepsilon(t_2), ..., Y_\varepsilon(t_n)$ are given by a linear transformation of the jointly normal r.v.'s $X(t_1), X(t_1 + \varepsilon)$, $X(t_2), X(t_2 + \varepsilon), ..., X(t_n), X(t_n + \varepsilon)$. It then follows by the result of Prob. 5.60 that $Y_\varepsilon(t_1), Y_\varepsilon(t_2), ..., Y_\varepsilon(t_n)$ are jointly normal r.v.'s, and hence $Y_\varepsilon(t)$ is a normal random process. Thus, we conclude that the m.s. derivative $X'(t)$, which is the limit of $Y_\varepsilon(t)$ as $\varepsilon \to 0$, is also a normal random process, since m.s. convergence implies convergence in probability (see Prob. 6.5).

6.11. Show that the m.s. integral of a random process $X(t)$ exists if the following integral exists:

$$\int_{t_0}^t \int_{t_0}^t R_X(\alpha, \beta)\, d\alpha\, d\beta$$

A m.s. integral of $X(t)$ is defined by [Eq. (6.8)]

$$Y(t) = \int_{t_0}^t X(\alpha)\, d\alpha = \underset{\Delta t_i \to 0}{\text{l.i.m.}} \sum_i X(t_i)\, \Delta t_i$$

Again using the Cauchy criterion, the m.s. integral $Y(t)$ of $X(t)$ exists if

$$\lim_{\Delta t_i, \Delta t_k \to 0} E\left\{ \left[\sum_i X(t_i)\, \Delta t_i - \sum_k X(t_k)\, \Delta t_k \right]^2 \right\} = 0 \tag{6.120}$$

As in the case of the m.s. derivative [Eq. (6.111)], expanding the square, we obtain

$$E\left\{ \left[\sum_i X(t_i)\, \Delta t_i - \sum_k X(t_k)\, \Delta t_k \right]^2 \right\}$$

$$= E\left[\sum_i \sum_k X(t_i)X(t_k)\, \Delta t_i\, \Delta t_k + \sum_i \sum_k X(t_i)X(t_k)\, \Delta t_i\, \Delta t_k - 2 \sum_i \sum_k X(t_i)X(t_k)\, \Delta t_i\, \Delta t_k \right]$$

$$= \sum_i \sum_k R_X(t_i, t_k)\, \Delta t_i\, \Delta t_k + \sum_i \sum_k R_X(t_i, t_k)\, \Delta t_i\, \Delta t_k - 2 \sum_i \sum_k R_X(t_i, t_k)\, \Delta t_i\, \Delta t_k$$

and Eq. (6.120) holds if

$$\lim_{\Delta t_i, \Delta t_k \to 0} \sum_i \sum_k R_X(t_i, t_k)\, \Delta t_i\, \Delta t_k$$

exists, or, equivalently,

$$\int_{t_0}^{t}\int_{t_0}^{t} R_X(\alpha, \beta)\, d\alpha\, d\beta$$

exists.

6.12. Let $X(t)$ be the Wiener process with parameter σ^2. Let

$$Y(t) = \int_0^t X(\alpha)\, d\alpha$$

(a) Find the mean and the variance of $Y(t)$.

(b) Find the autocorrelation function of $Y(t)$.

(a) By assumption 3 of the Wiener process (Sec. 5.7), that is, $E[X(t)] = 0$, we have

$$E[Y(t)] = E\left[\int_0^t X(\alpha)\, d\alpha\right] = \int_0^t E[X(\alpha)]\, d\alpha = 0 \tag{6.121}$$

Then

$$\mathrm{Var}[Y(t)] = E[Y^2(t)] = \int_0^t\int_0^t E[X(\alpha)X(\beta)]\, d\alpha\, d\beta$$

$$= \int_0^t\int_0^t R_X(\alpha, \beta)\, d\alpha\, d\beta$$

By Eq. (5.64), $R_X(\alpha, \beta) = \sigma^2 \min(\alpha, \beta)$; thus, referring to Fig. 6-3, we obtain

$$\mathrm{Var}[Y(t)] = \sigma^2 \int_0^t\int_0^t \min(\alpha, \beta)\, d\alpha\, d\beta$$

$$= \sigma^2 \int_0^t d\beta \int_0^\beta \alpha\, d\alpha + \sigma^2 \int_0^t d\alpha \int_0^\alpha \beta\, d\beta = \frac{\sigma^2 t^3}{3} \tag{6.122}$$

(b) Let $t > s \geq 0$ and write

$$Y(t) = \int_0^s X(\alpha)\, d\alpha + \int_s^t [X(\alpha) - X(s)]\, d\alpha + (t - s)X(s)$$

$$= Y(s) + \int_s^t [X(\alpha) - X(s)]\, d\alpha + (t - s)X(s)$$

Then, for $t > s \geq 0$,

$$R_Y(t, s) = E[Y(t)Y(s)]$$

$$= E[Y^2(s)] + \int_s^t E\{[X(\alpha) - X(s)]Y(s)\}\, d\alpha + (t - s)E[X(s)Y(s)] \tag{6.123}$$

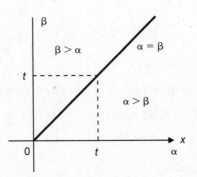

Fig. 6-3

Now by Eq. (6.122),

$$E[Y^2(s)] = \text{Var}[Y(s)] = \frac{\sigma^2 s^3}{3}$$

Using assumptions 1, 3, and 4 of the Wiener process (Sec. 5.7), and since $s \le \alpha \le t$, we have

$$\int_s^t E\{[X(\alpha) - X(s)]Y(s)\} \, d\alpha = \int_s^t E\left\{[X(\alpha) - X(s)]\int_0^s X(\beta)\, d\beta\right\} d\alpha$$

$$= \int_s^t \int_0^s E\{[X(\alpha) - X(s)][X(\beta) - X(0)]\} \, d\beta \, d\alpha$$

$$= \int_s^t \int_0^s E[X(\alpha) - X(s)]E[X(s) - X(0)] \, d\beta \, d\alpha = 0$$

Finally, for $0 \le \beta \le s$,

$$(t - s)E[X(s)Y(s)] = (t - s)\int_0^s E[X(s)]X(\beta)] \, d\beta$$

$$= (t - s)\int_0^s R_X(s, \beta) \, d\beta = (t - s)\int_0^s \sigma^2 \min(s, \beta) \, d\beta$$

$$= \sigma^2(t - s)\int_0^s \beta \, d\beta = \sigma^2(t - s)\frac{s^2}{2}$$

Substituting these results into Eq. (6.123), we get

$$R_Y(t, s) = \frac{\sigma^2 s^3}{3} + \sigma^2(t - s)\frac{s^2}{2} = \frac{1}{6}\sigma^2 s^3(3t - s)$$

Since $R_Y(t, s) = R_Y(s, t)$, we obtain

$$R_Y(t, s) = \begin{cases} \dfrac{1}{6}\sigma^2 s^2(3t - s) & t > s \ge 0 \\[2mm] \dfrac{1}{6}\sigma^2 t^2(3s - t) & s > t \ge 0 \end{cases} \tag{6.124}$$

Power Spectral Density

6.13. Verify Eqs. (6.13) and (6.14).

From Eq. (6.12),

$$R_X(\tau) = E[X(t)X(t + \tau)]$$

Setting $t + \tau = s$, we get

$$R_X(\tau) = E[X(s - \tau)X(s)] = E[X(s)X(s - \tau)] = R_X(-\tau)$$

Next, we have

$$E\{[X(t) \pm X(t + \tau)]^2\} \ge 0$$

Expanding the square, we have

$$E[X^2(t) \pm 2X(t)X(t + \tau) + X^2(r + \tau)] \ge 0$$

or $$E[X^2(t)] \pm 2E[X(t)X(t + \tau)] + E[X^2(t + \tau)] \ge 0$$

Thus, $$2R_X(0) \pm 2R_X(\tau) \ge 0$$

from which we obtain Eq. (6.14); that is,

$$R_X(0) \ge | R_X(\tau) |$$

6.14. Verify Eqs. (6.18) to (6.20).

By Eq. (6.17),

$$R_{XY}(-\tau) = E[X(t)Y(t - \tau)]$$

Setting $t - \tau = s$, we get

$$R_{XY}(-\tau) = E[X(s + \tau)Y(s)] = E[Y(s)X(s + \tau)] = R_{YX}(\tau)$$

Next, from the Cauchy-Schwarz inequality, Eq. (3.97) (Prob. 3.35), it follows that

$$\{E[X(t)Y(t + \tau)]\}^2 \le E[X^2(t)]E[Y^2(t + \tau)]$$

or

$$[R_{XY}(\tau)]^2 \le R_X(0)R_Y(0)$$

from which we obtain Eq. (6.19); that is,

$$|R_{XY}(\tau)| \le \sqrt{R_X(0)R_Y(0)}$$

Now

$$E\{[X(t) - Y(t + \tau)]^2\} \ge 0$$

Expanding the square, we have

$$E[X^2(t) - 2X(t)Y(t + \tau) + Y^2(t + \tau)] \ge 0$$

or

$$E[X^2(t)] - 2E[X(t)Y(t + \tau)] + E[Y^2(t + \tau)] \ge 0$$

Thus,

$$R_X(0) - 2R_{XY}(\tau) + R_Y(0) \ge 0$$

from which we obtain Eq. (6.20); that is,

$$R_{XY}(\tau) \le \frac{1}{2}[R_X(0) + R_Y(0)]$$

6.15. Two random processes $X(t)$ and $Y(t)$ are given by

$$X(t) = A \cos(\omega t + \Theta) \qquad Y(t) = A \sin(\omega t + \Theta)$$

where A and ω are constants and Θ is a uniform r.v. over $(0, 2\pi)$. Find the cross-correlation function of $X(t)$ and $Y(t)$ and verify Eq. (6.18).

From Eq. (6.17), the cross-correlation function of $X(t)$ and $Y(t)$ is

$$\begin{aligned}
R_{XY}(t, t + \tau) &= E[X(t)Y(t + \tau)] \\
&= E\{A^2 \cos(\omega t + \Theta)\sin[\omega(t + \tau) + \Theta]\} \\
&= \frac{A^2}{2} E[\sin(2\omega t + \omega\tau + 2\Theta) - \sin(-\omega\tau)] \\
&= \frac{A^2}{2} \sin\omega\tau = R_{XY}(\tau)
\end{aligned} \qquad (6.125)$$

Similarly,

$$\begin{aligned}
R_{YX}(t, t + \tau) &= E[Y(t)X(t + \tau)] \\
&= E\{A^2 \sin(\omega t + \Theta)\cos[\omega(t + \tau) + \Theta]\} \\
&= \frac{A^2}{2} E[\sin(2\omega t + \omega\tau + 2\Theta) + \sin(-\omega\tau)] \\
&= -\frac{A^2}{2} \sin\omega\tau = R_{YX}(\tau)
\end{aligned} \qquad (6.126)$$

From Eqs. (6.125) and (6.126), we see that

$$R_{XY}(-\tau) = \frac{A^2}{2}\sin \omega(-\tau) = -\frac{A^2}{2}\sin \omega\tau = R_{YX}(\tau)$$

which verifies Eq. (6.18).

6.16. Show that the power spectrum of a (real) random process $X(t)$ is real and verify Eq. (6.26).

From Eq. (6.23) and expanding the exponential, we have

$$S_X(\omega) = \int_{-\infty}^{\infty} R_X(\tau)\, e^{-j\omega\tau}\, d\tau$$

$$= \int_{-\infty}^{\infty} R_X(\tau)(\cos \omega\tau - j\sin \omega\tau)\, d\tau$$

$$= \int_{-\infty}^{\infty} R_X(\tau)\cos \omega\tau\, d\tau - j\int_{-\infty}^{\infty} R_X(\tau)\sin \omega\tau\, d\tau \qquad (6.127)$$

Since $R_X(-\tau) = R_X(\tau)$, $R_X(\tau)\cos \omega\tau$ is an even function of τ and $R_X(\tau)\sin \omega\tau$ is an odd function of τ, and hence the imaginary term in Eq. (6.127) vanishes and we obtain

$$S_X(\omega) = \int_{-\infty}^{\infty} R_X(\tau)\cos \omega\tau\, d\tau \qquad (6.128)$$

which indicates that $S_X(\omega)$ is real. Since $\cos(-\omega\tau) = \cos(\omega\tau)$, it follows that

$$S_X(-\omega) = S_X(\omega)$$

which indicates that the power spectrum of a real random process $X(t)$ is an even function of frequency.

6.17. Consider the random process

$$Y(t) = (-1)^{X(t)}$$

where $X(t)$ is a Poisson process with rate λ. Thus, $Y(t)$ starts at $Y(0) = 1$ and switches back and forth from $+1$ to -1 at random Poisson times T_i, as shown in Fig. 6-4. The process $Y(t)$ is known as the *semirandom telegraph signal* because its initial value $Y(0) = 1$ is not random.

(a) Find the mean of $Y(t)$.

(b) Find the autocorrelation function of $Y(t)$.

(a) We have

$$Y(t) = \begin{cases} 1 & \text{if } X(t) \text{ is even} \\ -1 & \text{if } X(t) \text{ is odd} \end{cases}$$

Thus, using Eq. (5.55), we have

$$P[Y(t) = 1] = P[X(t) = \text{even integer}]$$

$$= e^{-\lambda t}\left[1 + \frac{(\lambda t)^2}{2!} + \cdots\right] = e^{-\lambda t}\cosh \lambda t$$

$$P[Y(t) = -1] = P[X(t) = \text{odd integer}]$$

$$= e^{-\lambda t}\left[\lambda t + \frac{(\lambda t)^3}{3!} + \cdots\right] = e^{-\lambda t}\sinh \lambda t$$

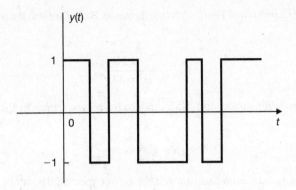

Fig. 6-4 Semirandom telegraph signal.

Hence, $$\mu_Y(t) = E[Y(t)] = (1)P[Y(t) = 1] + (-1)P[Y(t) = -1]$$
$$= e^{-\lambda t}(\cosh \lambda t - \sinh \lambda t) = e^{-2\lambda t} \qquad (6.129)$$

(b) Similarly, since $Y(t)Y(t + \tau) = 1$ if there are an even number of events in $(t, t + \tau)$ for $\tau > 0$ and $Y(t)Y(t + \tau) = -1$ if there are an odd number of events, then for $t > 0$ and $t + \tau > 0$,

$$R_Y(t, t + \tau) = E[Y(t)]Y(t + \tau)]$$

$$= (1) \sum_{n \text{ even}} e^{-\lambda \tau} \frac{(-\lambda \tau)^n}{n!} + (-1) \sum_{n \text{ odd}} e^{-\lambda \tau} \frac{(\lambda \tau)^n}{n!}$$

$$= e^{-\lambda \tau} \sum_{n=0}^{\infty} \frac{(-\lambda \tau)^n}{n!} = e^{-\lambda \tau} e^{-\lambda \tau} = e^{-2\lambda \tau}$$

which indicates that $R_Y(t, t + \tau) = R_Y(\tau)$, and by Eq. (6.13),

$$R_Y(\tau) = e^{-2\lambda |\tau|} \qquad (6.130)$$

Note that since $E[Y(t)]$ is not a constant, $Y(t)$ is not WSS.

6.18. Consider the random process

$$Z(t) = AY(t)$$

where $Y(t)$ is the semirandom telegraph signal of Prob. 6.17 and A is a r.v. independent of $Y(t)$ and takes on the values ± 1 with equal probability. The process $Z(t)$ is known as the *random telegraph signal*.

(a) Show that $Z(t)$ is WSS.

(b) Find the power spectral density of $Z(t)$.

(a) Since $E(A) = 0$ and $E(A^2) = 1$, the mean of $Z(t)$ is

$$\mu_Z(t) = E[Z(t)] = E(A)E[Y(t)] = 0 \qquad (6.131)$$

and the autocorrelation of $Z(t)$ is

$$R_Z(t, t + \tau) = E[A^2 Y(t)Y(t + \tau)] = E(A^2) E[Y(t)Y(t + \tau)] = R_Y(t, t + \tau)$$

Thus, using Eq. (6.130), we obtain

$$R_Z(t, t + \tau) = R_Z(\tau) = e^{-2\lambda |\tau|} \qquad (6.132)$$

Thus, we see that $Z(t)$ is WSS.

(b) Taking the Fourier transform of Eq. (6.132) (see Appendix B), we see that the power spectrum of $Z(t)$ is given by

$$S_Z(\omega) = \frac{4\lambda}{\omega^2 + 4\lambda^2} \tag{6.133}$$

6.19. Let $X(t)$ and $Y(t)$ be both zero-mean and WSS random processes. Consider the random process $Z(t)$ defined by

$$Z(t) = X(t) + Y(t)$$

(a) Determine the autocorrelation function and the power spectral density of $Z(t)$, (i) if $X(t)$ and $Y(t)$ are jointly WSS; (ii) if $X(t)$ and $Y(t)$ are orthogonal.

(b) Show that if $X(t)$ and $Y(t)$ are orthogonal, then the mean square of $Z(t)$ is equal to the sum of the mean squares of $X(t)$ and $Y(t)$.

(a) The autocorrelation of $Z(t)$ is given by

$$
\begin{aligned}
R_Z(t, s) = E[Z(t)Z(s)] &= E\{[X(t) + Y(t)][X(s) + Y(s)]\} \\
&= E[X(t)X(s)] + E[X(t)Y(s)] + E[Y(t)X(s)] + E[Y(t)Y(s)] \\
&= R_X(t, s) + R_{XY}(t, s) + R_{YX}(t, s) + R_Y(t, s)
\end{aligned}
$$

(i) If $X(t)$ and $Y(t)$ are jointly WSS, then we have

$$R_Z(\tau) = R_X(\tau) + R_{XY}(\tau) + R_{YX}(\tau) + R_Y(\tau)$$

where $\tau = s - t$. Taking the Fourier transform of the above expression, we obtain

$$S_Z(\omega) = S_X(\omega) + S_{XY}(\omega) + S_{YX}(\omega) + S_Y(\omega)$$

(ii) If $X(t)$ and $Y(t)$ are orthogonal [Eq. (6.21)],

$$R_{XY}(\tau) = R_{YX}(\tau) = 0$$

Then
$$R_Z(\tau) = R_X(\tau) + R_Y(\tau) \tag{6.134a}$$
$$S_Z(\omega) = S_X(\omega) + S_Y(\omega) \tag{6.134b}$$

(b) Setting $\tau = 0$ in Eq. (6.134a), and using Eq. (6.15), we get

$$E[Z^2(t)] = E[X^2(t)] + E[Y^2(t)]$$

which indicates that the mean square of $Z(t)$ is equal to the sum of the mean squares of $X(t)$ and $Y(t)$.

White Noise

6.20. Using the notion of generalized derivative, show that the generalized derivative $X'(t)$ of the Wiener process $X(t)$ is a white noise.

From Eq. (5.64),

$$R_X(t, s) = \sigma^2 \min(t, s)$$

and from Eq. (6.119) (Prob. 6.9), we have

$$\frac{\partial}{\partial s} R_X(t, s) = \sigma^2 u(t - s) \tag{6.135}$$

Now, using the δ function, the generalized derivative of a unit step function $u(t)$ is given by

$$\frac{d}{dt} u(t) = \delta(t)$$

Applying the above relation to Eq. (6.135), we obtain

$$\frac{\partial^2}{\partial t \, \partial s} R_X(t, s) = \sigma^2 \frac{\partial}{\partial t} u(t - s) = \sigma^2 \delta(t - s) \tag{6.136}$$

which is, by Eq. (6.116) (Prob. 6.7), the autocorrelation function of the generalized derivative $X'(t)$ of the Wiener process $X(t)$; that is,

$$R_X(t, s) = \sigma^2 \delta(t - s) = \sigma^2 \delta(\tau) \tag{6.137}$$

where $\tau = t - s$. Thus, by definition (6.43), we see that the generalized derivative $X'(t)$ of the Wiener process $X(t)$ is a white noise.

Recall that the Wiener process is a normal process and its derivative is also normal (see Prob. 6.10). Hence, the generalized derivative $X'(t)$ of the Wiener process is called *white normal* (or *white Gaussian*) *noise*.

6.21. Let $X(t)$ be a Poisson process with rate λ. Let

$$Y(t) = X(t) - \lambda t$$

Show that the generalized derivative $Y'(t)$ of $Y(t)$ is a white noise.

Since $Y(t) = X(t) - \lambda t$, we have formally

$$Y'(t) = X'(t) - \lambda \tag{6.138}$$

Then
$$E[Y'(t)] = E[X'(t) - \lambda] = E[X'(t)] - \lambda \tag{6.139}$$

$$\begin{aligned} R_{Y'}(t, s) &= E[Y'(t)Y'(s)] = E\{[X'(t) - \lambda][X'(s) - \lambda]\} \\ &= E[X'(t)X'(s) - \lambda X'(s) - \lambda X'(t) + \lambda^2] \\ &= E[X'(t)X'(s)] - \lambda E[X'(s)] - \lambda E[X'(t)] + \lambda^2 \end{aligned} \tag{6.140}$$

Now, from Eqs. (5.56) and (5.60), we have

$$E[X(t)] = \lambda t$$
$$R_X(t, s) = \lambda \min(t, s) + \lambda^2 ts$$

Thus,
$$E[X'(t)] = \lambda \qquad \text{and} \qquad E[X'(s)] = \lambda \tag{6.141}$$

and from Eqs. (6.7) and (6.137),

$$E[X'(t)X'(s)] = R_{X'}(t, s) = \frac{\partial^2 R_X(t, s)}{\partial t \, \partial s} = \lambda \delta(t - s) + \lambda^2 \tag{6.142}$$

Substituting Eq. (6.141) into Eq. (6.139), we obtain

$$E[Y'(t)] = 0 \tag{6.143}$$

Substituting Eqs. (6.141) and (6.142) into Eq. (6.140), we get

$$R_{Y'}(t, s) = \lambda \delta(t - s) \tag{6.144}$$

Hence, we see that $Y'(t)$ is a zero-mean WSS random process, and by definition (6.43), $Y'(t)$ is a white noise with $\sigma^2 = \lambda$. The process $Y'(t)$ is known as the *Poisson white noise*.

6.22. Let $X(t)$ be a white normal noise. Let

$$Y(t) = \int_0^t X(\alpha)\, d\alpha$$

(a) Find the autocorrelation function of $Y(t)$.

(b) Show that $Y(t)$ is the Wiener process.

(a) From Eq. (6.137) of Prob. 6.20,

$$R_X(t, s) = \sigma^2 \delta(t - s)$$

Thus, by Eq. (6.11), the autocorrelation function of $Y(t)$ is

$$
\begin{aligned}
R_Y(t, s) &= \int_0^t \int_0^s R_X(\alpha, \beta)\, d\beta\, d\alpha \\
&= \int_0^s \int_0^t \sigma^2 \delta(\alpha - \beta)\, d\alpha\, d\beta \\
&= \sigma^2 \int_0^s u(t - \beta)\, d\beta \\
&= \sigma^2 \int_0^{\min(t,\, s)} d\beta = \sigma^2 \min(t, s)
\end{aligned}
\tag{6.145}
$$

(b) Comparing Eq. (6.145) and Eq. (5.64), we see that $Y(t)$ has the same autocorrelation function as the Wiener process. In addition, $Y(t)$ is normal, since $X(t)$ is a normal process and $Y(0) = 0$. Thus, we conclude that $Y(t)$ is the Wiener process.

6.23. Let $Y(n) = X(n) + W(n)$, where $X(n) = A$ (for all n) and A is a r.v. with zero mean and variance σ_A^2, and $W(n)$ is a discrete-time white noise with average power σ^2. It is also assumed that $X(n)$ and $W(n)$ are independent.

(a) Show that $Y(n)$ is WSS.

(b) Find the power spectral density $S_Y(\Omega)$ of $Y(n)$.

(a) The mean of $Y(n)$ is

$$E[Y(n)] = E[X(n)] + E[W(n)] = E(A) + E[W(n)] = 0$$

The autocorrelation function of $Y(n)$ is

$$
\begin{aligned}
R_Y(n, n + k) &= E\{[X(n) + W(n)][X(n + k) + W(n + k)]\} \\
&= E[X(n)X(n + k)] + E[X(n)]E[W(n + k)] + E[W(n)]E[X(n + k)] + E[W(n)W(n + k)] \\
&= E(A^2) + R_W(k) = \sigma_A^2 + \sigma^2 \delta(k) = R_Y(k)
\end{aligned}
\tag{6.146}
$$

Thus, $Y(n)$ is WSS.

(b) Taking the Fourier transform of Eq. (6.146), we obtain

$$S_Y(\Omega) = 2\pi\sigma_A^2 \delta(\Omega) + \sigma^2 \qquad -\pi < \Omega < \pi \tag{6.147}$$

Response of Linear Systems to Random Inputs

6.24. Derive Eq. (6.58).

Using Eq. (6.56), we have

$$R_Y(t, s) = E[Y(t)Y(s)]$$

$$= E\left[\int_{-\infty}^{\infty} h(\alpha)X(t-\alpha)\,d\alpha \int_{-\infty}^{\infty} h(\beta)X(s-\beta)\,d\beta\right]$$

$$= \int_{-\infty}^{\infty}\int_{-\infty}^{\infty} h(\alpha)h(\beta)E[X(t-\alpha)X(s-\beta)]\,d\alpha\,d\beta$$

$$= \int_{-\infty}^{\infty}\int_{-\infty}^{\infty} h(\alpha)h(\beta)R_X(t-\alpha, s-\beta)\,d\alpha\,d\beta$$

6.25. Derive Eq. (6.63).

From Eq. (6.62), we have

$$R_Y(\tau) = \int_{-\infty}^{\infty}\int_{-\infty}^{\infty} h(\alpha)h(\beta)R_X(\tau+\alpha-\beta)\,d\alpha\,d\beta$$

Taking the Fourier transform of $R_Y(\tau)$, we obtain

$$S_Y(\omega) = \int_{-\infty}^{\infty} R_Y(\tau)\,e^{-j\omega\tau}\,d\tau = \int_{-\infty}^{\infty}\int_{-\infty}^{\infty}\int_{-\infty}^{\infty} h(\alpha)h(\beta)R_X(\tau+\alpha-\beta)\,e^{-j\omega\tau}\,d\alpha\,d\beta\,d\tau$$

Letting $\tau + \alpha - \beta = \lambda$, we get

$$S_Y(\omega) = \int_{-\infty}^{\infty}\int_{-\infty}^{\infty}\int_{-\infty}^{\infty} h(\alpha)h(\beta)R_X(\lambda)e^{-j\omega(\lambda-\alpha+\beta)}\,d\alpha\,d\beta\,d\lambda$$

$$= \int_{-\infty}^{\infty} h(\alpha)e^{j\omega\alpha}\,d\alpha \int_{-\infty}^{\infty} h(\beta)e^{-j\omega\beta}\,d\beta \int_{-\infty}^{\infty} R_X(\lambda)\,e^{-j\omega\lambda}\,d\lambda$$

$$= H(-\omega)H(\omega)S_X(\omega)$$

$$= H^*(\omega)H(\omega)S_X(\omega) = |H(\omega)|^2 S_X(\omega)$$

6.26. A WSS random process $X(t)$ with autocorrelation function

$$R_X(\tau) = e^{-a|\tau|}$$

where a is a real positive constant, is applied to the input of an LTI system with impulse response

$$h(t) = e^{-bt}u(t)$$

where b is a real positive constant. Find the autocorrelation function of the output $Y(t)$ of the system.

The frequency response $H(\omega)$ of the system is

$$H(\omega) = \mathscr{F}[h(t)] = \frac{1}{j\omega + b}$$

The power spectral density of $X(t)$ is

$$S_X(\omega) = \mathscr{F}[R_X(\tau)] = \frac{2a}{\omega^2 + a^2}$$

By Eq. (6.63), the power spectral density of $Y(t)$ is

$$S_Y(\omega) = |H(\omega)|^2 S_X(\omega) = \left(\frac{1}{\omega^2 + b^2}\right)\left(\frac{2a}{\omega^2 + a^2}\right)$$

$$= \frac{a}{(a^2 + b^2)b}\left(\frac{2b}{\omega^2 + b^2}\right) - \frac{b}{(a^2 - b^2)b}\left(\frac{2a}{\omega^2 + a^2}\right)$$

Taking the inverse Fourier transform of both sides of the above equation, we obtain

$$R_Y(\tau) = \frac{1}{(a^2 - b^2)b}(ae^{-b|\tau|} - be^{-a|\tau|})$$

6.27. Verify Eq. (6.25), that is, the power spectral density of any WSS process $X(t)$ is real and $S_X(\omega) \geq 0$.

The realness of $S_X(\omega)$ was shown in Prob. 6.16. Consider an ideal bandpass filter with frequency response (Fig. 6-5)

$$H(\omega) = \begin{cases} 1 & \omega_1 < |\omega| < \omega_2 \\ 0 & \text{otherwise} \end{cases}$$

with a random process $X(t)$ as its input.

From Eq. (6.63), it follows that the power spectral density $S_Y(\omega)$ of the output $Y(t)$ equals

$$S_Y(\omega) = \begin{cases} S_X(\omega) & \omega_1 < |\omega| < \omega_2 \\ 0 & \text{otherwise} \end{cases}$$

Hence, from Eq. (6.27), we have

$$E[Y^2(t)] = \frac{1}{2\pi}\int_{-\infty}^{\infty} S_Y(\omega)\,d\omega = 2\frac{1}{2\pi}\int_{\omega_1}^{\omega_2} S_X(\omega)\,d\omega \geq 0$$

which indicates that the area of $S_X(\omega)$ in any interval of ω is nonnegative. This is possible only if $S_X(\omega) \geq 0$ for every ω.

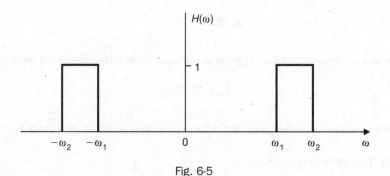

Fig. 6-5

6.28. Verify Eq. (6.72).

From Eq. (6.71), we have

$$R_Y(k) = \sum_{i=-\infty}^{\infty} \sum_{l=-\infty}^{\infty} h(i)h(l)R_X(k+i-l)$$

Taking the Fourier transform of $R_Y(k)$, we obtain

$$S_Y(\Omega) = \sum_{k=-\infty}^{\infty} R_Y(k)e^{-j\Omega k} = \sum_{k=-\infty}^{\infty}\sum_{i=-\infty}^{\infty}\sum_{l=-\infty}^{\infty} h(i)h(l)R_X(k+i-l)e^{-j\Omega k}$$

Letting $k + i - l = n$, we get

$$S_Y(\Omega) = \sum_{n=-\infty}^{\infty} \sum_{i=-\infty}^{\infty} \sum_{l=-\infty}^{\infty} h(i)h(l)R_X(n)e^{-j\Omega(n-i+l)}$$

$$= \sum_{i=-\infty}^{\infty} h(i)e^{j\Omega i} \sum_{l=-\infty}^{\infty} h(l)e^{-j\Omega l} \sum_{n=-\infty}^{\infty} R_X(n)e^{-j\Omega n}$$

$$= H(-\Omega)H(\Omega)S_X(\Omega)$$

$$= H^*(\Omega)H(\Omega)S_X(\Omega) = |H(\Omega)|^2 S_X(\Omega)$$

6.29. The discrete-time system shown in Fig. 6-6 consists of one unit delay element and one scalar multiplier ($a < 1$). The input $X(n)$ is discrete-time white noise with average power σ^2. Find the spectral density and average power of the output $Y(n)$.

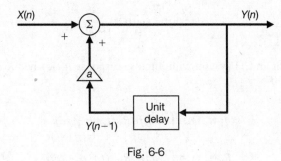

Fig. 6-6

From Fig. 6-6, $Y(n)$ and $X(n)$ are related by

$$Y(n) = aY(n-1) + X(n) \tag{6.148}$$

The impulse response $h(n)$ of the system is defined by

$$h(n) = ah(n-1) + \delta(n) \tag{6.149}$$

Solving Eq. (6.149), we obtain

$$h(n) = a^n u(n) \tag{6.150}$$

where $u(n)$ is the unit step sequence defined by

$$u(n) = \begin{cases} 1 & n \geq 0 \\ 0 & n < 0 \end{cases}$$

Taking the Fourier transform of Eq. (6.150), we obtain

$$H(\Omega) = \sum_{n=0}^{\infty} a^n e^{-j\Omega n} = \frac{1}{1 - ae^{-j\Omega}} \qquad a < 1, |\Omega| < \pi$$

Now, by Eq. (6.48),

$$S_X(\Omega) = \sigma^2 \qquad |\Omega| < \pi$$

and by Eq. (6.72), the power spectral density of $Y(n)$ is

$$S_Y(\Omega) = |H(\Omega)|^2 S_X(\Omega) = H(\Omega)H(-\Omega)S_X(\Omega)$$

$$= \frac{\sigma^2}{(1 - ae^{-j\Omega})(1 - ae^{j\Omega})}$$

$$= \frac{\sigma^2}{1 + a^2 - 2a\cos\Omega} \qquad |\Omega| < \pi \tag{6.151}$$

Taking the inverse Fourier transform of Eq. (6.151), we obtain

$$R_Y(k) = \frac{\sigma^2}{1 - a^2} a^{|k|}$$

Thus, by Eq. (6.33), the average power of $Y(n)$ is

$$E[Y^2(n)] = R_Y(0) = \frac{\sigma^2}{1 - a^2}$$

6.30. Let $Y(t)$ be the output of an LTI system with impulse response $h(t)$, when $X(t)$ is applied as input. Show that

(a) $$R_{XY}(t, s) = \int_{-\infty}^{\infty} h(\beta)R_X(t, s - \beta)\,d\beta \tag{6.152}$$

(b) $$R_Y(t, s) = \int_{-\infty}^{\infty} h(\alpha)R_{XY}(t - \alpha, s)\,d\alpha \tag{6.153}$$

(a) Using Eq. (6.56), we have

$$R_{XY}(t, s) = E[X(t)Y(s)] = E\left[X(t)\int_{-\infty}^{\infty} h(\beta)X(s - \beta)\,d\beta\right]$$

$$= \int_{-\infty}^{\infty} h(\beta)E[X(t)X(s - \beta)]\,d\beta = \int_{-\infty}^{\infty} h(\beta)R_X(t, s - \beta)\,d\beta$$

(b) Similarly,

$$R_Y(t, s) = E[Y(t)Y(s)] = E\left[\int_{-\infty}^{\infty} h(\alpha)X(t - \alpha)\,d\alpha\, Y(s)\right]$$

$$= \int_{-\infty}^{\infty} h(\alpha)E[X(t - \alpha)\,Y(s)]\,d\alpha = \int_{-\infty}^{\infty} h(\alpha)R_{XY}(t - \alpha,\ s)\,d\alpha$$

6.31. Let $Y(t)$ be the output of an LTI system with impulse response $h(t)$ when a WSS random process $X(t)$ is applied as input. Show that

(a) $S_{XY}(\omega) = H(\omega)S_X(\omega)$ (6.154)

(b) $S_Y(\omega) = H^*(\omega)S_{XY}(\omega)$ (6.155)

(a) If $X(t)$ is WSS, then Eq. (6.152) of Prob. 6.30 becomes

$$R_{XY}(t, s) = \int_{-\infty}^{\infty} h(\beta)R_X(s - t - \beta)\,d\beta \tag{6.156}$$

which indicates that $R_{XY}(t, s)$ is a function of the time difference $\tau = s - t$ only. Hence,

$$R_{XY}(\tau) = \int_{-\infty}^{\infty} h(\beta)R_X(\tau - \beta)\,d\beta \tag{6.157}$$

Taking the Fourier transform of Eq. (6.157), we obtain

$$S_{XY}(\omega) = \int_{-\infty}^{\infty} R_{XY}(\tau)e^{-j\omega\tau}\,d\tau = \int_{-\infty}^{\infty}\int_{-\infty}^{\infty} h(\beta)R_X(\tau-\beta)e^{-j\omega\tau}\,d\beta\,d\tau$$

$$= \int_{-\infty}^{\infty}\int_{-\infty}^{\infty} h(\beta)R_X(\lambda)e^{-j\omega(\lambda+\beta)}\,d\beta\,d\lambda$$

$$= \int_{-\infty}^{\infty} h(\beta)\,e^{-j\omega\beta}\,d\beta \int_{-\infty}^{\infty} R_X(\lambda)e^{-j\omega\lambda}\,d\lambda = H(\omega)S_X(\omega)$$

(b)   Similarly, if $X(t)$ is WSS, then by Eq. (6.156), Eq. (6.153) becomes

$$R_Y(t, s) = \int_{-\infty}^{\infty} h(\alpha)R_{XY}(s-t+\alpha)\,d\alpha$$

which indicates that $R_Y(t, s)$ is a function of the time difference $\tau = s - t$ only. Hence,

$$R_Y(\tau) = \int_{-\infty}^{\infty} h(\alpha)R_{XY}(\tau+\alpha)\,d\alpha \qquad (6.158)$$

Taking the Fourier transform of $R_Y(\tau)$, we obtain

$$S_Y(\omega) = \int_{-\infty}^{\infty} R_Y(\tau)e^{-j\omega\tau}\,d\tau = \int_{-\infty}^{\infty}\int_{-\infty}^{\infty} h(\alpha)R_{XY}(\tau+\alpha)e^{-j\omega\tau}\,d\alpha\,d\tau$$

$$= \int_{-\infty}^{\infty}\int_{-\infty}^{\infty} h(\alpha)R_{XY}(\lambda)e^{-j\omega(\lambda-\alpha)}\,d\alpha\,d\lambda$$

$$= \int_{-\infty}^{\infty} h(\alpha)\,e^{j\omega\alpha}\,d\alpha \int_{-\infty}^{\infty} R_{XY}(\lambda)e^{-j\omega\lambda}\,d\lambda$$

$$= H(-\omega)S_{XY}(\omega) = H^*(\omega)S_{XY}(\omega)$$

Note that from Eqs. (6.154) and (6.155), we obtain Eq. (6.63); that is,

$$S_Y(\omega) = H^*(\omega)S_{XY}(\omega) = H^*(\omega)H(\omega)S_X(\omega) = |H(\omega)|^2 S_X(\omega)$$

6.32. Consider a WSS process $X(t)$ with autocorrelation function $R_X(\tau)$ and power spectral density $S_X(\omega)$. Let $X'(t) = dX(t)/dt$. Show that

(a) $R_{XX'}(\tau) = \dfrac{d}{d\tau} R_X(\tau)$ $\qquad\qquad\qquad\qquad\qquad\qquad\qquad$ (6.159)

(b) $R_{X'}(\tau) = -\dfrac{d^2}{d\tau^2} R_X(\tau)$ $\qquad\qquad\qquad\qquad\qquad\qquad$ (6.160)

(c) $S_{X'}(\omega) = \omega^2 S_X(\omega)$ $\qquad\qquad\qquad\qquad\qquad\qquad\qquad$ (6.161)

(a)   If $X(t)$ is the input to a differentiator, then its output is $Y(t) = X'(t)$. The frequency response of a differentiator is known as $H(\omega) = j\omega$. Then from Eq. (6.154),

$$S_{XX'}(\omega) = H(\omega)S_X(\omega) = j\omega S_X(\omega)$$

Taking the inverse Fourier transform of both sides, we obtain

$$R_{XX'}(\tau) = \dfrac{d}{d\tau} R_X(\tau)$$

(b)   From Eq. (6.155),

$$S_{X'}(\omega) = H^*(\omega)S_{XX'}(\omega) = -j\omega S_{XX'}(\omega)$$

Again taking the inverse Fourier transform of both sides and using the result of part (a), we have

$$R_{X'}(\tau) = -\frac{d}{d\tau} R_{XX'}(\tau) = -\frac{d^2}{d\tau^2} R_X(\tau)$$

(c)   From Eq. (6.63),

$$S_{X'}(\omega) = |H(\omega)|^2 S_X(\omega) = |j\omega|^2 S_X(\omega) = \omega^2 S_X(\omega)$$

Note that Eqs. (6.159) and (6.160) were proved in Prob. 6.8 by a different method.

Fourier Series and Karhunen-Loéve Expansions

6.33. Verify Eqs. (6.80) and (6.81).

From Eq. (6.78),

$$X_n = \frac{1}{T} \int_0^T X(t) e^{-jn\omega_0 t} \, dt \qquad \omega_0 = 2\pi/T$$

Since $X(t)$ is WSS, $E[X(t)] = \mu_X$, and we have

$$E(X_n) = \frac{1}{T} \int_0^T E[X(t)] e^{-jn\omega_0 t} \, dt$$

$$= \mu_X \frac{1}{T} \int_0^T e^{-jn\omega_0 t} \, dt = \mu_X \delta(n)$$

Again using Eq. (6.78), we have

$$E(X_n X_m^*) = E\left[X_n \frac{1}{T} \int_0^T X^*(s) e^{jm\omega_0 s} \, ds \right]$$

$$= \frac{1}{T} \int_0^T E[X_n X^*(s)] e^{jm\omega_0 s} \, ds$$

Now

$$E[X_n X^*(s)] = E\left[\frac{1}{T} \int_0^T X(t) e^{-jn\omega_0 t} \, dt \, X^*(s) \right]$$

$$= \frac{1}{T} \int_0^T E[X(t) X^*(s)] e^{-jn\omega_0 t} \, dt$$

$$= \frac{1}{T} \int_0^T R_X(t - s) e^{-jn\omega_0 t} \, dt$$

Letting $t - s = \tau$, and using Eq. (6.76), we obtain

$$E[X_n X^*(s)] = \frac{1}{T} \int_0^T R_X(\tau) e^{-jn\omega_0(\tau + s)} \, d\tau$$

$$= \left\{ \frac{1}{T} \int_0^T R_X(\tau) e^{-jn\omega_0 \tau} \, d\tau \right\} e^{-jn\omega_0 s} = c_n e^{-jn\omega_0 s} \qquad (6.162)$$

Thus,

$$E(X_n X_m^*) = \frac{1}{T} \int_0^T c_n e^{-jn\omega_0 s} e^{jm\omega_0 s} \, ds$$

$$= c_n \frac{1}{T} \int_0^T e^{-j(n-m)\omega_0 s} \, ds = c_n \delta(n - m)$$

6.34. Let $\hat{X}(t)$ be the Fourier series representation of $X(t)$ shown in Eq. (6.77). Verify Eq. (6.79).

From Eq. (6.77), we have

$$E\left\{|X(t) - \hat{X}(t)|^2\right\} = E\left\{\left|X(t) - \sum_{n=-\infty}^{\infty} X_n e^{jn\omega_0 t}\right|^2\right\}$$

$$= E\left\{\left[X(t) - \sum_{n=-\infty}^{\infty} X_n e^{jn\omega_0 t}\right]\left[X^*(t) - \sum_{n=-\infty}^{\infty} X_n^* e^{-jn\omega_0 t}\right]\right\}$$

$$= E\left[|X(t)|^2\right] - \sum_{n=-\infty}^{\infty} E[X_n^* X(t)]e^{-jn\omega_0 t}$$

$$- \sum_{n=-\infty}^{\infty} E[X_n X^*(t)]e^{jn\omega_0 t} + \sum_{n=-\infty}^{\infty}\sum_{m=-\infty}^{\infty} E[X_n X_m^*]e^{j(n-m)\omega_0 t}$$

Now, by Eqs. (6.81) and (6.162), we have

$$E[X_n^* X(t)] = c_n^* e^{jn\omega_0 t}$$

$$E[X_n X^*(t)] = c_n e^{-jn\omega_0 t}$$

$$E(X_n X_m^*) = c_n \delta(n-m)$$

Using these results, finally we obtain

$$E\left\{|X(t) - \hat{X}(t)|^2\right\} = R_X(0) - \sum_{n=-\infty}^{\infty} c_n^* - \sum_{n=-\infty}^{\infty} c_n + \sum_{n=-\infty}^{\infty} c_n = 0$$

since each sum above equals $R_X(0)$ [see Eq. (6.75)].

6.35. Let $X(t)$ be m.s. periodic and represented by the Fourier series [Eq. (6.77)]

$$X(t) = \sum_{n=-\infty}^{\infty} X_n e^{jn\omega_0 t} \qquad \omega_0 = 2\pi/T_0$$

Show that

$$E[|X(t)|^2] = \sum_{n=-\infty}^{\infty} (E|X_n|^2) \tag{6.163}$$

From Eq. (6.81), we have

$$E(|X_n|^2) = E(X_n X_n^*) = c_n \tag{6.164}$$

Setting $\tau = 0$ in Eq. (6.75), we obtain

$$E[|X(t)|^2] = R_X(0) = \sum_{n=-\infty}^{\infty} c_n = \sum_{n=-\infty}^{\infty} E(|X_n|^2)$$

Equation (6.163) is known as *Parseval's theorem* for the Fourier series.

6.36. If a random process $X(t)$ is represented by a Karhunen-Loéve expansion [Eq. (6.82)]

$$X(t) = \sum_{n=1}^{\infty} X_n \phi_n(t) \qquad 0 < t < T$$

and X_n's are orthogonal, show that $\phi_n(t)$ must satisfy integral equation (6.86); that is,

$$\int_0^T R_X(t, s)\phi_n(s)\, ds = \lambda_n\phi_n(t) \qquad 0 < t < T$$

Consider

$$X(t)X_n^* = \sum_{m=1}^{\infty} X_m X_n^* \phi_m(t)$$

Then

$$E[X(t)X_n^*] = \sum_{m=1}^{\infty} E(X_m X_n^*)\phi_m(t) = E(|X_n|^2)\phi_n(t) \tag{6.165}$$

since X_n's are orthogonal; that is, $E(X_m X_n^*) = 0$ if $m \neq n$. But by Eq. (6.84),

$$\begin{aligned}
E[X(t)X_n^*] &= E\left[X(t)\int_0^T X^*(s)\phi_n(s)\, ds\right] \\
&= \int_0^T E[X(t)X^*(s)]\phi_n(s)\, ds \\
&= \int_0^T R_X(t, s)\phi_n(s)\, ds
\end{aligned} \tag{6.166}$$

Thus, equating Eqs. (6.165) and (6.166), we obtain

$$\int_0^T R_X(t, s)\phi_n(s)\, ds = E(|X_n|^2)\phi_n(t) = \lambda_n\phi_n(t)$$

where $\lambda_n = E(|X_n|^2)$.

6.37. Let $\hat{X}(t)$ be the Karhunen-Loéve expansion of $X(t)$ shown in Eq. (6.82). Verify Eq. (6.88).

From Eqs. (6.166) and (6.86), we have

$$E[X(t)X_m^*] = \int_0^T R_X(t, s)\phi_m(s)\, ds = \lambda_m\phi_m(t) \tag{6.167}$$

Now by Eqs. (6.83), (6.84), and (6.167) we obtain

$$\begin{aligned}
E[X_n X_m^*] &= E\left[\int_0^T X(t)\phi_n^*(t)\, dt\, X_m^*\right] = \int_0^T E[X(t)X_m^*]\phi_n^*(t)\, dt \\
&= \int_0^T \lambda_m\phi_m(t)\phi_n^*(t)\, dt = \lambda_m \int_0^T \phi_m(t)\phi_n^*(t)\, dt \\
&= \lambda_m\delta(m-n) = \lambda_n\delta(n-m)
\end{aligned} \tag{6.168}$$

6.38. Let $\hat{X}(t)$ be the Karhunen-Loéve expansion of $X(t)$ shown in Eq. (6.82). Verify Eq. (6.85).

From Eq. (6.82), we have

$$\begin{aligned}
E[|X(t) - \hat{X}(t)|^2] &= E\left\{\left|X(t) - \sum_{n=1}^{\infty} X_n\phi_n(t)\right|^2\right\} \\
&= E\left\{\left[X(t) - \sum_{n=1}^{\infty} X_n\phi_n(t)\right]\left[X^*(t) - \sum_{n=1}^{\infty} X_n^*\phi_n^*(t)\right]\right\} \\
&= E[|X(t)|^2] - \sum_{n=1}^{\infty} E[X(t)X_n^*]\phi_n^*(t) \\
&\quad - \sum_{n=1}^{\infty} E[X^*(t)X_n]\phi_n(t) + \sum_{n=1}^{\infty}\sum_{m=1}^{\infty} E(X_n X_m^*)\phi_n(t)\phi_m^*(t)
\end{aligned}$$

Using Eqs. (6.167) and (6.168), we have

$$E[|X(t) - \hat{X}(t)|^2] = R_X(t,t) - \sum_{n=1}^{\infty} \lambda_n \phi_n(t) \phi_n^*(t) - \sum_{n=1}^{\infty} \lambda_n^* \phi_n(t) \phi_n^*(t) + \sum_{n=1}^{\infty} \lambda_n \phi_n(t) \phi_n^*(t)$$

$$= 0$$

since by Mercer's theorem [Eq. (6.87)]

$$R_X(t,t) = \sum_{n=1}^{\infty} \lambda_n \phi_n(t) \phi_n^*(t)$$

and

$$\lambda_n = E(|X_n|^2) = \lambda_n^*$$

6.39. Find the Karhunen-Loéve expansion of the Wiener process $X(t)$.

From Eq. (5.64),

$$R_X(t,s) = \sigma^2 \min(t,s) = \begin{cases} \sigma^2 s & s < t \\ \sigma^2 t & s > t \end{cases}$$

Substituting the above expression into Eq. (6.86), we obtain

$$\sigma^2 \int_0^T \min(t,s) \phi_n(s)\, ds = \lambda_n \phi_n(t) \qquad 0 < t < T \tag{6.169}$$

or

$$\sigma^2 \int_0^t s\phi_n(s)\, ds + \sigma^2 t \int_t^T \phi_n(s)\, ds = \lambda_n \phi_n(t) \tag{6.170}$$

Differentiating Eq. (6.170) with respect to t, we get

$$\sigma^2 \int_t^T \phi_n(s)\, ds = \lambda_n \phi_n'(t) \tag{6.171}$$

Differentiating Eq. (6.171) with respect to t again, we obtain

$$\phi_n''(t) + \frac{\sigma^2}{\lambda_n} \phi_n(t) = 0 \tag{6.172}$$

A general solution of Eq. (6.172) is

$$\phi_n(t) = a_n \sin \omega_n t + b_n \cos \omega_n t \qquad \omega_n = \sigma/\sqrt{\lambda_n}$$

In order to determine the values of a_n, b_n, and λ_n (or ω_n), we need appropriate boundary conditions. From Eq. (6.170), we see that $\phi_n(0) = 0$. This implies that $b_n = 0$. From Eq. (6.171), we see that $\phi_n'(T) = 0$. This implies that

$$\omega_n = \frac{\sigma}{\sqrt{\lambda_n}} = \frac{(2n-1)\pi}{2T} = \frac{\left(n - \dfrac{1}{2}\right)\pi}{T} \qquad n = 1, 2, \dots$$

Therefore, the eigenvalues are given by

$$\lambda_n = \frac{\sigma^2 T^2}{\left(n - \dfrac{1}{2}\right)^2 \pi^2} \qquad n = 1, 2, \dots \tag{6.173}$$

The normalization requirement [Eq. (6.83)] implies that

$$\int_0^T (a_n \sin \omega_n t)^2 \, dt = \frac{a_n^2 T}{2} = 1 \rightarrow a_n = \sqrt{\frac{2}{T}}$$

Thus, the eigenfunctions are given by

$$\phi_n(t) = \sqrt{\frac{2}{T}} \sin\left(n - \frac{1}{2}\right)\frac{\pi}{T} t \qquad 0 < t < T \tag{6.174}$$

and the Karhunen-Loéve expansion of the Wiener process $X(t)$ is

$$X(t) = \sqrt{\frac{2}{T}} \sum_{n=1}^{\infty} X_n \sin\left(n - \frac{1}{2}\right)\frac{\pi}{T} t \qquad 0 < t < T \tag{6.175}$$

where X_n are given by

$$X_n = \sqrt{\frac{2}{T}} \int_0^{\infty} X(t) \sin\left(n - \frac{1}{2}\right)\frac{\pi}{T} t$$

and they are uncorrelated with variance λ_n.

6.40. Find the Karhunen-Loéve expansion of the white normal (or white Gaussian) noise $W(t)$.

From Eq. (6.43),

$$R_W(t, s) = \sigma^2 \delta(t - s)$$

Substituting the above expression into Eq. (6.86), we obtain

$$\sigma^2 \int_0^T \delta(t - s)\phi_n(s) \, ds = \lambda_n \phi_n(t) \qquad 0 < t < T$$

or [by Eq. (6.44)]

$$\sigma^2 \phi_n(t) = \lambda_n \phi_n(t) \tag{6.176}$$

which indicates that all $\lambda_n = \sigma^2$ and $\phi_n(t)$ are arbitrary. Thus, any complete orthogonal set $\{\phi_n(t)\}$ with corresponding eigenvalues $\lambda_n = \sigma^2$ can be used in the Karhunen-Loéve expansion of the white Gaussian noise.

Fourier Transform of Random Processes

6.41. Derive Eq. (6.94).

From Eq. (6.89),

$$\tilde{X}(\omega_1) = \int_{-\infty}^{\infty} X(t)e^{-j\omega_1 t} \, dt \qquad \tilde{X}(\omega_2) = \int_{-\infty}^{\infty} X(s)e^{-j\omega_2 s} \, ds$$

Then
$$R_{\tilde{X}}(\omega_1, \omega_2) = E[\tilde{X}(\omega_1)\tilde{X}^*(\omega_2)] = E\left[\int_{-\infty}^{\infty}\int_{-\infty}^{\infty} X(t)X^*(s)e^{-j(\omega_1 t - \omega_2 s)} \, dt \, ds\right]$$

$$= \int_{-\infty}^{\infty}\int_{-\infty}^{\infty} E[X(t)X^*(s)]e^{-j(\omega_1 t - \omega_2 s)} \, dt \, ds$$

$$= \int_{-\infty}^{\infty}\int_{-\infty}^{\infty} R_X(t, s)e^{-j[\omega_1 t + (-\omega_2)s]} \, dt \, ds = \tilde{R}_X(\omega_1, -\omega_2)$$

in view of Eq. (6.93).

6.42. Derive Eqs. (6.98) and (6.99).

Since $X(t)$ is WSS, by Eq. (6.93), and letting $t - s = \tau$, we have

$$\widetilde{R}_X(\omega_1, \omega_2) = \int_{-\infty}^{\infty}\int_{-\infty}^{\infty} R_X(t - s)e^{-j(\omega_1 t + \omega_2 s)}\, dt\, ds$$

$$= \int_{-\infty}^{\infty} R_X(\tau)e^{-j\omega_1\tau}\, dt \int_{-\infty}^{\infty} e^{-j(\omega_1 + \omega_2)s}\, ds$$

$$= S_X(\omega_1)\int_{-\infty}^{\infty} e^{-j(\omega_1 + \omega_2)s}\, ds$$

From the Fourier transform pair (Appendix B) $1 \leftrightarrow 2\pi\delta(\omega)$, we have

$$\int_{-\infty}^{\infty} e^{-j\omega t}\, dt = 2\pi\delta(\omega)$$

Hence, $$\widetilde{R}_X(\omega_1, \omega_2) = 2\pi S_X(\omega_1)\delta(\omega_1 + \omega_2)$$

Next, from Eq. (6.94) and the above result, we obtain

$$R_{\widetilde{X}}(\omega_1, \omega_2) = \widetilde{R}_X(\omega_1, -\omega_2) = 2\pi S_X(\omega_1)\delta(\omega_1 - \omega_2)$$

6.43. Let $\widetilde{X}(\omega)$ be the Fourier transform of a random process $X(t)$. If $\widetilde{X}(\omega)$ is a white noise with zero mean and autocorrelation function $q(\omega_1)\delta(\omega_1 - \omega_2)$, then show that $X(t)$ is WSS with power spectral density $q(\omega)/2\pi$.

By Eq. (6.91),

$$X(t) = \frac{1}{2\pi}\int_{-\infty}^{\infty}\widetilde{X}(\omega)e^{j\omega t}\, d\omega$$

Then $$E[X(t)] = \frac{1}{2\pi}\int_{-\infty}^{\infty} E[\widetilde{X}(\omega)]e^{j\omega t}\, d\omega = 0 \tag{6.177}$$

Assuming that $X(t)$ is a complex random process, we have

$$R_X(t, s) = E[X(t)X^*(s)] = E\left[\frac{1}{4\pi^2}\int_{-\infty}^{\infty}\int_{-\infty}^{\infty}\widetilde{X}(\omega_1)\widetilde{X}^*(\omega_2)e^{j(\omega_1 t - \omega_2 s)}\, d\omega_1\, d\omega_2\right]$$

$$= \frac{1}{4\pi^2}\int_{-\infty}^{\infty}\int_{-\infty}^{\infty} E[\widetilde{X}(\omega_1)\widetilde{X}^*(\omega_2)]e^{j(\omega_1 t - \omega_2 s)}\, d\omega_1\, d\omega_2$$

$$= \frac{1}{4\pi^2}\int_{-\infty}^{\infty}\int_{-\infty}^{\infty} q(\omega_1)\delta(\omega_1 - \omega_2)e^{j(\omega_1 t - \omega_2 s)}\, d\omega_1\, d\omega_2$$

$$= \frac{1}{4\pi^2}\int_{-\infty}^{\infty} q(\omega_1)e^{j\omega_1(t - s)}\, d\omega_1 \tag{6.178}$$

which depends only on $t - s = \tau$. Hence, we conclude that $X(t)$ is WSS. Setting $t - s = \tau$ and $\omega_1 = \omega$ in Eq. (6.178), we have

$$R_X(\tau) = \frac{1}{4\pi^2}\int_{-\infty}^{\infty} q(\omega)e^{j\omega\tau}\, d\omega = \frac{1}{2\pi}\int_{-\infty}^{\infty}\left[\frac{1}{2\pi}q(\omega)\right]e^{j\omega\tau}\, d\omega$$

$$= \frac{1}{2\pi}\int_{-\infty}^{\infty} S_X(\omega)e^{j\omega\tau}\, d\omega$$

in view of Eq. (6.24). Thus, we obtain $S_X(\omega) = q(\omega)/2\pi$.

6.44. Verify Eq. (6.104).

By Eq. (6.100),

$$\widetilde{X}(\Omega_1) = \sum_{n=-\infty}^{\infty} X(n)e^{-j\Omega_1 n} \qquad \widetilde{X}*(\Omega_2) = \sum_{m=-\infty}^{\infty} X*(m)e^{j\Omega_2 m}$$

Then

$$R_{\widetilde{X}}(\Omega_1, \Omega_2) = E[\widetilde{X}(\Omega_1)\widetilde{X}*(\Omega_2)] = \sum_{n=-\infty}^{\infty} \sum_{m=-\infty}^{\infty} E[X(n)X*(m)]e^{-j(\Omega_1 n - \Omega_2 m)}$$

$$= \sum_{n=-\infty}^{\infty} \sum_{m=-\infty}^{\infty} R_X(n, m)e^{-j[\Omega_1 n + (-\Omega_2)m]} = \widetilde{R}_X(\Omega_1, -\Omega_2)$$

in view of Eq. (6.103).

6.45. Derive Eqs. (6.105) and (6.106).

If $X(n)$ is WSS, then $R_X(n, m) = R_X(n - m)$. By Eq. (6.103), and letting $n - m = k$, we have

$$\widetilde{R}_X(\Omega_1, \Omega_2) = \sum_{n=-\infty}^{\infty} \sum_{m=-\infty}^{\infty} R_X(n - m)e^{-j(\Omega_1 n + \Omega_2 m)}$$

$$= \sum_{k=-\infty}^{\infty} R_X(k)e^{-j\Omega_1 k} \sum_{m=-\infty}^{\infty} e^{-j(\Omega_1 + \Omega_2)m}$$

$$= S_X(\Omega_1) \sum_{m=-\infty}^{\infty} e^{-j(\Omega_1 + \Omega_2)m}$$

From the Fourier transform pair (Appendix B) $x(n) = 1 \leftrightarrow 2\pi\delta(\Omega)$, we have

$$\sum_{m=-\infty}^{\infty} e^{-jm(\Omega_1 + \Omega_2)} = 2\pi\delta(\Omega_1 + \Omega_2)$$

Hence,

$$\widetilde{R}_X(\Omega_1, \Omega_2) = 2\pi S_X(\Omega_1)\delta(\Omega_1 + \Omega_2)$$

Next, from Eq. (6.104) and the above result, we obtain

$$R_{\widetilde{X}}(\Omega_1, \Omega_2) = \widetilde{R}_X(\Omega_1, -\Omega_2) = 2\pi S_X(\Omega_1)\delta(\Omega_1 - \Omega_2)$$

SUPPLEMENTARY PROBLEMS

6.46. Is the Poisson process $X(t)$ m.s. continuous?

6.47. Let $X(t)$ be defined by (Prob. 5.4)

$$X(t) = Y \cos \omega t \qquad t \geq 0$$

where Y is a uniform r.v. over $(0, 1)$ and ω is a constant.

(a) Is $X(t)$ m.s. continuous?

(b) Does $X(t)$ have a m.s. derivative?

6.48. Let $Z(t)$ be the random telegraph signal of Prob. 6.18.

    (*a*)   Is $Z(t)$ m.s. continuous?

    (*b*)   Does $Z(t)$ have a m.s. derivative?

6.49. Let $X(t)$ be a WSS random process, and let $X'(t)$ be its m.s. derivative. Show that $E[X(t)X'(t)] = 0$.

6.50. Let
$$Z(t) = \frac{2}{T} \int_t^{t+T/2} X(\alpha)\, d\alpha$$

where $X(t)$ is given by Prob. 6.47 with $\omega = 2\pi/T$.

    (*a*)   Find the mean of $Z(t)$.

    (*b*)   Find the autocorrelation function of $Z(t)$.

6.51. Consider a WSS random process $X(t)$ with $E[X(t)] = \mu_X$. Let

$$\langle X(t) \rangle_T = \frac{1}{T} \int_{-T/2}^{T/2} X(t)\, dt$$

The process $X(t)$ is said to be *ergodic in the mean* if

$$\underset{T \to \infty}{\text{l.i.m.}} \langle X(t) \rangle_T = E[X(t)] = \mu_X$$

Find $E\left[\langle X(t) \rangle_T\right]$.

6.52. Let $X(t) = A \cos(\omega_0 t + \Theta)$, where A and ω_0 are constants, Θ is a uniform r.v. over $(-\pi, \pi)$ (Prob. 5.20). Find the power spectral density of $X(t)$.

6.53. A random process $Y(t)$ is defined by

$$Y(t) = A X(t) \cos(\omega_c t + \Theta)$$

where A and ω_c are constants, Θ is a uniform r.v. over $(-\pi, \pi)$, and $X(t)$ is a zero-mean WSS random process with the autocorrelation function $R_X(\tau)$ and the power spectral density $S_X(\omega)$. Furthermore, $X(t)$ and Θ are independent. Show that $Y(t)$ is WSS, and find the power spectral density of $Y(t)$.

6.54. Consider a discrete-time random process defined by

$$X(n) = \sum_{i=1}^{m} a_i \cos(\Omega_i n + \Theta_i)$$

where a_i and Ω_i are real constants and Θ_i are independent uniform r.v.'s over $(-\pi, \pi)$.

    (*a*)   Find the mean of $X(n)$.

    (*b*)   Find the autocorrelation function of $X(n)$.

6.55. Consider a discrete-time WSS random process $X(n)$ with the autocorrelation function

$$R_X(k) = 10 e^{-0.5|k|}$$

Find the power spectral density of $X(n)$.

6.56. Let $X(t)$ and $Y(t)$ be defined by

$$X(t) = U \cos \omega_0 t + V \sin \omega_0 t$$
$$Y(t) = V \cos \omega_0 t - U \sin \omega_0 t$$

where ω_0 is constant and U and V are independent r.v.'s both having zero mean and variance σ^2.

(a) Find the cross-correlation function of $X(t)$ and $Y(t)$.

(b) Find the cross power spectral density of $X(t)$ and $Y(t)$.

6.57. Verify Eqs. (6.36) and (6.37).

6.58. Let $Y(t) = X(t) + W(t)$, where $X(t)$ and $W(t)$ are orthogonal and $W(t)$ is a white noise specified by Eq. (6.43) or (6.45). Find the autocorrelation function of $Y(t)$.

6.59. A zero-mean WSS random process $X(t)$ is called band-limited white noise if its spectral density is given by Find the autocorrelation function of $X(t)$.

$$S_X(\omega) = \begin{cases} N_0/2 & |\omega| \leq \omega_B \\ 0 & |\omega| > \omega_B \end{cases}$$

6.60. A WSS random process $X(t)$ is applied to the input of an LTI system with impulse response $h(t) = 3e^{-2t}u(t)$. Find the mean value of $Y(t)$ of the system if $E[X(t)] = 2$.

6.61. The input $X(t)$ to the *RC* filter shown in Fig. 6-7 is a white noise specified by Eq. (6.45). Find the mean-square value of $Y(t)$.

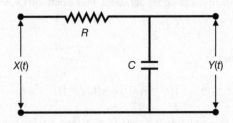

Fig. 6-7 *RC* filter.

6.62. The input $X(t)$ to a differentiator is the random telegraph signal of Prob. 6.18.

(a) Determine the power spectral density of the differentiator output.

(b) Find the mean-square value of the differentiator output.

6.63. Suppose that the input to the filter shown in Fig. 6-8 is a white noise specified by Eq. (6.45). Find the power spectral density of $Y(t)$.

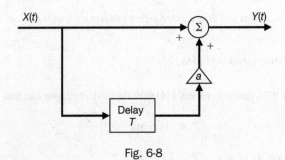

Fig. 6-8

6.64. Verify Eq. (6.67).

6.65. Suppose that the input to the discrete-time filter shown in Fig. 6-9 is a discrete-time white noise with average power σ^2. Find the power spectral density of $Y(n)$.

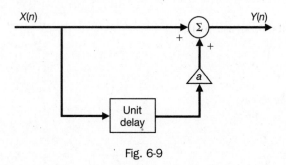

Fig. 6-9

6.66. Using the Karhunen-Loéve expansion of the Wiener process, obtain the Karhunen-Loéve expansion of the white normal noise.

6.67. Let $Y(t) = X(t) + W(t)$, where $X(t)$ and $W(t)$ are orthogonal and $W(t)$ is a white noise specified by Eq. (6.43) or (6.45). Let $\phi_n(t)$ be the eigenfunctions of the integral equation (6.86) and λ_n the corresponding eigenvalues.

   (a)   Show that $\phi_n(t)$ are also the eigenfunctions of the integral equation for the Karhunen-Loéve expansion of $Y(t)$ with $R_Y(t, s)$.

   (b)   Find the corresponding eigenvalues.

6.68. Suppose that

$$X(t) = \sum_n X_n \, e^{jn\omega_0 t}$$

where X_n are r.v.'s and ω_0 is a constant. Find the Fourier transform of $X(t)$.

6.69. Let $\tilde{X}(\omega)$ be the Fourier transform of a continuous-time random process $X(t)$. Find the mean of $\tilde{X}(\omega)$.

6.70. Let

$$\tilde{X}(\Omega) = \sum_{n=-\infty}^{\infty} X(n)e^{-j\Omega n}$$

where $E[X(n)] = 0$ and $E[X(n)X(k)] = \sigma_n^2 \, \delta(n - k)$. Find the mean and the autocorrelation function of $\tilde{X}(\Omega)$.

ANSWERS TO SUPPLEMENTARY PROBLEMS

6.46. *Hint:*   Use Eq. (5.60) and proceed as in Prob. 6.4.

   Yes.

6.47. *Hint:*   Use Eq. (5.87) of Prob. 5.12.

   (*a*) Yes;   (*b*) yes.

6.48. *Hint:*   Use Eq. (6.132) of Prob. 6.18.

   (*a*) Yes;   (*b*) no.

6.49. *Hint:*   Use Eqs. (6.13) [or (6.14)] and (6.117).

6.50. (*a*) $-\dfrac{1}{\pi}\sin \omega t$

   (*b*) $R_Z(t, s) = \dfrac{4}{3\pi^2}\sin \omega t \sin \omega s$

6.51. μ_X

6.52. $S_X(\omega) = \dfrac{A^2\pi}{2}[\delta(\omega - \omega_0) + \delta(\omega + \omega_0)]$

6.53. $S_Y(\omega) = \dfrac{A^2}{4}[S_X(\omega - \omega_c) + S_X(\omega + \omega_c)]$

6.54. (*a*) $E[X(n)] = 0$

   (*b*) $R_X(n, n + k) = \dfrac{1}{2}\displaystyle\sum_{i=1}^{m} a_i^2 \cos(\Omega_i k)$

6.55. $S_X(\Omega) = \dfrac{6.32}{1.368 - 1.213 \cos \Omega} \qquad -\pi < \Omega < \pi$

6.56. (*a*) $R_{XY}(t, t + \tau) = -\sigma^2 \sin \omega_0 \tau$

   (*b*) $S_{XY}(\omega) = j\sigma^2\pi[\delta(\omega - \omega_0) - \delta(\omega + \omega_0)]$

6.57. *Hint:*   Substitute Eq. (6.18) into Eq. (6.34).

6.58. $R_Y(t, s) = R_X(t, s) + \sigma^2\delta(t - s)$

6.59. $R_X(\tau) = \dfrac{N_0\,\omega_B}{2\pi}\dfrac{\sin \omega_B\tau}{\omega_B\tau}$

6.60. *Hint:*   Use Eq. (6.59).

   3

6.61. *Hint:*   Use Eqs. (6.64) and (6.65).

   $\sigma^2/(2RC)$

6.62. (*a*) $S_Y(\omega) = \dfrac{4\lambda\omega^2}{\omega^2 + 4\lambda^2}$

   (*b*) $E[Y^2(t)] = \infty$

6.63. $S_Y(\omega) = \sigma^2(1 + a^2 + 2a \cos \omega T)$

6.64. *Hint:* Proceed as in Prob. 6.24.

6.65. $S_Y(\Omega) = \sigma^2(1 + a^2 + 2a \cos \Omega)$

6.66. *Hint:* Take the derivative of Eq. (6.175) of Prob. 6.39.

$$\sqrt{\frac{2}{T}} \sum_{n=1}^{\infty} W_n \cos\left(n - \frac{1}{2}\right)\frac{\pi}{T}t \qquad 0 < t < T$$

where W_n are independent normal r.v.'s with the same variance σ^2.

6.67. *Hint:* Use the result of Prob. 6.58.

(b) $\lambda_n + \sigma^2$

6.68. $\widetilde{X}(\omega) = \sum_n 2\pi X_n \delta(\omega - n\omega_0)$

6.69. $\mathcal{F}[\mu_X(t)] = \int_{-\infty}^{\infty} \mu_X(t)e^{-j\omega t}\,dt \qquad$ where $\mu_X(t) = E[X(t)]$

6.70. $E[\widetilde{X}(\Omega)] = 0 \qquad R_{\widetilde{X}}(\Omega_1, \Omega_2) = \sum_{n=-\infty}^{\infty} \sigma_n{}^2 e^{-j(\Omega_1 - \Omega_2)n}$

CHAPTER 7

Estimation Theory

7.1 Introduction

In this chapter, we present a classical estimation theory. There are two basic types of estimation problems. In the first type, we are interested in estimating the parameters of one or more r.v.'s, and in the second type, we are interested in estimating the value of an inaccessible r.v. Y in terms of the observation of an accessible r.v. X.

7.2 Parameter Estimation

Let X be a r.v. with pdf $f(x)$ and $X_1, \ldots, X_n$ a set of n independent r.v.'s each with pdf $f(x)$. The set of r.v.'s $(X_1, \ldots, X_n)$ is called a *random sample* (or *sample vector*) of size n of X. Any real-valued function of a random sample $s(X_1, \ldots, X_n)$ is called a *statistic*.

Let X be a r.v. with pdf $f(x; \theta)$ which depends on an unknown parameter θ. Let $(X_1, \ldots, X_n)$ be a random sample of X. In this case, the joint pdf of $X_1, \ldots, X_n$ is given by

$$f(\mathbf{x}; \theta) = f(x_1, \ldots, x_n; \theta) = \prod_{i=1}^{n} f(x_i; \theta) \tag{7.1}$$

where $x_1, \ldots, x_n$ are the values of the observed data taken from the random sample.

An *estimator* of θ is any statistic $s(X_1, \ldots, X_n)$, denoted as

$$\Theta = s(X_1, \ldots, X_n) \tag{7.2}$$

For a particular set of observations $X_1 = x_1, \ldots, X_n = x_n$, the value of the estimator $s(x_1, \ldots, x_n)$ will be called an *estimate* of θ and denoted by $\hat{\theta}$. Thus, an estimator is a r.v. and an estimate is a particular realization of it. It is not necessary that an estimate of a parameter be one single value; instead, the estimate could be a *range* of values. Estimates which specify a single value are called *point estimates*, and estimates which specify a range of values are called *interval estimates*.

7.3 Properties of Point Estimators

A. Unbiased Estimators:

An estimator $\Theta = s(X_1, \ldots, X_n)$ is said to be an *unbiased* estimator of the parameter θ if

$$E(\Theta) = \theta \tag{7.3}$$

for all possible values of θ. If Θ is an unbiased estimator, then its mean square error is given by

$$E[(\Theta - \theta)^2] = E\{[\Theta - E(\Theta)]^2\} = \text{Var}(\Theta) \tag{7.4}$$

That is, its mean square error equals its variance.

B. Efficient Estimators:

An estimator Θ_1 is said to be a more *efficient* estimator of the parameter θ than the estimator Θ_2 if

1. Θ_1 and Θ_2 are both unbiased estimators of θ.
2. $\text{Var}(\Theta_1) < \text{Var}(\Theta_2)$.

The estimator $\Theta_{MV} = s(X_1, \ldots, X_n)$ is said to be a *most efficient* (or *minimum variance*) unbiased estimator of the parameter θ if

1.   It is an unbiased estimator of θ.
2. $\text{Var}(\Theta_{MV}) \le \text{Var}(\Theta)$ for all Θ.

C. Consistent Estimators:

The estimator Θ_n of θ based on a random sample of size n is said to be *consistent* if for any small $\varepsilon > 0$,

$$\lim_{n \to \infty} P\big(|\Theta_n - \theta| < \varepsilon\big) = 1 \tag{7.5}$$

or equivalently,

$$\lim_{n \to \infty} P\big(|\Theta_n - \theta| \ge \varepsilon\big) = 0 \tag{7.6}$$

The following two conditions are sufficient to define consistency (Prob. 7.5):

1. $\lim_{n \to \infty} E(\Theta_n) = \theta$ $\tag{7.7}$
2. $\lim_{n \to \infty} \text{Var}(\Theta_n) = 0$ $\tag{7.8}$

7.4 Maximum-Likelihood Estimation

Let $f(\mathbf{x}; \theta) = f(x_1, \ldots, x_n; \theta)$ denote the joint pmf of the r.v.'s $X_1, \ldots, X_n$ when they are discrete, and let it be their joint pdf when they are continuous. Let

$$L(\theta) = f(\mathbf{x}; \theta) = f(x_1, \ldots, x_n; \theta) \tag{7.9}$$

Now $L(\theta)$ represents the likelihood that the values $x_1, \ldots, x_n$ will be observed when θ is the true value of the parameter. Thus, $L(\theta)$ is often referred to as the *likelihood function* of the random sample. Let $\theta_{ML} = s(x_1, \ldots, x_n)$ be the maximizing value of $L(\theta)$; that is,

$$L(\theta_{ML}) = \max_{\theta} L(\theta) \tag{7.10}$$

Then the maximum-likelihood estimator of θ is

$$\Theta_{ML} = s(X_1, \ldots, X_n) \tag{7.11}$$

and θ_{ML} is the maximum-likelihood estimate of θ.

Since $L(\theta)$ is a product of either pmf's or pdf's, it will always be positive (for the range of possible values of θ). Thus, $\ln L(\theta)$ can always be defined, and in determining the maximizing value of θ, it is often useful to use the fact that $L(\theta)$ and $\ln L(\theta)$ have their maximum at the same value of θ. Hence, we may also obtain θ_{ML} by maximizing $\ln L(\theta)$.

7.5   Bayes' Estimation

Suppose that the unknown parameter θ is considered to be a r.v. having some fixed distribution or *prior* pdf $f(\theta)$. Then $f(x; \theta)$ is now viewed as a conditional pdf and written as $f(x|\theta)$, and we can express the joint pdf of the random sample $(X_1, ..., X_n)$ and θ as

$$f(x_1, ..., x_n, \theta) = f(x_1, ..., x_n | \theta) f(\theta) \tag{7.12}$$

and the marginal pdf of the sample is given by

$$f(x_1, ..., x_n) = \int_{R_\theta} f(x_1, ..., x_n, \theta) d\theta \tag{7.13}$$

where R_θ is the range of the possible value of θ. The other conditional pdf,

$$f(\theta|x_1, ..., x_n) = \frac{f(x_1, ..., x_n, \theta)}{f(x_1, ..., x_n)} = \frac{f(x_1, ..., x_n|\theta)f(\theta)}{f(x_1, ..., x_n)} \tag{7.14}$$

is referred to as the *posterior* pdf of θ. Thus, the prior pdf $f(\theta)$ represents our information about θ prior to the observation of the outcomes of $X_1, ..., X_n$, and the posterior pdf $f(\theta|x_1, ..., x_n)$ represents our information about θ after having observed the sample.

The conditional mean of θ, defined by

$$\theta_B = E(\theta|x_1, ..., x_n) = \int_{R_\theta} \theta f(\theta|x_1, ..., x_n) d\theta \tag{7.15}$$

is called the *Bayes' estimate* of θ, and

$$\Theta_B = E(\theta|X_1, ..., X_n) \tag{7.16}$$

is called the *Bayes' estimator* of θ.

7.6   Mean Square Estimation

In this section, we deal with the second type of estimation problem—that is, estimating the value of an inaccessible r.v. Y in terms of the observation of an accessible r.v. X. In general, the estimator $\hat{Y}$ of Y is given by a function of X, $g(X)$. Then $Y - \hat{Y} = Y - g(X)$ is called the *estimation error*, and there is a *cost* associated with this error, $C[Y - g(X)]$. We are interested in finding the function $g(X)$ that minimizes this cost. When X and Y are continuous r.v.'s, the *mean square* (m.s.) *error* is often used as the cost function,

$$C[Y - g(X)] = E\{[Y - g(X)]^2\} \tag{7.17}$$

It can be shown that the estimator of Y given by (Prob. 7.17),

$$\hat{Y} = g(X) = E(Y|X) \tag{7.18}$$

is the best estimator in the sense that the m.s. error defined by Eq. (7.17) is a minimum.

7.7 Linear Mean Square Estimation

Now consider the estimator $\hat{Y}$ of Y given by

$$\hat{Y} = g(X) = aX + b \tag{7.19}$$

We would like to find the values of a and b such that the m.s. error defined by

$$e = E[(Y - \hat{Y})^2] = E\{[Y - (aX + b)]^2\} \tag{7.20}$$

is minimum. We maintain that a and b must be such that (Prob. 7.20)

$$E\{[Y - (aX + b)]X\} = 0 \tag{7.21}$$

and a and b are given by

$$a = \frac{\sigma_{XY}}{\sigma_X^2} = \frac{\sigma_Y}{\sigma_X}\rho_{XY} \qquad b = \mu_Y - a\mu_X \tag{7.22}$$

and the minimum m.s. error e_m is (Prob. 7.22)

$$e_m = \sigma_Y^2(1 - \rho_{XY}^2) \tag{7.23}$$

where $\sigma_{XY} = \text{Cov}(X, Y)$ and ρ_{XY} is the correlation coefficient of X and Y. Note that Eq. (7.21) states that the optimum linear m.s. estimator $\hat{Y} = aX + b$ of Y is such that the estimation error $Y - \hat{Y} = Y - (aX + b)$ is orthogonal to the observation X. This is known as the *orthogonality principle*. The line $y = ax + b$ is often called a *regression line*.

Next, we consider the estimator $\hat{Y}$ of Y with a linear combination of the random sample $(X_1, \ldots, X_n)$ by

$$\hat{Y} = \sum_{i=1}^{n} a_i X_i \tag{7.24}$$

Again, we maintain that in order to produce the linear estimator with the minimum m.s. error, the coefficients a_i must be such that the following orthogonality conditions are satisfied (Prob. 7.35):

$$E\left[\left(Y - \sum_{i=1}^{n} a_i X_i\right) X_j\right] = 0 \qquad j = 1, \ldots, n \tag{7.25}$$

Solving Eq. (7.25) for a_i, we obtain

$$\mathbf{a} = R^{-1}\mathbf{b} \tag{7.26}$$

where

$$\mathbf{a} = \begin{bmatrix} a_1 \\ \vdots \\ a_n \end{bmatrix} \qquad \mathbf{b} = \begin{bmatrix} b_1 \\ \vdots \\ b_n \end{bmatrix} \qquad b_j = E(YX_j) \qquad R = \begin{bmatrix} R_{11} & \cdots & R_{1n} \\ \vdots & \ddots & \vdots \\ R_{n1} & \cdots & R_{nn} \end{bmatrix} \qquad R_{ij} = E(X_i X_j)$$

and R^{-1} is the inverse of R.

SOLVED PROBLEMS

Properties of Point Estimators

7.1. Let $(X_1, \ldots, X_n)$ be a random sample of X having unknown mean μ. Show that the estimator of μ defined by

$$M = \frac{1}{n} \sum_{i=1}^{n} X_i = \bar{X} \tag{7.27}$$

is an unbiased estimator of μ. Note that $\bar{X}$ is known as the *sample mean* (Prob. 4.64).

By Eq. (4.108),

$$E(M) = E\left(\frac{1}{n} \sum_{i=1}^{n} X_i\right) = \frac{1}{n} \sum_{i=1}^{n} E(X_i) = \frac{1}{n} \sum_{i=1}^{n} \mu = \frac{1}{n}(n\mu) = \mu$$

Thus, M is an unbiased estimator of μ.

7.2. Let $(X_1, \ldots, X_n)$ be a random sample of X having unknown mean μ and variance σ^2. Show that the estimator of σ^2 defined by

$$S^2 = \frac{1}{n} \sum_{i=1}^{n} (X_i - \bar{X})^2 \tag{7.28}$$

where $\bar{X}$ is the sample mean, is a biased estimator of σ^2.

By definition, we have

$$\sigma^2 = E\left[(X_i - \mu)^2\right]$$

Now

$$E(S^2) = E\left[\frac{1}{n} \sum_{i=1}^{n} (X_i - \bar{X})^2\right] = E\left\{\frac{1}{n} \sum_{i=1}^{n} [(X_i - \mu) - (\bar{X} - \mu)]^2\right\}$$

$$= E\left\{\frac{1}{n} \sum_{i=1}^{n} [(X_i - \mu)^2 - 2(X_i - \mu)(\bar{X} - \mu) + (\bar{X} - \mu)^2]\right\}$$

$$= E\left\{\frac{1}{n}\left[\sum_{i=1}^{n} (X_i - \mu)^2 - n(\bar{X} - \mu)^2\right]\right\}$$

$$= \frac{1}{n} \sum_{i=1}^{n} E[(X_i - \mu)^2] - E[(\bar{X} - \mu)^2] = \sigma^2 - \sigma_{\bar{X}}^2$$

By Eqs. (4.112) and (7.27), we have

$$\sigma_{\bar{X}}^2 = \text{Var}(\bar{X}) = \sum_{i=1}^{n} \frac{1}{n^2} \sigma^2 = \frac{1}{n} \sigma^2 \tag{7.29}$$

Thus,

$$E(S^2) = \sigma^2 - \frac{1}{n} \sigma^2 = \frac{n-1}{n} \sigma^2$$

which shows that S^2 is a biased estimator of σ^2.

7.3. Let $(X_1, \ldots, X_n)$ be a random sample of a Poisson r.v. X with unknown parameter λ.

(a) Show that

$$\Lambda_1 = \frac{1}{n} \sum_{i=1}^{n} X_i \quad \text{and} \quad \Lambda_2 = \frac{1}{2}(X_1 + X_2)$$

are both unbiased estimators of λ.

(b) Which estimator is more efficient?

(a) By Eqs. (2.50) and (4.132), we have

$$E(\Lambda_1) = \frac{1}{n} \sum_{i=1}^{n} E(X_i) = \frac{1}{n}(n\lambda) = \lambda$$

$$E(\Lambda_2) = \frac{1}{2}\left[E(X_1) + E(X_2)\right] = \frac{1}{2}(2\lambda) = \lambda$$

Thus, both estimators are unbiased estimators of λ.

(b) By Eqs. (2.51) and (4.136),

$$\operatorname{Var}(\Lambda_1) = \frac{1}{n^2} \sum_{i=1}^{n} \operatorname{Var}(X_i) = \frac{1}{n^2} \sum_{i=1}^{n} \operatorname{Var}(X_i) = \frac{1}{n^2}(n\lambda) = \frac{\lambda}{n}$$

$$\operatorname{Var}(\Lambda_2) = \frac{1}{4}(2\lambda) = \frac{\lambda}{2}$$

Thus, if $n > 2$, Λ_1 is a more efficient estimator of λ than Λ_2, since $\lambda/n < \lambda/2$.

7.4. Let $(X_1, \ldots, X_n)$ be a random sample of X with mean μ and variance σ^2. A linear estimator of μ is defined to be a linear function of $X_1, \ldots, X_n$, $l(X_1, \ldots, X_n)$. Show that the linear estimator defined by [Eq. (7.27)],

$$M = \frac{1}{n} \sum_{i=1}^{n} X_i = \bar{X}$$

is the most efficient linear unbiased estimator of μ.

Assume that

$$M_1 = l(X_1, \ldots, X_n) = \sum_{i=1}^{n} a_i X_i$$

is a linear unbiased estimator of μ with lower variance than M. Since M_1 is unbiased, we must have

$$E(M_1) = \sum_{i=1}^{n} a_i E(X_i) = \mu \sum_{i=1}^{n} a_i = \mu$$

which implies that $\sum_{i=1}^{n} a_i = 1$. By Eq. (4.136),

$$\operatorname{Var}(M) = \frac{1}{n}\sigma^2 \quad \text{and} \quad \operatorname{Var}(M_1) = \sigma^2 \sum_{i=1}^{n} a_i^2$$

By assumption,

$$\sigma^2 \sum_{i=1}^{n} a_i^2 < \frac{1}{n}\sigma^2 \qquad \text{or} \qquad \sum_{i=1}^{n} a_i^2 < \frac{1}{n} \tag{7.30}$$

Consider the sum

$$0 \leq \sum_{i=1}^{n} \left(a_i - \frac{1}{n}\right)^2 = \sum_{i=1}^{n} \left(a_i^2 - 2\frac{a_i}{n} + \frac{1}{n^2}\right)$$

$$= \sum_{i=1}^{n} a_i^2 - \frac{2}{n}\sum_{i=1}^{n} a_i + \frac{1}{n}$$

$$= \sum_{i=1}^{n} a_i^2 - \frac{1}{n}.$$

which, by assumption (7.30), is less than 0. This is impossible unless $a_i = 1/n$, implying that M is the most efficient linear unbiased estimator of μ.

7.5. Show that if

$$\lim_{n \to \infty} E(\Theta_n) = \theta \qquad \text{and} \qquad \lim_{n \to \infty} \text{Var}(\Theta_n) = 0$$

then the estimator Θ_n is consistent.

Using Chebyshev's inequality (2.116), we can write

$$P(|\Theta_n - \theta| \geq \varepsilon) \leq \frac{E[(\Theta_n - \theta)^2]}{\varepsilon^2} = \frac{1}{\varepsilon^2} E\{[\Theta_n - E(\Theta_n) + E(\Theta_n) - \theta]^2\}$$

$$= \frac{1}{\varepsilon^2} E\{[\Theta_n - E(\Theta_n)]^2 + [E(\Theta_n) - \theta]^2 + 2[\Theta_n - E(\Theta_n)][E(\Theta_n) - \theta]\}$$

$$= \frac{1}{\varepsilon^2} (\text{Var}(\Theta_n) + E\{[E(\Theta_n) - \theta]^2\} + 2E\{[\Theta_n - E(\Theta_n)][E(\Theta_n) - \theta]\})$$

Thus, if

$$\lim_{n \to \infty} E(\Theta_n) = \theta \qquad \text{and} \qquad \lim_{n \to \infty} \text{Var}(\Theta_n) = 0$$

then

$$\lim_{n \to \infty} P(|\Theta_n - \theta| \geq \varepsilon) = 0$$

that is, Θ_n is consistent [see Eq. (7.6)].

7.6. Let $(X_1, \ldots, X_n)$ be a random sample of a uniform r.v. X over $(0, a)$, where a is unknown. Show that

$$A = \max(X_1, X_2, \ldots, X_n) \tag{7.31}$$

is a consistent estimator of the parameter a.

If X is uniformly distributed over $(0, a)$, then from Eqs. (2.56), (2.57), and (4.122) of Prob. 4.36, the pdf of $Z = \max(X_1, \ldots, X_n)$ is

$$f_Z(z) = n f_X(z)[F_X(z)]^{n-1} = \frac{n}{a}\left(\frac{z}{a}\right)^{n-1} \qquad 0 < z < a \tag{7.32}$$

Thus,

$$E(A) = \int_0^a z f_z(z)\, dz = \frac{n}{a^n} \int_0^a z^n dz = \frac{n}{n+1} a$$

and

$$\lim_{n \to \infty} E(A) = a$$

Next,

$$E(A^2) = \int_0^a z^2 f_z(z)\, dz = \frac{n}{a^n} \int_0^a z^{n+1} dz = \frac{n}{n+2} a^2$$

$$\text{Var}(A) = E(A^2) - [E(A)]^2 = \frac{na^2}{n+2} - \frac{n^2 a^2}{(n+1)^2} = \frac{n}{(n+2)(n+1)^2} a^2$$

and

$$\lim_{n \to \infty} \text{Var}(A) = 0$$

Thus, by Eqs. (7.7) and (7.8), A is a consistent estimator of parameter a.

Maximum-Likelihood Estimation

7.7. Let $(X_1, \ldots, X_n)$ be a random sample of a binomial r.v. X with parameters (m, p), where m is assumed to be known and p unknown. Determine the maximum-likelihood estimator of p.

The likelihood function is given by [Eq. (2.36)]

$$L(p) = f(x_1, \ldots, x_n; p) = \binom{m}{x_1} p^{x_1} (1-p)^{(m-x_1)} \cdots \binom{m}{x_n} p^{x_n} (1-p)^{(m-x_n)}$$

$$= \binom{m}{x_1} \cdots \binom{m}{x_n} p^{\sum_{i=1}^n x_i} (1-p)^{(mn - \sum_{i=1}^n x_i)}$$

Taking the natural logarithm of the above expression, we get

$$\ln L(p) = \ln c + \left(\sum_{i=1}^n x_i \right) \ln p + \left(mn - \sum_{i=1}^n x_i \right) \ln(1-p)$$

where

$$c = \prod_{i=1}^n \binom{m}{x_i}$$

and

$$\frac{d}{dp} \ln L(p) = \frac{1}{p} \sum_{i=1}^n x_i - \frac{1}{1-p} \left(mn - \sum_{i=1}^n x_i \right)$$

Setting $d[\ln L(p)]/dp = 0$, the maximum-likelihood estimate $\hat{p}_{ML}$ of p is obtained as

$$\frac{1}{\hat{P}_{ML}} \sum_{i=1}^n x_i = \frac{1}{1 - \hat{P}_{ML}} \left(mn - \sum_{i=1}^n x_i \right)$$

or

$$\hat{P}_{ML} = \frac{1}{mn} \sum_{i=1}^n x_i \tag{7.33}$$

Hence, the maximum-likelihood estimator of p is given by

$$P_{ML} = \frac{1}{mn} \sum_{i=1}^n X_i = \frac{1}{m} \overline{X} \tag{7.34}$$

7.8. Let $(X_1, \ldots, X_n)$ be a random sample of a Poisson r.v. with unknown parameter λ. Determine the maximum-likelihood estimator of λ.

The likelihood function is given by [Eq. (2.48)]

$$L(\lambda) = f(x_1, \ldots, x_n; \lambda) = \prod_{i=1}^n \frac{e^{-\lambda} \lambda^{x_i}}{x_i!} = \frac{e^{-n\lambda} \lambda^{\sum_{i=1}^n x_i}}{x_1! \cdots x_n!}$$

Thus,
$$\ln L(\lambda) = -n\lambda + \ln \lambda \sum_{i=1}^{n} x_i - \ln c$$

where
$$c = \prod_{i=1}^{n} (x_i !)$$

and
$$\frac{d}{d\lambda} \ln L(\lambda) = -n + \frac{1}{\lambda} \sum_{i=1}^{n} x_i$$

Setting $d[\ln L(\lambda)]/d\lambda = 0$, the maximum-likelihood estimate $\hat{\lambda}_{ML}$ of λ is obtained as

$$\hat{\lambda}_{ML} = \frac{1}{n} \sum_{i=1}^{n} x_i \tag{7.35}$$

Hence, the maximum-likelihood estimator of λ is given by

$$\Lambda_{ML} = s(X_1, \ldots, X_n) = \frac{1}{n} \sum_{i=1}^{n} X_i = \overline{X} \tag{7.36}$$

7.9. Let $(X_1, \ldots, X_n)$ be a random sample of an exponential r.v. X with unknown parameter λ. Determine the maximum-likelihood estimator of λ.

The likelihood function is given by [Eq. (2.60)]

$$L(\lambda) = f(x_1, \ldots, x_n; \lambda) = \prod_{i=1}^{n} \lambda e^{-\lambda x_i} = \lambda^n e^{-\lambda \sum_{i=1}^{n} x_i}$$

Thus,
$$\ln L(\lambda) = n \ln \lambda - \lambda \sum_{i=1}^{n} x_i$$

and
$$\frac{d}{d\lambda} \ln L(\lambda) = \frac{n}{\lambda} - \sum_{i=1}^{n} x_i$$

Setting $d[\ln L(\lambda)]/d\lambda = 0$, the maximum-likelihood estimate $\hat{\lambda}_{ML}$ of λ is obtained as

$$\hat{\lambda}_{ML} = \frac{n}{\sum_{i=1}^{n} x_i} \tag{7.37}$$

Hence, the maximum-likelihood estimator of λ is given by

$$\Lambda_{ML} = s(X_1, \ldots, X_n) = \frac{n}{\sum_{i=1}^{n} X_i} = \frac{1}{\overline{X}} \tag{7.38}$$

7.10. Let $(X_1, \ldots, X_n)$ be a random sample of a normal random r.v. X with unknown mean μ and unknown variance σ^2. Determine the maximum-likelihood estimators of μ and σ^2.

The likelihood function is given by [Eq. (2.71)]

$$L(\mu, \sigma) = f(x_1, \ldots, x_n; \mu, \sigma) = \prod_{i=1}^{n} \frac{1}{\sqrt{2\pi}\sigma} \exp\left[-\frac{1}{2\sigma^2}(x_i - \mu)^2\right]$$

$$= \left(\frac{1}{2\pi}\right)^{n/2} \frac{1}{\sigma^n} \exp\left[-\frac{1}{2\sigma^2} \sum_{i=1}^{n} (x_i - \mu)^2\right]$$

Thus,
$$\ln L(\mu, \sigma) = -\frac{n}{2}\ln(2\pi) - n\ln\sigma - \frac{1}{2\sigma^2}\sum_{i=1}^{n}(x_i - \mu)^2$$

In order to find the values of μ and σ maximizing the above, we compute

$$\frac{\partial}{\partial\mu}\ln L(\mu, \sigma) = \frac{1}{\sigma^2}\sum_{i=1}^{n}(x_i - \mu)$$

$$\frac{\partial}{\partial\sigma}\ln L(\mu, \sigma) = -\frac{n}{\sigma} + \frac{1}{\sigma^3}\sum_{i=1}^{n}(x_i - \mu)^2$$

Equating these equations to zero, we get

$$\sum_{i=1}^{n}(x_i - \hat{\mu}_{ML}) = 0$$

and

$$\frac{1}{\hat{\sigma}_{ML}^{3}}\sum_{i=1}^{n}(x_i - \hat{\mu}_{ML})^2 = \frac{n}{\hat{\sigma}_{ML}}$$

Solving for $\hat{\mu}_{ML}$ and $\hat{\sigma}_{ML}$, the maximum-likelihood estimates of μ and σ^2 are given, respectively, by

$$\hat{\mu}_{ML} = \frac{1}{n}\sum_{i=1}^{n}x_i \tag{7.39}$$

$$\hat{\sigma}_{ML}^{2} = \frac{1}{n}\sum_{i=1}^{n}(x_i - \hat{\mu}_{ML})^2 \tag{7.40}$$

Hence, the maximum-likelihood estimators of μ and σ^2 are given, respectively, by

$$M_{ML} = \frac{1}{n}\sum_{i=1}^{n}X_i = \bar{X} \tag{7.41}$$

$$S_{ML}^{2} = \frac{1}{n}\sum_{i=1}^{n}(X_i - \bar{X})^2 \tag{7.42}$$

Bayes' Estimation

7.11. Let $(X_1, \ldots, X_n)$ be the random sample of a Bernoulli r.v. X with pmf given by [Eq. (2.32)]

$$f(x; p) = p^x(1 - p)^{1-x} \qquad x = 0, 1 \tag{7.43}$$

where $p, 0 \le p \le 1$, is unknown. Assume that p is a uniform r.v. over $(0, 1)$. Find the Bayes' estimator of p.

The prior pdf of p is the uniform pdf; that is,

$$f(p) = 1 \qquad 0 < p < 1$$

The posterior pdf of p is given by

$$f(p \,|\, x_1, \ldots, x_n) = \frac{f(x_1, \ldots, x_n, p)}{f(x_1, \ldots, x_n)}$$

Then, by Eq. (7.12),

$$f(x_1, \ldots, x_n, p) = f(x_1, \ldots, x_n \,|\, p)f(p)$$

$$= p^{\sum_{i=1}^{n}x_i}(1 - p)^{n - \sum_{i=1}^{n}x_i} = p^m(1 - p)^{n-m}$$

where $m = \sum_{i=1}^{n} x_i$, and by Eq. (7.13),

$$f(x_1, ..., x_n) = \int_0^1 f(x_1, ..., x_n, p)\, dp = \int_0^1 p^m (1-p)^{n-m}\, dp$$

Now, from calculus, for integers m and k, we have

$$\int_0^1 p^m (1-p)^k\, dp = \frac{m!\, k!}{(m+k+1)!} \qquad (7.44)$$

Thus, by Eq. (7.14), the posterior pdf of p is

$$f(p \mid x_1, ..., x_n) = \frac{f(x_1, ..., x_n, p)}{f(x_1, ..., x_n)} = \frac{(n+1)!\, p^m (1-p)^{n-m}}{m!\,(n-m)!}$$

and by Eqs. (7.15) and (7.44),

$$\begin{aligned}
E(p \mid x_1, ..., x_n) &= \int_0^1 p f(p \mid x_1, ..., x_n)\, dp \\
&= \frac{(n+1)!}{m!\,(n-m)!} \int_0^1 p^{m+1}(1-p)^{n-m}\, dp \\
&= \frac{(n+1)!}{m!\,(n-m)!}\, \frac{(m+1)!\,(n-m)!}{(n+2)!} \\
&= \frac{m+1}{n+2} = \frac{1}{n+2}\left(\sum_{i=1}^{n} x_i + 1 \right)
\end{aligned}$$

Hence, by Eq. (7.16), the Bayes' estimator of p is

$$P_B = E(p \mid X_1, ..., X_n) = \frac{1}{n+2}\left(\sum_{i=1}^{n} X_i + 1 \right) \qquad (7.45)$$

7.12. Let $(X_1, ..., X_n)$ be a random sample of an exponential r.v. X with unknown parameter λ. Assume that λ is itself to be an exponential r.v. with parameter α. Find the Bayes' estimator of λ.

The assumed prior pdf of λ is [Eq. (2.48)]

$$f(\lambda) = \begin{cases} \alpha e^{-\alpha \lambda} & \alpha, \lambda > 0 \\ 0 & \text{otherwise} \end{cases}$$

Now

$$f(x_1, ..., x_n \mid \lambda) = \prod_{i=1}^{n} \lambda e^{-\lambda x_i} = \lambda^n e^{-\lambda \sum_{i=1}^{n} x_i} = \lambda^n e^{-m\lambda}$$

where $m = \sum_{i=1}^{n} x_i$. Then, by Eqs. (7.12) and (7.13),

$$\begin{aligned}
f(x_1, ..., x_n) &= \int_0^\infty f(x_1, ..., x_n \mid \lambda) f(\lambda)\, d\lambda \\
&= \int_0^\infty \lambda^n e^{-m\lambda} \alpha e^{-\alpha \lambda}\, d\lambda \\
&= \alpha \int_0^\infty \lambda^n e^{-(\alpha + m)\lambda}\, d\lambda = \alpha\, \frac{n!}{(\alpha + m)^{n+1}}
\end{aligned}$$

By Eq. (7.14), the posterior pdf of λ is given by

$$f(\lambda \mid x_1, ..., x_n) = \frac{f(x_1, ..., x_n \mid \lambda) f(\lambda)}{f(x_1, ..., x_n)} = \frac{(\alpha + m)^{n+1} \lambda^n e^{-(\alpha + m)\lambda}}{n!}$$

Thus, by Eq. (7.15), the Bayes' estimate of λ is

$$
\begin{aligned}
\hat{\lambda}_B = E(\lambda \mid x_1, \ldots, x_n) &= \int_0^\infty \lambda f(\lambda \mid x_1, \ldots, x_n)\, d\lambda \\
&= \frac{(\alpha+m)^{n+1}}{n!} \int_0^\infty \lambda^{n+1} e^{-(\alpha+m)\lambda}\, d\lambda \\
&= \frac{(\alpha+m)^{n+1}}{n!} \frac{(n+1)!}{(\alpha+m)^{n+2}} \\
&= \frac{n+1}{\alpha+m} = \frac{n+1}{\alpha + \sum\limits_{i=1}^{n} x_i}
\end{aligned}
\tag{7.46}
$$

and the Bayes' estimator of λ is

$$
\Lambda_B = \frac{n+1}{\alpha + \sum\limits_{i=1}^{n} X_i} = \frac{n+1}{\alpha + n\overline{X}}
\tag{7.47}
$$

7.13. Let $(X_1, \ldots, X_n)$ be a random sample of a normal r.v. X with unknown mean μ and variance 1. Assume that μ is itself to be a normal r.v. with mean 0 and variance 1. Find the Bayes' estimator of μ.

The assumed prior pdf of μ is

$$
f(\mu) = \frac{1}{\sqrt{2\pi}} e^{-\mu^2/2}
$$

Then by Eq. (7.12),

$$
\begin{aligned}
f(x_1, \ldots, x_n, \mu) &= f(x_1, \ldots, x_n \mid \mu) f(\mu) \\
&= \frac{1}{(2\pi)^{n/2}} \exp\left[-\sum_{i=1}^{n} \frac{(x_i - \mu)^2}{2} \right] \frac{1}{\sqrt{2\pi}} e^{-\mu^2/2} \\
&= \frac{\exp\left(-\sum_{i=1}^{n} \frac{x_i^2}{2} \right)}{(2\pi)^{(n+1)/2}} \exp\left[-\frac{(n+1)}{2}\left(\mu^2 - \frac{2\mu}{n+1} \sum_{i=1}^{n} x_i \right) \right] \\
&= \frac{\exp\left(-\sum_{i=1}^{n} \frac{x_i^2}{2} \right) \exp\left[\frac{1}{2(n+1)}\left(\sum_{i=1}^{n} x_i \right)^2 \right]}{(2\pi)^{(n+1)/2}} \exp\left[-\frac{(n+1)}{2}\left(\mu - \frac{1}{n+1} \sum_{i=1}^{n} x_i \right)^2 \right]
\end{aligned}
$$

Then, by Eq. (7.14), the posterior pdf of μ is given by

$$
\begin{aligned}
f(\mu \mid x_1, \ldots, x_n) &= \frac{f(x_1, \ldots, x_n, \mu)}{\int_{-\infty}^{\infty} f(x_1, \ldots, x_n, \mu)\, d\mu} \\
&= C \exp\left[-\frac{(n+1)}{2}\left(\mu - \frac{1}{n+1} \sum_{i=1}^{n} x_i \right)^2 \right]
\end{aligned}
\tag{7.48}
$$

where $C = C(x_1, \ldots, x_n)$ is independent of μ. However, Eq. (7.48) is just the pdf of a normal r.v. with mean

$$
\frac{1}{n+1}\left(\sum_{i=1}^{n} x_i \right)
$$

and variance

$$\frac{1}{n+1}$$

Hence, the conditional distribution of μ given $x_1, \ldots, x_n$ is the normal distribution with mean

$$\frac{1}{n+1}\left(\sum_{i=1}^{n} x_i\right)$$

and variance

$$\frac{1}{n+1}$$

Thus, the Bayes' estimate of μ is given by

$$\hat{\mu}_B = E(\mu \mid x_1, \ldots, x_n) = \frac{1}{n+1}\sum_{i=1}^{n} x_i \tag{7.49}$$

and the Bayes' estimator of μ is

$$M_B = \frac{1}{n+1}\sum_{i=1}^{n} X_i = \frac{n}{n+1}\bar{X} \tag{7.50}$$

7.14. Let $(X_1, \ldots, X_n)$ be a random sample of a r.v. X with pdf $f(x; \theta)$, where θ is an unknown parameter. The statistics L and U determine a $100(1 - \alpha)$ percent *confidence interval* (L, U) for the parameter θ if

$$P(L < \theta < U) \geq 1 - \alpha \qquad 0 < \alpha < 1 \tag{7.51}$$

and $1 - \alpha$ is called the *confidence coefficient*. Find L and U if X is a normal r.v. with known variance σ^2 and mean μ is an unknown parameter.

If $X = N(\mu; \sigma^2)$, then

$$Z = \frac{\bar{X} - \mu}{\sigma/\sqrt{n}} \qquad \text{where} \qquad \bar{X} = \frac{1}{n}\sum_{i=1}^{n} X_i$$

is a standard normal r.v., and hence for a given α we can find a number $z_{\alpha/2}$ from Table A (Appendix A) such that

$$P\left(-z_{\alpha/2} < \frac{\bar{X} - \mu}{\sigma/\sqrt{n}} < z_{\alpha/2}\right) = 1 - \alpha \tag{7.52}$$

For example, if $1 - \alpha = 0.95$, then $z_{\alpha/2} = z_{0.025} = 1.96$, and if $1 - \alpha = 0.9$, then $z_{\alpha/2} = z_{0.05} = 1.645$. Now, recalling that $\sigma > 0$, we have the following equivalent inequality relationships;

$$-z_{\alpha/2} < \frac{\bar{X} - \mu}{\sigma/\sqrt{n}} < z_{\alpha/2}$$

$$-z_{\alpha/2}(\sigma/\sqrt{n}) < \bar{X} - \mu < z_{\alpha/2}(\sigma/\sqrt{n})$$

$$-\bar{X} - z_{\alpha/2}(\sigma/\sqrt{n}) < -\mu < -\bar{X} + z_{\alpha/2}(\sigma/\sqrt{n})$$

and

$$\bar{X} + z_{\alpha/2}(\sigma/\sqrt{n}) > \mu > \bar{X} - z_{\alpha/2}(\sigma/\sqrt{n})$$

Thus, we have

$$P[\bar{X} + z_{\alpha/2}(\sigma/\sqrt{n}) < \mu < \bar{X} + z_{\alpha/2}(\sigma/\sqrt{n})] = 1 - \alpha \tag{7.53}$$

and so

$$L = \bar{X} - z_{\alpha/2}(\sigma/\sqrt{n}) \qquad \text{and} \qquad U = \bar{X} + z_{\alpha/2}(\sigma/\sqrt{n}) \tag{7.54}$$

7.15. Consider a normal r.v. with variance 1.66 and unknown mean μ. Find the 95 percent confidence interval for the mean based on a random sample of size 10.

As shown in Prob. 7.14, for $1 - \alpha = 0.95$, we have $z_{\alpha/2} = z_{0.025} = 1.96$ and

$$z_{\alpha/2}(\sigma/\sqrt{n}) = 1.96(\sqrt{1.66}/\sqrt{10}) = 0.8$$

Thus, by Eq. (7.54), the 95 percent confidence interval for μ is

$$(\bar{X} - 0.8, \bar{X} + 0.8)$$

Mean Square Estimation

7.16. Find the m.s. estimate of a r.v. Y by a constant c.

By Eq. (7.17), the m.s. error is

$$e = E[(Y - c)^2] = \int_{-\infty}^{\infty}(y - c)^2 f(y)\, dy \tag{7.55}$$

Clearly the m.s. error e depends on c, and it is minimum if

$$\frac{de}{dc} = -2\int_{-\infty}^{\infty}(y - c)f(y)\, dy = 0$$

or

$$c\int_{-\infty}^{\infty} f(y)\, dy = c = \int_{-\infty}^{\infty} yf(y)\, dy$$

Thus, we conclude that the m.s. estimate c of Y is given by

$$\hat{y} = c = \int_{-\infty}^{\infty} yf(y)\, dy = E(Y) \tag{7.56}$$

7.17. Find the m.s. estimator of a r.v. Y by a function $g(X)$ of the r.v. X.

By Eq. (7.17), the m.s. error is

$$e = E\{[Y - g(X)]^2\} = \int_{-\infty}^{\infty}\int_{-\infty}^{\infty}[y - g(x)]^2 f(x, y)\, dx\, dy$$

Since $f(x, y) = f(y\,|\,x) f(x)$, we can write

$$e = \int_{-\infty}^{\infty} f(x)\left\{\int_{-\infty}^{\infty}[y - g(x)]^2 f(y\,|\,x)\, dy\right\} dx \tag{7.57}$$

Since the integrands above are positive, the m.s. error e is minimum if the inner integrand,

$$\int_{-\infty}^{\infty}[y - g(x)]^2 f(y\,|\,x)\, dy \tag{7.58}$$

is minimum for every x. Comparing Eq. (7.58) with Eq. (7.55) (Prob. 7.16), we see that they are the same form if c is changed to $g(x)$ and $f(y)$ is changed to $f(y|x)$. Thus, by the result of Prob. 7.16 [Eq. (7.56)], we conclude that the m.s. estimate of Y is given by

$$\hat{y} = g(x) = \int_{-\infty}^{\infty} yf(y|x)\,dy = E(Y|x) \tag{7.59}$$

Hence, the m.s. estimator of Y is

$$\hat{Y} = g(X) = E(Y|X) \tag{7.60}$$

7.18. Find the m.s. error if $g(x) = E(Y|x)$ is the m.s. estimate of Y.

As we see from Eq. (3.58), the conditional mean $E(Y|x)$ of Y, given that $X = x$, is a function of x, and by Eq. (4.39),

$$E[E(Y|X)] = E(Y) \tag{7.61}$$

Similarly, the conditional mean $E[g(X, Y)|x]$ of $g(X, Y)$, given that $X = x$, is a function of x. It defines, therefore, the function $E[g(X, Y)|X]$ of the r.v. X. Then

$$
\begin{aligned}
E\{E[g(X, Y)|X]\} &= \int_{-\infty}^{\infty}\left[\int_{-\infty}^{\infty} g(x, y)f(y|x)\,dy\right]f(x)\,dx \\
&= \int_{-\infty}^{\infty}\int_{-\infty}^{\infty} g(x, y)f(y|x)\,f(x)\,dx\,dy \\
&= \int_{-\infty}^{\infty}\int_{-\infty}^{\infty} g(x, y)f(x, y)\,dx\,dy = E[g(X, Y)]
\end{aligned} \tag{7.62}
$$

Note that Eq. (7.62) is the generalization of Eq. (7.61). Next, we note that

$$E[g_1(X)g_2(Y)|x] = E[g_1(x)g_2(Y)|x] = g_1(x)E[g_2(Y)|x] \tag{7.63}$$

Then by Eqs. (7.62) and (7.63), we have

$$E[g_1(X)g_2(Y)] = E\{E[g_1(X)g_2(Y)|X]\} = E\{g_1(X)E(g_2(Y)|X)\} \tag{7.64}$$

Now, setting $g_1(X) = g(X)$ and $g_2(Y) = Y$ in Eq. (7.64), and using Eq. (7.18), we obtain

$$E[g(X)Y] = E[g(X)E(Y|X)] = E[g^2(X)]$$

Thus, the m.s. error is given by

$$
\begin{aligned}
e &= E\{[Y - g(X)]^2\} = E(Y^2) - 2E[g(X)Y] + E[g^2(X)] \\
&= E(Y^2) - E[g^2(X)]
\end{aligned} \tag{7.65}
$$

7.19. Let $Y = X^2$ and X be a uniform r.v. over $(-1, 1)$. Find the m.s. estimator of Y in terms of X and its m.s. error.

By Eq. (7.18), the m.s. estimate of Y is given by

$$g(x) = E(Y|x) = E(X^2|X = x) = x^2$$

Hence, the m.s. estimator of Y is

$$\hat{Y} = X^2 \tag{7.66}$$

The m.s. error is

$$e = E\{[Y - g(X)]^2\} = E\{[X^2 - X^2]^2\} = 0 \tag{7.67}$$

Linear Mean Square Estimation

7.20. Derive the orthogonality principle (7.21) and Eq. (7.22).

By Eq. (7.20), the m.s. error is

$$e(a, b) = E\{[Y - (aX + b)]^2\}$$

Clearly, the m.s. error e is a function of a and b, and it is minimum if $\partial e/\partial a = 0$ and $\partial e/\partial b = 0$. Now

$$\frac{\partial e}{\partial a} = E\{2[Y - (aX + b)](-X)\} = -2E\{[Y - (aX + b)]X\}$$

$$\frac{\partial e}{\partial b} = E\{2[Y - (aX + b)](-1)\} = -2E\{[Y - (aX + b)]\}$$

Setting $\partial e/\partial a = 0$ and $\partial e/\partial b = 0$, we obtain

$$E\{[Y - (aX + b)]X\} = 0 \qquad\qquad (7.68)$$

$$E[Y - (aX + b)] = 0 \qquad\qquad (7.69)$$

Note that Eq. (7.68) is the orthogonality principle (7.21).

Rearranging Eqs. (7.68) and (7.69), we get

$$E(X^2)a + E(X)b = E(XY)$$

$$E(X)a + b = E(Y)$$

Solving for a and b, we obtain Eq. (7.22); that is,

$$a = \frac{E(XY) - E(X)E(Y)}{E(X^2) - [E(X)]^2} = \frac{\sigma_{XY}}{\sigma_X^2} = \frac{\sigma_Y}{\sigma_X}\rho_{XY}$$

$$b = E(Y) - aE(X) = \mu_Y - a\mu_X$$

where we have used Eqs. (2.31), (3.51), and (3.53).

7.21. Show that m.s. error defined by Eq. (7.20) is minimum when Eqs. (7.68) and (7.69) are satisfied.

Assume that $\hat{Y} = cX + d$, where c and d are arbitrary constants. Then

$$\begin{aligned}
e(c, d) &= E\{[Y - (cX + d)]^2\} = E\{[Y - (aX + b) + (a - c)X + (b - d)]^2\} \\
&= E\{[Y - (aX + b)]^2\} + E\{[(a - c)X + (b - d)]^2\} \\
&\quad + 2(a - c)E\{[Y - (aX + b)]X\} + 2(b - d)E\{[Y - (aX + b)]\} \\
&= e(a, b) + E\{[(a - c)X + (b - d)]^2\} \\
&\quad + 2(a - c)E\{[Y - (aX + b)]X\} + 2(b - d)E\{[Y - (aX + b)]\}
\end{aligned}$$

The last two terms on the right-hand side are zero when Eqs. (7.68) and (7.69) are satisfied, and the second term on the right-hand side is positive if $a \neq c$ and $b \neq d$. Thus, $e(c, d) \geq e(a, b)$ for any c and d. Hence, $e(a, b)$ is minimum.

7.22. Derive Eq. (7.23).

By Eqs. (7.68) and (7.69), we have

$$R\{[Y - (aX + b)]aX\} = 0 = E\{[Y - (aX + b)]b\}$$

Then $\qquad e_m = e(a, b) = E\{[Y - (aX + b)]^2\} = E\{[Y - (aX + b)][Y - (aX + b)]\}$

$$= E\{[Y - (aX + b)]Y\} = E(Y^2) - aE(XY) - bE(Y)$$

Using Eqs. (2.31), (3.51), and (3.53), and substituting the values of a and b [Eq. (7.22)] in the above expression, the minimum m.s. error is

$$e_m = \sigma_Y^2 + \mu_Y^2 - a(\sigma_{XY} + \mu_X\mu_Y) - (\mu_Y - a\mu_X)\mu_Y$$

$$= \sigma_Y^2 - a\sigma_{XY} = \sigma_Y^2 - \frac{\sigma_{XY}^2}{\sigma_X^2} = \sigma_Y^2\left(1 - \frac{\sigma_{XY}^2}{\sigma_X^2\sigma_Y^2}\right) = \sigma_Y^2(1 - \rho_{XY}^2)$$

which is Eq. (7.23).

7.23. Let $Y = X^2$, and let X be a uniform r.v. over $(-1, 1)$ (see Prob. 7.19). Find the linear m.s. estimator of Y in terms of X and its m.s. error.

The linear m.s. estimator of Y in terms of X is

$$\hat{Y} = aX + b$$

where a and b are given by [Eq. (7.22)]

$$a = \frac{\sigma_{XY}}{\sigma_X^2} \qquad b = \mu_Y - a\mu_X$$

Now, by Eqs. (2.58) and (2.56),

$$\mu_X = E(X) = 0$$

$$E(XY) = E(XX^2) = E(X^3) = \frac{1}{2}\int_{-1}^{1} x^3\, dx = 0$$

By Eq. (3.51),

$$\sigma_{XY} = \text{Cov}(XY) = E(XY) - E(X)E(Y) = 0$$

Thus, $a = 0$ and $b = E(Y)$, and the linear m.s. estimator of Y is

$$\hat{Y} = b = E(Y) \tag{7.70}$$

and the m.s. error is

$$e = E\{[Y - E(Y)]^2\} = \sigma_Y^2 \tag{7.71}$$

7.24. Find the minimum m.s. error estimator of Y in terms of X when X and Y are jointly normal r.v.'s.

By Eq. (7.18), the minimum m.s. error estimator of Y in terms of X is

$$\hat{Y} = E(Y|X)$$

Now, when X and Y are jointly normal, by Eq. (3.108) (Prob. 3.51), we have

$$E(Y|x) = \rho_{XY}\frac{\sigma_Y}{\sigma_X}x + \mu_Y - \rho_{XY}\frac{\sigma_Y}{\sigma_X}\mu_X$$

Hence, the minimum m.s. error estimator of Y is

$$\hat{Y} = E(Y|X) = \rho_{XY}\frac{\sigma_Y}{\sigma_X}X + \mu_Y - \rho_{XY}\frac{\sigma_Y}{\sigma_X}\mu_X \tag{7.72}$$

Comparing Eq. (7.72) with Eqs. (7.19) and (7.22), we see that for jointly normal r.v.'s the linear m.s. estimator is the minimum m.s. error estimator.

SUPPLEMENTARY PROBLEMS

7.25. Let $(X_1, \ldots, X_n)$ be a random sample of X having unknown mean μ and variance σ^2. Show that the estimator of σ^2 defined by

$$S_1^2 = \frac{1}{n-1} \sum_{i=1}^{n} (X_i - \bar{X})^2$$

where $\bar{X}$ is the sample mean, is an unbiased estimator of σ^2. Note that S_1^2 is often called the *sample variance*.

7.26. Let $(X_1, \ldots, X_n)$ be a random sample of X having known mean μ and unknown variance σ^2. Show that the estimator of σ^2 defined by

$$S_0^2 = \frac{1}{n} \sum_{i=1}^{n} (X_i - \mu)^2$$

is an unbiased estimator of σ^2.

7.27. Let $(X_1, \ldots, X_n)$ be a random sample of a binomial r.v. X with parameter (m, p), where p is unknown. Show that the maximum-likelihood estimator of p given by Eq. (7.34) is unbiased.

7.28. Let $(X_1, \ldots, X_n)$ be a random sample of a Bernoulli r.v. X with pmf $f(x; p) = p^x (1 - p)^{1-x}$, $x = 0, 1$, where p, $0 \le p \le 1$, is unknown. Find the maximum-likelihood estimator of p.

7.29. The values of a random sample, 2.9, 0.5, 1.7, 4.3, and 3.2, are obtained from a r.v. X that is uniformly distributed over the unknown interval (a, b). Find the maximum-likelihood estimates of a and b.

7.30. In analyzing the flow of traffic through a drive-in bank, the times (in minutes) between arrivals of 10 customers are recorded as 3.2, 2.1, 5.3, 4.2, 1.2, 2.8, 6.4, 1.5, 1.9, and 3.0. Assuming that the interarrival time is an exponential r.v. with parameter λ, find the maximum likelihood estimate of λ.

7.31. Let $(X_1, \ldots, X_n)$ be a random sample of a normal r.v. X with known mean μ and unknown variance σ^2. Find the maximum likelihood estimator of σ^2.

7.32. Let $(X_1, \ldots, X_n)$ be the random sample of a normal r.v. X with mean μ and variance σ^2, where μ is unknown. Assume that μ is itself to be a normal r.v. with mean μ_1 and variance σ_1^2. Find the Bayes' estimate of μ.

7.33. Let $(X_1, \ldots, X_n)$ be the random sample of a normal r.v. X with variance 100 and unknown μ. What sample size n is required such that the width of 95 percent confidence interval is 5?

7.34. Find a constant a such that if Y is estimated by aX, the m.s. error is minimum, and also find the minimum m.s. error e_m.

7.35. Derive Eqs. (7.25) and (7.26).

ANSWERS TO SUPPLEMENTARY PROBLEMS

7.25. *Hint:* Show that $S_1^2 = \dfrac{n}{n-1} S^2$, and use Eq. (7.29).

7.26. *Hint:* Proceed as in Prob. 7.2.

7.27. *Hint:* Use Eq. (2.38).

7.28. $P_{ML} = \dfrac{1}{n}\displaystyle\sum_{i=1}^{n} X_i = \bar{X}$

7.29. $\hat{a}_{ML} = \min_i x_i = 0.5, \qquad \hat{b}_{ML} = \max_i x_i = 4.3$

7.30. $\hat{\lambda}_{ML} = \dfrac{1}{3.16}$

7.31. $S_{ML}{}^2 = \dfrac{1}{n}\displaystyle\sum_{i=1}^{n}(X_i - \mu)^2$

7.32. $\hat{\mu}_B = \left(\dfrac{\mu_1}{\sigma_1{}^2} + \dfrac{n\bar{x}}{\sigma^2}\right)\Big/\left(\dfrac{1}{\sigma_1{}^2} + \dfrac{n}{\sigma^2}\right) \qquad \bar{x} = \dfrac{1}{n}\displaystyle\sum_{i=1}^{n} x_i$

7.33. $n = 62$

7.34. $a = E(XY)/E(X^2) \qquad e_m = E(Y^2) - [E(XY)]^2/[E(X)]^2$

7.35. *Hint:* Proceed as in Prob. 7.20.

CHAPTER 8

Decision Theory

8.1 Introduction

There are many situations in which we have to make decisions based on observations or data that are random variables. The theory behind the solutions for these situations is known as *decision theory* or *hypothesis testing*. In communication or radar technology, decision theory or hypothesis testing is known as (signal) detection theory. In this chapter we present a brief review of the binary decision theory and various decision tests.

8.2 Hypothesis Testing

A. Definitions:

A *statistical hypothesis* is an assumption about the probability law of r.v.'s. Suppose we observe a random sample $(X_1, \ldots, X_n)$ of a r.v. X whose pdf $f(\mathbf{x}; \theta) = f(\mathbf{x}_1, \ldots, x_n; \theta)$ depends on a parameter θ. We wish to test the assumption $\theta = \theta_0$ against the assumption $\theta = \theta_1$. The assumption $\theta = \theta_0$ is denoted by H_0 and is called the *null hypothesis*. The assumption $\theta = \theta_1$ is denoted by H_1 and is called the *alternative hypothesis*.

$$H_0: \quad \theta = \theta_0 \quad \text{(Null hypothesis)}$$
$$H_1: \quad \theta = \theta_1 \quad \text{(Alternative hypothesis)}$$

A hypothesis is called *simple* if all parameters are specified exactly. Otherwise it is called *composite*. Thus, suppose $H_0: \theta = \theta_0$ and $H_1: \theta \neq \theta_0$; then H_0 is simple and H_1 is composite.

B. Hypothesis Testing and Types of Errors:

Hypothesis testing is a decision process establishing the validity of a hypothesis. We can think of the decision process as dividing the observation space R^n (*Euclidean n-space*) into two regions R_0 and R_1. Let $\mathbf{x} = (x_1, \ldots, x_n)$ be the observed vector. Then if $\mathbf{x} \in R_0$, we will decide on H_0; if $\mathbf{x} \in R_1$, we decide on H_1. The region R_0 is known as the *acceptance region* and the region R_1 as the *rejection* (or *critical*) *region* (since the null hypothesis is rejected). Thus, with the observation vector (or data), one of the following four actions can happen:

1. H_0 true; accept H_0
2. H_0 true; reject H_0 (or accept H_1)
3. H_1 true; accept H_1
4. H_1 true; reject H_1 (or accept H_0)

The first and third actions correspond to correct decisions, and the second and fourth actions correspond to errors. The errors are classified as

1. Type I error: Reject H_0 (or accept H_1) when H_0 is true.
2. Type II error: Reject H_1 (or accept H_0) when H_1 is true.

Let P_I and P_{II} denote, respectively, the probabilities of Type I and Type II errors:

$$P_I = P(D_1 | H_0) = P(\mathbf{x} \in R_1; H_0) \tag{8.1}$$

$$P_{II} = P(D_0 | H_1) = P(\mathbf{x} \in R_0; H_1) \tag{8.2}$$

where D_i ($i = 0, 1$) denotes the event that the decision is made to accept H_i. P_I is often denoted by α and is known as the *level of significance,* and P_{II} is denoted by β and $(1 - \beta)$ is known as the *power of the test.* Note that since α and β represent probabilities of events from the same decision problem, they are not independent of each other or of the sample size n. It would be desirable to have a decision process such that both α and β will be small. However, in general, a decrease in one type of error leads to an increase in the other type for a fixed sample size (Prob. 8.4). The only way to simultaneously reduce both types of errors is to increase the sample size (Prob. 8.5). One might also attach some relative importance (or cost) to the four possible courses of action and minimize the total cost of the decision (see Sec. 8.3D).

The probabilities of correct decisions (actions 1 and 3) may be expressed as

$$P(D_0 | H_0) = P(\mathbf{x} \in R_0; H_0) \tag{8.3}$$

$$P(D_1 | H_1) = P(\mathbf{x} \in R_1; H_1) \tag{8.4}$$

In radar signal detection, the two hypotheses are

$$H_0: \quad \text{No target exists}$$

$$H_1: \quad \text{Target is present}$$

In this case, the probability of a Type I error $P_I = P(D_1 | H_0)$ is often referred to as the *false-alarm* probability (denoted by P_F), the probability of a Type II error $P_{II} = P(D_0 | H_1)$ as the *miss* probability (denoted by P_M), and $P(D_1 | H_1)$ as the *detection* probability (denoted by P_D). The cost of failing to detect a target cannot be easily determined. In general we set a value of P_F which is acceptable and seek a decision test that constrains P_F to this value while maximizing P_D (or equivalently minimizing P_M). This test is known as the *Neyman-Pearson* test (see Sec. 8.3C).

8.3   Decision Tests

A.   Maximum-Likelihood Test:

Let $\mathbf{x}$ be the observation vector and $P(\mathbf{x} | H_i)$, $i = 0.1$, denote the probability of observing $\mathbf{x}$ given that H_i was true. In the *maximum-likelihood* test, the decision regions R_0 and R_1 are selected as

$$R_0 = \{\mathbf{x}: P(\mathbf{x} | H_0) > P(\mathbf{x} | H_1)\}$$
$$R_1 = \{\mathbf{x}: P(\mathbf{x} | H_0) < P(\mathbf{x} | H_1)\} \tag{8.5}$$

Thus, the maximum-likelihood test can be expressed as

$$d(\mathbf{x}) = \begin{cases} H_0 & \text{if } P(\mathbf{x} | H_0) > P(\mathbf{x} | H_1) \\ H_1 & \text{if } P(\mathbf{x} | H_0) < P(\mathbf{x} | H_1) \end{cases} \tag{8.6}$$

The above decision test can be rewritten as

$$\frac{P(\mathbf{x} | H_1)}{P(\mathbf{x} | H_0)} \underset{H_0}{\overset{H_1}{\gtrless}} 1 \tag{8.7}$$

If we define the likelihood ratio $\Lambda(\mathbf{x})$ as

$$\Lambda(\mathbf{x}) = \frac{P(\mathbf{x} \mid H_1)}{P(\mathbf{x} \mid H_0)} \tag{8.8}$$

then the maximum-likelihood test (8.7) can be expressed as

$$\Lambda(\mathbf{x}) \underset{H_0}{\overset{H_1}{\gtrless}} 1 \tag{8.9}$$

which is called the *likelihood ratio test,* and 1 is called the *threshold value* of the test.

Note that the likelihood ratio $\Lambda(\mathbf{x})$ is also often expressed as

$$\Lambda(\mathbf{x}) = \frac{f(\mathbf{x} \mid H_1)}{f(\mathbf{x} \mid H_0)} \tag{8.10}$$

B. MAP Test:

Let $P(H_i \mid \mathbf{x})$, $i = 0, 1$, denote the probability that H_i was true given a particular value of $\mathbf{x}$. The conditional probability $P(H_i \mid \mathbf{x})$ is called *a posteriori* (or posterior) probability, that is, a probability that is computed after an observation has been made. The probability $P(H_i)$, $i = 0, 1$, is called *a priori* (or prior) probability. In the *maximum a posteriori* (MAP) test, the decision regions R_0 and R_1 are selected as

$$R_0 = \{\mathbf{x} \colon P(H_0 \mid \mathbf{x}) > P(H_1 \mid \mathbf{x})\} \tag{8.11}$$
$$R_1 = \{\mathbf{x} \colon P(H_0 \mid \mathbf{x}) < P(H_1 \mid \mathbf{x})\}$$

Thus, the MAP test is given by

$$d(\mathbf{x}) = \begin{cases} H_0 & \text{if } P(H_0 \mid \mathbf{x}) > P(H_1 \mid \mathbf{x}) \\ H_1 & \text{if } P(H_0 \mid \mathbf{x}) < P(H_1 \mid \mathbf{x}) \end{cases} \tag{8.12}$$

which can be rewritten as

$$\frac{P(H_1 \mid \mathbf{x})}{P(H_0 \mid \mathbf{x})} \underset{H_0}{\overset{H_1}{\gtrless}} 1 \tag{8.13}$$

Using Bayes' rule [Eq. (1.58)], Eq. (8.13) reduces to

$$\frac{P(\mathbf{x} \mid H_1)P(H_1)}{P(\mathbf{x} \mid H_0)P(H_0)} \underset{H_0}{\overset{H_1}{\gtrless}} 1 \tag{8.14}$$

Using the likelihood ratio $\Lambda(\mathbf{x})$ defined in Eq. (8.8), the MAP test can be expressed in the following likelihood ratio test as

$$\Lambda(\mathbf{x}) \underset{H_0}{\overset{H_1}{\gtrless}} \eta = \frac{P(H_0)}{P(H_1)} \tag{8.15}$$

where $\eta = P(H_0)/P(H_1)$ is the threshold value for the MAP test. Note that when $P(H_0) = P(H_1)$, the maximum-likelihood test is also the MAP test.

C. Neyman-Pearson Test:

As mentioned, it is not possible to simultaneously minimize both $\alpha(= P_\text{I})$ and $\beta(= P_\text{II})$. The Neyman-Pearson test provides a workable solution to this problem in that the test minimizes β for a given level of α. Hence, the Neyman-Pearson test is the test which maximizes the power of the test $1 - \beta$ for a given level of significance α. In the

Neyman-Pearson test, the critical (or rejection) region R_1 is selected such that $1 - \beta = 1 - P(D_0|H_1) = P(D_1|H_1)$ is maximum subject to the constraint $\alpha = P(D_1|H_0) = \alpha_0$. This is a classical problem in optimization: maximizing a function subject to a constraint, which can be solved by the use of Lagrange multiplier method. We thus construct the objective function

$$J = (1 - \beta) - \lambda(\alpha - \alpha_0) \tag{8.16}$$

where $\lambda \geq 0$ is a Lagrange multiplier. Then the critical region R_1 is chosen to maximize J. It can be shown that the Neyman-Pearson test can be expressed in terms of the likelihood ratio test as (Prob. 8.8)

$$\Lambda(\mathbf{x}) \underset{H_0}{\overset{H_1}{\gtrless}} \eta = \lambda \tag{8.17}$$

where the threshold value η of the test is equal to the Lagrange multiplier λ, which is chosen to satisfy the constraint $\alpha = \alpha_0$.

D. Bayes' Test:

Let C_{ij} be the cost associated with (D_i, H_j), which denotes the event that we accept H_i when H_j is true. Then the average cost, which is known as the *Bayes' risk,* can be written as

$$\bar{C} = C_{00} P(D_0, H_0) + C_{10} P(D_1, H_0) + C_{01} P(D_0, H_1) + C_{11} P(D_1, H_1) \tag{8.18}$$

where $P(D_i, H_j)$ denotes the probability that we accept H_i when H_j is true. By Bayes' rule (1.42), we have

$$\bar{C} = C_{00} P(D_0|H_0)P(H_0) + C_{10} P(D_1|H_0)P(H_0) + C_{01}P(D_0|H_1)P(H_1) + C_{11}P(D_1|H_1)P(H_1) \tag{8.19}$$

In general, we assume that

$$C_{10} > C_{00} \qquad \text{and} \qquad C_{01} > C_{11} \tag{8.20}$$

since it is reasonable to assume that the cost of making an incorrect decision is higher than the cost of making a correct decision. The test that minimizes the average cost $\bar{C}$ is called the Bayes' test, and it can be expressed in terms of the likelihood ratio test as (Prob. 8.10)

$$\Lambda(\mathbf{x}) \underset{H_0}{\overset{H_1}{\gtrless}} \eta = \frac{(C_{10} - C_{00})P(H_0)}{(C_{01} - C_{11})P(H_1)} \tag{8.21}$$

Note that when $C_{10} - C_{00} = C_{01} - C_{11}$, the Bayes' test (8.21) and the MAP test (8.15) are identical.

E. Minimum Probability of Error Test:

If we set $C_{00} = C_{11} = 0$ and $C_{01} = C_{10} = 1$ in Eq. (8.18), we have

$$\bar{C} = P(D_1, H_0) + P(D_0, H_1) = P_e \tag{8.22}$$

which is just the probability of making an incorrect decision. Thus, in this case, the Bayes' test yields the minimum probability of error, and Eq. (8.21) becomes

$$\Lambda(\mathbf{x}) \underset{H_0}{\overset{H_1}{\gtrless}} \eta = \frac{P(H_0)}{P(H_1)} \tag{8.23}$$

We see that the minimum probability of error test is the same as the MAP test.

F. Minimax Test:

We have seen that the Bayes' test requires the a priori probabilities $P(H_0)$ and $P(H_1)$. Frequently, these probabilities are not known. In such a case, the Bayes' test cannot be applied, and the following minimax (min-max) test may be used. In the minimax test, we use the Bayes' test which corresponds to the least favorable $P(H_0)$ (Prob. 8.12). In the minimax test, the critical region R_1^* is defined by

$$\max_{P(H_0)} \bar{C}[P(H_0), R_1^*] = \min_{R_1} \max_{P(H_0)} \bar{C}[P(H_0), R_1] < \max_{P(H_0)} \bar{C}[P(H_0), R_1] \tag{8.24}$$

for all $R_1 \neq R_1^*$. In other words, R_1^* is the critical region which yields the minimum Bayes' risk for the least favorable $P(H_0)$. Assuming that the minimization and maximization operations are interchangeable, then we have

$$\min_{R_1} \max_{P(H_0)} \bar{C}[P(H_0), R_1] = \max_{P(H_0)} \min_{R_1} \bar{C}[P(H_0), R_1] \tag{8.25}$$

The minimization of $\bar{C}[P(H_0), R_1]$ with respect to R_1 is simply the Bayes' test, so that

$$\min_{R_1} \bar{C}[P(H_0), R_1] = C^*[P(H_0)] \tag{8.26}$$

where $C^*[P(H_0)]$ is the minimum Bayes' risk associated with the a priori probability $P(H_0)$. Thus, Eq. (8.25) states that we may find the minimax test by finding the Bayes' test for the least favorable $P(H_0)$, that is, the $P(H_0)$ which maximizes $\bar{C}[P(H_0)]$.

SOLVED PROBLEMS

Hypothesis Testing

8.1. Suppose a manufacturer of memory chips observes that the probability of chip failure is $p = 0.05$. A new procedure is introduced to improve the design of chips. To test this new procedure, 200 chips could be produced using this new procedure and tested. Let r.v. X denote the number of these 200 chips that fail. We set the test rule that we would accept the new procedure if $X \leq 5$. Let

$$H_0: \quad p = 0.05 \qquad \text{(No change hypothesis)}$$
$$H_1: \quad p < 0.05 \qquad \text{(Improvement hypothesis)}$$

Find the probability of a Type I error.

If we assume that these tests using the new procedure are independent and have the same probability of failure on each test, then X is a binomial r.v. with parameters $(n, p) = (200, p)$. We make a Type I error if $X \leq 5$ when in fact $p = 0.05$. Thus, using Eq. (2.37), we have

$$P_I = P(D_1 | H_0) = P(X \leq 5; p = 0.05)$$

$$= \sum_{k=0}^{5} \binom{200}{k} (0.05)^k (0.95)^{200-k}$$

Since n is rather large and p is small, these binomial probabilities can be approximated by Poisson probabilities with $\lambda = np = 200(0.05) = 10$ (see Prob. 2.43). Thus, using Eq. (2.119), we obtain

$$P_I \approx \sum_{k=0}^{5} e^{-10} \frac{10^k}{k!} = 0.067$$

Note that H_0 is a simple hypothesis but H_1 is a composite hypothesis.

8.2. Consider again the memory chip manufacturing problem of Prob. 8.1. Now let

$$H_0: \quad p = 0.05 \qquad \text{(No change hypothesis)}$$
$$H_1: \quad p = 0.02 \qquad \text{(Improvement hypothesis)}$$

Again our rule is, we would reject the new procedure if $X > 5$. Find the probability of a Type II error.

Now both hypotheses are simple. We make a Type II error if $X > 5$ when in fact $p = 0.02$. Hence, by Eq. (2.37),

$$P_{\text{II}} = P(D_0 | H_1) = P(X > 5; p = 0.02)$$
$$= \sum_{k=6}^{\infty} \binom{200}{k} (0.02)^k (0.98)^{200-k}$$

Again using the Poisson approximation with $\lambda = np = 200(0.02) = 4$, we obtain

$$P_{\text{II}} \approx 1 - \sum_{k=0}^{5} e^{-4} \frac{4^k}{k!} = 0.215$$

8.3. Let $(X_1, \ldots, X_n)$ be a random sample of a normal r.v. X with mean μ and variance 100. Let

$$H_0: \quad \mu = 50$$
$$H_1: \quad \mu = \mu_1 \; (>50)$$

and sample size $n = 25$. As a decision procedure, we use the rule to reject H_0 if $\bar{x} \geq 52$, where $\bar{x}$ is the value of the sample mean $\bar{X}$ defined by Eq. (7.27).

(a) Find the probability of rejecting H_0: $\mu = 50$ as a function of $\mu \, (> 50)$.

(b) Find the probability of a Type I error α.

(c) Find the probability of a Type II error β (i) when $\mu_1 = 53$ and (ii) when $\mu_1 = 55$.

(a) Since the test calls for the rejection of H_0: $\mu = 50$ when $\bar{x} \geq 52$, the probability of rejecting H_0 is given by

$$g(\mu) = P(\bar{X} \geq 52; \mu) \qquad (8.27)$$

Now, by Eqs. (4.136) and (7.27), we have

$$\text{Var}(\bar{X}) = \sigma_{\bar{X}}^2 = \frac{1}{n}\sigma^2 = \frac{100}{25} = 4$$

Thus, $\bar{X}$ is $N(\mu; 4)$, and using Eq. (2.74), we obtain

$$g(\mu) = P\left(\frac{\bar{X} - \mu}{2} \geq \frac{52 - \mu}{2}; \mu\right) = 1 - \Phi\left(\frac{52 - \mu}{2}\right) \qquad \mu \geq 50 \qquad (8.28)$$

The function $g(\mu)$ is known as the *power function of the test,* and the value of $g(\mu)$ at $\mu = \mu_1$, $g(\mu_1)$, is called the *power* at μ_1.

(b) Note that the power at $\mu = 50$, $g(50)$, is the probability of rejecting H_0: $\mu = 50$ when H_0 is true—that is, a Type I error. Thus, using Table A (Appendix A), we obtain

$$\alpha = P_1 = g(50) = 1 - \Phi\left(\frac{52 - 50}{2}\right) = 1 - \Phi(1) = 0.1587$$

(c)   Note that the power at $\mu = \mu_1$, $g(\mu_1)$, is the probability of rejecting H_0: $\mu = 50$ when $\mu = \mu_1$. Thus, $1 - g(\mu_1)$ is the probability of accepting H_0 when $\mu = \mu_1$—that is, the probability of a Type II error β.

(i)   Setting $\mu = \mu_1 = 53$ in Eq. (8.28) and using Table A (Appendix A), we obtain

$$\beta = P_{II} = 1 - g(53) = \Phi\left(\frac{52-53}{2}\right) = \Phi\left(-\frac{1}{2}\right) = 1 - \Phi\left(\frac{1}{2}\right) = 0.3085$$

(ii)   Similarly, for $\mu = \mu_1 = 55$ we obtain

$$\beta = P_{II} = 1 - g(55) = \Phi\left(\frac{52-55}{2}\right) = \Phi\left(-\frac{3}{2}\right) = 1 - \Phi\left(\frac{3}{2}\right) = 0.0668$$

Notice that clearly, the probability of a Type II error depends on the value of μ_1.

8.4.   Consider the binary decision problem of Prob. 8.3. We modify the decision rule such that we reject H_0 if $x \geq c$.

(a)   Find the value of c such that the probability of a Type I error $\alpha = 0.05$.

(b)   Find the probability of a Type II error β when $\mu_1 = 55$ with the modified decision rule.

(a)   Using the result of part (b) in Prob. 8.3, c is selected such that [see Eq. (8.27)]

$$\alpha = g(50) = P(\bar{X} \geq c; \mu = 50) = 0.05$$

However, when $\mu = 50$, $\bar{X} = N(50; 4)$, and [see Eq. (8.28)]

$$g(50) = P\left(\frac{\bar{X}-50}{2} \geq \frac{c-50}{2}; \mu = 50\right) = 1 - \Phi\left(\frac{c-50}{2}\right) = 0.05$$

From Table A (Appendix A), we have $\Phi(1.645) = 0.95$. Thus,

$$\frac{c-50}{2} = 1.645 \quad \text{and} \quad c = 50 + 2(1.645) = 53.29$$

(b)   The power function $g(\mu)$ with the modified decision rule is

$$g(\mu) = P(\bar{X} \geq 53.29; \mu) = P\left(\frac{\bar{X}-\mu}{2} \geq \frac{53.29-\mu}{2}; \mu\right) = 1 - \Phi\left(\frac{53.29-\mu}{2}\right)$$

Setting $\mu = \mu_1 = 55$ and using Table A (Appendix A), we obtain

$$\beta = P_{II} = 1 - g(55) = \Phi\left(\frac{53.29-55}{2}\right) = \Phi(-0.855)$$
$$= 1 - \Phi(0.855) = 0.1963$$

Comparing with the results of Prob. 8.3, we notice that with the change of the decision rule, α is reduced from 0.1587 to 0.05, but β is increased from 0.0668 to 0.1963.

8.5.   Redo Prob. 8.4 for the case where the sample size $n = 100$.

(a)   With $n = 100$, we have

$$\text{Var}(\bar{X}) = \sigma_{\bar{X}}^2 = \frac{1}{n}\sigma^2 = \frac{100}{100} = 1$$

As in part (a) of Prob. 8.4, c is selected so that

$$\alpha = g(50) = P(\bar{X} \geq c; \mu = 50) = 0.05$$

Since $\bar{X} = N(50; 1)$, we have

$$g(50) = P\left(\frac{\bar{X} - 50}{1} \geq \frac{c - 50}{1}; \mu = 50\right) = 1 - \Phi(c - 50) = 0.05$$

Thus, $$c - 50 = 1.645 \quad \text{and} \quad c = 51.645$$

(b) The power function is

$$g(\mu) = P(\bar{X} \geq 51.645; \mu)$$

$$= P\left(\frac{\bar{X} - \mu}{1} \geq \frac{51.645 - \mu}{1}; \mu\right) = 1 - \Phi(51.645 - \mu)$$

Setting $\mu = \mu_1 = 55$ and using Table A (Appendix A), we obtain

$$\beta = P_{II} = 1 - g(55) = \Phi(51.645 - 55) = \Phi(-3.355) \approx 0.0004$$

Notice that with sample size $n = 100$, both α and β have decreased from their respective original values of 0.1587 and 0.0668 when $n = 25$.

Decision Tests

8.6. In a simple binary communication system, during every T seconds, one of two possible signals $s_0(t)$ and $s_1(t)$ is transmitted. Our two hypotheses are

$$H_0: \quad s_0(t) \text{ was transmitted.}$$
$$H_1: \quad s_1(t) \text{ was transmitted.}$$

We assume that

$$s_0(t) = 0 \quad \text{and} \quad s_1(t) = 1 \quad 0 < t < T$$

The communication channel adds noise $n(t)$, which is a zero-mean normal random process with variance 1. Let $x(t)$ represent the received signal:

$$x(t) = s_i(t) + n(t) \quad i = 0, 1$$

We observe the received signal $x(t)$ at some instant during each signaling interval. Suppose that we received an observation $x = 0.6$.

(a) Using the maximum likelihood test, determine which signal is transmitted.

(b) Find P_I and P_{II}.

(a) The received signal under each hypothesis can be written as

$$H_0: \quad x = n$$
$$H_1: \quad x = 1 + n$$

Then the pdf of x under each hypothesis is given by

$$f(x|H_0) = \frac{1}{\sqrt{2\pi}} e^{-x^2/2}$$

$$f(x|H_1) = \frac{1}{\sqrt{2\pi}} e^{-(x-1)^2/2}$$

The likelihood ratio is then given by

$$\Lambda(x) = \frac{f(x|H_1)}{f(x|H_0)} = e^{(x-1/2)}$$

By Eq. (8.9), the maximum likelihood test is

$$e^{(x-1/2)} \underset{H_0}{\overset{H_1}{\gtrless}} 1$$

Taking the natural logarithm of the above expression, we get

$$x - \frac{1}{2} \underset{H_0}{\overset{H_1}{\gtrless}} 0 \qquad \text{or} \qquad x \underset{H_0}{\overset{H_1}{\gtrless}} \frac{1}{2}$$

Since $x = 0.6 > \frac{1}{2}$, we determine that signal $s_1(t)$ was transmitted.

(b)   The decision regions are given by

$$R_0 = \left\{ x : x < \frac{1}{2} \right\} = \left(-\infty, \frac{1}{2} \right) \qquad R_1 = \left\{ x : x > \frac{1}{2} \right\} = \left(\frac{1}{2}, \infty \right)$$

Then by Eqs. (8.1) and (8.2) and using Table A (Appendix A), we obtain

$$P_{\text{I}} = P(D_1|H_0) = \int_{R_1} f(x|H_0)\, dx = \frac{1}{\sqrt{2\pi}} \int_{1/2}^{\infty} e^{-x^2/2}\, dx = 1 - \Phi\left(\frac{1}{2}\right) = 0.3085$$

$$P_{\text{II}} = P(D_0|H_1) = \int_{R_0} f(x|H_1)\, dx = \frac{1}{\sqrt{2\pi}} \int_{-\infty}^{1/2} e^{-(x-1)^2/2}\, dx$$

$$= \frac{1}{\sqrt{2\pi}} \int_{-\infty}^{-1/2} e^{-y^2/2}\, dy = \Phi\left(-\frac{1}{2}\right) = 0.3085$$

8.7. In the binary communication system of Prob. 8.6, suppose that $P(H_0) = \frac{2}{3}$ and $P(H_1) = \frac{1}{3}$.

(a)   Using the MAP test, determine which signal is transmitted when $x = 0.6$.

(b)   Find P_{I} and P_{II}.

(a)   Using the result of Prob. 8.6 and Eq. (8.15), the MAP test is given by

$$e^{(x-1/2)} \underset{H_0}{\overset{H_1}{\gtrless}} \frac{P(H_0)}{P(H_1)} = 2$$

Taking the natural logarithm of the above expression, we get

$$x - \frac{1}{2} \underset{H_0}{\overset{H_1}{\gtrless}} \ln 2 \qquad \text{or} \qquad x \underset{H_0}{\overset{H_1}{\gtrless}} \frac{1}{2} + \ln 2 = 1.193$$

Since $x = 0.6 < 1.193$, we determine that signal $s_0(t)$ was transmitted.

(b)   The decision regions are given by

$$R_0 = \{x: x < 1.193\} = (-\infty, 1.193)$$
$$R_1 = \{x: x > 1.193\} = (1.193, \infty)$$

Thus, by Eqs. (8.1) and (8.2) and using Table A (Appendix A), we obtain

$$P_{\text{I}} = P(D_1 | H_0) = \int_{R_1} f(x | H_0)\, dx = \frac{1}{\sqrt{2\pi}} \int_{1.193}^{\infty} e^{-x^2/2}\, dx = 1 - \Phi(1.193) = 0.1164$$

$$P_{\text{II}} = P(D_0 | H_1) = \int_{R_0} f(x | H_1)\, dx = \frac{1}{\sqrt{2\pi}} \int_{-\infty}^{1.193} e^{-(x-1)^2/2}\, dx$$

$$= \frac{1}{\sqrt{2\pi}} \int_{-\infty}^{0.193} e^{-y^2/2}\, dy = \Phi(0.193) = 0.5765$$

8.8.   Derive the Neyman-Pearson test, Eq. (8.17).

From Eq. (8.16), the objective function is

$$J = (1 - \beta) - \lambda(\alpha - \alpha_0) = P(D_1 | H_1) - \lambda[P(D_1 | H_0) - \alpha_0] \tag{8.29}$$

where λ is an undetermined Lagrange multiplier which is chosen to satisfy the constraint $\alpha = \alpha_0$. Now, we wish to choose the critical region R_1 to maximize J. Using Eqs. (8.1) and (8.2), we have

$$J = \int_{R_1} f(\mathbf{x} | H_1)\, d\mathbf{x} - \lambda \left[\int_{R_1} f(\mathbf{x} | H_0)\, d\mathbf{x} - \alpha_0 \right]$$

$$= \int_{R_1} [f(\mathbf{x} | H_1) - \lambda f(\mathbf{x} | H_0)]\, d\mathbf{x} + \lambda \alpha_0 \tag{8.30}$$

To maximize J by selecting the critical region R_1, we select $\mathbf{x} \in R_1$ such that the integrand in Eq. (8.30) is positive. Thus, R_1 is given by

$$R_1 = \{\mathbf{x}: [f(\mathbf{x} | H_1) - \lambda\, f(\mathbf{x} | H_0)] > 0\}$$

and the Neyman-Pearson test is given by

$$\Lambda(\mathbf{x}) = \frac{f(\mathbf{x} | H_1)}{f(\mathbf{x} | H_0)} \underset{H_0}{\overset{H_1}{\gtrless}} \lambda$$

and λ is determined such that the constraint

$$\alpha = P_{\text{I}} = P(D_1 | H_0) = \int_{R_1} f(\mathbf{x} | H_0)\, d\mathbf{x} = \alpha_0$$

is satisfied.

8.9.   Consider the binary communication system of Prob. 8.6 and suppose that we require that $\alpha = P_{\text{I}} = 0.25$.

(a)   Using the Neyman-Pearson test, determine which signal is transmitted when $x = 0.6$.

(b)   Find P_{II}.

(a)   Using the result of Prob. 8.6 and Eq. (8.17), the Neyman-Pearson test is given by

$$e^{(x-1/2)} \underset{H_0}{\overset{H_1}{\gtrless}} \lambda$$

Taking the natural logarithm of the above expression, we get

$$x - \frac{1}{2} \underset{H_0}{\overset{H_1}{\gtrless}} \ln \lambda \qquad \text{or} \qquad x \underset{H_0}{\overset{H_1}{\gtrless}} \frac{1}{2} + \ln \lambda$$

The critical region R_1 is thus

$$R_1 = \left\{ x : x > \frac{1}{2} + \ln \lambda \right\}$$

Now we must determine λ such that $\alpha = P_{\mathrm{I}} = P(D_1 | H_0) = 0.25$. By Eq. (8.1), we have

$$P_{\mathrm{I}} = P(D_1 | H_0) = \int_{R_1} f(x | H_0)\, dx = \frac{1}{\sqrt{2\pi}} \int_{1/2 + \ln \lambda}^{\infty} e^{-x^2/2}\, dx = 1 - \Phi\left(\frac{1}{2} + \ln \lambda \right)$$

Thus,
$$1 - \Phi\left(\frac{1}{2} + \ln \lambda \right) = 0.25 \qquad \text{or} \qquad \Phi\left(\frac{1}{2} + \ln \lambda \right) = 0.75$$

From Table A (Appendix A), we find that $\Phi(0.674) = 0.75$. Thus,

$$\frac{1}{2} + \ln \lambda = 0.674 \rightarrow \lambda = 1.19$$

Then the Neyman-Pearson test is

$$x \underset{H_0}{\overset{H_1}{\gtrless}} 0.674$$

Since $x = 0.6 < 0.674$, we determine that signal $s_0(t)$ was transmitted.

(b) By Eq. (8.2), we have

$$P_{\mathrm{II}} = P(D_0 | H_1) = \int_{R_0} f(x | H_1)\, dx = \frac{1}{\sqrt{2\pi}} \int_{-\infty}^{0.674} e^{-(x-1)^2/2}\, dx$$

$$= \frac{1}{\sqrt{2\pi}} \int_{-\infty}^{-0.326} e^{-y^2/2}\, dy = \Phi(-0.326) = 0.3722$$

8.10. Derive Eq. (8.21).

By Eq. (8.19), the Bayes' risk is

$$\overline{C} = C_{00}\, P(D_0 | H_0) P(H_0) + C_{10}\, P(D_1 | H_0) P(H_0) + C_{01} P(D_0 | H_1) P(H_1) + C_{11} P(D_1 | H_1) P(H_1)$$

Now we can express

$$P(D_i | H_j) = \int_{R_i} f(\mathbf{x} | H_j)\, d\mathbf{x} \qquad i = 0,1;\ j = 0,1 \tag{8.31}$$

Then $\overline{C}$ can be expressed as

$$\overline{C} = C_{00}\, P(H_0) \int_{R_0} f(\mathbf{x} | H_0)\, d\mathbf{x} + C_{10}\, P(H_0) \int_{R_1} f(\mathbf{x} | H_0)\, d\mathbf{x}$$
$$+ C_{01} P(H_1) \int_{R_0} f(\mathbf{x} | H_1)\, d\mathbf{x} + C_{11} P(H_1) \int_{R_1} f(\mathbf{x} | H_1)\, d\mathbf{x} \tag{8.32}$$

Since $R_0 \cup R_1 = S$ and $R_0 \cap R_1 = \varphi$, we can write

$$\int_{R_0} f(\mathbf{x} | H_j)\, d\mathbf{x} = \int_{S} f(\mathbf{x} | H_j)\, d\mathbf{x} - \int_{R_1} f(\mathbf{x} | H_j)\, d\mathbf{x} = 1 - \int_{R_1} f(\mathbf{x} | H_j)\, d\mathbf{x}$$

Then Eq. (8.32) becomes

$$\overline{C} = C_{00} P(H_0) + C_{01} P(H_1) + \int_{R_1} \{ [(C_{10} - C_{00}) P(H_0) f(\mathbf{x} | H_0)] - [(C_{01} - C_{11}) P(H_1) f(\mathbf{x} | H_1)] \}\, d\mathbf{x}$$

The only variable in the above expression is the critical region R_1. By the assumptions [Eq. (8.20)] $C_{10} > C_{00}$ and $C_{01} > C_{11}$, the two terms inside the brackets in the integral are both positive. Thus, $\bar{C}$ is minimized if R_1 is chosen such that

$$(C_{01} - C_{11})P(H_1)\, f(\mathbf{x}|H_1) > (C_{10} - C_{00})P(H_0)\, f(\mathbf{x}|H_0)$$

for all $\mathbf{x} \in R_1$. That is, we decide to accept H_1 if

$$(C_{01} - C_{11})P(H_1)\, f(\mathbf{x}|H_1) > (C_{10} - C_{00})P(H_0)\, f(\mathbf{x}|H_0)$$

In terms of the likelihood ratio, we obtain

$$\Lambda(\mathbf{x}) = \frac{f(\mathbf{x}|H_1)}{f(\mathbf{x}|H_0)} \underset{H_0}{\overset{H_1}{\gtrless}} \frac{(C_{10} - C_{00})P(H_0)}{(C_{01} - C_{11})P(H_1)}$$

which is Eq. (8.21).

8.11. Consider a binary decision problem with the following conditional pdf's:

$$f(x|H_0) = \frac{1}{2}e^{-|x|}$$

$$f(x|H_1) = e^{-2|x|}$$

The Bayes' costs are given by

$$C_{00} = C_{11} = 0 \qquad C_{01} = 2 \qquad C_{10} = 1$$

(a) Determine the Bayes' test if $P(H_0) = \frac{2}{3}$ and the associated Bayes' risk.

(b) Repeat (a) with $P(H_0) = \frac{1}{2}$.

(a) The likelihood ratio is

$$\Lambda(x) = \frac{f(x|H_1)}{f(x|H_0)} = \frac{e^{-2|x|}}{\frac{1}{2}e^{-|x|}} = 2e^{-|x|} \tag{8.33}$$

By Eq. (8.21), the Bayes' test is given by

$$2e^{-|x|} \underset{H_0}{\overset{H_1}{\gtrless}} \frac{(1-0)\frac{2}{3}}{(2-0)\frac{1}{3}} = 1 \qquad \text{or} \qquad e^{-|x|} \underset{H_0}{\overset{H_1}{\gtrless}} \frac{1}{2}$$

Taking the natural logarithm of both sides of the last expression yields

$$|x| \underset{H_0}{\overset{H_1}{\lessgtr}} -\ln\left(\frac{1}{2}\right) = 0.693$$

Thus, the decision regions are given by

$$R_0 = \{x : |x| > 0.693\} \qquad R_1 = \{x : |x| < 0.693\}$$

Then

$$P_1 = P(D_1|H_0) = \int_{-0.693}^{0.693} \frac{1}{2}e^{-|x|}\,dx = 2\int_0^{0.693} \frac{1}{2}e^{-x}\,dx = 0.5$$

$$P_{\mathrm{II}} = P(D_0|H_1) = \int_{-\infty}^{-0.693} e^{2x}\,dx + \int_{0.693}^{\infty} e^{-2x}\,dx = 2\int_{0.693}^{\infty} e^{-2x}\,dx = 0.25$$

and by Eq. (8.19), the Bayes' risk is

$$\bar{C} = P(D_1|H_0)P(H_0) + 2P(D_0|H_1)P(H_1) = (0.5)\left(\frac{2}{3}\right) + 2(0.25)\left(\frac{1}{3}\right) = 0.5$$

(b)   The Bayes' test is

$$2e^{-|x|} \underset{H_0}{\overset{H_1}{\gtrless}} \frac{(1-0)\dfrac{1}{2}}{(2-0)\dfrac{1}{2}} = \frac{1}{2} \qquad \text{or} \qquad e^{-|x|} \underset{H_0}{\overset{H_1}{\gtrless}} \frac{1}{4}$$

Again, taking the natural logarithm of both sides of the last expression yields

$$|x| \underset{H_0}{\overset{H_1}{\lessgtr}} -\ln\left(\frac{1}{4}\right) = 1.386$$

Thus, the decision regions are given by

$$R_0 = \{x : |x| > 1.386\} \qquad R_1 = \{x : |x| < 1.386\}$$

Then

$$P_1 = P(D_1|H_0) = 2\int_0^{1.386} \frac{1}{2}e^{-x}\,dx = 0.75$$

$$P_{\text{II}} = P(D_0|H_1) = 2\int_{1.386}^{\infty} e^{-2x}\,dx = 0.0625$$

and by Eq. (8.19), the Bayes' risk is

$$\bar{C} = (0.75)\left(\frac{1}{2}\right) + 2(0.0625)\left(\frac{1}{2}\right) = 0.4375$$

8.12.   Consider the binary decision problem of Prob. 8.11 with the same Bayes' costs. Determine the minimax test.

From Eq. (8.33), the likelihood ratio is

$$\Lambda(x) = \frac{f(x|H_1)}{f(x|H_0)} = 2e^{-|x|}$$

In terms of $P(H_0)$, the Bayes' test [Eq. (8.21)] becomes

$$2e^{-|x|} \underset{H_0}{\overset{H_1}{\gtrless}} \frac{1}{2}\frac{P(H_0)}{1-P(H_0)} \qquad \text{or} \qquad e^{-|x|} \underset{H_0}{\overset{H_1}{\gtrless}} \frac{1}{4}\frac{P(H_0)}{1-P(H_0)}$$

Taking the natural logarithm of both sides of the last expression yields

$$|x| \underset{H_0}{\overset{H_1}{\lessgtr}} \ln\frac{4[1-P(H_0)]}{P(H_0)} = \delta \tag{8.34}$$

For $P(H_0) > 0.8$, δ becomes negative, and we always decide H_0. For $P(H_0) \le 0.8$, the decision regions are

$$R_0 = \{x : |x| > \delta\} \qquad R_1 = \{x : |x| < \delta\}$$

Then, by setting $C_{00} = C_{11} = 0$, $C_{01} = 2$, and $C_{10} = 1$ in Eq. (8.19), the minimum Bayes' risk $\bar{C}^*$ can be expressed as a function of $P(H_0)$ as

$$\bar{C}^*[P(H_0)] = P(H_0) \int_{-\delta}^{\delta} \frac{1}{2} e^{-|x|} \, dx + 2[1 - P(H_0)] \left[\int_{-\infty}^{-\delta} e^{2x} \, dx + \int_{\delta}^{\infty} e^{-2x} \, dx \right]$$

$$= P(H_0) \int_0^{\delta} e^{-x} \, dx + 4[1 - P(H_0)] \int_{\delta}^{\infty} e^{-2x} \, dx$$

$$= P(H_0)(1 - e^{-\delta}) + 2[1 - P(H_0)]e^{-2\delta} \tag{8.35}$$

From the definition of δ [Eq. (8.34)], we have

$$e^{\delta} = \frac{4[1 - P(H_0)]}{P(H_0)}$$

Thus,

$$e^{-\delta} = \frac{P(H_0)}{4[1 - P(H_0)]} \quad \text{and} \quad e^{-2\delta} = \frac{P^2(H_0)}{16[1 - P(H_0)]^2}$$

Substituting these values into Eq. (8.35), we obtain

$$\bar{C}^*[P(H_0)] = \frac{8P(H_0) - 9P^2(H_0)}{8[1 - P(H_0)]}$$

Now the value of $P(H_0)$ which maximizes $\bar{C}^*$ can be obtained by setting $d\bar{C}^*[P(H_0)]/dP(H_0)$ equal to zero and solving for $P(H_0)$. The result yields $P(H_0) = \frac{2}{3}$. Substituting this value into Eq. (8.34), we obtain the following minimax test:

$$|x| \underset{H_0}{\overset{H_1}{\gtrless}} \ln \frac{4\left(1 - \dfrac{2}{3}\right)}{\dfrac{2}{3}} = \ln 2 = 0.69$$

8.13. Suppose that we have n observations X_i, $i = 1, \dots, n$, of radar signals, and X_i are normal iid r.v.'s under each hypothesis. Under H_0, X_i have mean μ_0 and variance σ^2, while under H_1, X_i have mean μ_1 and variance σ^2, and $\mu_1 > \mu_0$. Determine the maximum likelihood test.

By Eq. (2.71) for each X_i, we have

$$f(x_i | H_0) = \frac{1}{\sqrt{2\pi}\sigma} \exp\left[-\frac{1}{2\sigma^2} (x_i - \mu_0)^2 \right]$$

$$f(x_i | H_1) = \frac{1}{\sqrt{2\pi}\sigma} \exp\left[-\frac{1}{2\sigma^2} (x_i - \mu_1)^2 \right]$$

Since the X_i are independent, we have

$$f(\mathbf{x} | H_0) = \prod_{i=1}^{n} f(x_i | H_0) = \frac{1}{\sqrt{2\pi}\sigma} \exp\left[-\frac{1}{2\sigma^2} \sum_{i=1}^{n} (x_i - \mu_0)^2 \right]$$

$$f(\mathbf{x} | H_1) = \prod_{i=1}^{n} f(x_i | H_1) = \frac{1}{\sqrt{2\pi}\sigma} \exp\left[-\frac{1}{2\sigma^2} \sum_{i=1}^{n} (x_i - \mu_1)^2 \right]$$

With $\mu_1 - \mu_0 > 0$, the likelihood ratio is then given by

$$\Lambda(\mathbf{x}) = \frac{f(\mathbf{x} | H_1)}{f(\mathbf{x} | H_0)} = \exp\left\{ \frac{1}{2\sigma^2} \left[\sum_{i=1}^{n} 2(\mu_1 - \mu_0)x_i - n(\mu_1^2 - \mu_0^2) \right] \right\}$$

Hence, the maximum likelihood test is given by

$$\exp\left\{\frac{1}{2\sigma^2}\left[\sum_{i=1}^{n} 2(\mu_1 - \mu_0)x_i - n(\mu_1^2 - \mu_0^2)\right]\right\} \underset{H_0}{\overset{H_1}{\gtrless}} 1$$

Taking the natural logarithm of both sides of the above expression yields

$$\frac{\mu_1 - \mu_0}{\sigma^2} \sum_{i=1}^{n} x_i \underset{H_0}{\overset{H_1}{\gtrless}} \frac{n(\mu_1^2 - \mu_0^2)}{2\sigma^2}$$

(8.36)

or

$$\frac{1}{n}\sum_{i=1}^{n} x_i \underset{H_0}{\overset{H_1}{\gtrless}} \frac{\mu_1 + \mu_0}{2}$$

Equation (8.36) indicates that the statistic

$$s(X, \ldots, X_n) = \frac{1}{n}\sum_{i=1}^{n} X_i = \bar{X}$$

provides enough information about the observations to enable us to make a decision. Thus, it is called the *sufficient statistic* for the maximum likelihood test.

8.14. Consider the same observations X_i, $i = 1, \ldots, n$, of radar signals as in Prob. 8.13, but now, under H_0, X_i have zero mean and variance σ_0^2, while under H_1, X_i have zero mean and variance σ_1^2, and $\sigma_1^2 > \sigma_0^2$. Determine the maximum likelihood test.

In a similar manner as in Prob. 8.13, we obtain

$$f(\mathbf{x}|H_0) = \frac{1}{(2\pi\sigma_0^2)^{n/2}} \exp\left(-\frac{1}{2\sigma_0^2}\sum_{i=1}^{n} x_i^2\right)$$

$$f(\mathbf{x}|H_1) = \frac{1}{(2\pi\sigma_1^2)^{n/2}} \exp\left(-\frac{1}{2\sigma_1^2}\sum_{i=1}^{n} x_i^2\right)$$

With $\sigma_1^2 - \sigma_0^2 > 0$, the likelihood ratio is

$$\Lambda(\mathbf{x}) = \frac{f(\mathbf{x}|H_1)}{f(\mathbf{x}|H_0)} = \left(\frac{\sigma_0}{\sigma_1}\right)^n \exp\left[\left(\frac{\sigma_1^2 - \sigma_0^2}{2\sigma_0^2\sigma_1^2}\right)\sum_{i=1}^{n} x_i^2\right]$$

and the maximum likelihood test is

$$\left(\frac{\sigma_0}{\sigma_1}\right)^n \exp\left[\left(\frac{\sigma_1^2 - \sigma_0^2}{2\sigma_0^2\sigma_1^2}\right)\sum_{i=1}^{n} x_i^2\right] \underset{H_0}{\overset{H_1}{\gtrless}} 1$$

Taking the natural logarithm of both sides of the above expression yields

$$\sum_{i=1}^{n} x_i^2 \underset{H_0}{\overset{H_1}{\gtrless}} n\left[\ln\left(\frac{\sigma_1}{\sigma_0}\right)\right]\left(\frac{2\sigma_0^2\sigma_1^2}{\sigma_1^2 - \sigma_0^2}\right)$$

(8.37)

Note that in this case,

$$s(X_1, \ldots, X_n) = \sum_{i=1}^{n} X_i^2$$

is the sufficient statistic for the maximum likelihood test.

8.15. In the binary communication system of Prob. 8.6, suppose that we have n independent observations $X_i = X(t_i)$, $i = 1, \ldots, n$, where $0 < t_1 < \cdots < t_n \leq T$.

(a)   Determine the maximum likelihood test.

(b)   Find P_I and P_{II} for $n = 5$ and $n = 10$.

(a)   Setting $\mu_0 = 0$ and $\mu_1 = 1$ in Eq. (8.36), the maximum likelihood test is

$$\frac{1}{n} \sum_{i=1}^{n} x_i \underset{H_0}{\overset{H_1}{\gtrless}} \frac{1}{2}$$

(b)   Let

$$Y = \frac{1}{n} \sum_{i=1}^{n} X_i$$

Then by Eqs. (4.132) and (4.136), and the result of Prob. 5.60, we see that Y is a normal r.v. with zero mean and variance $1/n$ under H_0, and is a normal r.v. with mean 1 and variance $1/n$ under H_1. Thus,

$$P_I = P(D_1 | H_0) = \int_{R_1} f_Y(y | H_0) \, dy = \frac{\sqrt{n}}{\sqrt{2\pi}} \int_{1/2}^{\infty} e^{-(n/2)y^2} \, dy$$

$$= \frac{1}{\sqrt{2\pi}} \int_{\sqrt{n}/2}^{\infty} e^{-z^2/2} \, dz = 1 - \Phi(\sqrt{n}/2)$$

$$P_{II} = P(D_0 | H_1) = \int_{R_0} f_Y(y | H_1) \, dy = \frac{\sqrt{n}}{\sqrt{2\pi}} \int_{1/2}^{\infty} e^{-(n/2)(y-1)^2} \, dy$$

$$= \frac{1}{\sqrt{2\pi}} \int_{-\infty}^{-\sqrt{n}/2} e^{-z^2/2} \, dz = \Phi(-\sqrt{n}/2) = 1 - \Phi(\sqrt{n}/2)$$

Note that $P_I = P_{II}$. Using Table A (Appendix A), we have

$$P_I = P_{II} = 1 - \Phi(1.118) = 0.1318 \qquad \text{for } n = 5$$
$$P_I = P_{II} = 1 - \Phi(1.581) = 0.057 \qquad \text{for } n = 10$$

8.16. In the binary communication system of Prob. 8.6, suppose that $s_0(t)$ and $s_1(t)$ are arbitrary signals and that n observations of the received signal $x(t)$ are made. Let n samples of $s_0(t)$ and $s_1(t)$ be represented, respectively, by

$$\mathbf{s}_0 = [s_{01}, s_{02}, \ldots, s_{0n}]^T \qquad \text{and} \qquad \mathbf{s}_1 = [s_{11}, s_{12}, \ldots, s_{1n}]^T$$

where T denotes "transpose of." Determine the MAP test.

For each X_i, we can write

$$f(x_i | H_0) = \frac{1}{\sqrt{2\pi}} \exp\left[-\frac{1}{2}(x_i - s_{0i})^2\right]$$

$$f(x_i | H_1) = \frac{1}{\sqrt{2\pi}} \exp\left[-\frac{1}{2}(x_i - s_{1i})^2\right]$$

Since the noise components are independent, we have

$$f(\mathbf{x} | H_j) = \prod_{i=1}^{n} f(x_i | H_j) \qquad j = 0, 1$$

and the likelihood ratio is given by

$$\Lambda(\mathbf{x}) = \frac{f(\mathbf{x}|H_1)}{f(\mathbf{x}|H_0)} = \frac{\displaystyle\prod_{i=1}^{n} \exp\left[-\frac{1}{2}(x_i - s_{1i})^2\right]}{\displaystyle\prod_{i=1}^{n} \exp\left[-\frac{1}{2}(x_i - s_{0i})^2\right]}$$

$$= \exp\left[\sum_{i=1}^{n}(s_{1i} - s_{0i})x_i - \frac{1}{2}(s_{1i}^2 - s_{0i}^2)\right]$$

Thus, by Eq. (8.15), the MAP test is given by

$$\exp\left[\sum_{i=1}^{n}(s_{1i} - s_{0i})x_i - \frac{1}{2}(s_{1i}^2 - s_{0i}^2)\right] \underset{H_0}{\overset{H_1}{\gtrless}} \eta = \frac{P(H_0)}{P(H_1)}$$

Taking the natural logarithm of both sides of the above expression yields

$$\sum_{i=1}^{n}(s_{1i} - s_{0i})x_i \underset{H_0}{\overset{H_1}{\gtrless}} \ln\left[\frac{P(H_0)}{P(H_1)}\right] + \frac{1}{2}(s_{1i}^2 - s_{0i}^2) \tag{8.38}$$

SUPPLEMENTARY PROBLEMS

8.17. Let $(X_1, \ldots, X_n)$ be a random sample of a Bernoulli r.v. X with pmf

$$f(x\,;p) = p^x(1-p)^{1-x} \qquad x = 0, 1$$

where it is known that $0 < p \le \frac{1}{2}$. Let

$$H_0: p = \frac{1}{2}$$

$$H_1: p = p_1 \left(< \frac{1}{2}\right)$$

and $n = 20$. As a decision test, we use the rule to reject H_0 if $\sum_{i=1}^{n} x_i \le 6$.

(a) Find the power function $g(p)$ of the test.

(b) Find the probability of a Type I error α.

(c) Find the probability of a Type II error β (i) when $p_1 = \frac{1}{4}$ and (ii) when $p_1 = \frac{1}{10}$.

8.18. Let $(X_1, \ldots, X_n)$ be a random sample of a normal r.v. X with mean μ and variance 36. Let

$$H_0: \quad \mu = 50$$

$$H_1: \quad \mu = 55$$

As a decision test, we use the rule to accept H_0 if $\bar{x} < 53$, where $\bar{x}$ is the value of the sample mean.

(a) Find the expression for the critical region R_1.

(b) Find α and β for $n = 16$.

8.19. Let $(X_1, \ldots, X_n)$ be a random sample of a normal r.v. X with mean μ and variance 100. Let

$$H_0: \quad \mu = 50$$
$$H_1: \quad \mu = 55$$

As a decision test, we use the rule that we reject H_0 if $\bar{x} \geq c$. Find the value of c and sample size n such that $\alpha = 0.025$ and $\beta = 0.05$.

8.20. Let X be a normal r.v. with zero mean and variance σ^2. Let

$$H_0: \quad \sigma^2 = 1$$
$$H_1: \quad \sigma^2 = 4$$

Determine the maximum likelihood test.

8.21. Consider the binary decision problem of Prob. 8.20. Let $P(H_0) = \frac{2}{3}$ and $P(H_1) = \frac{1}{3}$. Determine the MAP test.

8.22. Consider the binary communication system of Prob. 8.6.

    (a) Construct a Neyman-Pearson test for the case where $\alpha = 0.1$.

    (b) Find β.

8.23. Consider the binary decision problem of Prob. 8.11. Determine the Bayes' test if $P(H_0) = 0.25$ and the Bayes' costs are

$$C_{00} = C_{11} = 0 \qquad C_{01} = 1 \qquad C_{10} = 2$$

ANSWERS TO SUPPLEMENTARY PROBLEMS

8.17. (a) $g(p) = \displaystyle\sum_{k=0}^{6} \binom{20}{k} p^k (1-p)^{20-k} \qquad 0 < p \leq \frac{1}{2}$

    (b) $\alpha = 0.0577$;   (c) (i) $\beta = 0.2142$,   (ii) $\beta = 0.0024$

8.18. (a) $R_1 = \{(x_1, \ldots, x_N); \bar{x} \geq 53\} \qquad$ where $\bar{x} = \dfrac{1}{n}\displaystyle\sum_{n=1}^{n} x_i$

    (b) $\alpha = 0.0228, \beta = 0.0913$

8.19. $c = 52.718, n = 52$

8.20. $|x| \underset{H_0}{\overset{H_1}{\gtrless}} 1.36$

8.21. $|x| \underset{H_0}{\overset{H_1}{\gtrless}} 1.923$

8.22. (a) $|x| \underset{H_0}{\overset{H_1}{\gtrless}} 1.282$;   (b) $\beta = 0.6111$

8.23. $|x| \underset{H_0}{\overset{H_1}{\lessgtr}} 1.10$

CHAPTER 9

Queueing Theory

9.1 Introduction

Queueing theory deals with the study of queues (waiting lines). Queues abound in practical situations. The earliest use of queueing theory was in the design of a telephone system. Applications of queueing theory are found in fields as seemingly diverse as traffic control, hospital management, and time-shared computer system design. In this chapter, we present an elementary queueing theory.

9.2   Queueing Systems

A.   Description:

A simple queueing system is shown in Fig. 9-1. Customers arrive randomly at an average rate of λ_a. Upon arrival, they are served without delay if there are available servers; otherwise, they are made to wait in the queue until it is their turn to be served. Once served, they are assumed to leave the system. We will be interested in determining such quantities as the average number of customers in the system, the average time a customer spends in the system, the average time spent waiting in the queue, and so on.

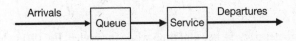

Fig. 9-1 A simple queueing system.

The description of any queueing system requires the specification of three parts:

1.   The arrival process
2.   The service mechanism, such as the number of servers and service-time distribution
3.   The queue discipline (for example, first-come, first-served)

B.   Classification:

The notation $A/B/s/K$ is used to classify a queueing system, where A specifies the type of arrival process, B denotes the service-time distribution, s specifies the number of servers, and K denotes the capacity of the system, that is, the maximum number of customers that can be accommodated. If K is not specified, it is assumed that the capacity of the system is unlimited. For example, an M/M/2 queueing system (M stands for Markov) is one with Poisson arrivals (or exponential interarrival time distribution), exponential service-time distribution, and 2 servers. An M/G/1 queueing system has Poisson arrivals, general service-time distribution, and a single server. A special case is the M/D/1 queueing system, where D stands for constant (deterministic) service time. Examples of queueing systems with limited capacity are telephone systems with limited trunks, hospital emergency rooms with limited beds, and airline terminals with limited space in which to park aircraft for loading and unloading. In each case, customers who arrive when the system is saturated are denied entrance and are lost.

C. Useful Formulas:

Some basic quantities of queueing systems are

L: the average number of customers in the system
L_q: the average number of customers waiting in the queue
L_s: the average number of customers in service
W: the average amount of time that a customer spends in the system
W_q: the average amount of time that a customer spends waiting in the queue
W_s: the average amount of time that a customer spends in service

Many useful relationships between the above and other quantities of interest can be obtained by using the following basic cost identity:

Assume that entering customers are required to pay an entrance fee (according to some rule) to the system. Then we have

$$\text{Average rate at which the system earns} = \lambda_a \times \text{average amount an entering customer pays} \qquad (9.1)$$

where λ_a is the average arrival rate of entering customers

$$\lambda_a = \lim_{t \to \infty} \frac{X(t)}{t}$$

and $X(t)$ denotes the number of customer arrivals by time t.

If we assume that each customer pays \$1 per unit time while in the system, Eq. (9.1) yields

$$L = \lambda_a W \qquad (9.2)$$

Equation (9.2) is sometimes known as *Little's formula*.

Similarly, if we assume that each customer pays \$1 per unit time while in the queue, Eq. (9.1) yields

$$L_q = \lambda_a W_q \qquad (9.3)$$

If we assume that each customer pays \$1 per unit time while in service, Eq. (9.1) yields

$$L_s = \lambda_a W_s \qquad (9.4)$$

Note that Eqs. (9.2) to (9.4) are valid for almost all queueing systems, regardless of the arrival process, the number of servers, or queueing discipline.

9.3 Birth-Death Process

We say that the queueing system is in state S_n if there are n customers in the system, including those being served. Let $N(t)$ be the Markov process that takes on the value n when the queueing system is in state S_n with the following assumptions:

1. If the system is in state S_n, it can make transitions only to S_{n-1} or S_{n+1}, $n \geq 1$; that is, either a customer completes service and leaves the system or, while the present customer is still being serviced, another customer arrives at the system; from S_0, the next state can only be S_1.

2. If the system is in state S_n at time t, the probability of a transition to S_{n+1} in the time interval $(t, t + \Delta t)$ is $a_n \Delta t$. We refer to a_n as the *arrival* parameter (or the *birth* parameter).

3. If the system is in state S_n at time t, the probability of a transition to S_{n-1} in the time interval $(t, t + \Delta t)$ is $d_n \, \Delta t$. We refer to dn as the *departure* parameter (or the *death* parameter).

The process $N(t)$ is sometimes referred to as the *birth-death* process.

Let $p_n(t)$ be the probability that the queueing system is in state S_n at time t; that is,

$$p_n(t) = P\{N(t) = n\} \tag{9.5}$$

Then we have the following fundamental recursive equations for $N(t)$ (Prob. 9.2):

$$p'_n(t) = -(a_n + d_n)p_n(t) + a_{n-1}p_{n-1}(t) + d_{n+1}p_{n+1}(t) \qquad n \geq 1$$

$$p'_0(t) = -(a_0 + d_0)p_0(t) + d_1 p_1(t) \tag{9.6}$$

Assume that in the steady state we have

$$\lim_{t \to \infty} p_n(t) = p_n \tag{9.7}$$

and setting $p'_0(t)$ and $p'_n(t) = 0$ in Eqs. (9.6), we obtain the following steady-state recursive equation:

$$(a_n + d_n)p_n = a_{n-1}p_{n-1} + d_{n+1}p_{n+1} \qquad n \geq 1 \tag{9.8}$$

and for the special case with $d_0 = 0$,

$$a_0 p_0 = d_1 p_1 \tag{9.9}$$

Equations (9.8) and (9.9) are also known as the steady-state *equilibrium equations*. The state transition diagram for the birth-death process is shown in Fig. 9-2.

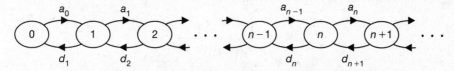

Fig. 9-2 State transition diagram for the birth-death process.

Solving Eqs. (9.8) and (9.9) in terms of p_0, we obtain

$$p_1 = \frac{a_0}{d_1} p_0$$

$$p_2 = \frac{a_0 a_1}{d_1 d_2} p_0 \tag{9.10}$$

$$\vdots$$

$$p_n = \frac{a_0 a_1 \cdots a_{n-1}}{d_1 d_2 \cdots d_n} p_0$$

where p_0 can be determined from the fact that

$$\sum_{n=0}^{\infty} p_n = \left(1 + \frac{a_0}{d_1} + \frac{a_0 a_1}{d_1 d_2} + \cdots\right)p_0 = 1 \tag{9.11}$$

provided that the summation in parentheses converges to a finite value.

9.4   The M/M/1 Queueing System

In the M/M/1 queueing system, the arrival process is the Poisson process with rate λ (the mean arrival rate) and the service time is exponentially distributed with parameter μ (the mean service rate). Then the process $N(t)$ describing the state of the M/M/1 queueing system at time t is a birth-death process with the following state independent parameters:

$$a_n = \lambda \qquad n \geq 0 \qquad d_n = \mu \qquad n \geq 1 \qquad\qquad (9.12)$$

Then from Eqs. (9.10) and (9.11), we obtain (Prob. 9.3)

$$p_0 = 1 - \frac{\lambda}{\mu} = 1 - \rho \qquad\qquad (9.13)$$

$$p_n = \left(1 - \frac{\lambda}{\mu}\right)\left(\frac{\lambda}{\mu}\right)^n = (1 - \rho)\rho^n \qquad\qquad (9.14)$$

where $\rho = \lambda/\mu < 1$, which implies that the server, on the average, must process the customers faster than their average arrival rate; otherwise, the queue length (the number of customers waiting in the queue) tends to infinity. The ratio $\rho = \lambda/\mu$ is sometimes referred to as the *traffic intensity* of the system. The traffic intensity of the system is defined as

$$\text{Traffic intensity} = \frac{\text{mean service time}}{\text{mean interarrival time}} = \frac{\text{mean arrival rate}}{\text{mean service rate}}$$

The average number of customers in the system is given by (Prob. 9.4)

$$L = \frac{\rho}{1 - \rho} = \frac{\lambda}{\mu - \lambda} \qquad\qquad (9.15)$$

Then, setting $\lambda_a = \lambda$ in Eqs. (9.2) to (9.4), we obtain (Prob. 9.5)

$$W = \frac{1}{\mu - \lambda} = \frac{1}{\mu(1 - \rho)} \qquad\qquad (9.16)$$

$$W_q = \frac{\lambda}{\mu(\mu - \lambda)} = \frac{\rho}{\mu(1 - \rho)} \qquad\qquad (9.17)$$

$$L_q = \frac{\lambda^2}{\mu(\mu - \lambda)} = \frac{\rho^2}{1 - \rho} \qquad\qquad (9.18)$$

9.5   The M/M/s Queueing System

In the M/M/s queueing system, the arrival process is the Poisson process with rate λ and each of the s servers has an exponential service time with parameter μ. In this case, the process $N(t)$ describing the state of the M/M/s queueing system at time t is a birth-death process with the following parameters:

$$a_n = \lambda \qquad n \geq 0 \qquad d_n = \begin{cases} n\mu & 0 < n < s \\ s\mu & n \geq s \end{cases} \qquad\qquad (9.19)$$

Note that the departure parameter d_n is state dependent. Then, from Eqs. (9.10) and (9.11) we obtain (Prob. 9.10)

$$p_0 = \left[\sum_{n=0}^{s-1} \frac{(s\rho)}{n!} + \frac{(s\rho)^s}{s!(1 - \rho)}\right]^{-1} \qquad\qquad (9.20)$$

$$p_n = \begin{cases} \dfrac{(s\rho)^n}{n!} \, p_0 & n < s \\[2ex] \dfrac{\rho^n s^s}{s!} \, p_0 & n \geq s \end{cases} \tag{9.21}$$

where $\rho = \lambda/(s\mu) < 1$. Note that the ratio $\rho = \lambda/(s\mu)$ is the traffic intensity of the M/M/s queueing system. The average number of customers in the system and the average number of customers in the queue are given, respectively, by (Prob. 9.12)

$$L = \frac{\lambda}{\mu} + \frac{\rho(sp)^s}{s!(1-\rho)^2} \, p_0 \tag{9.22}$$

$$L_q = \frac{\rho(s\rho)^s}{s!(1-\rho)^2} \, p_0 = L - \frac{\lambda}{\mu} \tag{9.23}$$

By Eqs. (9.2) and (9.3), the quantities W and W_q are given by

$$W = \frac{L}{\lambda} \tag{9.24}$$

$$W_q = \frac{L_q}{\lambda} = W - \frac{1}{\mu} \tag{9.25}$$

9.6 The M/M/1/K Queueing System

In the M/M/1/K queueing system, the capacity of the system is limited to K customers. When the system reaches its capacity, the effect is to reduce the arrival rate to zero until such time as a customer is served to again make queue space available. Thus, the M/M/1/K queueing system can be modeled as a birth-death process with the following parameters:

$$a_n = \begin{cases} \lambda & 0 \leq n < K \\ 0 & n \geq K \end{cases} \qquad d_n = \mu \qquad n \geq 1 \tag{9.26}$$

Then, from Eqs. (9.10) and (9.11) we obtain (Prob. 9.14)

$$p_0 = \frac{1 - (\lambda/\mu)}{1 - (\lambda/\mu)^{K+1}} = \frac{1-\rho}{1-\rho^{K+1}} \qquad \rho \neq 1 \tag{9.27}$$

$$p_n = \left(\frac{\lambda}{\mu}\right)^n p_0 = \frac{(1-\rho)\rho^n}{1-\rho^{K+1}} \qquad n = 1, \ldots, K \tag{9.28}$$

where $\rho = \lambda/\mu$. It is important to note that it is no longer necessary that the traffic intensity $\rho = \lambda/\mu$ be less than 1. Customers are denied service when the system is in state K. Since the fraction of arrivals that actually enter the system is $1 - p_K$, the effective arrival rate is given by

$$\lambda_e = \lambda(1 - p_K) \tag{9.29}$$

The average number of customers in the system is given by (Prob. 9.15)

$$L = \rho \, \frac{1 - (K+1)\rho^K + K\rho^{K+1}}{(1-\rho)(1-\rho^{K+1})} \qquad \rho = \frac{\lambda}{\mu} \tag{9.30}$$

Then, setting $\lambda_a = \lambda_e$ in Eqs. (9.2) to (9.4), we obtain

$$W = \frac{L}{\lambda_e} = \frac{L}{\lambda(1 - p_K)} \tag{9.31}$$

$$W_q = W - \frac{1}{\mu} \tag{9.32}$$

$$L_q = \lambda_e W_q = \lambda(1 - p_K)W_q \tag{9.33}$$

9.7 The M/M/*s*/*K* Queueing System

Similarly, the M/M/*s*/*K* queueing system can be modeled as a birth-death process with the following parameters:

$$a_n = \begin{cases} \lambda & 0 \le n < K \\ 0 & n \ge K \end{cases} \qquad d_n = \begin{cases} n\mu & 0 < n < s \\ s\mu & n \ge s \end{cases} \tag{9.34}$$

Then, from Eqs. (9.10) and (9.11), we obtain (Prob. 9.17)

$$p_0 = \left[\sum_{n=0}^{s-1} \frac{(s\rho)^n}{n!} + \frac{(s\rho)^s}{s!} \left(\frac{1 - \rho^{K-s+1}}{1 - \rho} \right) \right]^{-1} \tag{9.35}$$

$$p_n = \begin{cases} \dfrac{(s\rho)^n}{n!} p_0 & n < s \\ \dfrac{\rho^n s^s}{s!} p_0 & s \le n \le K \end{cases} \tag{9.36}$$

where $\rho = \lambda/(s\mu)$. Note that the expression for p_n is identical in form to that for the M/M/*s* system, Eq. (9.21). They differ only in the p_0 term. Again, it is not necessary that $\rho = \lambda/(s\mu)$ be less than 1. The average number of customers in the queue is given by (Prob. 9.18)

$$L_q = p_0 \frac{\rho(\rho s)^s}{s!(1 - \rho)^2} \{1 - [1 + (1 - \rho)(K - s)]\rho^{K-s}\} \tag{9.37}$$

The average number of customers in the system is

$$L = L_q + \frac{\lambda_e}{\mu} = L_q + \frac{\lambda}{\mu}(1 - p_K) \tag{9.38}$$

The quantities W and W_q are given by

$$W = \frac{L}{\lambda_e} = L_q + \frac{1}{\mu} \tag{9.39}$$

$$W_q = \frac{L_q}{\lambda_e} = \frac{L_q}{\lambda(1 - p_K)} \tag{9.40}$$

SOLVED PROBLEMS

9.1. Deduce the basic cost identity (9.1).

Let T be a fixed large number. The amount of money earned by the system by time T can be computed by multiplying the average rate at which the system earns by the length of time T. On the other hand, it can also be computed by multiplying the average amount paid by an entering customer by the average number of customers entering by time T, which is equal to $\lambda_a T$, where λ_a is the average arrival rate of entering customers. Thus, we have

Average rate at which the system earns $\times T$ = average amount paid by an entering customer $\times (\lambda_a T)$

Dividing both sides by T (and letting $T \to \infty$), we obtain Eq. (9.1).

9.2. Derive Eq. (9.6).

From properties 1 to 3 of the birth-death process $N(t)$, we see that at time $t + \Delta t$ the system can be in state S_n in three ways:

1. By being in state S_n at time t and no transition occurring in the time interval $(t, t + \Delta t)$. This happens with probability $(1 - a_n \Delta t)(1 - d_n \Delta t) \approx 1 - (a_n + d_n)\Delta t$ [by neglecting the second-order effect $a_n d_n (\Delta t)^2$].

2. By being in state S_{n-1} at time t and a transition to S_n occurring in the time interval $(t, t + \Delta t)$. This happens with probability $a_{n-1}\Delta t$.

3. By being in state S_{n+1} at time t and a transition to S_n occurring in the time interval $(t, t + \Delta t)$. This happens with probability $d_{n+1}\Delta t$.

Let
$$p_i(t) = P[N(t) = i]$$

Then, using the Markov property of $N(t)$, we obtain

$$p_n(t + \Delta t) = [1 - (a_n + d_n)\Delta t]p_n(t) + a_{n-1}\Delta t\, p_{n-1}(t) + d_{n+1}\Delta t\, p_{n+1}(t) \qquad n \geq 1$$
$$p_0(t + \Delta t) = [1 - (a_0 + d_0)\Delta t]p_0(t) + d_1 \Delta t\, p_1(t)$$

Rearranging the above relations

$$\frac{p_n(t + \Delta t) - p_n(t)}{\Delta t} = -(a_n + d_n)p_n(t) + a_{n-1}p_{n-1}(t) + d_{n+1}p_{n+1}(t) \qquad n \geq 1$$
$$\frac{p_0(t + \Delta t) - p_0(t)}{\Delta t} = -(a_0 + d_0)p_0(t) + d_1 p_1(t)$$

Letting $\Delta t \to 0$, we obtain

$$p'_n(t) = -(a_n + d_n)p_n(t) + a_{n-1}p_{n-1}(t) + d_{n+1}p_{n+1}(t) \qquad n \geq 1$$

$$p'_0(t) = -(a_0 + d_0)p_0(t) + d_1 p_1(t)$$

9.3. Derive Eqs. (9.13) and (9.14).

Setting $a_n = \lambda$, $d_0 = 0$, and $d_n = \mu$ in Eq. (9.10), we get

$$p_1 = \frac{\lambda}{\mu} p_0 = \rho p_0$$

$$p_2 = \left(\frac{\lambda}{\mu}\right)^2 p_0 = \rho^2 p_0$$

$$\vdots$$

$$p_n = \left(\frac{\lambda}{\mu}\right)^n p_0 = \rho^n p_0$$

where p_0 is determined by equating

$$\sum_{n=0}^{\infty} p_n = p_0 \sum_{n=0}^{\infty} \rho^n = p_0 \frac{1}{1-\rho} = 1 \quad |\rho| < 1$$

from which we obtain

$$p_0 = 1 - \rho = 1 - \frac{\lambda}{\mu}$$

$$p_n = \left(\frac{\lambda}{\mu}\right)^n p_0 = (1-\rho)\rho^n = \left(1 - \frac{\lambda}{\mu}\right)\left(\frac{\lambda}{\mu}\right)^n$$

9.4. Derive Eq. (9.15).

Since p_n is the steady-state probability that the system contains exactly n customers, using Eq. (9.14), the average number of customers in the M/M/1 queueing system is given by

$$L = \sum_{n=0}^{\infty} n p_n = \sum_{n=0}^{\infty} n(1-\rho)\rho^n = (1-\rho)\sum_{n=0}^{\infty} n\rho^n \tag{9.41}$$

where $\rho = \lambda/\mu < 1$. Using the algebraic identity

$$\sum_{n=0}^{\infty} n x^n = \frac{x}{(1-x)^2} \quad |x| < 1 \tag{9.42}$$

we obtain

$$L = \frac{\rho}{1-\rho} = \frac{\lambda/\mu}{1-\lambda/\mu} = \frac{\lambda}{\mu-\lambda}$$

9.5. Derive Eqs. (9.16) to (9.18).

Since $\lambda_a = \lambda$, by Eqs. (9.2) and (9.15), we get

$$W = \frac{L}{\lambda} = \frac{1}{\mu-\lambda} = \frac{1}{\mu(1-\rho)}$$

which is Eq. (9.16). Next, by definition,

$$W_q = W - W_s \tag{9.43}$$

where $W_s = 1/\mu$, that is, the average service time. Thus,

$$W_q = \frac{1}{\mu-\lambda} - \frac{1}{\mu} = \frac{\lambda}{\mu(\mu-\lambda)} = \frac{\rho}{\mu(1-\rho)}$$

which is Eq. (9.17). Finally, by Eq. (9.3),

$$L_q = \lambda W_q = \frac{\lambda^2}{\mu(\mu-\lambda)} = \frac{\rho^2}{1-\rho}$$

which is Eq. (9.18).

9.6. Let W_a denote the amount of time an arbitrary customer spends in the M/M/1 queueing system. Find the distribution of W_a.

We have

$$P\{W_a \le a\} = \sum_{n=0}^{\infty} P\{W_a \le a \,|\, n \text{ in the system when the customer arrives}\}$$

$$\times P\{n \text{ in the system when the customer arrives}\} \tag{9.44}$$

where n is the number of customers in the system. Now consider the amount of time W_a that this customer will spend in the system when there are already n customers when he or she arrives. When $n = 0$, then $W_a = W_{s(a)}$, that is, the service time. When $n \ge 1$, there will be one customer in service and $n - 1$ customers waiting in line ahead of this customer's arrival. The customer in service might have been in service for some time, but because of the memoryless property of the exponential distribution of the service time, it follows that (see Prob. 2.58) the arriving customer would have to wait an exponential amount of time with parameter μ for this customer to complete service. In addition, the customer also would have to wait an exponential amount of time for each of the other $n - 1$ customers in line. Thus, adding his or her own service time, the amount of time W_a that the customer will spend in the system when there are already n customers when he or she arrives is the sum of $n + 1$ iid exponential r.v.'s with parameter μ. Then by Prob. 4.39, we see that this r.v. is a gamma r.v. with parameters $(n + 1, \mu)$. Thus, by Eq. (2.65),

$$P\{W_a \le a \,|\, n \text{ in the system when customer arrives}\} = \int_0^a \mu e^{-\mu t} \frac{(\mu t)^n}{n!} \, dt$$

From Eq. (9.14),

$$P\{n \text{ in the system when customer arrives}\} = p_n = \left(1 - \frac{\lambda}{\mu}\right)\left(\frac{\lambda}{\mu}\right)^n$$

Hence, by Eq. (9.44),

$$F_{W_a} = P\{W_a \le a\} = \sum_{n=0}^{\infty} \left[\int_0^a \mu e^{-\mu t} \frac{(\mu t)^n}{n!}\left(1 - \frac{\lambda}{\mu}\right)\left(\frac{\lambda}{\mu}\right)^n dt\right]$$

$$= \int_0^a (\mu - \lambda)e^{-\mu t} \sum_{n=0}^{\infty} \frac{(\lambda t)^n}{n!} \, dt$$

$$= \int_0^a (\mu - \lambda)e^{-(\mu - \lambda)t} \, dt = 1 - e^{-(\mu - \lambda)a} \tag{9.45}$$

Thus, by Eq. (2.61), W_a is an exponential r.v. with parameter $\mu - \lambda$. Note that from Eq. (2.62), $E(W_a) = 1/(\mu - \lambda)$, which agrees with Eq. (9.16), since $W = E(W_a)$.

9.7. Customers arrive at a watch repair shop according to a Poisson process at a rate of one per every 10 minutes, and the service time is an exponential r.v. with mean 8 minutes.

(a) Find the average number of customers L, the average time a customer spends in the shop W, and the average time a customer spends in waiting for service W_q.

(b) Suppose that the arrival rate of the customers increases 10 percent. Find the corresponding changes in L, W, and W_q.

(a) The watch repair shop service can be modeled as an M/M/1 queueing system with $\lambda = \frac{1}{10}, \mu = \frac{1}{8}$. Thus, from Eqs. (9.15), (9.16), and (9.43), we have

$$L = \frac{\lambda}{\mu - \lambda} = \frac{\frac{1}{10}}{\frac{1}{8} - \frac{1}{10}} = 4$$

$$W = \frac{1}{\mu - \lambda} = \frac{1}{\frac{1}{8} - \frac{1}{10}} = 40 \text{ minutes}$$

$$W_q = W - W_s = 40 - 8 = 32 \text{ minutes}$$

(b)   Now $\lambda = \frac{1}{9}$, $\mu = \frac{1}{8}$. Then

$$L = \frac{\lambda}{\mu - \lambda} = \frac{\frac{1}{9}}{\frac{1}{8} - \frac{1}{9}} = 8$$

$$W = \frac{1}{\mu - \lambda} = \frac{1}{\frac{1}{8} - \frac{1}{9}} = 72 \text{ minutes}$$

$$W_q = W - W_s = 72 - 8 = 64 \text{ minutes}$$

It can be seen that an increase of 10 percent in the customer arrival rate doubles the average number of customers in the system. The average time a customer spends in queue is also doubled.

9.8.   A drive-in banking service is modeled as an M/M/1 queueing system with customer arrival rate of 2 per minute. It is desired to have fewer than 5 customers line up 99 percent of the time. How fast should the service rate be?

From Eq. (9.14),

$$P\{5 \text{ or more customers in the system}\} = \sum_{n=5}^{\infty} p_n = \sum_{n=5}^{\infty} (1 - \rho)\rho^n = \rho^5 \qquad \rho = \frac{\lambda}{\mu}$$

In order to have fewer than 5 customers line up 99 percent of the time, we require that this probability be less than 0.01. Thus,

$$\rho^5 = \left(\frac{\lambda}{\mu}\right)^5 \le 0.01$$

from which we obtain

$$\mu^5 \ge \frac{\lambda^5}{0.01} = \frac{2^5}{0.01} = 3200 \qquad \text{or} \qquad \mu \ge 5.024$$

Thus, to meet the requirements, the average service rate must be at least 5.024 customers per minute.

9.9.   People arrive at a telephone booth according to a Poisson process at an average rate of 12 per hour, and the average time for each call is an exponential r.v. with mean 2 minutes.

(a)   What is the probability that an arriving customer will find the telephone booth occupied?

(b)   It is the policy of the telephone company to install additional booths if customers wait an average of 3 or more minutes for the phone. Find the average arrival rate needed to justify a second booth.

(a)   The telephone service can be modeled as an M/M/1 queueing system with $\lambda = \frac{1}{5}$, $\mu = \frac{1}{2}$, and $\rho = \lambda/\mu = \frac{2}{5}$. The probability that an arriving customer will find the telephone occupied is $P(L > 0)$, where L is the average number of customers in the system. Thus, from Eq. (9.13),

$$P(L > 0) = 1 - p_0 = 1 - (1 - \rho) = \rho = \frac{2}{5} = 0.4$$

(b)   From Eq. (9.17),

$$W_q = \frac{\lambda}{\mu(\mu - \lambda)} = \frac{\lambda}{0.5(0.5 - \lambda)} \ge 3$$

from which we obtain $\lambda \geq 0.3$ per minute. Thus, the required average arrival rate to justify the second booth is 18 per hour.

9.10. Derive Eqs. (9.20) and (9.21).

From Eqs. (9.19) and (9.10), we have

$$p_n = p_0 \prod_{k=0}^{n-1} \frac{\lambda}{(k+1)\mu} = p_0 \left(\frac{\lambda}{\mu}\right)^n \frac{1}{n!} \qquad n < s \tag{9.46}$$

$$p_n = p_0 \prod_{k=0}^{s-1} \frac{\lambda}{(k+1)\mu} \prod_{k=s}^{n-1} \frac{\lambda}{s\mu} = p_0 \left(\frac{\lambda}{\mu}\right)^n \frac{1}{s!s^{n-s}} \qquad n \geq s \tag{9.47}$$

Let $\rho = \lambda/(s\mu)$. Then Eqs. (9.46) and (9.47) can be rewritten as

$$p_n = \begin{cases} \dfrac{(s\rho)^n}{n!} p_0 & n < s \\[2mm] \dfrac{\rho^n s^s}{s!} p_0 & n \geq s \end{cases}$$

which is Eq. (9.21). From Eq. (9.11), p_0 is obtained by equating

$$\sum_{n=0}^{\infty} p_n = p_0 \left[\sum_{n=0}^{s-1} \frac{(s\rho)^n}{n!} + \sum_{n=s}^{\infty} \frac{\rho^n s^s}{s!} \right] = 1$$

Using the summation formula

$$\sum_{n=k}^{\infty} x^n = \frac{x^k}{1-x} \qquad |x| < 1 \tag{9.48}$$

we obtain Eq. (9.20); that is,

$$p_0 = \left[\sum_{n=0}^{s-1} \frac{(s\rho)^n}{n!} + \frac{s^s}{s!} \sum_{n=s}^{\infty} \rho^n \right]^{-1} = \left[\sum_{n=0}^{s-1} \frac{(s\rho)^n}{n!} + \frac{(s\rho)^s}{s!(1-\rho)} \right]^{-1}$$

provided $\rho = \lambda/(s\mu) < 1$.

9.11. Consider an M/M/s queueing system. Find the probability that an arriving customer is forced to join the queue.

An arriving customer is forced to join the queue when all servers are busy—that is, when the number of customers in the system is equal to or greater than s. Thus, using Eqs. (9.20) and (9.21), we get

$$P(\text{a customer is forced to join queue}) = \sum_{n=s}^{\infty} p_n = p_0 \frac{s^s}{s!} \sum_{n=s}^{\infty} \rho^n = p_0 \frac{(s\rho)^s}{s!(1-\rho)}$$

$$= \frac{\dfrac{(s\rho)^s}{s!(1-\rho)}}{\displaystyle\sum_{n=0}^{s-1} \frac{(s\rho)^n}{n!} + \frac{(s\rho)^s}{s!(1-\rho)}} \tag{9.49}$$

Equation (9.49) is sometimes referred to as *Erlang's delay* (or *C*) formula and denoted by $C(s, \lambda/\mu)$. Equation (9.49) is widely used in telephone systems and gives the probability that no trunk (server) is available for an incoming call (arriving customer) in a system of *s* trunks.

9.12. Derive Eqs. (9.22) and (9.23).

Equation (9.21) can be rewritten as

$$
p_n = \begin{cases} \dfrac{(s\rho)^n}{n!} \, p_0 & n \le s \\[2mm] \dfrac{\rho^n s^s}{s!} \, p_0 & n > s \end{cases}
$$

Then the average number of customers in the system is

$$
L = E(N) = \sum_{n=0}^{\infty} n p_n = p_0 \left[\sum_{n=0}^{s} n \frac{(s\rho)^n}{n!} + \sum_{n=s+1}^{\infty} n \frac{\rho^n s^s}{s!} \right]
$$

$$
= p_0 \left[s\rho \sum_{n=1}^{s} \frac{(s\rho)^{n-1}}{(n-1)!} + \frac{s^s}{s!} \sum_{n=s+1}^{\infty} n\rho^n \right]
$$

$$
= p_0 \left[s\rho \sum_{n=0}^{s-1} \frac{(s\rho)^n}{n!} + \frac{s^s}{s!} \left(\sum_{n=0}^{\infty} n\rho^n - \sum_{n=0}^{s} n\rho^n \right) \right]
$$

Using the summation formulas,

$$
\sum_{n=0}^{\infty} n x^n = \frac{x}{(1-x)^2} \qquad |x| < 1 \tag{9.50}
$$

$$
\sum_{n=0}^{k} n x^n = \frac{x[kx^{k+1} - (k+1)x^k + 1]}{(1-x)^2} \qquad |x| < 1 \tag{9.51}
$$

and Eq. (9.20), we obtain

$$
L = p_0 \left(s\rho \sum_{n=0}^{s-1} \frac{(s\rho)^n}{n!} + \frac{s^s}{s!} \left\{ \frac{\rho}{(1-\rho)^2} - \frac{\rho[s\rho^{s+1} - (s+1)\rho^s + 1]}{(1-\rho)^2} \right\} \right)
$$

$$
= p_0 \left\{ s\rho \left[\sum_{n=0}^{s-1} \frac{(s\rho)^n}{n!} + \frac{(s\rho)^s}{s!(1-\rho)} \right] + \frac{\rho(\rho s)^s}{s!(1-\rho)^2} \right\}
$$

$$
= p_0 \left[s\rho \frac{1}{p_0} + \frac{\rho(s\rho)^s}{s!(1-\rho)^2} \right]
$$

$$
= s\rho + \frac{\rho(s\rho)^s}{s!(1-\rho)^2} \, p_0 = \frac{\lambda}{\mu} + \frac{\rho(s\rho)^s}{s!(1-\rho)^2} \, p_0
$$

Next, using Eqs. (9.21) and (9.50), the average number of customers in the queue is

$$
L_q = \sum_{n=s}^{\infty} (n-s) p_n = \sum_{n=s}^{\infty} (n-s) \frac{\rho^n s^s}{s!} \, p_0
$$

$$
= p_0 \frac{(s\rho)^s}{s!} \sum_{n=s}^{\infty} (n-s) \rho^{n-s} = p_0 \frac{(s\rho)^s}{s!} \sum_{k=0}^{\infty} k \rho^k
$$

$$
= \frac{\rho(s\rho)^s}{s!(1-\rho)^2} \, p_0 = L - \frac{\lambda}{\mu}
$$

9.13. A corporate computing center has two computers of the same capacity. The jobs arriving at the center are of two types, internal jobs and external jobs. These jobs have Poisson arrival times with rates 18 and 15 per hour, respectively. The service time for a job is an exponential r.v. with mean 3 minutes.

(a) Find the average waiting time per job when one computer is used exclusively for internal jobs and the other for external jobs.

(b) Find the average waiting time per job when two computers handle both types of jobs.

(a) When the computers are used separately, we treat them as two M/M/1 queueing systems. Let W_{q1} and W_{q2} be the average waiting time per internal job and per external job, respectively. For internal jobs, $\lambda_1 = \frac{18}{60} = \frac{3}{10}$ and $\mu_1 = \frac{1}{3}$. Then, from Eq. (9.16),

$$W_{q1} = \frac{\frac{3}{10}}{\frac{1}{3}\left(\frac{1}{3} - \frac{3}{10}\right)} = 27 \text{ min}$$

For external jobs, $\lambda_2 = \frac{15}{60} = \frac{1}{4}$ and $\mu_2 = \frac{1}{3}$, and

$$W_{q2} = \frac{\frac{1}{4}}{\frac{1}{3}\left(\frac{1}{3} - \frac{1}{4}\right)} = 9 \text{ min}$$

(b) When two computers handle both types of jobs, we model the computing service as an M/M/2 queueing system with

$$\lambda = \frac{18 + 15}{60} = \frac{11}{20}, \qquad \mu = \frac{1}{3} \qquad \rho = \frac{\lambda}{2\mu} = \frac{33}{40}$$

Now, substituting $s = 2$ in Eqs. (9.20), (9.22), (9.24), and (9.25), we get

$$p_0 = \left[1 + 2\rho + \frac{(2\rho)^2}{2(1-\rho)}\right]^{-1} = \frac{1-\rho}{1+\rho} \tag{9.52}$$

$$L = 2\rho + \frac{4\rho^3}{2(1-\rho)^2}\left(\frac{1-\rho}{1+\rho}\right) = \frac{2\rho}{1-\rho^2} \tag{9.53}$$

$$W_q = \frac{L}{\lambda} - \frac{1}{\mu} = \frac{1}{\lambda}\frac{2\rho}{1-\rho^2} - \frac{1}{\mu} \tag{9.54}$$

Thus, from Eq. (9.54), the average waiting time per job when both computers handle both types of jobs is given by

$$W_q = \frac{2\left(\frac{33}{40}\right)}{\frac{11}{20}\left[1 - \left(\frac{33}{40}\right)^2\right]} = 6.39 \text{ min}$$

From these results, we see that it is more efficient for both computers to handle both types of jobs.

9.14. Derive Eqs. (9.27) and (9.28).

From Eqs. (9.26) and (9.10), we have

$$p_n = \left(\frac{\lambda}{\mu}\right)^n p_0 = \rho^n p_0 \qquad 0 \le n \le K \tag{9.55}$$

From Eq. (9.11), p_0 is obtained by equating

$$\sum_{n=0}^{K} p_n = p_0 \sum_{n=0}^{K} \rho^n = 1$$

Using the summation formula

$$\sum_{n=0}^{K} x^n = \frac{1 - x^{K+1}}{1 - x} \tag{9.56}$$

we obtain

$$p_0 = \frac{1 - \rho}{1 - \rho^{K+1}} \qquad \text{and} \qquad p_n = \frac{(1 - \rho)\rho^n}{1 - \rho^{K+1}}$$

Note that in this case, there is no need to impose the condition that $\rho = \lambda / \mu < 1$.

9.15. Derive Eq. (9.30).

Using Eqs. (9.28) and (9.51), the average number of customers in the system is given by

$$L = E(N) = \sum_{n=0}^{K} n p_n = \frac{1 - \rho}{1 - \rho^{K+1}} \sum_{n=0}^{K} n \rho^n$$

$$= \frac{1 - \rho}{1 - \rho^{K+1}} \left\{ \frac{\rho[K\rho^{K+1} - (K+1)\rho^K + 1]}{(1 - \rho)^2} \right\}$$

$$= \rho \frac{1 - (K+1)\rho^K + K\rho^{K+1}}{(1 - \rho)(1 - \rho^{K+1})}$$

9.16. Consider the M/M/1/K queueing system. Show that

$$L_q = L - (1 - p_0) \tag{9.57}$$

$$W_q = \frac{1}{\mu} L \tag{9.58}$$

$$W = \frac{1}{\mu}(L + 1) \tag{9.59}$$

In the M/M/1/K queueing system, the average number of customers in the system is

$$L = E(N) = \sum_{n=0}^{K} n p_n \qquad \text{and} \qquad \sum_{n=0}^{K} p_n = 1$$

The average number of customers in the queue is

$$L_q = E(N_q) = \sum_{n=1}^{K} (n-1) p_n = \sum_{n=0}^{K} n p_n - \sum_{n=1}^{K} p_n = L - (1 - p_0)$$

A customer arriving with the queue in state S_n has a wait time T_q that is the sum of n independent exponential r.v.'s, each with parameter μ. The expected value of this sum is n/μ [Eq. (4.132)]. Thus, the average amount of time that a customer spends waiting in the queue is

$$W_q = E(T_q) = \sum_{n=0}^{K} \frac{n}{\mu} p_n = \frac{1}{\mu} \sum_{n=0}^{K} n p_n = \frac{1}{\mu} L$$

Similarly, the amount of time that a customer spends in the system is

$$W = E(T) = \sum_{n=1}^{K} \frac{(n+1)}{\mu} p_n = \frac{1}{\mu} \left(\sum_{n=0}^{K} n p_n + \sum_{n=0}^{K} p_n \right) = \frac{1}{\mu}(L+1)$$

Note that Eqs. (9.57) to (9.59) are equivalent to Eqs. (9.31) to (9.33) (Prob. 9.27).

9.17. Derive Eqs. (9.35) and (9.36).

As in Prob. 9.10, from Eqs. (9.34) and (9.10), we have

$$p_n = p_0 \prod_{k=0}^{n-1} \frac{\lambda}{(k+1)\mu} = p_0 \left(\frac{\lambda}{\mu} \right)^n \frac{1}{n!} \qquad\qquad n < s \qquad\qquad (9.60)$$

$$p_n = p_0 \prod_{k=0}^{s-1} \frac{\lambda}{(k+1)\mu} \sum_{k=s}^{n-1} \frac{\lambda}{s\mu} = p_0 \left(\frac{\lambda}{\mu} \right)^n \frac{1}{s! s^{n-s}} \qquad s \leq n \leq K \qquad (9.61)$$

Let $\rho = \lambda /(s\mu)$. Then Eqs. (9.60) and (9.61) can be rewritten as

$$p_n = \begin{cases} \dfrac{(s\rho)^n}{n!} p_0 & n < s \\[3mm] \dfrac{\rho^n s^s}{s!} p_0 & s \leq n \leq K \end{cases}$$

which is Eq. (9.36). From Eq. (9.11), p_0 is obtained by equating

$$\sum_{n=0}^{K} p_n = p_0 \left[\sum_{n=0}^{s-1} \frac{(s\rho)^n}{n!} + \sum_{n=s}^{K} \frac{\rho^n s^s}{s!} \right] = 1$$

Using the summation formula (9.56), we obtain

$$\begin{aligned}
p_0 &= \left[\sum_{n=0}^{s-1} \frac{(s\rho)^n}{n!} + \frac{s^s}{s!} \sum_{n=s}^{K} \rho^n \right]^{-1} \\[3mm]
&= \left[\sum_{n=0}^{s-1} \frac{(s\rho)^n}{n!} + \frac{s^s}{s!} \left(\sum_{n=0}^{K} \rho^n - \sum_{n=0}^{s-1} \rho^n \right) \right]^{-1} \\[3mm]
&= \left[\sum_{n=0}^{s-1} \frac{(s\rho)^n}{n!} + \frac{(s\rho)^s (1 - \rho^{K-s+1})}{s!(1-\rho)} \right]^{-1}
\end{aligned}$$

which is Eq. (9.35).

9.18. Derive Eq. (9.37).

Using Eq. (9.36) and (9.51), the average number of customers in the queue is given by

$$\begin{aligned}
L_q &= \sum_{n=s}^{K} (n-s) p_n = p_0 \frac{s^s}{s!} \sum_{n=s}^{K} (n-s)\rho^n \\[3mm]
&= p_0 \frac{(s\rho)^s}{s!} \sum_{n=s}^{K} (n-s)\rho^{n-s} = p_0 \frac{(s\rho)^s}{s!} \sum_{m=0}^{K-s} m\rho^m \\[3mm]
&= p_0 \frac{(s\rho)^s}{s!} \frac{\rho[(K-s)\rho^{K-s+1} - (K-s+1)\rho^{K-s} + 1]}{(1-\rho)^2} \\[3mm]
&= p_0 \frac{\rho(s\rho)^s}{s!(1-\rho)^2} \{1 - [1 + (1-\rho)(K-s)]\rho^{K-s}\}
\end{aligned}$$

9.19. Consider an M/M/s/s queueing system. Find the probability that all servers are busy.

Setting $K = s$ in Eqs. (9.60) and (9.61), we get

$$p_n = p_0 \left(\frac{\lambda}{\mu} \right)^n \frac{1}{n!} \qquad\qquad 0 \le n \le s \qquad\qquad (9.62)$$

and p_0 is obtained by equating

$$\sum_{n=0}^{s} p_n = p_0 \left[\sum_{n=0}^{s} \left(\frac{\lambda}{\mu} \right)^n \frac{1}{n!} \right] = 1$$

Thus,
$$p_0 = \left[\sum_{n=0}^{s} \left(\frac{\lambda}{\mu} \right)^n \frac{1}{n!} \right]^{-1} \qquad\qquad (9.63)$$

The probability that all servers are busy is given by

$$p_s = p_0 \left(\frac{\lambda}{\mu} \right)^s \frac{1}{s!} = \frac{(\lambda / \mu)^s / s!}{\displaystyle\sum_{n=0}^{s} (\lambda / \mu)^n / n!} \qquad\qquad (9.64)$$

Note that in an M/M/s/s queueing system, if an arriving customer finds that all servers are busy, the customer will turn away and is lost. In a telephone system with s trunks, p_s is the portion of incoming calls which will receive a busy signal. Equation (9.64) is often referred to as *Erlang's loss* (or *B*) formula and is commonly denoted as $B(s, \lambda / \mu)$.

9.20. An air freight terminal has four loading docks on the main concourse. Any aircraft which arrive when all docks are full are diverted to docks on the back concourse. The average aircraft arrival rate is 3 aircraft per hour. The average service time per aircraft is 2 hours on the main concourse and 3 hours on the back concourse.

(*a*) Find the percentage of the arriving aircraft that are diverted to the back concourse.

(*b*) If a holding area which can accommodate up to 8 aircraft is added to the main concourse, find the percentage of the arriving aircraft that are diverted to the back concourse and the expected delay time awaiting service.

(*a*) The service system at the main concourse can be modeled as an M/M/s/s queueing system with $s = 4$, $\lambda = 3$, $\mu = \frac{1}{2}$, and $\lambda / \mu = 6$. The percentage of the arriving aircraft that are diverted to the back concourse is

$$100 \times P(\text{all docks on the main concourse are full})$$

From Eq. (9.64).

$$P(\text{all docks on the main concourse are full}) = p_4 = \frac{6^4 / 4!}{\displaystyle\sum_{n=0}^{4} (6^n / n!)} = \frac{54}{115} \approx 0.47$$

Thus, the percentage of the arriving aircraft that are diverted to the back concourse is about 47 percent.

(*b*) With the addition of a holding area for 8 aircraft, the service system at the main concourse can now be modeled as an M/M/s/K queueing system with $s = 4$, $K = 12$, and $\rho = \lambda / (s\mu) = 1.5$. Now, from Eqs. (9.35) and (9.36),

$$p_0 = \left[\sum_{n=0}^{3} \frac{6^n}{n!} + \frac{6^4}{4!} \left(\frac{1 - 1.5^9}{1 - 1.5} \right) \right]^{-1} \approx 0.00024$$

$$p_{12} = \frac{1.5^{12} 4^4}{4!} p_0 \approx 0.332$$

Thus, about 33.2 percent of the arriving aircraft will still be diverted to the back concourse.

Next, from Eq. (9.37), the average number of aircraft in the queue is

$$L_q = 0.00024 \frac{1.5(6^4)}{4!(1-1.5)^2} \{1 - [1 + (1-1.5)8](1.5)^8\} \approx 6.0565$$

Then, from Eq. (9.40), the expected delay time waiting for service is

$$W_q = \frac{L_q}{\lambda(1 - p_{12})} = \frac{6.0565}{3(1 - 0.332)} \approx 3.022 \text{ hours}$$

Note that when the 2-hour service time is added, the total expected processing time at the main concourse will be 5.022 hours compared to the 3-hour service time at the back concourse.

SUPPLEMENTARY PROBLEMS

9.21. Customers arrive at the express checkout lane in a supermarket in a Poisson process with a rate of 15 per hour. The time to check out a customer is an exponential r.v. with mean of 2 minutes.

    (*a*)   Find the average number of customers present.

    (*b*)   What is the expected idle delay time experienced by a customer?

    (*c*)   What is the expected time for a customer to clear a system?

9.22. Consider an M/M/1 queueing system. Find the probability of finding at least k customers in the system.

9.23. In a university computer center, 80 jobs an hour are submitted on the average. Assuming that the computer service is modeled as an M/M/1 queueing system, what should the service rate be if the average turnaround time (time at submission to time of getting job back) is to be less than 10 minutes?

9.24. The capacity of a communication line is 2000 bits per second. The line is used to transmit 8-bit characters, and the total volume of expected calls for transmission from many devices to be sent on the line is 12,000 characters per minute. Find (*a*) the traffic intensity, (*b*) the average number of characters waiting to be transmitted, and (*c*) the average transmission (including queueing delay) time per character.

9.25. A bank counter is currently served by two tellers. Customers entering the bank join a single queue and go to the next available teller when they reach the head of the line. On the average, the service time for a customer is 3 minutes, and 15 customers enter the bank per hour. Assuming that the arrivals process is Poisson and the service time is an exponential r.v., find the probability that a customer entering the bank will have to wait for service.

9.26. A post office has three clerks serving at the counter. Customers arrive on the average at the rate of 30 per hour, and arriving customers are asked to form a single queue. The average service time for each customer is 3 minutes. Assuming that the arrivals process is Poisson and the service time is an exponential r.v., find (*a*) the probability that all the clerks will be busy, (*b*) the average number of customers in the queue, and (*c*) the average length of time customers have to spend in the post office.

9.27. Show that Eqs. (9.57) to (9.59) and Eqs. (9.31) to (9.33) are equivalent.

9.28. Find the average number of customers L in the M/M/1/K queueing system when $\lambda = \mu$.

9.29. A gas station has one diesel fuel pump for trucks only and has room for three trucks (including one at the pump). On the average, trucks arrive at the rate of 4 per hour, and each truck takes 10 minutes to service. Assume that the arrivals process is Poisson and the service time is an exponential r.v.

    (a)   What is the average time for a truck from entering to leaving the station?

    (b)   What is the average time for a truck to wait for service?

    (c)   What percentage of the truck traffic is being turned away?

9.30. Consider the air freight terminal service of Prob. 9.20. How many additional docks are needed so that at least 80 percent of the arriving aircraft can be served in the main concourse with the addition of holding area?

ANSWERS TO SUPPLEMENTARY PROBLEMS

9.21. (a) 1;   (b) 2 min;   (c) 4 min

9.22. $\rho^k = (\lambda/\mu)^k$

9.23. 1.43 jobs per minute

9.24. (a) 0.8;   (b) 3.2;   (c) 20 ms

9.25. 0.205

9.26. (a) 0.237;   (b) 0.237;   (c) 3.947 min

9.27. *Hint:* Use Eq. (9.29).

9.28. $K/2$

9.29. (a) 20.15 min;   (b) 10.14 min;   (c) 12.3 percent

9.30. 4

CHAPTER 10

Information Theory

10.1 Introduction

Information theory provides a quantitative measure of the information contained in message signals and allows us to determine the capacity of a communication system to transfer this information from source to destination. In this chapter we briefly explore some basic ideas involved in information theory.

10.2 Measure of Information

A. Information Sources:

An information source is an object that produces an event, the outcome of which is selected at random according to a probability distribution. A practical source in a communication system is a device that produces messages, and it can be either analog or discrete. In this chapter we deal mainly with the discrete sources, since analog sources can be transformed to discrete sources through the use of sampling and quantization techniques. A discrete information source is a source that has only a finite set of symbols as possible outputs. The set of source symbols is called the *source alphabet*, and the elements of the set are called *symbols* or *letters*.

Information sources can be classified as having memory or being memoryless. A source with memory is one for which a current symbol depends on the previous symbols. A memoryless source is one for which each symbol produced is independent of the previous symbols.

A *discrete memoryless source* (DMS) can be characterized by the list of the symbols, the probability assignment to these symbols, and the specification of the rate of generating these symbols by the source.

B. Information Content of a Discrete Memoryless Source:

The amount of information contained in an event is closely related to its uncertainty. Messages containing knowledge of high probability of occurrence convey relatively little information. We note that if an event is certain (that is, the event occurs with probability 1), it conveys zero information.

Thus, a mathematical measure of information should be a function of the probability of the outcome and should satisfy the following axioms:

1. Information should be proportional to the uncertainty of an outcome.
2. Information contained in independent outcomes should add.

1. Information Content of a Symbol:

Consider a DMS, denoted by X, with alphabet $\{x_1, x_2, \ldots, x_m\}$. The *information content* of a symbol x_i, denoted by $I(x_i)$, is defined by

$$I(x_i) = \log_b \frac{1}{P(x_i)} = -\log_b P(x_i) \tag{10.1}$$

where $P(x_i)$ is the probability of occurrence of symbol x_i. Note that $I(x_i)$ satisfies the following properties:

$$I(x_i) = 0 \qquad \text{for} \qquad P(x_i) = 1 \tag{10.2}$$

$$I(x_i) \geq 0 \tag{10.3}$$

$$I(x_i) > I(x_j) \quad \text{if} \qquad P(x_i) < P(x_j) \tag{10.4}$$

$$I(x_i x_j) = I(x_i) + I(x_j) \qquad \text{if } x_i \text{ and } x_j \text{ are independent} \tag{10.5}$$

The unit of $I(x_i)$ is the bit (*binary unit*) if $b = 2$, Hartley or decit if $b = 10$, and nat (*natural unit*) if $b = e$. It is standard to use $b = 2$. Here the unit bit (abbreviated "b") is a measure of information content and is not to be confused with the term *bit* meaning "binary digit." The conversion of these units to other units can be achieved by the following relationships.

$$\log_2 a = \frac{\ln a}{\ln 2} = \frac{\log a}{\log 2} \tag{10.6}$$

2. Average Information or Entropy:

In a practical communication system, we usually transmit long sequences of symbols from an information source. Thus, we are more interested in the average information that a source produces than the information content of a single symbol.

The mean value of $I(x_i)$ over the alphabet of source X with m different symbols is given by

$$H(X) = E[I(x_i)] = \sum_{i=1}^{m} P(x_i) I(x_i)$$

$$= -\sum_{i=1}^{m} P(x_i) \log_2 P(x_i) \quad \text{b/symbol} \tag{10.7}$$

The quantity $H(X)$ is called the *entropy* of source X. It is a measure of the *average information content per source symbol*. The source entropy $H(X)$ can be considered as the average amount of uncertainty within source X that is resolved by use of the alphabet.

Note that for a binary source X that generates independent symbols 0 and 1 with equal probability, the source entropy $H(X)$ is

$$H(X) = -\frac{1}{2} \log_2 \frac{1}{2} - \frac{1}{2} \log_2 \frac{1}{2} = 1 \quad \text{b/symbol} \tag{10.8}$$

The source entropy $H(X)$ satisfies the following relation:

$$0 \leq H(X) \leq \log_2 m \tag{10.9}$$

where m is the size (number of symbols) of the alphabet of source X (Prob. 10.5). The lower bound corresponds to no uncertainty, which occurs when one symbol has probability $P(x_i) = 1$ while $P(x_j) = 0$ for $j \neq i$, so X emits the same symbol x_i all the time. The upper bound corresponds to the maximum uncertainty which occurs when $P(x_i) = 1/m$ for all i—that is, when all symbols are equally likely to be emitted by X.

3. Information Rate:

If the time rate at which source X emits symbols is r (symbols/s), the *information rate* R of the source is given by

$$R = rH(X) \qquad \text{b/s} \tag{10.10}$$

10.3   Discrete Memoryless Channels

A.   Channel Representation:

A communication channel is the path or medium through which the symbols flow to the receiver. A *discrete memoryless channel* (DMC) is a statistical model with an input X and an output Y (Fig. 10-1). During each unit of the time (signaling interval), the channel accepts an input symbol from X, and in response it generates an output symbol from Y. The channel is "discrete" when the alphabets of X and Y are both finite. It is "memoryless" when the current output depends on only the current input and not on any of the previous inputs.

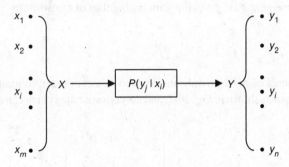

Fig. 10-1 Discrete memoryless channel.

A diagram of a DMC with m inputs and n outputs is illustrated in Fig. 10-1. The input X consists of input symbols $x_1, x_2, ..., x_m$. The a priori probabilities of these source symbols $P(x_i)$ are assumed to be known. The output Y consists of output symbols $y_1, y_2, ..., y_n$. Each possible input-to-output path is indicated along with a conditional probability $P(y_j|x_i)$, where $P(y_j|x_i)$ is the conditional probability of obtaining output y_j given that the input is x_i, and is called a *channel transition probability*.

B.   Channel Matrix:

A channel is completely specified by the complete set of transition probabilities. Accordingly, the channel of Fig. 10-1 is often specified by the matrix of transition probabilities $[P(Y\,|\,X)]$, given by

$$[P(Y|X)] = \begin{bmatrix} P(y_1|x_1) & P(y_2|x_1) & \cdots & P(y_n|x_1) \\ P(y_1|x_2) & P(y_2|x_2) & \cdots & P(y_n|x_2) \\ \cdots & \cdots & \cdots & \cdots \\ P(y_1|x_m) & P(y_2|x_m) & \cdots & P(y_n|x_m) \end{bmatrix} \tag{10.11}$$

The matrix $[P(Y\,|\,X)]$ is called the *channel matrix*. Since each input to the channel results in some output, each row of the channel matrix must sum to unity; that is,

$$\sum_{j=1}^{n} P(y_j|x_i) = 1 \quad \text{for all } i \tag{10.12}$$

Now, if the input probabilities $P(X)$ are represented by the row matrix

$$[P(X)] = [P(x_1) \quad P(x_2) \quad ... \quad P(x_m)] \tag{10.13}$$

and the output probabilities $P(Y)$ are represented by the row matrix

$$[P(Y)] = [P(y_1) \quad P(y_2) \quad ... \quad P(y_n)] \tag{10.14}$$

then
$$[P(Y)] = [P(X)][P(Y\,|\,X)] \tag{10.15}$$

If $P(X)$ is represented as a diagonal matrix

$$[P(X)]_d = \begin{bmatrix} P(x_1) & 0 & \dots & 0 \\ 0 & P(x_2) & \dots & 0 \\ \dots & \dots & \dots & \dots \\ 0 & 0 & \dots & P(x_m) \end{bmatrix} \tag{10.16}$$

then $$[P(X, Y)] = [P(X)]_d[P(Y \mid X)] \tag{10.17}$$

where the (i, j) element of matrix $[P(X, Y)]$ has the form $P(x_i, y_j)$. The matrix $[P(X, Y)]$ is known as the *joint probability matrix,* and the element $P(x_i, y_j)$ is the joint probability of transmitting x_i and receiving y_j.

C. Special Channels:

1. Lossless Channel:

A channel described by a channel matrix with only one nonzero element in each column is called a *lossless channel.* An example of a lossless channel is shown in Fig. 10-2, and the corresponding channel matrix is shown in Eq. (10.18).

$$[P(Y \mid X)] = \begin{bmatrix} \dfrac{3}{4} & \dfrac{1}{4} & 0 & 0 & 0 \\ 0 & 0 & \dfrac{1}{3} & \dfrac{2}{3} & 0 \\ 0 & 0 & 0 & 0 & 1 \end{bmatrix} \tag{10.18}$$

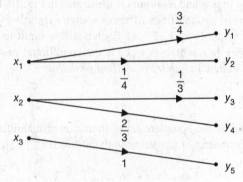

Fig. 10-2 Lossless channel.

It can be shown that in the lossless channel no source information is lost in transmission. [See Eq. (10.35) and Prob. 10.10.]

2. Deterministic Channel:

A channel described by a channel matrix with only one nonzero element in each row is called a *deterministic channel.* An example of a deterministic channel is shown in Fig. 10-3, and the corresponding channel matrix is shown in Eq. (10.19).

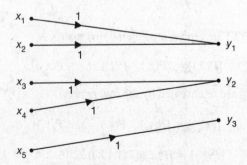

Fig. 10-3 Deterministic channel.

$$[P(Y\,|\,X)] = \begin{bmatrix} 1 & 0 & 0 \\ 1 & 0 & 0 \\ 0 & 1 & 0 \\ 0 & 1 & 0 \\ 0 & 0 & 1 \end{bmatrix} \tag{10.19}$$

Note that since each row has only one nonzero element, this element must be unity by Eq. (10.12). Thus, when a given source symbol is sent in the deterministic channel, it is clear which output symbol will be received.

3. Noiseless Channel:

A channel is called *noiseless* if it is both lossless and deterministic. A noiseless channel is shown in Fig. 10-4. The channel matrix has only one element in each row and in each column, and this element is unity. Note that the input and output alphabets are of the same size; that is, $m = n$ for the noiseless channel.

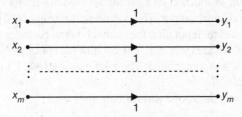

Fig. 10-4 Noiseless channel.

4. Binary Symmetric Channel:

The *binary symmetric channel* (BSC) is defined by the channel diagram shown in Fig. 10-5, and its channel matrix is given by

$$[P(Y\,|\,X)] = \begin{bmatrix} 1-p & p \\ p & 1-p \end{bmatrix} \tag{10.20}$$

The channel has two inputs ($x_1 = 0, x_2 = 1$) and two outputs ($y_1 = 0, y_2 = 1$). The channel is symmetric because the probability of receiving a 1 if a 0 is sent is the same as the probability of receiving a 0 if a 1 is sent. This common transition probability is denoted by p.

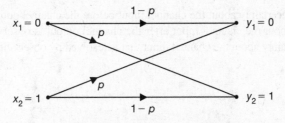

Fig. 10-5 Binary symmetrical channel.

10.4 Mutual Information

A. Conditional and Joint Entropies:

Using the input probabilities $P(x_i)$, output probabilities $P(y_j)$, transition probabilities $P(y_j\,|\,x_i)$, and joint probabilities $P(x_i, y_j)$, we can define the following various entropy functions for a channel with m inputs and n outputs:

$$H(X) = -\sum_{i=1}^{m} P(x_i)\log_2 P(x_i) \tag{10.21}$$

$$H(Y) = -\sum_{j=1}^{n} P(y_j)\log_2 P(y_j) \tag{10.22}$$

$$H(X|Y) = -\sum_{j=1}^{n}\sum_{i=1}^{m} P(x_i, y_j)\log_2 P(x_i|y_j) \tag{10.23}$$

$$H(Y|X) = -\sum_{j=1}^{n}\sum_{i=1}^{m} P(x_i, y_j)\log_2 P(y_j|x_i) \tag{10.24}$$

$$H(X, Y) = -\sum_{j=1}^{n}\sum_{i=1}^{m} P(x_i, y_j)\log_2 P(x_i, y_j) \tag{10.25}$$

These entropies can be interpreted as follows: $H(X)$ is the average uncertainty of the channel input, and $H(Y)$ is the average uncertainty of the channel output. The conditional entropy $H(X|Y)$ is a measure of the average uncertainty remaining about the channel input after the channel output has been observed. And $H(X|Y)$ is sometimes called the *equivocation* of X with respect to Y. The conditional entropy $H(Y|X)$ is the average uncertainty of the channel output given that X was transmitted. The joint entropy $H(X, Y)$ is the average uncertainty of the communication channel as a whole.

Two useful relationships among the above various entropies are

$$H(X, Y) = H(X|Y) + H(Y) \tag{10.26}$$

$$H(X, Y) = H(Y|X) + H(X) \tag{10.27}$$

Note that if X and Y are independent, then

$$H(X|Y) = H(X) \tag{10.28}$$

$$H(Y|X) = H(Y) \tag{10.29}$$

$$H(X, Y) = H(X) + H(Y) \tag{10.30}$$

B. Mutual Information:

The *mutual information* $I(X; Y)$ of a channel is defined by

$$I(X; Y) = H(X) - H(X|Y) \qquad \text{b/symbol} \tag{10.31}$$

Since $H(X)$ represents the uncertainty about the channel input before the channel output is observed and $H(X|Y)$ represents the uncertainty about the channel input after the channel output is observed, the mutual information $I(X; Y)$ represents the uncertainty about the channel input that is resolved by observing the channel output.

Properties of $I(X; Y)$:

1. $I(X; Y) = I(Y; X)$ (10.32)
2. $I(X; Y) \geq 0$ (10.33)
3. $I(X; Y) = H(Y) - H(Y|X)$ (10.34)
4. $I(X; Y) = H(X) + H(Y) - H(X, Y)$ (10.35)
5. $I(X; Y) \geq 0$ if X and Y are independent   (10.36)

Note that from Eqs. (10.31), (10.33), and (10.34) we have

6. $H(X) \geq H(X|Y)$ (10.37)
7. $H(Y) \geq H(Y|X)$ (10.38)

with equality if X and Y are independent.

C.　Relative Entropy:

The *relative entropy* between two pmf's $p(x_i)$, and $q(x_i)$ on X is defined as

$$D(p \mathbin{/\mkern-4mu/} q) = \sum_i p(x_i) \log_2 \frac{p(x_i)}{q(x_i)} \tag{10.39}$$

The relative entropy, also known as Kullback-Leibler *divergence*, measures the "closeness" of one distribution from another. It can be shown that (Prob. 10.15)

$$D(p \mathbin{/\mkern-4mu/} q) \ge 0 \tag{10.40}$$

and equal zero if $p(x_i) = q(x_i)$. Note that, in general, it is not symmetric, that is, $D(p \mathbin{/\mkern-4mu/} q) \ne D(q \mathbin{/\mkern-4mu/} p)$.

The mutual information $I(X; Y)$ can be expressed as the relative entropy between the joint distribution $p_{XY}(x, y)$ and the product of distribution $p_X(x)\, p_Y(y)$; that is,

$$I(X; Y) = D(p_{XY}(x, y) \mathbin{/\mkern-4mu/} p_x(x)\, p_Y(y)) \tag{10.41}$$

$$= \sum_{x_i} \sum_{y_j} p_{XY}(x_i, y_j) \log_2 \frac{p_{XY}(x_i, y_j)}{p_X(x_i)\, p_Y(y_j)} \tag{10.42}$$

10.5　Channel Capacity

A.　Channel Capacity per Symbol C_s:

The *channel capacity per symbol* of a DMC is defined as

$$C_s = \max_{\{P(x_i)\}} I(X;Y) \quad \text{b/symbol} \tag{10.43}$$

where the maximization is over all possible input probability distributions $\{P(x_i)\}$ on X. Note that the channel capacity C_s is a function of only the channel transition probabilities that define the channel.

B.　Channel Capacity per Second C:

If r symbols are being transmitted per second, then the maximum rate of transmission of information per second is rC_s. This is the *channel capacity per second* and is denoted by C (b/s):

$$C = rC_s \quad \text{b/s} \tag{10.44}$$

C.　Capacities of Special Channels:

1.　Lossless Channel:

For a lossless channel, $H(X \mid Y) = 0$ (Prob. 10.12) and

$$I(X; Y) = H(X) \tag{10.45}$$

Thus, the mutual information (information transfer) is equal to the input (source) entropy, and no source information is lost in transmission. Consequently, the channel capacity per symbol is

$$C_s = \max_{\{P(x_i)\}} H(X) = \log_2 m \tag{10.46}$$

where m is the number of symbols in X.

2. Deterministic Channel:

For a deterministic channel, $H(Y \mid X) = 0$ for all input distributions $P(x_i)$, and

$$I(X; Y) = H(Y) \tag{10.47}$$

Thus, the information transfer is equal to the output entropy. The channel capacity per symbol is

$$C_s = \max_{\{P(x_i)\}} H(Y) = \log_2 n \tag{10.48}$$

where n is the number of symbols in Y.

3. Noiseless Channel:

Since a noiseless channel is both lossless and deterministic, we have

$$I(X; Y) = H(X) = H(Y) \tag{10.49}$$

and the channel capacity per symbol is

$$C_s = \log_2 m = \log_2 n \tag{10.50}$$

4. Binary Symmetric Channel:

For the BSC of Fig. 10-5, the mutual information is (Prob. 10.20)

$$I(X; Y) = H(Y) + p \log_2 p + (1 - p) \log_2(1 - p) \tag{10.51}$$

and the channel capacity per symbol is

$$C_s = 1 + p \log_2 p + (1 - p) \log_2(1 - p) \tag{10.52}$$

10.6 Continuous Channel

In a continuous channel an information source produces a continuous signal $x(t)$. The set of possible signals is considered as an ensemble of waveforms generated by some ergodic random process. It is further assumed that $x(t)$ has a finite bandwidth so that $x(t)$ is completely characterized by its periodic sample values. Thus, at any sampling instant, the collection of possible sample values constitutes a continuous random variable X described by its probability density function $f_X(x)$.

A. Differential Entropy:

The average amount of information per sample value of $x(t)$ is measured by

$$H(X) = -\int_{-\infty}^{\infty} f_X(x) \log_2 f_X(x) \, dx \qquad \text{b/sample} \tag{10.53}$$

The entropy defined by Eq. (10.53) is known as the *differential entropy* of a continuous r.v. X with pdf $f_X(x)$. Note that as opposed to the discrete case (see Eq. (10.9)), the differential entropy can be negative (see Prob. 10.24).

Properties of Differential Entropy:

    (1) $H(X + c) = H(X)$ (10.54)

    Translation does not change the differential entropy. Equation (10.54) follows directly from the definition Eq. (10.53).

    (2) $H(aX) = H(X) + \log_2 |a|$ (10.55)

    Equation (10.55) is proved in Prob. 10.29.

B. Mutual Information:

The average mutual information in a continuous channel is defined (by analogy with the discrete case) as

$$I(X; Y) = H(X) - H(X \mid Y) \tag{10.56}$$

or

$$I(X; Y) = H(Y) - H(Y \mid X) \tag{10.57}$$

where

$$H(Y) = -\int_{-\infty}^{\infty} f_Y(y) \log_2 f_Y(y) \, dy \tag{10.58}$$

$$H(X \mid Y) = -\int_{-\infty}^{\infty}\int_{-\infty}^{\infty} f_{XY}(x, y) \log_2 f_X(x \mid y) \, dx \, dy \tag{10.59}$$

$$H(Y \mid X) = -\int_{-\infty}^{\infty}\int_{-\infty}^{\infty} f_{XY}(x, y) \log_2 f_Y(y \mid x) \, dx \, dy \tag{10.60}$$

C. Relative Entropy:

Similar to the discrete case, the relative entropy (or Kullback-Leibler divergence) between two pdf's $f_X(x)$ and $g_X(x)$ on continuous r.v. X is defined by

$$D(f \mathbin{/\!/} g) = \int_{-\infty}^{\infty} f_X(x) \log_2 \left(\frac{f_X(x)}{g_X(x)} \right) dx \tag{10.61}$$

From this definition, we can express the average mutual information $I(X; Y)$ as

$$I(X; Y) = D(f_{XY}(x, y) \mathbin{/\!/} f_X(x) f_Y(y)) \tag{10.62}$$

$$= \int_{-\infty}^{\infty}\int_{-\infty}^{\infty} f_{XY}(x_i, y_j) \log_2 \frac{f_{XY}(x_i, y_j)}{f_X(x_i) f_Y(y_j)} \, dx \, dy \tag{10.63}$$

D. Properties of Differential Entropy, Relative Entropy, and Mutual Information:

1. $D(f \mathbin{/\!/} g) \geq 0$ (10.64)

with equality iff $f = g$. (See Prob. 10.30.)

2. $I(X; Y) \geq 0$ (10.65)
3. $H(X) \geq H(X \mid Y)$ (10.66)
4. $H(Y) \geq H(Y \mid X)$ (10.67)

with equality iff X and Y are independent.

10.7 Additive White Gaussian Noise Channel

A. Additive White Gaussian Noise Channel:

An additive white Gaussian noise (AWGN) channel is depicted in Fig. 10-6.

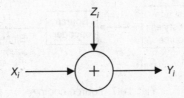

Fig. 10-6 Gaussian channel.

This is a time discrete channel with output Y_i at time i, where Y_i is the sum of the input X_i and the noise Z_i.

$$Y_i = X_i + Z_i \tag{10.68}$$

The noise Z_i is drawn i.i.d. from a Gaussian distribution with zero mean and variance N. The noise Z_i is assumed to be independent of the input signal X_i. We also assume that the average power of input signal is finite. Thus,

$$E(X_i^2) \leq S \tag{10.69}$$
$$E(Z_i^2) = N \tag{10.70}$$

The capacity C_s of an AWGN channel is given by (Prob. 10.31)

$$C_s = \max_{\{f_X(x)\}} I(X;Y) = \frac{1}{2}\log_2\left(1 + \frac{S}{N}\right) \tag{10.71}$$

where S/N is the signal-to-noise ratio at the channel output.

B. Band-Limited Channel:

A common model for a communication channel is a band-limited channel with additive white Gaussian noise. This is a continuous-time channel.

Nyquist Sampling Theorem

Suppose a signal $x(t)$ is band-limited to B; namely, the Fourier transform of the signal $x(t)$ is 0 for all frequencies greater than B(Hz). Then the signal $x(t)$ is completely determined by its periodic sample values taken at the Nyquist rate $2B$ samples. (For the proof of this sampling theorem, see any Fourier transform text.)

C. Capacity of the Continuous-Time Gaussian Channel:

If the channel bandwidth B(Hz) is fixed, then the output $y(t)$ is also a band-limited signal. Then the capacity C(b/s) of the continuous-time AWGN channel is given by (see Prob. 10.32)

$$C = B\log_2\left(1 + \frac{S}{N}\right) \quad \text{b/s} \tag{10.72}$$

Equation (10.72) is known as the *Shannon-Hartley law*.

The Shannon-Hartley law underscores the fundamental role of bandwidth and signal-to-noise ratio in communication. It also shows that we can exchange increased bandwidth for decreased signal power (Prob. 10.35) for a system with given capacity C.

10.8 Source Coding

A conversion of the output of a DMS into a sequence of binary symbols (binary code word) is called *source coding*. The device that performs this conversion is called the *source encoder* (Fig. 10-7).

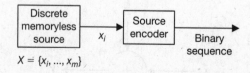

Fig. 10-7 Source coding.

An objective of source coding is to minimize the average bit rate required for representation of the source by reducing the redundancy of the information source.

A. Code Length and Code Efficiency:

Let X be a DMS with finite entropy $H(X)$ and an alphabet $\{x_1, \ldots, x_m\}$ with corresponding probabilities of occurrence $P(x_i)(i = 1, \ldots, m)$. Let the binary code word assigned to symbol x_i by the encoder have length n_i, measured in bits. The length of a code word is the number of binary digits in the code word. The average code word length L, per source symbol, is given by

$$L = \sum_{i=1}^{m} P(x_i)n_i \tag{10.73}$$

The parameter L represents the average number of bits per source symbol used in the source coding process.
   The *code efficiency* η is defined as

$$\eta = \frac{L_{min}}{L} \tag{10.74}$$

where L_{min} is the minimum possible value of L. When η approaches unity, the code is said to be *efficient*.
   The *code redundancy* γ is defined as

$$\gamma = 1 - \eta \tag{10.75}$$

B. Source Coding Theorem:

The source coding theorem states that for a DMS X with entropy $H(X)$, the average code word length L per symbol is bounded as (Prob. 10.39)

$$L \geq H(X) \tag{10.76}$$

and further, L can be made as close to $H(X)$ as desired for some suitably chosen code.
   Thus, with $L_{min} = H(X)$, the code efficiency can be rewritten as

$$\eta = \frac{H(X)}{L} \tag{10.77}$$

C. Classification of Codes:

Classification of codes is best illustrated by an example. Consider Table 10-1 where a source of size 4 has been encoded in binary codes with symbol 0 and 1.

TABLE 10-1 Binary Codes

x_i	CODE 1	CODE 2	CODE 3	CODE 4	CODE 5	CODE 6
x_1	00	00	0	0	0	1
x_2	01	01	1	10	01	01
x_3	00	10	00	110	011	001
x_4	11	11	11	111	0111	0001

1. Fixed-Length Codes:

A *fixed-length* code is one whose code word length is fixed. Code 1 and code 2 of Table 10-1 are fixed-length codes with length 2.

2. Variable-Length Codes:

A *variable-length* code is one whose code word length is not fixed. All codes of Table 10-1 except codes 1 and 2 are variable-length codes.

3. Distinct Codes:

A code is *distinct* if each code word is distinguishable from other code words. All codes of Table 10-1 except code 1 are distinct codes—notice the codes for x_1 and x_3.

4. Prefix-Free Codes:

A code in which no code word can be formed by adding code symbols to another code word is called a *prefix-free code*. Thus, in a prefix-free code no code word is a *prefix* of another. Codes 2, 4, and 6 of Table 10-1 are prefix-free codes.

5. Uniquely Decodable Codes:

A distinct code is *uniquely decodable* if the original source sequence can be reconstructed perfectly from the encoded binary sequence. Note that code 3 of Table 10-1 is not a uniquely decodable code. For example, the binary sequence 1001 may correspond to the source sequences $x_2 x_3 x_2$ or $x_2 x_1 x_1 x_2$. A sufficient condition to ensure that a code is uniquely decodable is that no code word is a prefix of another. Thus, the prefix-free codes 2, 4, and 6 are uniquely decodable codes. Note that the prefix-free condition is not a necessary condition for unique decodability. For example, code 5 of Table 10-1 does not satisfy the prefix-free condition, and yet it is uniquely decodable since the bit 0 indicates the beginning of each code word of the code.

6. Instantaneous Codes:

A uniquely decodable code is called an *instantaneous code* if the end of any code word is recognizable without examining subsequent code symbols. The instantaneous codes have the property previously mentioned that no code word is a prefix of another code word. For this reason, prefix-free codes are sometimes called instantaneous codes.

7. Optimal Codes:

A code is said to be *optimal* if it is instantaneous and has minimum average length L for a given source with a given probability assignment for the source symbols.

D. Kraft Inequality:

Let X be a DMS with alphabet $\{x_i\}$ ($i = 1, 2, ..., m$). Assume that the length of the assigned binary code word corresponding to x_i is n_i.

A necessary and sufficient condition for the existence of an instantaneous binary code is

$$K = \sum_{i=1}^{m} 2^{-n_i} \leq 1 \tag{10.78}$$

which is known as the *Kraft inequality*. (See Prob. 10.43.)

Note that the Kraft inequality assures us of the existence of an instantaneously decodable code with code word lengths that satisfy the inequality. But it does not show us how to obtain these code words, nor does it say that any code that satisfies the inequality is automatically uniquely decodable (Prob. 10.38).

10.9 Entropy Coding

The design of a variable-length code such that its average code word length approaches the entropy of the DMS is often referred to as *entropy coding*. This section presents two examples of entropy coding.

A. Shannon-Fano Coding:

An efficient code can be obtained by the following simple procedure, known as *Shannon-Fano algorithm:*

1. List the source symbols in order of decreasing probability.
2. Partition the set into two sets that are as close to equiprobable as possible, and assign 0 to the upper set and 1 to the lower set.
3. Continue this process, each time partitioning the sets with as nearly equal probabilities as possible until further partitioning is not possible.

An example of Shannon-Fano encoding is shown in Table 10-2. Note that in Shannon-Fano encoding the ambiguity may arise in the choice of approximately equiprobable sets. (See Prob. 10.46.) Note also that Shannon-Fano coding results in suboptimal code.

TABLE 10-2 Shannon-Fano Encoding

x_i	$P(x_i)$	STEP 1	STEP 2	STEP 3	STEP 4	CODE 5
x_1	0.30	0	0			00
x_2	0.25	0	1			01
x_3	0.20	1	0			10
x_4	0.12	1	1	0		110
x_5	0.08	1	1	1	0	1110
x_6	0.05	1	1	1	1	1111

$$H(X) = 2.36 \text{ b/symbol}$$
$$L = 2.38 \text{ b/symbol}$$
$$\eta = H(X)/L = 0.99$$

B. Huffman Encoding:

In general, Huffman encoding results in an optimum code. Thus, it is the code that has the highest efficiency (Prob. 10.47). The Huffman encoding procedure is as follows:

1. List the source symbols in order of decreasing probability.
2. Combine the probabilities of the two symbols having the lowest probabilities, and reorder the resultant probabilities; this step is called reduction 1. The same procedure is repeated until there are two ordered probabilities remaining.
3. Start encoding with the last reduction, which consists of exactly two ordered probabilities. Assign 0 as the first digit in the code words for all the source symbols associated with the first probability; assign 1 to the second probability.
4. Now go back and assign 0 and 1 to the second digit for the two probabilities that were combined in the previous reduction step, retaining all assignments made in Step 3.
5. Keep regressing this way until the first column is reached.

An example of Huffman encoding is shown in Table 10-3.

$$H(X) = 2.36 \text{ b/symbol}$$
$$L = 2.38 \text{ b/symbol}$$
$$\eta = 0.99$$

TABLE 10-3 Huffman Encoding

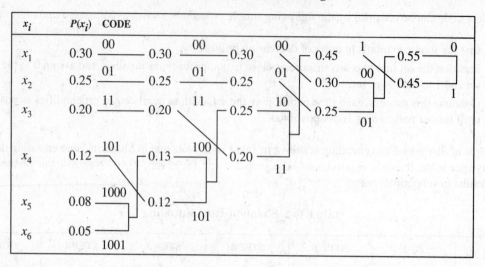

Note that the Huffman code is not unique depending on Huffman tree and the labeling (see Prob. 10.33).

SOLVED PROBLEMS

Measure of Information

10.1. Consider event E occurred when a random experiment is performed with probability p. Let $I(p)$ be the information content (or surprise measure) of event E, and assume that it satisfies the following axioms:

1. $I(p) \geq 0$
2. $I(1) = 0$
3. $I(p) > I(q)$ if $p < q$
4. $I(p)$ is a continuous function of p.
5. $I(p\,q) = I(p) + I(q)$ $0 < p \leq 1, 0 < q \leq 1$

Then show that $I(p)$ can be expressed as

$$I(p) = -C \log_2 p \tag{10.79}$$

where C is an arbitrary positive integer.

From Axiom 5, we have

$$I(p^2) = I(p\,p) = I(p) + I(p) = 2\,I(p)$$

and by induction, we have

$$I(p^m) = m\,I(p) \tag{10.80}$$

Also, for any integer n, we have

$$I(p) = I(p^{1/n} \cdots p^{1/n}) = n\,I(p^{1/n})$$

and

$$I\!\left(p^{1/n}\right) = \frac{1}{n} I(p) \tag{10.81}$$

Thus, from Eqs. (10.80) and (10.81), we have

$$I\left(p^{m/n}\right) = \frac{m}{n}I(p)$$

or

$$I(p^r) = r\,I(p)$$

where r is any positive rational number. Then by Axiom 4

$$I(p^{\alpha}) = \alpha\,I(p) \qquad\qquad (10.82)$$

where α is any nonnegative number.

Let $\alpha = -\log_2 p$ $(0 < p \le 1)$. Then $p = (1/2)^{\alpha}$ and from Eq. (10.82), we have

$$I(p) = I((1/2)^{\alpha}) = \alpha\,I(1/2) = -I(1/2)\log_2 p = -C\log_2 p$$

where $C = I(1/2) > I(1) = 0$ by Axioms 2 and 3. Setting $C = 1$, we have $I(p) = -\log_2 p$ bits.

10.2. Verify Eq. (10.5); that is,

$$I(x_i x_j) = I(x_i) + I(x_j) \qquad \text{if } x_i \text{ and } x_j \text{ are independent}$$

If x_i and x_j are independent, then by Eq. (3.22)

$$P(x_i x_j) = P(x_i)\,P(x_j)$$

By Eq. (10.1)

$$I(x_i x_j) = \log\frac{1}{P(x_i x_j)} = \log\frac{1}{P(x_i)P(x_j)}$$

$$= \log\frac{1}{P(x_i)} + \log\frac{1}{P(x_j)}$$

$$= I(x_i) + I(x_j)$$

10.3. A DMS X has four symbols x_1, x_2, x_3, x_4 with probabilities $P(x_1) = 0.4$, $P(x_2) = 0.3$, $P(x_3) = 0.2$, $P(x_4) = 0.1$.

(a) Calculate $H(X)$.

(b) Find the amount of information contained in the messages $x_1 x_2 x_1 x_3$ and $x_4 x_3 x_3 x_2$, and compare with the $H(X)$ obtained in part (a).

(a)

$$H(X) = -\sum_{i=1}^{4} p(x_i)\log_2[P(x_i)]$$

$$= -0.4\log_2 0.4 - 0.3\log_2 0.3/ -0.2\log_2 0.2 - 0.1\log_2 0.1$$

$$= 1.85 \text{ b/symbol}$$

(b)

$$P(x_1 x_2 x_1 x_3) = (0.4)(0.3)(0.4)(0.2) = 0.0096$$

$$I(x_1 x_2 x_1 x_3) = -\log_2 0.0096 = 6.70 \text{ b/symbol}$$

Thus,

$$I(x_1 x_2 x_1 x_3) < 7.4\ [=4H(X)]\ \text{b/symbol}$$

$$P(x_4 x_3 x_3 x_2) = (0.1)(0.2)^2(0.3) = 0.0012$$

$$I(x_4 x_3 x_3 x_2) = -\log_2 0.0012 = 9.70\text{b/symbol}$$

Thus,

$$I(x_4 x_3 x_3 x_2) > 7.4\ [= 4H(X)]\ \text{b/symbol}$$

10.4. Consider a binary memoryless source X with two symbols x_1 and x_2. Show that $H(X)$ is maximum when both x_1 and x_2 are equiprobable.

Let $P(x_1) = \alpha$. $P(x_2) = 1 - \alpha$.

$$H(X) = -\alpha \log_2 \alpha - (1 - \alpha) \log_2 (1 - \alpha) \tag{10.83}$$

$$\frac{dH(X)}{d\alpha} = \frac{d}{d\alpha} [-\alpha \log_2 \alpha - (1 - \alpha) \log_2 (1 - \alpha)]$$

Using the relation

$$\frac{d}{dx} \log_b y = \frac{1}{y} \log_b e \frac{dy}{dx}$$

we obtain

$$\frac{dH(X)}{d\alpha} = -\log_2 \alpha + \log_2 (1 - \alpha) = \log_2 \frac{1 - \alpha}{\alpha}$$

The maximum value of $H(X)$ requires that

$$\frac{dH(X)}{d\alpha} = 0$$

that is,

$$\frac{1 - \alpha}{\alpha} = 1 \rightarrow \alpha = \frac{1}{2}$$

Note that $H(X) = 0$ when $\alpha = 0$ or 1. When $P(x_1) = P(x_2) = \frac{1}{2}$, $H(X)$ is maximum and is given by

$$H(X) = \frac{1}{2} \log_2 2 + \frac{1}{2} \log_2 2 = 1 \quad \text{b/symbol} \tag{10.84}$$

10.5. Verify Eq. (10.9); that is,

$$0 \le H(X) \le \log_2 m$$

where m is the size of the alphabet of X.

Proof of the lower bound: Since $0 \le P(x_i) \le 1$,

$$\frac{1}{P(x_i)} \ge 1 \quad \text{and} \quad \log_2 \frac{1}{P(x_i)} \ge 0$$

Then it follows that

$$p(x_i) \log_2 \frac{1}{P(x_i)} \ge 0$$

Thus,

$$H(X) = \sum_{i=1}^m P(x_i) \log_2 \frac{1}{P(x_i)} \ge 0 \tag{10.85}$$

Next, we note that

$$P(x_i) \log_2 \frac{1}{P(x_i)} = 0$$

if and only if $P(x_i) = 0$ or 1. Since

$$\sum_{i=1}^m P(x_i) = 1$$

when $P(x_i) = 1$, then $P(x_j) = 0$ for $j \neq i$. Thus, only in this case, $H(X) = 0$.

Proof of the upper bound: Consider two probability distributions $\{P(x_i) = P_i\}$ and $\{Q(x_i) = Q_i\}$ on the alphabet $\{x_i\}$, $i = 1, 2, \ldots, m$, such that

$$\sum_{i=1}^{m} P_i = 1 \quad \text{and} \quad \sum_{i=1}^{m} Q_i = 1 \tag{10.86}$$

Using Eq. (10.6), we have

$$\sum_{i=1}^{m} P_i \log_2 \frac{Q_i}{P_i} = \frac{1}{\ln 2} \sum_{i=1}^{m} P_i \ln \frac{Q_i}{P_i}$$

Next, using the inequality

$$\ln \alpha \leq \alpha - 1 \qquad \alpha \geq 0 \tag{10.87}$$

and noting that the equality holds only if $\alpha = 1$, we get

$$\sum_{i=1}^{m} P_i \ln \frac{Q_i}{P_i} \leq \sum_{i=1}^{m} P_i \left(\frac{Q_i}{P_i} - 1 \right) = \sum_{i=1}^{m} (Q_i - P_i)$$

$$= \sum_{i=1}^{m} Q_i - \sum_{i=1}^{m} P_i = 0 \tag{10.88}$$

by using Eq. (10.86). Thus,

$$\sum_{i=1}^{m} P_i \log_2 \frac{Q_i}{P_i} \leq 0 \tag{10.89}$$

where the equality holds only if $Q_i = P_i$ for all i. Setting

$$Q_i = \frac{1}{m} \qquad i = 1, 2, \ldots, m \tag{10.90}$$

we obtain

$$\sum_{i=1}^{m} P_i \log_2 \frac{1}{P_i m} = -\sum_{i=1}^{m} P_i \log_2 P_i - \sum_{i=1}^{m} P_i \log_2 m$$

$$= H(X) - \log_2 m \sum_{i=1}^{m} P_i \tag{10.91}$$

$$= H(X) - \log_2 m \leq 0$$

Hence,
$$H(X) \leq \log_2 m$$

and the equality holds only if the symbols in X are equiprobable, as in Eq. (10.90).

10.6. Find the discrete probability distribution of X, $p_X(x_i)$, which maximizes information entropy $H(X)$.

From Eqs. (10.7) and (2.17), we have

$$H(X) = -\sum_{i=1}^{m} p_X(x_i) \ln p_X(x_i) \tag{10.92}$$

$$\sum_{i=1}^{m} p_X(x_i) = 1 \tag{10.93}$$

Thus, the problem is the maximization of Eq. (10.92) with constraint Eq. (10.93). Then using the method of Lagrange multipliers, we set Lagrangian J as

$$J = -\sum_{i=1}^{m} p_X(x_i) \ln p_X(x_i) + \lambda \left(\sum_{i=1}^{m} p_X(x_i) - 1 \right)$$

(10.94)

where λ is the Lagrangian multiplier. Taking the derivative of J with respect to $p_X(x_i)$ and setting equal to zero, we obtain

$$\frac{\partial J}{\partial p_X(x_i)} = -\ln p_X(x_i) - 1 + \lambda = 0$$

and

$$\ln p_X(x_i) = \lambda - 1 \Rightarrow p_X(x_i) = e^{\lambda - 1}$$

(10.95)

This shows that all $p_x(x_i)$ are equal (because they depend on λ only) and using the constraint Eq. (10.93), we obtain $p_x(x_i) = 1/m$. Hence, the uniform distribution is the distribution with the maximum entropy. (cf. Prob. 10.5.)

10.7. A high-resolution black-and-white TV picture consists of about 2×10^6 picture elements and 16 different brightness levels. Pictures are repeated at the rate of 32 per second. All picture elements are assumed to be independent, and all levels have equal likelihood of occurrence. Calculate the average rate of information conveyed by this TV picture source.

$$H(X) = -\sum_{i=1}^{16} \frac{1}{16} \log_2 \frac{1}{16} = 4 \text{ b/element}$$

$$r = 2(10^6)(32) = 64(10^6) \text{ elements/s}$$

Hence, by Eq. (10.10)

$$R = rH(X) = 64(10^6)(4) = 256(10^6) \text{ b/s} = 256 \text{ Mb/s}$$

10.8. Consider a telegraph source having two symbols, dot and dash. The dot duration is 0.2 s. The dash duration is 3 times the dot duration. The probability of the dot's occurring is twice that of the dash, and the time between symbols is 0.2 s. Calculate the information rate of the telegraph source.

$$P(\text{dot}) = 2P(\text{dash})$$
$$P(\text{dot}) + P(\text{dash}) = 3P(\text{dash}) = 1$$

Thus,

$$P(\text{dash}) = \frac{1}{3} \quad \text{and} \quad P(\text{dot}) = \frac{2}{3}$$

By Eq. (10.7)

$$H(X) = -P(\text{dot}) \log_2 P(\text{dot}) - P(\text{dash}) \log_2 P(\text{dash})$$
$$= 0.667(0.585) + 0.333(1.585) = 0.92 \text{ b/symbol}$$

$$t_{\text{dot}} = 0.2\text{s} \qquad t_{\text{dash}} = 0.6\text{s} \qquad t_{\text{space}} = 0.2 \text{ s}$$

Thus, the average time per symbol is

$$T_s = P(\text{dot})t_{\text{dot}} + P(\text{dash})t_{\text{dash}} + t_{\text{space}} = 0.5333 \text{ s/symbol}$$

and the average symbol rate is

$$r = \frac{1}{T_S} = 1.875 \quad \text{symbols/s}$$

Thus, the average information rate of the telegraph source is

$$R = rH(X) = 1.875(0.92) = 1.725 \text{ b/s}$$

Discrete Memoryless Channels

10.9. Consider a binary channel shown in Fig. 10-8.

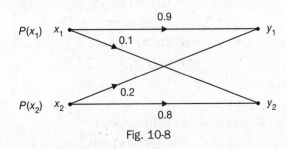

Fig. 10-8

(*a*) Find the channel matrix of the channel.

(*b*) Find $P(y_1)$ and $P(y_2)$ when $P(x_1) = P(x_2) = 0.5$.

(*c*) Find the joint probabilities $P(x_1, y_2)$ and $P(x_2, y_1)$ when $P(x_1) = P(x_2) = 0.5$.

(*a*) Using Eq. (10.11), we see the channel matrix is given by

$$[P(Y|X)] = \begin{bmatrix} P(y_1|x_1) & P(y_2|x_1) \\ P(y_1|x_2) & P(y_2|x_2) \end{bmatrix} = \begin{bmatrix} 0.9 & 0.1 \\ 0.2 & 0.8 \end{bmatrix}$$

(*b*) Using Eqs. (10.13), (10.14), and (10.15), we obtain

$$[P(Y)] = [P(X)][P(Y|X)]$$

$$= [0.5 \quad 0.5] \begin{bmatrix} 0.9 & 0.1 \\ 0.2 & 0.8 \end{bmatrix}$$

$$= [0.55 \quad 0.45] = [P(y_1) P(y_2)]$$

Hence, $P(y_1) = 0.55$ and $P(y_2) = 0.45$.

(*c*) Using Eqs. (10.16) and (10.17), we obtain

$$[P(X,Y)] = [P(X)]_d [P(Y|X)]$$

$$= \begin{bmatrix} 0.5 & 0 \\ 0 & 0.5 \end{bmatrix} \begin{bmatrix} 0.9 & 0.1 \\ 0.2 & 0.8 \end{bmatrix}$$

$$= \begin{bmatrix} 0.45 & 0.05 \\ 0.1 & 0.4 \end{bmatrix} = \begin{bmatrix} P(x_1, y_1) & P(x_1, y_2) \\ P(x_2, y_1) & P(x_2, y_2) \end{bmatrix}$$

Hence, $P(x_1, y_2) = 0.05$ and $P(x_2, y_1) = 0.1$.

10.10. Two binary channels of Prob. 10.9 are connected in a cascade, as shown in Fig. 10-9.

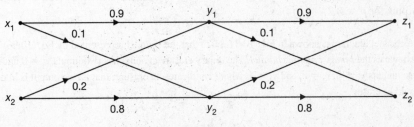

Fig. 10-9

(*a*) Find the overall channel matrix of the resultant channel, and draw the resultant equivalent channel diagram.

(*b*) Find $P(z_1)$ and $P(z_2)$ when $P(x_1) = P(x_2) = 0.5$.

(*a*) By Eq. (10.15)

$$[P(Y)] = [P(X)][P(Y\,|\,X)]$$
$$[P(Z)] = [P(Y)][P(Z\,|\,Y)]$$
$$= [P(X)][P(Y\,|\,X)][P(Z\,|\,Y)]$$
$$= [P(X)][P(Z\,|\,X)]$$

Thus, from Fig. 10-9

$$[P(Z\,|\,X)] = [P(Y\,|\,X)][P(Z\,|\,Y)]$$
$$= \begin{bmatrix} 0.9 & 0.1 \\ 0.2 & 0.8 \end{bmatrix}\begin{bmatrix} 0.9 & 0.1 \\ 0.2 & 0.8 \end{bmatrix} = \begin{bmatrix} 0.83 & 0.17 \\ 0.34 & 0.66 \end{bmatrix}$$

The resultant equivalent channel diagram is shown in Fig. 10-10.

(*b*)
$$[P(Z)] = [P(X)][P(Z\,|\,X)]$$
$$= [0.5 \quad 0.5]\begin{bmatrix} 0.83 & 0.17 \\ 0.34 & 0.66 \end{bmatrix} = [0.585 \quad 0.415]$$

Hence, $P(z_1) = 0.585$ and $P(z_2) = 0.415$.

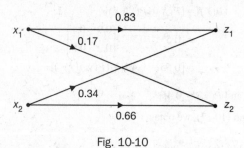

Fig. 10-10

10.11. A channel has the following channel matrix:

$$[P(Y\,|\,X)] = \begin{bmatrix} 1-p & p & 0 \\ 0 & p & 1-p \end{bmatrix} \tag{10.96}$$

(*a*) Draw the channel diagram.

(*b*) If the source has equally likely outputs, compute the probabilities associated with the channel outputs for $p = 0.2$.

(*a*) The channel diagram is shown in Fig. 10-11. Note that the channel represented by Eq. (10.96) (see Fig. 10-11) is known as the *binary erasure channel*. The binary erasure channel has two inputs $x_1 = 0$ and $x_2 = 1$ and three outputs $y_1 = 0$, $y_2 = e$, and $y_3 = 1$, where e indicates an erasure; that is, the output is in doubt, and it should be erased.

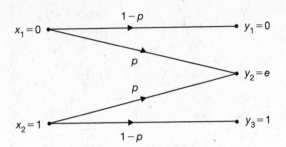

Fig. 10-11 Binary erasure channel.

(b)   By Eq. (10.15)

$$[P(Y)] = [0.5 \quad 0.5] \begin{bmatrix} 0.8 & 0.2 & 0 \\ 0 & 0.2 & 0.8 \end{bmatrix}$$

$$= [0.4 \quad 0.2 \quad 0.4]$$

Thus, $P(y_1) = 0.4$, $P(y_2) = 0.2$, and $P(y_3) = 0.4$.

Mutual Information

10.12. For a lossless channel show that

$$H(X \mid Y) = 0 \tag{10.97}$$

When we observe the output y_j in a lossless channel (Fig. 10-2), it is clear which x_i was transmitted; that is,

$$P(x_i \mid y_j) = 0 \text{ or } 1 \tag{10.98}$$

Now by Eq. (10.23)

$$H(X \mid Y) = \sum_{j=1}^{n} \sum_{i=1}^{m} P(x_i, y_j) \log_2 P(x_i \mid y_j)$$

$$= -\sum_{j=1}^{n} P(y_j) \sum_{i=1}^{m} P(x_i \mid y_j) \log_2 P(x_i \mid y_j) \tag{10.99}$$

Note that all the terms in the inner summation are zero because they are in the form of $1 \times \log_2 1$ or $0 \times \log_2 0$. Hence, we conclude that for a lossless channel

$$H(X \mid Y) = 0$$

10.13. Consider a noiseless channel with m input symbols and m output symbols (Fig. 10-4). Show that

$$H(X) = H(Y) \tag{10.100}$$

and

$$H(Y \mid X) = 0 \tag{10.101}$$

For a noiseless channel the transition probabilities are

$$P(y_j \mid x_i) = \begin{cases} 1 & i = j \\ 0 & i \neq j \end{cases} \tag{10.102}$$

Hence,
$$P(x_i, y_j) = P(y_j|x_i)P(x_i) = \begin{cases} P(x_i) & i = j \\ 0 & i \neq j \end{cases} \tag{10.103}$$

and
$$P(y_j) = \sum_{i=1}^{m} P(x_i, y_j) = P(x_j) \tag{10.104}$$

Thus, by Eqs. (10.7) and (10.104)

$$H(Y) = -\sum_{j=1}^{m} P(y_j)\log_2 P(y_j)$$

$$= -\sum_{i=1}^{m} P(x_i)\log_2 P(x_i) = H(X)$$

Next, by Eqs. (10.24), (10.102), and (10.103)

$$H(Y|X) = -\sum_{j=1}^{m} \sum_{i=1}^{m} P(x_i, y_j)\log_2 P(y_j|x_i)$$

$$= -\sum_{i=1}^{m} P(x_i) \sum_{j=1}^{m} \log_2 P(y_1|x_i)$$

$$= -\sum_{i=1}^{m} P(x_i)\log_2 1 = 0$$

10.14. Verify Eq. (10.26); that is,

$$H(X, Y) = H(X|Y) + H(Y)$$

From Eqs. (3.34) and (3.37)

$$P(x_i, y_j) = P(x_i|y_j)P(y_j)$$

and
$$\sum_{i=1}^{m} P(x_i, y_j) = P(y_j)$$

So by Eq. (10.25) and using Eqs. (10.22) and (10.23), we have

$$H(X,Y) = -\sum_{j=1}^{n} \sum_{i=1}^{m} P(x_i, y_j)\log P(x_i, y_j)$$

$$= -\sum_{j=1}^{n} \sum_{i=1}^{m} P(x_i, y_j)\log\left[P(x_i|y_j)P(y_j)\right]$$

$$= -\sum_{j=1}^{n} \sum_{i=1}^{m} P(x_i, y_j)\log P(x_i|y_j)$$

$$\quad -\sum_{j=1}^{n}\left[\sum_{i=1}^{m} P(x_i, y_j)\right]\log P(y_j)$$

$$= H(X|Y) - \sum_{j=1}^{n} P(y_j)\log P(y_j)$$

$$= H(X|Y) + H(Y)$$

10.15. Verify Eq. (10.40); that is,

$$D(p /\!/ q) \geq 0$$

By definition of $D(p /\!/ q)$, Eq. (10.39), we have

$$D(p /\!/ q) = \sum_i p(x_i) \log_2 \frac{p(x_i)}{q(x_i)} = E\left(-\log_2 \frac{q(x_i)}{p(x_i)}\right) \tag{10.105}$$

Since minus the logarithm is convex, then by Jensen's inequality (Eq. 4.40), we obtain

$$D(p /\!/ q) = E\left(-\log_2 \frac{q(x_i)}{p(x_i)}\right) \geq -\log_2 E\left(\frac{q(x_i)}{p(x_i)}\right)$$

Now

$$-\log_2 E\left(\frac{q(x_i)}{p(x_i)}\right) = -\log_2\left(\sum_i p(x_i) \frac{q(x_i)}{p(x_i)}\right) = -\log_2\left(\sum_i q(x_i)\right) = -\log_2 1 = 0$$

Thus, $D(p /\!/ q) \geq 0$. Next, when $p(x_i) = q(x_i)$, then $\log_2(q/p) = \log_2 1 = 0$, and we have $D(p /\!/ q) = 0$. On the other hand, if $D(p /\!/ q) = 0$, then $\log_2(q/p) = 0$, which implies that $q/p = 1$; that is, $p(x_i) = q(x_i)$. Thus, we have shown that $D(p /\!/ q) \geq 0$ and equality holds iff $p(x_i) = q(x_i)$.

10.16. Verify Eq. (10.41); that is,

$$I(X; Y) = D(p_{XY}(x, y) /\!/ p_X(x) p_Y(y))$$

From Eqs. (10.31) and using Eqs. (10.21) and (10.23), we obtain

$$
\begin{aligned}
I(X; Y) &= H(X) - H(X \mid Y) \\
&= -\sum_i p_X(x_i) \log_2 p(x_i) + \sum_i \sum_j p_{XY}(x_i, y_j) \log_2 p_{X|Y}(x_i \mid y_j) \\
&= -\sum_i \left[\sum_j p_{XY}(x_i, y_j)\right] \log_2 p_X(x_i) + \sum_i \sum_j p_{XY}(x_i, y_j) \log_2 p_{X|Y}(x_i \mid y_j) \\
&= \sum_i \sum_j p_{XY}(x_i, y_j) \log_2 \frac{p_{X|Y}(x_i \mid y_j)}{p_X(x_i)} \\
&= \sum_i \sum_j p_{XY}(x_i, y_j) \log_2 \frac{p_{XY}(x_i, y_j)}{p_X(x_i) p_Y(y_j)} = D\left(p_{XY}(x, y) /\!/ p_X(x) p_Y(y)\right)
\end{aligned}
$$

10.17. Verify Eq. (10.32); that is,

$$I(X; Y) = I(Y; X)$$

Since $p_{XY}(x, y) = p_{YX}(y, x)$, $p_X(x) p_Y(y) = p_Y(y) p_X(x)$, by Eq. (10.41), we have

$$
\begin{aligned}
I(X; Y) &= D(p_{XY}(x, y) /\!/ p_X(x) p_Y(y)) \\
&= D(p_{YX}(y, x) /\!/ p_Y(y) p_X(x)) = I(Y; X)
\end{aligned}
$$

10.18. Verify Eq. (10.33); that is

$$I(X; Y) \geq 0$$

From Eqs. (10.41) and (10.40), we have

$$I(X; Y) = D(p_{XY}(x, y) /\!/ p_X(x) p_Y(y)) \geq 0$$

10.19. Using Eq. (10.42) verify Eq. (10.35); that is

$$I(X; Y) = H(X) + H(Y) - H(X, Y)$$

From Eq. (10.42) we have

$$
\begin{aligned}
I(X;Y) &= D\big(p_{XY}(x, y)//p_X(x)p_Y(y)\big) \\
&= \sum_x \sum_y p_{XY}(x, y)\log_2 \frac{p_{XY}(x, y)}{p_X(x)p_Y(y)} \\
&= \sum_x \sum_y p_{XY}(x, y)\big(\log_2 p_{XY}(x, y) - \log_2 p_X(x) - \log_2 p_Y(y)\big) \\
&= -H(X,Y) - \sum_x \sum_y p_{XY}(x, y)\log_2 p_X(x) - \sum_x \sum_y p_{XY}(x, y)\log_2 p_Y(y) \\
&= -H(X,Y) - \sum_x \log_2 p_X(x)\Big(\sum_y p_{XY}(x, y)\Big) - \sum_y \log_2 p_Y(y)\Big(\sum_x p_{XY}(x, y)\Big) \\
&= -H(X,Y) - \sum_x p_X(x)\log_2 p_X(x) - \sum_y p_Y(y)\log_2 p_Y(y) \\
&= -H(X,Y) + H(X) + H(Y) \\
&= H(X) + H(Y) - H(X, Y)
\end{aligned}
$$

10.20. Consider a BSC (Fig. 10-5) with $P(x_1) = \alpha$.

(a) Show that the mutual information $I(X; Y)$ is given by

$$I(X; Y) = H(Y) + p \log_2 p + (1 - p) \log_2 (1 - p) \tag{10.106}$$

(b) Calculate $I(X; Y)$ for $\alpha = 0.5$ and $p = 0.1$.

(c) Repeat (b) for $\alpha = 0.5$ and $p = 0.5$, and comment on the result.

Figure 10-12 shows the diagram of the BSC with associated input probabilities.

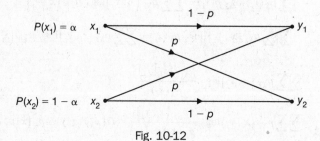

Fig. 10-12

(a) Using Eqs. (10.16), (10.17), and (10.20), we have

$$
\begin{aligned}
[P(X,Y)] &= \begin{bmatrix} \alpha & 0 \\ 0 & 1-\alpha \end{bmatrix}\begin{bmatrix} 1-p & p \\ p & 1-p \end{bmatrix} \\
&= \begin{bmatrix} \alpha(1-p) & \alpha p \\ (1-\alpha)p & (1-\alpha)(1-p) \end{bmatrix} = \begin{bmatrix} P(x_1, y_1) & P(x_1, y_2) \\ P(x_2, y_1) & P(x_2, y_2) \end{bmatrix}
\end{aligned}
$$

Then by Eq. (10.24)

$$
\begin{aligned}
H(Y|X) &= -P(x_1, y_1)\log_2 P(y_1|x_1) - P(x_1, y_2)\log_2 P(y_2|x_2) \\
&\quad - P(x_2, y_1)\log_2 P(y_1|x_2) - P(x_2, y_2)\log_2 P(y_2|x_2) \\
&= -\alpha(1-p)\log_2(1-p) - \alpha p\log_2 p \\
&\quad - (1-\alpha)p\log_2 p - (1-\alpha)(1-p)\log_2(1-p) \\
&= -p\log_2 p - (1-p)\log_2(1-p) \tag{10.107}
\end{aligned}
$$

Hence, by Eq. (10.31)

$$I(X; Y) = H(Y) - H(Y \mid X)$$
$$= H(Y) + p \log_2 p + (1 - p) \log_2 (1 - p)$$

(b) When $\alpha = 0.5$ and $p = 0.1$, by Eq. (10.15)

$$[P(Y)] = \begin{bmatrix} 0.5 & 0.5 \end{bmatrix} \begin{bmatrix} 0.9 & 0.1 \\ 0.1 & 0.9 \end{bmatrix} = \begin{bmatrix} 0.5 & 0.5 \end{bmatrix}$$

Thus, $P(y_1) = P(y_2) = 0.5$.

By Eq. (10.22)

$$H(Y) = - P(y_1) \log_2 P(y_1) - P(y_2) \log_2 P(y_2)$$
$$= -0.5 \log_2 0.5 - 0.5 \log_2 0.5 = 1$$
$$p \log_2 p + (1 - p) \log_2 (1 - p) = 0.1 \log_2 0.1 + 0.9 \log_2 0.9$$
$$= -0.469$$

Thus, $I(X; Y) = 1 - 0.469 = 0.531$

(c) When $\alpha = 0.5$ and $p = 0.5$,

$$[P(Y)] = \begin{bmatrix} 0.5 & 0.5 \end{bmatrix} \begin{bmatrix} 0.5 & 0.5 \\ 0.5 & 0.5 \end{bmatrix} = \begin{bmatrix} 0.5 & 0.5 \end{bmatrix}$$
$$H(Y) = 1$$

$$p \log_2 p + (1 - p) \log_2 (1 - p) = 0.5 \log_2 0.5 + 0.5 \log_2 0.5$$
$$= -1$$

Thus, $I(X; Y) = 1 - 1 = 0$

Note that in this case ($p = 0.5$) no information is being transmitted at all. An equally acceptable decision could be made by dispensing with the channel entirely and "flipping a coin" at the receiver. When $I(X; Y) = 0$, the channel is said to be *useless*.

Channel Capacity

10.21. Verify Eq. (10.46); that is,

$$C_s = \log_2 m$$

where C_s is the channel capacity of a lossless channel and m is the number of symbols in X.

For a lossless channel [Eq. (10.97), Prob. 10.12]

$$H(X \mid Y) = 0$$

Then by Eq. (10.31)

$$I(X; Y) = H(X) - H(X \mid Y) = H(X) \tag{10.108}$$

Hence, by Eqs. (10.43) and (10.9)

$$C_s = \max_{\{P(X)\}} I(X; Y) = \max_{\{P(x_i)\}} H(X) = \log_2 m$$

10.22. Verify Eq. (10.52); that is,

$$C_s = 1 + p \log_2 p + (1 - p) \log_2 (1 - p)$$

where C_s is the channel capacity of a BSC (Fig. 10-5).

By Eq. (10.106) (Prob. 10.20) the mutual information $I(X; Y)$ of a BSC is given by

$$I(X; Y) = H(Y) + p \log_2 p + (1 - p) \log_2 (1 - p)$$

which is maximum when $H(Y)$ is maximum. Since the channel output is binary, $H(Y)$ is maximum when each output has a probability of 0.5 and is achieved for equally likely inputs [Eq. (10.9)]. For this case $H(Y) = 1$, and the channel capacity is

$$C_s = \max_{\{P(X)\}} I(X; Y) = 1 + p \log_2 p + (1 - p) \log_2 (1 - p)$$

10.23. Find the channel capacity of the binary erasure channel of Fig. 10-13 (Prob. 10.11).

Let $P(x_1) = \alpha$. Then $P(x_2) = 1 - \alpha$. By Eq. (10.96)

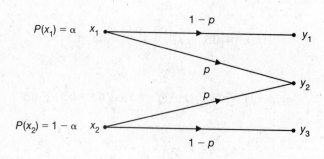

Fig. 10-13

$$[P(Y|X)] = \begin{bmatrix} 1-p & p & 0 \\ 0 & p & 1-p \end{bmatrix} = \begin{bmatrix} P(y_1|x_1) & P(y_2|x_1) & P(y_3|x_1) \\ P(y_1|x_2) & P(y_2|x_2) & P(y_3|x_2) \end{bmatrix}$$

By Eq. (10.15)

$$[P(Y)] = \begin{bmatrix} \alpha & 1-\alpha \end{bmatrix} \begin{bmatrix} 1-p & p & 0 \\ 0 & p & 1-p \end{bmatrix}$$

$$= [\alpha(1-p) \quad p \quad (1-\alpha)(1-p)]$$

$$= [P(y_1) \quad P(y_2) \quad P(y_3)]$$

By Eq. (10.17)

$$[P(X, Y)] = \begin{bmatrix} \alpha & 0 \\ 0 & 1-\alpha \end{bmatrix} \begin{bmatrix} 1-p & p & 0 \\ 0 & p & 1-p \end{bmatrix}$$

$$= \begin{bmatrix} \alpha(1-p) & \alpha p & 0 \\ 0 & (1-\alpha)p & (1-\alpha)(1-p) \end{bmatrix}$$

$$= \begin{bmatrix} P(x_1, y_1) & P(x_1, y_2) & P(x_1, y_3) \\ P(x_2, y_1) & P(x_2, y_2) & P(x_2, y_3) \end{bmatrix}$$

In addition, from Eqs. (10.22) and (10.24) we can calculate

$$H(Y) = -\sum_{j=1}^{3} P(y_j) \log_2 P(y_j)$$
$$= -\alpha(1-p)\log_2 \alpha(1-p) - p \log_2 p - (1-\alpha)(1-p)\log_2[(1-\alpha)(1-p)]$$
$$= (1-p)[-\alpha \log_2 \alpha - (1-\alpha)\log_2(1-\alpha)]$$
$$\qquad -p \log_2 p - (1-p)\log_2(1-p) \qquad\qquad (10.109)$$

$$H(Y|X) = -\sum_{j=1}^{3}\sum_{i=1}^{2} P(x_1, y_j)\log_2 P(y_j|x_i)$$
$$= -\alpha(1-p)\log_2(1-p) - \alpha p \log_2 p$$
$$\qquad -(1-\alpha)p \log_2 p - (1-\alpha)(1-p)\log_2(1-p)$$
$$= -p \log_2 p - (1-p)\log_2(1-p) \qquad\qquad (10.110)$$

Thus, by Eqs. (10.34) and (10.83)

$$I(X;Y) = H(Y) - H(Y|X)$$
$$= (1-p)[-\alpha \log_2 \alpha - (1-\alpha)\log_2(1-\alpha)]$$
$$= (1-p)H(X) \qquad\qquad (10.111)$$

And by Eqs. (10.43) and (10.84)

$$C_s = \max_{\{P(X)\}} I(X;Y) = \max_{\{P(x_1)\}} (1-p)H(X) = (1-p)\max_{\{P(x_1)\}} H(X) = 1-p \qquad (10.112)$$

Continuous Channel

10.24. Find the differential entropy $H(X)$ of the uniformly distributed random variable X with probability density function

$$f_X(x) = \begin{cases} \dfrac{1}{a} & 0 \le x \le a \\ 0 & \text{otherwise} \end{cases}$$

for (a) $a = 1$, (b) $a = 2$, and (c) $a = \frac{1}{2}$.

By Eq. (10.53)

$$H(X) = -\int_{-\infty}^{\infty} f_X(x)\log_2 f_X(x)\,dx$$
$$= -\int_{0}^{a} \frac{1}{a}\log_2 \frac{1}{a}\,dx = \log_2 a \qquad\qquad (10.113)$$

(a) $a = 1, H(X) = \log_2 1 = 0$

(b) $a = 2, H(X) \log_2 2 = 1$

(c) $a = \frac{1}{2}, H(X) = \log_2 \frac{1}{2} = -\log_2 2 = -1$

Note that the differential entropy $H(X)$ is not an absolute measure of information, and unlike discrete entropy, differential entropy can be negative.

10.25. Find the probability density function $f_X(x)$ of X for which differential entropy $H(X)$ is maximum.

Let the support of $f_X(x)$ be (a, b). [Note that the *support* of $f_X(x)$ is the region where $f_X(x) > 0$.] Now

$$H(X) = -\int_{a}^{b} f_X(x)\ln f_X(x)\,dx \qquad\qquad (10.114)$$

$$\int_{a}^{b} f_X(x)\,dx = 1 \qquad\qquad (10.115)$$

Using Lagrangian multipliers technique, we let

$$J = -\int_a^b f_X(x) \ln f_X(x)\, dx + \lambda \left(\int_a^b f_X(x)\, dx - 1 \right) \tag{10.116}$$

where λ is the Lagrangian multiplier. Taking the functional derivative of J with respect to $f_X(x)$ and setting equal to zero, we obtain

$$\frac{\partial J}{\partial f_X(x)} = -\ln f_X(x) - 1 + \lambda = 0 \tag{10.117}$$

or
$$\ln f_X(x) = \lambda - 1 \Rightarrow f_X(x) = e^{\lambda - 1} \tag{10.118}$$

so $f_X(x)$ is a constant. Then, by constraint Eq. (10.115), we obtain

$$f_X(x) = \frac{1}{b-a} \qquad a < x < b \tag{10.119}$$

Thus, the uniform distribution results in the maximum differential entropy.

10.26. Let X be $N(0; \sigma^2)$. Find its differential entropy.

The differential entropy in nats is expressed as

$$H(X) = -\int_{-\infty}^{\infty} f_X(x) \ln f_X(x)\, dx \tag{10.120}$$

By Eq. (2.71), the pdf of $N(0, \sigma^2)$ is

$$f_X(x) = \frac{1}{\sqrt{2\pi\sigma^2}} e^{-x^2/(2\sigma^2)} \tag{10.121}$$

Then

$$\ln f_X(x) = -\frac{x^2}{2\sigma^2} - \ln \sqrt{2\pi\sigma^2} \tag{10.122}$$

Thus, we obtain

$$
\begin{aligned}
H(X) &= -\int_{-\infty}^{\infty} f_X(x) \left(-\frac{x^2}{2\sigma^2} - \ln\sqrt{2\pi\sigma^2} \right) dx \\
&= \frac{1}{2\sigma^2} E(X^2) + \frac{1}{2}\ln(2\pi\sigma^2) = \frac{1}{2} + \frac{1}{2}\ln(2\pi\sigma^2) \\
&= \frac{1}{2}\ln e + \frac{1}{2}\ln(2\pi\sigma^2) = \frac{1}{2}\ln(2\pi e\sigma^2) \quad \text{nats/sample}
\end{aligned}
\tag{10.123}
$$

Changing the base of the logarithm, we have

$$H(X) = \frac{1}{2}\log_2(2\pi e\sigma^2) \qquad \text{bits/sample} \tag{10.124}$$

10.27. Let $(X_1, \ldots, X_n)$ be an n-variate r.v. defined by Eq. (3.92). Find the differential entropy of $(X_1, \ldots, X_n)$.

From Eq. (3.92) the joint pdf of $(X_1, \ldots, X_n)$ is given by

$$f_{\mathbf{X}}(\mathbf{x}) = \frac{1}{(2\pi)^{n/2} |\mathbf{K}|^{1/2}} \exp\left[-\frac{1}{2}(\mathbf{x} - \mu)^T \mathbf{K}^{-1}(\mathbf{x} - \mu) \right] \tag{10.125}$$

where

$$\boldsymbol{\mu} = E(\mathbf{X}) = \begin{bmatrix} \mu_1 \\ \vdots \\ \mu_n \end{bmatrix} = \begin{bmatrix} E(X_1) \\ \vdots \\ E(X_n) \end{bmatrix} \tag{10.126}$$

$$\mathbf{K} = \begin{bmatrix} \sigma_{11} & \cdots & \sigma_{1n} \\ \vdots & \ddots & \vdots \\ \sigma_{n1} & \cdots & \sigma_{nn} \end{bmatrix} = \left[\sigma_{ij} \right]_{n \times n} \qquad \sigma_{ij} = \mathrm{Cov}\left(X_i, X_j\right) \tag{10.127}$$

Then, we obtain

$$
\begin{aligned}
H(\mathbf{X}) &= -\int f_{\mathbf{X}}(\mathbf{x}) \ln f_{\mathbf{X}}(\mathbf{x}) d\mathbf{x} \\
&= -\int f_{\mathbf{X}}(\mathbf{x}) \left[-\frac{1}{2}(\mathbf{x}-\boldsymbol{\mu})^T \mathbf{K}^{-1}(\mathbf{x}-\boldsymbol{\mu}) - \ln (2\pi)^{n/2} |\mathbf{K}|^{1/2} \right] d\mathbf{x} \\
&= \frac{1}{2} E\left[\sum_{i,j} (X_i - u_i)(\mathbf{K}^{-1})_{ij} \left(X_j - \mu_j\right) \right] + \frac{1}{2} \ln (2\pi)^n |\mathbf{K}| \\
&= \frac{1}{2} E\left[\sum_{i,j} (X_i - u_i)(X_j - \mu_j)\left(\mathbf{K}^{-1}\right)_{ij} \right] + \frac{1}{2} \ln (2\pi)^n |\mathbf{K}| \\
&= \frac{1}{2} \sum_{i,j} E\left[(X_i - \mu_i)(X_j - \mu_j) \right] (\mathbf{K}^{-1})_{ij} + \frac{1}{2} \ln (2\pi)^n |\mathbf{K}| \\
&= \frac{1}{2} \sum_j \sum_i \mathbf{K}_{ji}\, (\mathbf{K}^{-1})_{ij} + \frac{1}{2} \ln (2\pi)^n |\mathbf{K}| \\
&= \frac{1}{2} \sum_j (\mathbf{K}\mathbf{K}^{-1})_{jj} + \frac{1}{2} \ln (2\pi)^n |\mathbf{K}| \\
&= \frac{1}{2} \sum_j \mathrm{I}_{jj} + \frac{1}{2} \ln (2\pi)^n |\mathbf{K}| \\
&= \frac{n}{2} + \frac{1}{2} \ln (2\pi)^n |\mathbf{K}| \\
&= \frac{1}{2} \ln (2\pi e)^n |\mathbf{K}| \qquad \text{nats/sample} \tag{10.128} \\
&= \frac{1}{2} \log_2 (2\pi e)^n |\mathbf{K}| \qquad \textit{bits / sample} \tag{10.129}
\end{aligned}
$$

10.28. Find the probability density function $f_X(x)$ of X with zero mean and variance σ^2 for which differential entropy $H(X)$ is maximum.

The differential entropy in nats is expressed as

$$H(X) = -\int_{-\infty}^{\infty} f_X(x) \ln f_X(x) \, dx \tag{10.130}$$

The constraints are

$$\int_{-\infty}^{\infty} f_X(x) \, dx = 1 \tag{10.131}$$

$$\int_{-\infty}^{\infty} f_X(x) \, x \, dx = 0 \tag{10.132}$$

$$\int_{-\infty}^{\infty} f_X(x) \, x^2 dx = \sigma^2 \tag{10.133}$$

Using the method of Lagrangian multipliers, and setting Lagrangian as

$$J = -\int_{-\infty}^{\infty} f_X(x) \ln f_X(x)\, dx + \lambda_0 \left(\int_{-\infty}^{\infty} f_X(x)\, dx - 1 \right)$$
$$+ \lambda_1 \left(\int_{-\infty}^{\infty} f_X(x) x\, dx \right) + \lambda_2 \left(\int_{-\infty}^{\infty} f_X(x) x^2\, dx - \sigma^2 \right)$$

and taking the functional derivative of J with respect to $f_X(x)$ and setting equal to zero, we obtain

$$\frac{\partial J}{\partial f_X(x)} = -\ln f_X(x) - 1 + \lambda_0 + \lambda_1 x + \lambda_2 x^2 = 0$$

$$\ln f_X(x) = \lambda_0 - 1 + \lambda_1 x + \lambda_2 x^2$$

or
$$f_X(x) = e^{\lambda_0 - 1 + \lambda_1 x + \lambda_2 x^2} = C(\lambda_0) e^{\lambda_1 x + \lambda_2 x^2} \tag{10.134}$$

It is obvious from the generic form of the exponential family restricted to the second order polynomials that they cover Gaussian distribution only. Thus, we obtain $N(0; \sigma^2)$ and

$$f_X(x) = \frac{1}{\sqrt{2\pi}\sigma} e^{-x^2/(2\sigma^2)} \qquad -\infty < x < \infty \tag{10.135}$$

10.29. Verify Eq. (10.55); that is,

$$H(aX) = H(X) + \log_2 |a|$$

Let $Y = aX$. Then by Eq. (4.86)

$$f_Y(y) = \frac{1}{|a|} f_X\left(\frac{y}{a}\right)$$

and

$$H(aX) = H(Y) = -\int_{-\infty}^{\infty} f_Y(y) \log_2 f_Y(Y)\, dy$$

$$= -\int_{-\infty}^{\infty} \frac{1}{|a|} f_X\left(\frac{y}{a}\right) \log_2 \left(\frac{1}{|a|} f_X\left(\frac{y}{a}\right) \right) dy$$

After a change of variables $y / a = x$, we have

$$H(aX) = -\int_{-\infty}^{\infty} f_X(x) \log_2 f_X(x)\, dx - \log_2 \frac{1}{|a|} \int_{-\infty}^{\infty} f_X(x)\, dx$$

$$= -H(X) + \log_2 |a|$$

10.30. Verify Eq. (10.64); that is,

$$D(f \,//\, g) \geq 0$$

By definition (10.61), we have

$$-D(f \,//\, g) = \int_{-\infty}^{\infty} f_X(x) \log_2 \left(\frac{g_X(x)}{f_X(x)} \right) dx$$

$$\leq \log_2 \int_{-\infty}^{\infty} f_X(x) \left(\frac{g_X(x)}{f_X(x)} \right) dx \qquad \text{(by Jensen's inequality (4.40))}$$

$$= \log_2 \int_{-\infty}^{\infty} g_X(x)\, dx = \log_2 1 = 0$$

Thus, $D(f \,//\, g) \geq 0$. We have equality in Jensen's inequality which occurs iff $f = g$.

Additive White Gaussian Noise Channel

10.31. Verify Eq. (10.71); that is,

$$C_s = \max_{\{f_X(x)\}} I(X;Y) = \frac{1}{2}\log_2\left(1 + \frac{S}{N}\right)$$

From Eq. (10.68), $Y = X + Z$. Then by Eq. (10.57) we have

$$
\begin{aligned}
I(X;Y) &= H(Y) - H(Y|X) = H(Y) - H(X+Z|X) \\
&= H(Y) - H(Z|X) = H(Y) - H(Z)
\end{aligned}
\tag{10.136}
$$

since Z is independent of X.

Now, from Eq. (10.124) (Prob. 10.26) and setting $\sigma^2 = N$, we have

$$H(Z) = \frac{1}{2}\log_2(2\pi e N) \tag{10.137}$$

Since X and Z are independent and $E(Z) = 0$, we have

$$E(Y^2) = E\left[(X+Z)^2\right] = E(X^2) + 2E(X)E(Z) + E(Z^2) = S + N \tag{10.138}$$

Given $E(Y^2) = S + N$, the differential entropy of Y is bounded by $\frac{1}{2}\log_2 2\pi e(S+N)$, since the Gaussian distribution maximizes the differential entropy for a given variance (see Prob. 10.28). Applying this result to bound the mutual information, we obtain

$$
\begin{aligned}
I(X;Y) &= H(Y) - H(z) \\
&\leq \frac{1}{2}\log_2 2\pi e\,(S+N) - \frac{1}{2}\log_2 2\pi e N \\
&= \frac{1}{2}\log_2\left(\frac{2\pi e(S+N)}{2\pi e N}\right) = \frac{1}{2}\log_2\left(1 + \frac{S}{N}\right)
\end{aligned}
$$

Hence, the capacity C_s of the AWGN channel is

$$C_s = \max_{\{f_X(x)\}} I(X;Y) = \frac{1}{2}\log_2\left(1 + \frac{S}{N}\right)$$

10.32. Verify Eq. (10.72); that is,

$$C = B\log_2\left(1 + \frac{S}{N}\right) \qquad \text{bits/second}$$

Assuming that the channel bandwidth is B Hz, then by Nyquist sampling theorem we can represent both the input and output by samples taken $1/(2B)$ seconds apart. Each of the input samples is corrupted by the noise to produce the corresponding output sample. Since the noise is white and Gaussian, each of the noise samples is an i.i.d. Gaussian r.v.. If the noise has power spectral density $N_0/2$ and bandwidth B, then the noise has power $(N_0/2)\,2B = N_0 B$ and each of the $2BT$ noise samples in time t has variance $N_0BT/(2BT) = N_0/2$. Now the capacity of the discrete time Gaussian channel is given by (Eq. (10.71))

$$C_s = \frac{1}{2}\log_2\left(1 + \frac{S}{N}\right) \qquad \text{bits/sample}$$

Let the channel be used over the time interval $[0, T]$. Then the power per sample is $ST / (2BT) = S / (2B)$, the noise variance per sample is $N_0 / 2$, and hence the capacity per sample is

$$C_s = \frac{1}{2} \log_2 \left(1 + \frac{S/(2B)}{N_0/2}\right) = \frac{1}{2} \log_2 \left(1 + \frac{S}{N_0 B}\right) \qquad \text{bits/sample} \tag{10.139}$$

Since there are $2B$ samples each second, the capacity of the channel can be rewritten as

$$C = B \log_2 \left(1 + \frac{S}{N_0 B}\right) = B \log_2 \left(1 + \frac{S}{N}\right) \qquad \text{bits/second}$$

where $N = N_0 B$ is the total noise power.

10.33. Show that the channel capacity of an ideal AWGN channel with infinite bandwidth is given by

$$C_\infty = \frac{1}{\ln 2} \frac{S}{\eta} \approx 1.44 \frac{S}{\eta} \qquad \text{b/s} \tag{10.140}$$

where S is the average signal power and $\eta / 2$ is the power spectral density of white Gaussian noise.

The noise power N is given by $N = \eta B$. Thus, by Eq. (10.72)

$$C = B \log_2 \left(1 + \frac{S}{\eta B}\right)$$

Let $S/(\eta B) = \lambda$. Then

$$C = \frac{S}{\eta \lambda} \log_2 (1 + \lambda) = \frac{1}{\ln 2} \frac{S}{\eta} \frac{\ln(1 + \lambda)}{\lambda} \tag{10.141}$$

Now

$$C_\infty = \lim_{B \to \infty} B \log_2 \left(1 + \frac{S}{\eta B}\right)$$

$$= \frac{1}{\ln 2} \frac{S}{\eta} \lim_{\lambda \to 0} \frac{\ln(1 + \lambda)}{\lambda}$$

Since $\lim_{\lambda \to 0} [\ln(1 + \lambda)]/\lambda = 1$, we obtain

$$C_\infty = \frac{1}{\ln 2} \frac{S}{\eta} \approx 1.44 \frac{S}{\eta} \qquad \text{b/s}$$

Note that Eq. (10.140) can be used to estimate upper limits on the performance of any practical communication system whose transmission channel can be approximated by the AWGN channel.

10.34. Consider an AWGN channel with 4-kHz bandwidth and the noise power spectral density $\eta / 2 = 10^{-12}$ W/Hz. The signal power required at the receiver is 0.1 mW. Calculate the capacity of this channel.

$$B = 4000 \text{ Hz} \qquad S = 0.1(10^{-3}) \text{ W}$$

$$N = \eta B = 2(10^{-12})(4000) = 8(10^{-9}) \text{ W}$$

Thus,

$$\frac{S}{N} = \frac{0.1(10^{-3})}{8(10^{-9})} = 1.25(10^4)$$

And by Eq. (10.72)

$$C = B \log_2 \left(1 + \frac{S}{N}\right)$$

$$= 4000 \log_2 [1 + 1.25(10^4)] = 54.44(10^3) \text{ b/s}$$

10.35. An analog signal having 4-kHz bandwidth is sampled at 1.25 times the Nyquist rate, and each sample is quantized into one of 256 equally likely levels. Assume that the successive samples are statistically independent.

(*a*) What is the information rate of this source?

(*b*) Can the output of this source be transmitted without error over an AWGN channel with a bandwidth of 10 kHz and an S/N ratio of 20 dB?

(*c*) Find the S/N ratio required for error-free transmission for part (*b*).

(*d*) Find the bandwidth required for an AWGN channel for error-free transmission of the output of this source if the S/N ratio is 20 dB.

(*a*)
$$f_M = 4(10^3)\text{Hz}$$

$$\text{Nyquist rate} = 2f_M = 8(10^3) \text{ samples/s}$$

$$r = 8(10^3)(1.25) = 10^4 \text{ samples/s}$$

$$H(X) = \log_2 256 = 8 \text{ b/sample}$$

By Eq (10.10) the information rate R of the source is

$$R = rH(X) = 10^4(8) \text{ b/s} = 80 \text{ kb/s}$$

(*b*) By Eq. (10.72)

$$C = B \log_2 \left(1 + \frac{S}{N}\right) = 10^4 \log_2 (1 + 10^2) = 66.6(10^3) \text{ b/s}$$

Since $R > C$, error-free transmission is not possible.

(*c*) The required S/N ratio can be found by

$$C = 10^4 \log_2 \left(1 + \frac{S}{N}\right) \geq 8(10^4)$$

or
$$\log_2 \left(1 + \frac{S}{N}\right) \geq 8$$

or
$$1 + \frac{S}{N} \geq 2^8 = 256 \rightarrow \frac{S}{N} \geq 255 \quad (= 24.1 \text{ dB})$$

Thus, the required S/N ratio must be greater than or equal to 24.1 dB for error-free transmission.

(*d*) The required bandwidth B can be found by

$$C = B \log_2 (1 + 100) \geq 8(10^4)$$

or
$$B \geq \frac{8(10^4)}{\log_2 (1 + 100)} = 1.2(10^4)\text{Hz} = 12 \text{ kHz}$$

and the required bandwidth of the channel must be greater than or equal to 12 kHz.

Source Coding

10.36. Consider a DMS X with two symbols x_1 and x_2 and $P(x_1) = 0.9$, $P(x_2) = 0.1$. Symbols x_1 and x_2 are encoded as follows (Table 10-4):

TABLE 10-4

x_i	$P(x_i)$	CODE
x_1	0.9	0
x_2	0.1	1

Find the efficiency η and the redundancy γ of this code.

By Eq. (10.73) the average code length L per symbol is

$$L = \sum_{i=1}^{2} P(x_i)n_i = (0.9)(1) + (0.1)(1) = 1 \text{ b}$$

By Eq. (10.7)

$$H(X) = -\sum_{i=1}^{2} P(x_i) \log_2 P(x_i)$$

$$= -0.9 \log_2 0.9 - 0.1 \log_2 0.1 = 0.469 \text{ b/symbol}$$

Thus, by Eq. (10.77) the code efficiency η is

$$\eta = \frac{H(X)}{L} = 0.469 = 46.9\%$$

By Eq. (10.75) the code redundancy γ is

$$\gamma = 1 - \eta = 0.531 = 53.1\%$$

10.37. The second-order extension of the DMS X of Prob. 10.25, denoted by X^2, is formed by taking the source symbols two at a time. The coding of this extension is shown in Table 10-5. Find the efficiency η and the redundancy γ of this extension code.

TABLE 10-5

a_i	$P(a_i)$	CODE
$a_1 = x_1x_1$	0.81	0
$a_2 = x_1x_2$	0.09	10
$a_3 = x_2x_1$	0.09	110
$a_4 = x_2x_2$	0.01	111

$$L = \sum_{i=1}^{4} P(a_i)n_i = 0.81(1) + 0.09(2) + 0.09(3) + 0.01(3)$$

$$= 1.29 \text{ b/symbol}$$

The entropy of the second-order extension of X, $H(X^2)$, is given by

$$H(X^2) = \sum_{i=1}^{4} P(a_i) \log_2 P(a_i)$$

$$= -0.81 \log_2 0.81 - 0.09 \log_2 0.09 - 0.09 \log_2 0.09 - 0.01 \log_2 0.01$$

$$= 0.938 \text{ b/symbol}$$

Therefore, the code efficiency η is

$$\eta = \frac{H(X^2)}{L} = \frac{0.938}{1.29} = 0.727 = 72.7\%$$

and the code redundancy γ is

$$\gamma = 1 - \eta = 0.273 = 27.3\%$$

Note that $H(X^2) = 2H(X)$.

10.38. Consider a DMS X with symbols x_i, $i = 1, 2, 3, 4$. Table 10-6 lists four possible binary codes.

TABLE 10-6

x_i	CODE A	CODE B	CODE C	CODE D
x_1	00	0	0	0
x_2	01	10	11	100
x_3	10	11	100	110
x_4	11	110	110	111

(a) Show that all codes except code B satisfy the Kraft inequality.

(b) Show that codes A and D are uniquely decodable but codes B and C are not uniquely decodable.

(a) From Eq. (10.78) we obtain the following:

For code A: $n_1 = n_2 = n_3 = n_4 = 2$

$$K = \sum_{i=1}^{4} 2^{-n_i} = \frac{1}{4} + \frac{1}{4} + \frac{1}{4} + \frac{1}{4} = 1$$

For code B: $n_1 = 1 \quad n_2 = n_3 = 2 \quad n_4 = 3$

$$K = \sum_{i=1}^{4} 2^{-n_i} = \frac{1}{2} + \frac{1}{4} + \frac{1}{4} + \frac{1}{8} = 1\frac{1}{8} > 1$$

For code C: $n_1 = 1 \quad n_2 = 2 \quad n_3 = n_4 = 3$

$$K = \sum_{i=1}^{4} 2^{-n_i} = \frac{1}{2} + \frac{1}{4} + \frac{1}{8} + \frac{1}{8} = 1$$

For code D: $n_1 = 1 \quad n_2 = n_3 = n_4 = 3$

$$K = \sum_{i=1}^{4} 2^{-n_i} = \frac{1}{2} + \frac{1}{8} + \frac{1}{8} + \frac{1}{8} = \frac{7}{8} < 1$$

All codes except code B satisfy the Kraft inequality.

(b)   Codes A and D are prefix-free codes. They are therefore uniquely decodable. Code B does not satisfy the Kraft inequality, and it is not uniquely decodable. Although code C does satisfy the Kraft inequality, it is not uniquely decodable. This can be seen by the following example: Given the binary sequence 0110110. This sequence may correspond to the source sequences $x_1 x_2 x_1 x_4$ or $x_1 x_4 x_4$.

10.39. Verify Eq. (10.76); that is,

$$L \geq H(X)$$

where L is the average code word length per symbol and $H(X)$ is the source entropy.

From Eq. (10.89) (Prob. 10.5), we have

$$\sum_{i=1}^{m} P_i \log_2 \frac{Q_i}{P_i} \leq 0$$

where the equality holds only if $Q_i = P_i$. Let

$$Q_i = \frac{2^{-n_i}}{K} \tag{10.142}$$

where

$$K = \sum_{i=1}^{m} 2^{-n_i} \tag{10.143}$$

which is defined in Eq. (10.78). Then

$$\sum_{i=1}^{m} Q_i = \frac{1}{K} \sum_{i=1}^{m} 2^{-n_i} = 1 \tag{10.144}$$

and

$$\sum_{i=1}^{m} P_i \log_2 \frac{2^{-n_i}}{KP_i} = \sum_{i=1}^{m} P_i \left(\log_2 \frac{1}{P_i} - n_i - \log_2 K \right)$$

$$= -\sum_{i=1}^{m} P_i \log_2 P_i - \sum_{i=1}^{m} P_i n_i - (\log_2 K) \sum_{i=1}^{m} P_i \tag{10.145}$$

$$= H(X) - L - \log_2 K \leq 0$$

From the Kraft inequality (10.78) we have

$$\log_2 K \leq 0 \tag{10.146}$$

Thus,
$$H(X) - L \leq \log_2 K \leq 0 \tag{10.147}$$

or
$$L \geq H(X)$$

The equality holds when $K = 1$ and $P_i = Q_i$.

10.40. Let X be a DMS with symbols x_i and corresponding probabilities $P(x_i) = P_i, i = 1, 2, \ldots, m$. Show that for the optimum source encoding we require that

$$K = \sum_{i=1}^{m} 2^{-n_i} = 1 \tag{10.148}$$

and
$$n_i = \log_2 \frac{1}{P_i} = I_i \tag{10.149}$$

where n_i is the length of the code word corresponding to x_i and I_i is the information content of x_i.

From the result of Prob. 10.39, the optimum source encoding with $L = H(X)$ requires $K = 1$ and $P_i = Q_i$. Thus, by Eqs. (10.143) and (10.142)

$$K = \sum_{i=1}^{m} 2^{-n_i} = 1 \tag{10.150}$$

and

$$P_i = Q_i = 2^{-n_i} \tag{10.151}$$

Hence,

$$n_i = -\log_2 P_i = \log_2 \frac{1}{P_i} = I_i$$

Note that Eq. (10.149) implies the following commonsense principle: Symbols that occur with high probability should be assigned shorter code words than symbols that occur with low probability.

10.41. Consider a DMS X with symbols x_i and corresponding probabilities $P(x_i) = P_i, i = 1, 2, \ldots, m$. Let n_i be the length of the code word for x_i such that

$$\log_2 \frac{1}{P_i} \leq n_i \leq \log_2 \frac{1}{P_i} + 1 \tag{10.152}$$

Show that this relationship satisfies the Kraft inequality (10.78), and find the bound on K in Eq. (10.78).

Equation (10.152) can be rewritten as

$$-\log_2 P_i \leq n_i \leq -\log_2 P_i + 1 \tag{10.153}$$

or $$\log_2 P_i \geq -n_i \geq \log_2 P_i - 1$$

Then $$2^{\log_2 P_i} \geq 2^{-n_i} \geq 2^{\log_2 P_i} 2^{-1}$$

or $$P_i \geq 2^{-n_i} \geq \frac{1}{2} P_i \tag{10.154}$$

Thus, $$\sum_{i=1}^{m} P_i \geq \sum_{i=1}^{m} 2^{-n_i} \geq \frac{1}{2} \sum_{i=1}^{m} P_i \tag{10.155}$$

or $$1 \geq \sum_{i=1}^{m} 2^{-n_i} \geq \frac{1}{2} \tag{10.156}$$

which indicates that the Kraft inequality (10.78) is satisfied, and the bound on K is

$$\frac{1}{2} \leq K \leq 1 \tag{10.157}$$

10.42. Consider a DMS X with symbols x_i and corresponding probabilities $P(x_i) = P_i, i = 1, 2, \ldots, m$. Show that a code constructed in agreement with Eq. (10.152) will satisfy the following relation:

$$H(X) \leq L \leq H(X) + 1 \tag{10.158}$$

where $H(X)$ is the source entropy and L is the average code word length.

Multiplying Eq. (10.153) by P_i and summing over i yields

$$-\sum_{i=1}^{m} P_i \log_2 P_i \leq \sum_{i=1}^{m} n_i P_i \leq \sum_{i=1}^{m} P_i (-\log_2 P_i + 1) \tag{10.159}$$

Now
$$\sum_{i=1}^{m} P_i(-\log_2 P_i + 1) = -\sum_{i=1}^{m} P_i \log_2 P_i + \sum_{i=1}^{m} P_i$$
$$= H(X) + 1$$

Thus, Eq. (10.159) reduces to

$$H(X) \le L \le H(X) + 1$$

10.43. Verify Kraft inequality Eq. (10.78); that is,

$$K = \sum_{i=1}^{m} 2^{-n_i} \le 1$$

Consider a binary tree representing the code words; (Fig. 10-14). This tree extends downward toward infinity. The path down the tree is the sequence of symbols $(0, 1)$, and each leaf of the tree with its unique path corresponds to a code word. Since an instantaneous code is a prefix-free code, each code word eliminates its descendants as possible code words.

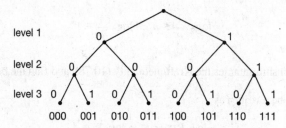

Fig. 10-14　Binary tree.

Let $n_{\max}$ be the length of the longest code word. Of all the possible nodes at a level of $n_{\max}$, some may be code words, some may be descendants of code words, and some may be neither. A code word of level n_i has $2^{n_{\max} - n_i}$ descendants at level $n_{\max}$. The total number of possible leaf nodes at level $n_{\max}$ is $2^{n_{\max}}$. Hence, summing over all code words, we have

$$\sum_{i=1}^{m} 2^{n_{\max} - n_i} \le 2^{n_{\max}}$$

Dividing through by $2^{n_{\max}}$, we obtain

$$K = \sum_{i=1}^{m} 2^{-n_i} \le 1$$

Conversely, given any set of code words with length n_i $(i = 1, \ldots, m)$ which satisfy the inequality, we can always construct a tree. First, order the code word lengths according to increasing length, then construct code words in terms of the binary tree introduced in Fig. 10.14.

10.44. Show that every source alphabet $X = \{x_1, \ldots, x_m\}$ has a binary prefix code.

Given source symbols $x_1, \ldots, x_m$, choose the code length n_i such that $2^{n_i} \ge m$; that is, $n_i \ge \log_2 m$. Then

$$\sum_{i=1}^{m} 2^{-n_i} \le \sum_{i=1}^{m} \frac{1}{m} = m\left(\frac{1}{m}\right) = 1$$

Thus, Kraft's inequality is satisfied and there is a prefix code.

Entropy Coding

10.45. A DMS X has four symbols x_1, x_2, x_3, and x_4 with $P(x_1) = \frac{1}{2}$, $P(x_2) = \frac{1}{4}$, and $P(x_3) = P(x_4) = \frac{1}{8}$. Construct a Shannon-Fano code for X; show that this code has the optimum property that $n_i = I(x_i)$ and that the code efficiency is 100 percent.

The Shannon-Fano code is constructed as follows (see Table 10-7):

TABLE 10-7

x_i	$P(x_i)$	STEP 1	STEP 2	STEP 3	CODE
x_1	$\frac{1}{2}$	0			0
x_2	$\frac{1}{4}$	1	0		10
x_3	$\frac{1}{8}$	1	1	0	110
x_4	$\frac{1}{8}$	1	1	1	111

$$I(x_1) = -\log_2 \frac{1}{2} = 1 = n_1 \quad I(x_2) = -\log_2 \frac{1}{4} = 2 = n_2$$

$$I(x_3) = -\log_2 \frac{1}{8} = 3 = n_3 \quad I(x_4) = -\log_2 \frac{1}{8} = 3 = n_4$$

$$H(X) = \sum_{i=1}^{4} P(x_i)I(x_i) = \frac{1}{2}(1) + \frac{1}{4}(2) + \frac{1}{8}(3) + \frac{1}{8}(3) = 1.75$$

$$L = \sum_{i=1}^{4} P(x_i)n_i = \frac{1}{2}(1) + \frac{1}{4}(2) + \frac{1}{8}(3) + \frac{1}{8}(3) = 1.75$$

$$\eta = \frac{H(X)}{L} = 1 = 100\%$$

10.46. A DMS X has five equally likely symbols.

(a) Construct a Shannon-Fano code for X, and calculate the efficiency of the code.

(b) Construct another Shannon-Fano code and compare the results.

(c) Repeat for the Huffman code and compare the results.

(a) A Shannon-Fano code [by choosing two approximately equiprobable (0.4 versus 0.6) sets] is constructed as follows (see Table 10-8):

TABLE 10-8

x_i	$P(x_i)$	STEP 1	STEP 2	STEP 3	CODE
x_1	0.2	0	0		00
x_2	0.2	0	1		01
x_3	0.2	1	0		10
x_4	0.2	1	1	0	110
x_5	0.2	1	1	1	111

$$H(X) = -\sum_{i=1}^{5} P(x_i) \log_2 P(x_i) = 5(-0.2 \log_2 0.2) = 2.32$$

$$L = \sum_{i=1}^{5} P(x_i) n_i = 0.2(2 + 2 + 2 + 3 + 3) = 2.4$$

The efficiency η is

$$\eta = \frac{H(X)}{L} = \frac{2.32}{2.4} = 0.967 = 96.7\%$$

(*b*)   Another Shannon-Fano code [by choosing another two approximately equiprobable (0.6 versus 0.4) sets] is constructed as follows (see Table 10-9):

TABLE 10-9

x_i	$P(x_i)$	STEP 1	STEP 2	STEP 3	CODE
x_1	0.2	0	0		00
x_2	0.2	0	1	0	010
x_3	0.2	0	1	1	011
x_4	0.2	1	0		10
x_5	0.2	1	1		11

$$L = \sum_{i=1}^{5} P(x_i) n_i = 0.2(2 + 3 + 3 + 2 + 2) = 2.4$$

Since the average code word length is the same as that for the code of part (*a*), the efficiency is the same.

(*c*)   The Huffman code is constructed as follows (see Table 10-10):

$$L = \sum_{i=1}^{5} P(x_i) n_i = 0.2(2 + 3 + 3 + 2 + 2) = 2.4$$

Since the average code word length is the same as that for the Shannon-Fano code, the efficiency is also the same.

TABLE 10-10

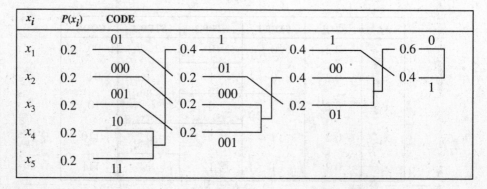

10.47. A DMS X has five symbols x_1, x_2, x_3, x_4, and x_5 with $P(x_1) = 0.4$, $P(x_2) = 0.19$, $P(x_3) = 0.16$, $P(x_4) = 0.15$, and $P(x_5) = 0.1$.

(a) Construct a Shannon-Fano code for X, and calculate the efficiency of the code.

(b) Repeat for the Huffman code and compare the results.

(a) The Shannon-Fano code is constructed as follows (see Table 10-11):

TABLE 10-11

x_i	$P(x_i)$	STEP 1	STEP 2	STEP 3	CODE
x_1	0.4	0	0		00
x_2	0.19	0	1		01
x_3	0.16	1	0		10
x_4	0.15	1	1	0	110
x_5	0.1	1	1	1	111

$$H(X) = -\sum_{i=1}^{5} P(x_i) \log_2 P(x_i)$$

$$= -0.4 \log_2 0.4 - 0.19 \log_2 0.19 - 0.16 \log_2 0.16$$

$$- 0.15 \log_2 0.15 - 0.1 \log_2 0.1$$

$$= 2.15$$

$$L = \sum_{i=1}^{5} P(x_i) n_i$$

$$= 0.4(2) + 0.19(2) + 0.16(2) + 0.15(3) + 0.1(3) = 2.25$$

$$\eta = \frac{H(X)}{L} = \frac{2.15}{2.25} = 0.956 = 95.6\%$$

(b) The Huffman code is constructed as follows (see Table 10-12):

$$L = \sum_{i=1}^{5} P(x_i) n_i$$

$$= 0.4(1) + (0.19 + 0.16 + 0.15 + 0.1)(3) = 2.2$$

TABLE 10-12

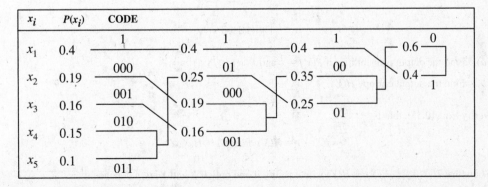

$$\eta = \frac{H(X)}{L} = \frac{2.15}{2.2} = 0.977 = 97.7\%$$

The average code word length of the Huffman code is shorter than that of the Shannon-Fano code, and thus the efficiency is higher than that of the Shannon-Fano code.

SUPPLEMENTARY PROBLEMS

10.48. Consider a source X that produces five symbols with probabilities $\frac{1}{2}, \frac{1}{4}, \frac{1}{8}, \frac{1}{16}$, and $\frac{1}{16}$. Determine the source entropy $H(X)$.

10.49. Calculate the average information content in the English language, assuming that each of the 26 characters in the alphabet occurs with equal probability.

10.50. Two BSCs are connected in cascade, as shown in Fig. 10-15.

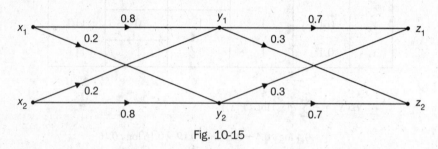

Fig. 10-15

(a) Find the channel matrix of the resultant channel.

(b) Find $P(z_1)$ and $P(z_2)$ if $P(x_1) = 0.6$ and $P(x_2) = 0.4$.

10.51. Consider the DMC shown in Fig. 10-16.

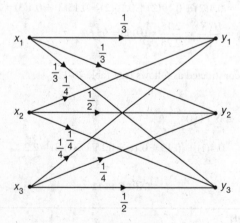

Fig. 10-16

(a) Find the output probabilities if $P(x_1) = \frac{1}{2}$ and $P(x_2) = P(x_3) = \frac{1}{4}$.

(b) Find the output entropy $H(Y)$.

10.52. Verify Eq. (10.35), that is,

$$I(X; Y) = H(X) + H(Y) - H(X, Y)$$

10.53. Show that $H(X, Y) \le H(X) + H(Y)$ with equality if and only if X and Y are independent.

10.54. Show that for a deterministic channel

$$H(Y \mid X) = 0$$

10.55. Consider a channel with an input X and an output Y. Show that if X and Y are statistically independent, then $H(X \mid Y) = H(X)$ and $I(X; Y) = 0$.

10.56. A channel is described by the following channel matrix.

(a) Draw the channel diagram.

(b) Find the channel capacity.

$$\begin{bmatrix} \frac{1}{2} & \frac{1}{2} & 0 \\ 0 & 0 & 1 \end{bmatrix}$$

10.57. Let X be a random variable with probability density function $f_X(x)$, and let $Y = aX + b$, where a and b are constants. Find $H(Y)$ in terms of $H(X)$.

10.58. Show that $H(X + c) = H(X)$, where c is a constant.

10.59. Show that $H(X) \geq H(X \mid Y)$, and $H(Y) \geq H(Y \mid X)$.

10.60. Verify Eq. (10.30), that is, $H(X, Y) = H(X) + H(Y)$, if X and Y are independent.

10.61. Find the pdf $f_X(x)$ of a continuous r.v X with $E(X) = \mu$ which maximizes the differential entropy $H(X)$.

10.62. Calculate the capacity of AWGN channel with a bandwidth of 1 MHz and an S/N ratio of 40 dB.

10.63. Consider a DMS X with m equiprobable symbols $x_i, i = 1, 2, \ldots, m$.

(a) Show that the use of a fixed-length code for the representation of x_i is most efficient.

(b) Let n_0 be the fixed code word length. Show that if $n_0 = \log_2 m$, then the code efficiency is 100 percent.

10.64. Construct a Huffman code for the DMS X of Prob. 10.45, and show that the code is an optimum code.

10.65. A DMS X has five symbols x_1, x_2, x_3, x_4, and x_5 with respective probabilities 0.2, 0.15, 0.05, 0.1, and 0.5.

(a) Construct a Shannon-Fano code for X, and calculate the code efficiency.

(b) Repeat (a) for the Huffman code.

10.66. Show that the Kraft inequality is satisfied by the codes of Prob. 10.46.

ANSWERS TO SUPPLEMENTARY PROBLEMS

10.48. 1.875 b/symbol

10.49. 4.7 b/character

10.50. (a) $\begin{bmatrix} 0.62 & 0.38 \\ 0.38 & 0.62 \end{bmatrix}$

(b) $P(z_1) = 0.524, \quad P(z_2) = 0.476$

10.51. (a) $P(y_1) = 7/24, P(y_2) = 17/48$, and $P(y_3) = 17/48$

(b) 1.58 b/symbol

10.52. *Hint:* Use Eqs. (10.31) and (10.26).

10.53. *Hint:* Use Eqs. (10.33) and (10.35).

10.54. *Hint:* Use Eq. (10.24), and note that for a deterministic channel $P(y_j \mid x_i)$ are either 0 or 1.

10.55. *Hint:* Use Eqs. (3.32) and (3.37) in Eqs. (10.23) and (10.31).

10.56. (*a*) See Fig. 10-17.

   (*b*) 1 b/symbol

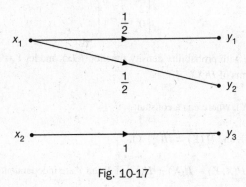

Fig. 10-17

10.57. $H(Y) = H(X) + \log_2 a$

10.58. *Hint:* Let $Y = X + c$ and follow Prob. 10.29.

10.59. *Hint:* Use Eqs. (10.31), (10.33), and (10.34).

10.60. *Hint:* Use Eqs. (10.28), (10.29), and (10.26).

10.61. $f_X(x) = \dfrac{1}{\mu} e^{-x/\mu}$ $\qquad x \geq 0$

10.62. 13.29 Mb/s

10.63. *Hint:* Use Eqs. (10.73) and (10.76).

10.64. Symbols: x_1 x_2 x_3 x_4

       Code: 0   10   110   111

10.65. (*a*) Symbols: x_1 x_2 x_3 x_4 x_5

           Code: 10   110   1111   1110   0

           Code efficiency $\eta = 98.6$ percent.

   (*b*) Symbols: x_1 x_2 x_3 x_4 x_5

           Code: 11   100   1011   1010   0

           Code efficiency $\eta = 98.6$ percent.

10.66. *Hint:* Use Eq. (10.78)

Normal Distribution

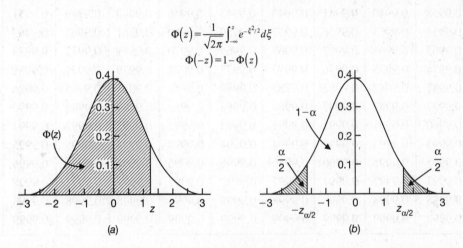

$$\Phi(z) = \frac{1}{\sqrt{2\pi}} \int_{-\infty}^{z} e^{-\xi^2/2} d\xi$$

$$\Phi(-z) = 1 - \Phi(z)$$

Fig. A

TABLE A Normal Distribution $\Phi(z)$

z	0.00	0.01	0.02	0.03	0.04	0.05	0.06	0.07	0.08	0.09
0.0	0.5000	0.5040	0.5080	0.5120	0.5160	0.5199	0.5239	0.5279	0.5319	0.5359
0.1	0.5399	0.5438	0.5478	0.5517	0.5557	0.5596	0.5636	0.5675	0.5714	0.5753
0.2	0.5793	0.5832	0.5871	0.5910	0.5948	0.5987	0.6026	0.6064	0.6103	0.6141
0.3	0.6179	0.6217	0.6255	0.6293	0.6331	0.6368	0.6406	0.6443	0.6480	0.6517
0.4	0.6554	0.6591	0.6628	0.6664	0.6700	0.6736	0.6772	0.6808	0.6844	0.6879
0.5	0.6915	0.6950	0.6985	0.7019	0.7054	0.7088	0.7123	0.7157	0.7190	0.7224
0.6	0.7257	0.7291	0.7324	0.7357	0.7389	0.7422	0.7454	0.7486	0.7517	0.7549
0.7	0.7580	0.7611	0.7642	0.7673	0.7703	0.7734	0.7764	0.7794	0.7823	0.7852
0.8	0.7881	0.7910	0.7939	0.7967	0.7995	0.8023	0.8051	0.8078	0.8106	0.8133
0.9	0.8159	0.8186	0.8212	0.8238	0.8364	0.8289	0.8315	0.8340	0.8365	0.8389
1.0	0.8413	0.8438	0.8461	0.8485	0.8508	0.8531	0.8554	0.8577	0.8599	0.8621
1.1	0.8643	0.8665	0.8686	0.8708	0.8729	0.8749	0.8770	0.8790	0.8810	0.8830
1.2	0.8849	0.8869	0.8888	0.8907	0.8925	0.8944	0.8962	0.8980	0.8997	0.9015
1.3	0.9032	0.9049	0.9066	0.9082	0.9099	0.9115	0.9131	0.9147	0.9162	0.9177
1.4	0.9192	0.9207	0.9222	0.9236	0.9251	0.9265	0.9279	0.9292	0.9306	0.9319

TABLE A—*Continued*

z	0.00	0.01	0.02	0.03	0.04	0.05	0.06	0.07	0.08	0.09
1.5	0.9332	0.9345	0.9357	0.9370	0.9382	0.9394	0.9406	0.9418	0.9429	0.9441
1.6	0.9452	0.9463	0.9474	0.9484	0.9495	0.9505	0.9515	0.9525	0.9535	0.9545
1.7	0.9554	0.9564	0.9573	0.9582	0.9591	0.9599	0.9608	0.9616	0.9625	0.9633
1.8	0.9641	0.9649	0.9656	0.9664	0.9671	0.9678	0.9686	0.9693	0.9699	0.9706
1.9	0.9713	0.9719	0.9726	0.9732	0.9738	0.9744	0.9750	0.9756	0.9761	0.9767
2.0	0.9772	0.9778	0.9783	0.9788	0.9793	0.9798	0.9803	0.9808	0.9812	0.9817
2.1	0.9821	0.9826	0.9830	0.9834	0.9838	0.9842	0.9846	0.9850	0.9854	0.9857
2.2	0.9861	0.9864	0.9868	0.9871	0.9875	0.9878	0.9881	0.9884	0.9887	0.9890
2.3	0.9893	0.9896	0.9898	0.9901	0.9904	0.9906	0.9909	0.9911	0.9913	0.9916
2.4	0.9918	0.9920	0.9922	0.9925	0.9927	0.9929	0.9931	0.9932	0.9934	0.9936
2.5	0.9938	0.9940	0.9941	0.9943	0.9945	0.9946	0.9948	0.9949	0.9951	0.9952
2.6	0.9953	0.9955	0.9956	0.9957	0.9959	0.9960	0.9961	0.9962	0.9963	0.9964
2.7	0.9965	0.9966	0.9967	0.9968	0.9969	0.9970	0.9971	0.9972	0.9973	0.9974
2.8	0.9974	0.9975	0.9976	0.9977	0.9977	0.9078	0.9979	0.9979	0.9980	0.9981
2.9	0.9981	0.9982	0.9982	0.9983	0.9984	0.9984	0.9085	0.9985	0.9986	0.9986
3.0	0.9987	0.9987	0.9987	0.9988	0.9988	0.9989	0.9989	0.9989	0.9990	0.9990
3.1	0.9990	0.9991	0.9991	0.9991	0.9992	0.9992	0.9992	0.9992	0.9993	0.9993
3.2	0.9993	0.9993	0.9994	0.9994	0.9994	0.9994	0.9994	0.9995	0.9995	0.9995
3.3	0.9995	0.9995	0.9996	0.9996	0.9996	0.9996	0.9996	0.9996	0.9996	0.9997
3.4	0.9997	0.9997	0.9997	0.9997	0.9997	0.9997	0.9997	0.9997	0.9998	0.9998
3.5	0.9998	0.9998	0.9998	0.9998	0.9998	0.9998	0.9998	0.9998	0.9998	0.9998
3.6	0.9998	0.9999	0.9999	0.9999	0.9999	0.9999	0.9999	0.9999	0.9999	0.9999

The material below refers to Fig. A.

α	0.2	0.1	0.05	0.025	0.01	0.005
$z_{\alpha/2}$	1.282	1.645	1.960	2.240	2.576	2.807

APPENDIX B

Fourier Transform

B.1 Continuous-Time Fourier Transform

Definition:

$$X(\omega) = \int_{-\infty}^{\infty} x(t)e^{-j\omega t}\, dt \qquad x(t) = \frac{1}{2\pi}\int_{-\infty}^{\infty} X(\omega)e^{j\omega t}\, d\omega$$

TABLE B-1 Properties of the Continuous-Time Fourier Transform

PROPERTY	SIGNAL	FOURIER TRANSFORM
	$x(t)$	$X(\omega)$
	$x_1(t)$	$X_1(\omega)$
	$x_2(t)$	$X_2(\omega)$
Linearity	$a_1 x_1(t) + a_2 x_2(t)$	$a_1 X_1(\omega) + a_2 X_2(\omega)$
Time shifting	$x(t - t_0)$	$e^{-j\omega t_0}X(\omega)$
Frequency shifting	$e^{j\omega_0 t}x(t)$	$X(\omega - \omega_0)$
Time scaling	$x(at)$	$\dfrac{1}{\lvert a \rvert}X\left(\dfrac{\omega}{a}\right)$
Time reversal	$x(-t)$	$X(-\omega)$
Duality	$X(t)$	$2\pi x(-\omega)$
Time differentiation	$\dfrac{dx(t)}{dt}$	$j\omega X(\omega)$
Frequency differentiation	$(-jt)x(t)$	$\dfrac{dX(\omega)}{d\omega}$
Integration	$\displaystyle\int_{-\infty}^{t} x(\tau)\, d\tau$	$\pi X(0)\delta(\omega) + \dfrac{1}{j\omega} + X(\omega)$
Convolution	$x_1(t) * x_2(t) = \displaystyle\int_{-\infty}^{\infty} x_1(\tau)x_2(t - \tau)\, d\tau$	$X_1(\omega)X_2(\omega)$
Multiplication	$x_1(t)x_2(t)$	$\dfrac{1}{2\pi}X_1(\omega) * X_2(\omega)$ $= \dfrac{1}{2\pi}\displaystyle\int_{-\infty}^{\infty} X_1(\lambda)X_2(\omega - \lambda)\, d\lambda$
Real signal	$x(t) = x_e(t) + x_0(t)$	$X(\omega) = A(\omega) + jB(\omega)$ $X(-\omega) = X^*(\omega)$
Even component	$x_e(t)$	$\mathrm{Re}\{X(\omega)\} = A(\omega)$
Odd component	$x_0(t)$	$j\,\mathrm{Im}\{X(\omega)\} = jB(\omega)$
Parseval's theorem	$\displaystyle\int_{-\infty}^{\infty} \lvert x(t) \rvert^2\, dt = \dfrac{1}{2\pi}\int_{-\infty}^{\infty} \lvert X(\omega) \rvert^2\, d\omega$	

TABLE B-2 Common Continuous-Time Fourier Transform Pairs

	$x(t)$	$X(\omega)$
1.	$\delta(t)$	1
2.	$\delta(t - t_0)$	$e^{-j\omega t_0}$
3.	1	$2\pi\delta(\omega)$
4.	$e^{j\omega_0 t}$	$2\pi\delta(\omega - \omega_0)$
5.	$\cos \omega_0 t$	$\pi[\delta(\omega - \omega_0) + \delta(\omega + \omega_0)]$
6.	$\sin \omega_0 t$	$-j\pi[\delta(\omega - \omega_0) - \delta(\omega + \omega_0)]$
7.	$u(t) = \begin{cases} 1 & t > 0 \\ 0 & t < 0 \end{cases}$	$\pi\delta(\omega) + \dfrac{1}{j\omega}$
8.	$e^{-at}u(t) \quad a > 0$	$\dfrac{1}{j\omega + a}$
9.	$te^{-at}u(t) \quad a > 0$	$\dfrac{1}{(j\omega + a)^2}$
10.	$e^{-a|t|} \quad a > 0$	$\dfrac{2a}{a^2 + \omega^2}$
11.	$\dfrac{1}{a^2 + t^2}$	$e^{-a|\omega|}$
12.	$e^{-at^2} \quad a > 0$	$\sqrt{\dfrac{\pi}{a}}\, e^{-\omega^2/4a}$
13.	$p_a(t) = \begin{cases} 1 & |t| < a \\ 0 & |t| > a \end{cases}$	$2a\dfrac{\sin \omega a}{\omega a}$
14.	$\dfrac{\sin at}{\pi t}$	$p_a(\omega) = \begin{cases} 1 & |\omega| < a \\ 0 & |\omega| > a \end{cases}$
15.	$\mathrm{sgn}\, t = \begin{cases} 1 & t > 0 \\ -1 & t < 0 \end{cases}$	$\dfrac{2}{j\omega}$
16.	$\displaystyle\sum_{k=-\infty}^{\infty} \delta(t - kT)$	$\omega_0 \displaystyle\sum_{k=-\infty}^{\infty} \delta(\omega - k\omega_0),\ \omega_0 = \dfrac{2\pi}{T}$

B.2 Discrete-Time Fourier Transform

Definition:

$$X(\Omega) = \sum_{n=-\infty}^{\infty} x(n)e^{-j\Omega n} \qquad x(n) = \frac{1}{2\pi}\int_{-\pi}^{\pi} X(\Omega)e^{j\Omega n}\, d\Omega$$

TABLE B-3 Properties of the Discrete-Time Fourier Transform

PROPERTY	SEQUENCE	FOURIER TRANSFORM
	$x(n)$	$X(\Omega)$
	$x_1(n)$	$X_1(\Omega)$
	$x_2(n)$	$X_2(\Omega)$
Periodicity	$x(n)$	$X(\Omega + 2\pi) = X(\Omega)$
Linearity	$a_1 x_1(n) + a_2 x_2(n)$	$a_1 X_1(\Omega) + a_2 X_2(\Omega)$
Time shifting	$x(n - n_0)$	$e^{-j\Omega n_0} X(\Omega)$

TABLE B-3—*Continued*

PROPERTY	SEQUENCE	FOURIER TRANSFORM
Frequency shifting	$e^{j\Omega_0 n} x(n)$	$X(\Omega - \Omega_0)$
Time reversal	$x(-n)$	$X(-\Omega)$
Frequency differentiation	$nx(n)$	$j\dfrac{dX(\Omega)}{d\Omega}$
Accumulation	$\displaystyle\sum_{k=-\infty}^{n} x(k)$	$\pi X(0)\delta(\Omega) + \dfrac{1}{1-e^{-j\Omega}}X(\Omega)$
Convolution	$x_1(n) * x_2(n) = \displaystyle\sum_{k=-\infty}^{n} x_1(k)x_2(n-k)$	$X_1(\Omega)X_2(\Omega)$
Multiplication	$x_1(n)x_2(n)$	$\dfrac{1}{2\pi}X_1(\Omega) \otimes X_2(\Omega)$ $= \dfrac{1}{2\pi}\displaystyle\int_{-\pi}^{\pi} X_1(\lambda)X_2(\Omega-\lambda)d\lambda$
Real sequence	$x(n) = x_e(n) + x_0(n)$	$X(\Omega) = A(\Omega) + jB(\Omega)$ $X(-\Omega) = X^*(\Omega)$
Even component	$x_e(n)$	$\mathrm{Re}\{X(\Omega) = A(\Omega)$
Odd component	$x_0(n)$	$j\,\mathrm{Im}\{X(\Omega)\} = jB(\Omega)$
Parseval's theorem	$\displaystyle\sum_{n=-\infty}^{\infty} |x(n)|^2 = \dfrac{1}{2\pi}\int_{-\pi}^{\pi} |X(\Omega)|^2\, d\Omega$	

TABLE B-4 Common Discrete-Time Fourier Transform Pairs

	$x[n]$	$X(\Omega)$
1.	$\delta(n) = \begin{cases} 1 & n=0 \\ 0 & n\neq 0 \end{cases}$	1
2.	$\delta(n-n_0)$	$e^{-j\Omega n_0}$
3.	$x(n) = 1$	$2\pi\delta(\Omega)$
4.	$e^{j\Omega_0 n}$	$2\pi\delta(\Omega - \Omega_0)$
5.	$\cos\Omega_0 n$	$\pi[\delta(\Omega - \Omega_0) + \delta(\Omega + \Omega_0)]$
6.	$\sin\Omega_0 n$	$-j\pi[\delta(\Omega - \Omega_0) - \delta(\Omega + \Omega_0)]$
7.	$u(n) = \begin{cases} 1 & n\geq 0 \\ 0 & n<0 \end{cases}$	$\pi\delta(\Omega) + \dfrac{1}{1-e^{-j\Omega}}$
8.	$a^n u(n) \quad |a|<1$	$\dfrac{1}{1-ae^{-j\Omega}}$
9.	$(n+1)a^n u(n) \quad |a|<1$	$\dfrac{1}{(1-ae^{-j\Omega})^2}$
10.	$a^{|n|} \quad |a|<1$	$\dfrac{1-a^2}{1-2a\cos\Omega + a^2}$
11.	$x(n) = \begin{cases} 1 & |n|\leq N_1 \\ 0 & |n|>N_1 \end{cases}$	$\dfrac{\sin\left[\Omega\left(N_1 + \dfrac{1}{2}\right)\right]}{\sin(\Omega/2)}$
12.	$\dfrac{\sin W_n}{\pi n} \quad 0<W<\pi$	$X(\Omega) = \begin{cases} 1 & 0\leq|\Omega|\leq W \\ 0 & W<|\Omega|\leq\pi \end{cases}$
13.	$\displaystyle\sum_{k=-\infty}^{\infty}\delta(n-kN_0)$	$\Omega_0\displaystyle\sum_{k=-\infty}^{\infty}\delta(\Omega - k\Omega_0) \quad \Omega_0 = \dfrac{2\pi}{N_0}$

INDEX